Noncommutative Geometry and Cayley-smooth Orders

PURE AND APPLIED MATHEMATICS

A Program of Monographs, Textbooks, and Lecture Notes

MONOGRAPHS AND TEXTBOOKS IN PURE AND APPLIED MATHEMATICS

Recent Titles

Noncommutative Geometry and Cayley-smooth Orders

Lieven Le Bruyn

Universiteit Antwerpen
Belgium

CRC Press

Taylor & Francis Group
Boca Raton London New York

CRC Press is an imprint of the
Taylor & Francis Group, an **informa** business

A CHAPMAN & HALL BOOK

CRC Press
Taylor & Francis Group
6000 Broken Sound Parkway NW, Suite 300
Boca Raton, FL 33487-2742

First issued in paperback 2019

© 2008 by Taylor & Francis Group, LLC
CRC Press is an imprint of Taylor & Francis Group, an Informa business

No claim to original U.S. Government works

ISBN-13: 978-1-4200-6422-3 (hbk)
ISBN-13: 978-0-367-38870-6 (pbk)

Library of Congress Cataloging-in-Publication Data

Le Bruyn, Lieven, 1958-
    Noncommutative geometry and Cayley-smooth orders / Lieven Le Bruyn.
      p. cm. -- (Pure and applied mathematics ; 290)
    Includes bibliographical references and index.
    ISBN 978-1-4200-6422-3 (alk. paper)
    1. Noncommutative differential geometry. 2. Cayley-Hamilton theorem. I. Title. II. Series.

QC20.7.D52L4 2007
512'.55--dc22
                                    2007019964

Visit the Taylor & Francis Web site at
http://www.taylorandfrancis.com

and the CRC Press Web site at
http://www.crcpress.com

This book is dedicated to the women in my life
Simonne Stevens (1926-2004), Ann, Gitte & Bente

Contents

Contents

Preface

This book explains the theory of *Cayley-smooth orders* in central simple algebras over functionfields of varieties. In particular, we will describe the étale local structure of such orders as well as their central singularities and finite dimensional representations. There are two major motivations to study Cayley-smooth orders.

A first application is the construction of partial desingularizations of (commutative) singularities from noncommutative algebras. This approach is summarized in the introduction and can be read independently, modulo technical details and proofs, which are deferred to the main body of the book. A second motivation stems from *noncommutative algebraic geometry* as developed by Joachim Cuntz, Daniel Quillen, Maxim Kontsevich, Michael Kapranov and others. One studies *formally smooth algebras* or *quasi-free algebras* (in this book we will call them *Quillen-smooth algebras*), which are huge, non-Noetherian algebras, the free associative algebras being the archetypical examples. One attempts to study these algebras via their finite dimensional representations which, in turn, are controlled by associated Cayley-smooth algebras. In the final two chapters, we will give an introduction to this fast developing theory.

This book is based on a series of courses given since 1999 in the "advanced master program on noncommutative geometry" organized by the NOncommutative Geometry (NOG) project, sponsored by the European Science Foundation (ESF). As the participating students came from different countries there was a need to include background information on a variety of topics including invariant theory, algebraic geometry, central simple algebras and the representation theory of quivers. In this book, these prerequisites are covered in chapters 1 to 4.

Chapters 1 and 2 contain the invariant theoretic description of orders and their centers, due to Michael Artin and Claudio Procesi. Chapter 3 contains an introduction to étale topology and its use in noncommutative algebra, in particular to the study of Azumaya algebras and to the local description of algebras via Luna slices. Chapter 4 collects the necessary material on representations of quivers, including the description of their indecomposable roots, due to Victor Kac, the determination of dimension vectors of simple representations, and results on general quiver representations, due to Aidan Schofield. The results in these chapters are due to many people and the presentation is influenced by a variety of sources.

Chapters 5 and 6 contain the main results on Cayley-smooth orders. In chapter 5, we describe the étale local structure of a Cayley-smooth order in a semisimple representation and classify the associated central singularity up to smooth equivalence. This is done by associating to a semisimple representation a combinatorial gadget, a *marked quiver setting*, which encodes the tangent-space information to the noncommutative manifold in the cluster of points determined by the simple factors of the representation. In chapter 6 we will describe the nullcone of these marked quiver representations and relate them to the study of all isomorphism classes of n-dimensional representations of a Cayley-smooth order.

Introduction

Ever since the dawn of noncommutative algebraic geometry in the mid-seventies, see for example the work of P. Cohn [21], J. Golan [38], C. Procesi [86], F. Van Oystaeyen and A. Verschoren [103],[105], it has been ring theorists' hope that this theory might one day be relevant to commutative geometry, in particular to the study of singularities and their resolutions.

Over the last decade, noncommutative algebras have been used to construct canonical (partial) resolutions of quotient singularities. That is, take a finite group G acting on $\mathbb{C}^d$ freely away from the origin then its orbit-space $\mathbb{C}^d/G$ is an isolated singularity. Resolutions $Y \longrightarrow \mathbb{C}^d/G$ have been constructed using the skew group algebra

$$\mathbb{C}[x_1, \ldots, x_d] \# G$$

which is an order with center $\mathbb{C}[\mathbb{C}^d/G] = \mathbb{C}[x_1, \ldots, x_d]^G$ or deformations of it.

In dimension $d = 2$ (the case of Kleinian singularities) this gives us minimal resolutions via the connection with the preprojective algebra, see for example [27]. In dimension $d = 3$, the skew group algebra appears via the superpotential and commuting matrices setting (in the physics literature) or via the McKay quiver, see for example [23]. If G is Abelian one obtains from this study crepant resolutions but for general G one obtains at best partial resolutions with conifold singularities remaining. In dimension $d > 3$ the situation is unclear at this moment.

Usually, skew group algebras and their deformations are studied via homological methods as they are Serre-smooth orders, see for example [102]. In this book, we will follow a different approach.

We want to find a noncommutative explanation for the omnipresence of conifold singularities in partial resolutions of three-dimensional quotient singularities. One may argue that they have to appear because they are somehow the nicest singularities. But then, what is the corresponding list of "nice" singularities in dimension four? or five, six...?

The results contained in this book suggest that the nicest partial resolutions of $\mathbb{C}^4/G$ should only contain singularities that are either polynomials over the conifold or one of the following three types

$$\frac{\mathbb{C}[[a,b,c,d,e,f]]}{(ae - bd, af - cd, bf - ce)} \qquad \frac{\mathbb{C}[[a,b,c,d,e]]}{(abc - de)} \qquad \frac{\mathbb{C}[[a,b,c,d,e,f,g,h]]}{I}$$

where I is the ideal of all 2×2 minors of the matrix

$$\begin{bmatrix} a & b & c & d \\ e & f & g & h \end{bmatrix}$$

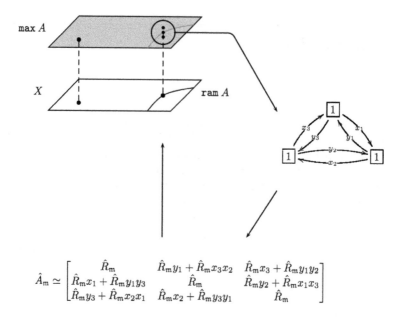

$$\hat{A}_{\mathfrak{m}} \simeq \begin{bmatrix} \hat{R}_{\mathfrak{m}} & \hat{R}_{\mathfrak{m}}y_1 + \hat{R}_{\mathfrak{m}}x_3x_2 & \hat{R}_{\mathfrak{m}}x_3 + \hat{R}_{\mathfrak{m}}y_1y_2 \\ \hat{R}_{\mathfrak{m}}x_1 + \hat{R}_{\mathfrak{m}}y_1y_3 & \hat{R}_{\mathfrak{m}} & \hat{R}_{\mathfrak{m}}y_2 + \hat{R}_{\mathfrak{m}}x_1x_3 \\ \hat{R}_{\mathfrak{m}}y_3 + \hat{R}_{\mathfrak{m}}x_2x_1 & \hat{R}_{\mathfrak{m}}x_2 + \hat{R}_{\mathfrak{m}}y_3y_1 & \hat{R}_{\mathfrak{m}} \end{bmatrix}$$

FIGURE I.1:   Local structure of Cayley-smooth orders

In dimension $d = 5$ there is another list of ten new specific singularities that
will appear; in dimension $d = 6$ another 63 new ones appear and so on.

How do we arrive at these specific lists? The hope is that any quotient
singularity $X = \mathbb{C}^d/G$ has associated to it a "nice" order A with center
$R = \mathbb{C}[X]$ such that there is a stability structure θ such that the scheme of
all θ-semistable representations of A is a smooth variety (all these terms will
be explained in the main body of the book). If this is the case, the associated
moduli space will be a partial resolution

$$\texttt{moduli}_\alpha^\theta\ A \longrightarrow X = \mathbb{C}^d/G$$

and has a sheaf of Cayley-smooth orders $\mathcal{A}$ over it, allowing us to control its
singularities in a combinatorial way as depicted in figure I.1.

If A is a Cayley-smooth order over $R = \mathbb{C}[X]$ then its noncommutative
variety $\texttt{max}\ A$ of maximal twosided ideals is birational to X away from the
ramification locus. If P is a point of the ramification locus $\texttt{ram}\ A$ then there is
a finite cluster of infinitesimally nearby noncommutative points lying over it.
The local structure of the noncommutative variety $\texttt{max}\ A$ near this cluster can
be summarized by a (marked) quiver setting (Q, α), which in turn allows us to
compute the étale local structure of A and R in P. The central singularities
that appear in this way have been classified in [14] (see also section 5.8) up to
smooth equivalence giving us the small lists of singularities mentioned before.

In this introduction we explain this noncommutative approach to the desingularization project of commutative singularities. Proofs and more details will be given in the following chapters.

I.1 Noncommutative algebra

Let me begin by trying to motivate why one might be interested in noncommutative algebra if you want to understand quotient singularities and their resolutions. Suppose we have a finite group G acting on d-dimensional affine space $\mathbb{C}^d$ such that this action is free away from the origin. Then the orbit-space, the so called *quotient singularity* $\mathbb{C}^d/G$, is an isolated singularity

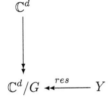

and we want to construct "minimal" or "canonical" resolutions (so called crepant resolutions) of this singularity. In his Bourbaki talk [89] Miles Reid asserts that McKay correspondence follows from a much more general principle

Miles Reid's Principle: Let M be an algebraic manifold, G a group of automorphisms of M, and $Y \longrightarrow X$ a resolution of singularities of $X = M/G$. Then the answer to any well-posed question about the geometry of Y is the G-equivariant geometry of M.

Applied to the case of quotient singularities, the content of his slogan is that the G-equivariant geometry of $\mathbb{C}^d$ already *knows* about the crepant resolution $Y \longrightarrow \mathbb{C}^d/G$. Let us change this principle slightly: assume we have an affine variety M on which a reductive group (we will take PGL_n) acts with algebraic quotient variety $M/PGL_n \simeq \mathbb{C}^d/G$

$$\mathbb{C}^d$$
$$\downarrow$$
$$M \longrightarrow M/PGL_n \simeq \mathbb{C}^d/G \xleftarrow{\;res\;} Y$$

then, in favorable situations, we can argue that the PGL_n-equivariant geometry of M knows about good resolutions Y. One of the key lessons to be learned from this book is that PGL_n-equivariant geometry of M is *roughly* equivalent to the study of a certain noncommutative algebra over $\mathbb{C}^d/G$. In fact, an *order* in a central simple algebra of dimension n^2 over the function field of the quotient singularity. Hence, if we know of *good* orders over $\mathbb{C}^d/G$, we might get our hands on "good" resolutions Y by noncommutative methods.

We will work in the following, quite general, setting:

- X will be a *normal* affine variety, possibly having singularities.

- R will be the coordinate ring $\mathbb{C}[X]$ of X.

- K will be the function field $\mathbb{C}(X)$ of X.

If you are only interested in quotient singularities, you should replace X by $\mathbb{C}^d/G$, R by the invariant ring $\mathbb{C}[x_1,\ldots,x_d]^G$ and K by the invariant field $\mathbb{C}(x_1,\ldots,x_d)^G$ in all statements below.

Our goal will be to construct lots of R-orders A in a central simple K-algebra Σ.

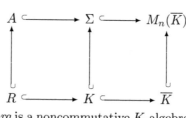

A *central simple algebra* is a noncommutative K-algebra Σ with center $Z(\Sigma) = K$ such that over the algebraic closure $\overline{K}$ of K we obtain full $n \times n$ matrices

$$\Sigma \otimes_K \overline{K} \simeq M_n(\overline{K})$$

(more details will be given in section 3.2). There are plenty such central simple K-algebras:

Example I.1

For any nonzero functions $f, g \in K^*$, the *cyclic algebra*

$$\Sigma = (f,g)_n \qquad \text{defined by} \qquad (f,g)_n = \frac{K\langle x,y\rangle}{(x^n - f, y^n - g, yx - qxy)}$$

with q is a primitive n-th root of unity, is a central simple K-algebra of dimension n^2. Often, $(f,g)_n$ will even be a *division algebra*, that is a noncommutative algebra such that every nonzero element has an inverse.

For example, this is always the case when $E = K[x]$ is a (commutative) field extension of dimension n and if g has order n in the quotient $K^*/N_{E/K}(E^*)$ where $N_{E/K}$ is the *norm map* of E/K. ◻

Fix a central simple K-algebra Σ, then an R-*order* A *in* Σ is a subalgebras $A \subset \Sigma$ with center $Z(A) = R$ such that A is finitely generated as an R-module and contains a K-basis of Σ, that is

$$A \otimes_R K \simeq \Sigma$$

The classic reference for orders is Irving Reiner's book [90] but it is somewhat outdated and focuses mainly on the one-dimensional case. With this book we hope to remedy this situation somewhat.

Example I.2
In the case of quotient singularities $X = \mathbb{C}^d/G$ a natural choice of R-order might be the *skew group ring*: $\mathbb{C}[x_1, \ldots, x_d]\#G$, which consists of all formal sums $\sum_{g \in G} r_g \# g$ with multiplication defined by

$$(r\#g)(r'\#g') = r\phi_g(r')\#gg'$$

where ϕ_g is the action of g on $\mathbb{C}[x_1, \ldots, x_d]$. The center of the skew group algebra is easily verified to be the ring of G-invariants

$$R = \mathbb{C}[\mathbb{C}^d/G] = \mathbb{C}[x_1, \ldots, x_d]^G$$

Further, one can show that $\mathbb{C}[x_1, \ldots, x_d]\#G$ is an R-order in $M_n(K)$ with n the order of G. Later we will give another description of the skew group algebra in terms of the McKay-quiver setting and the variety of commuting matrices. ▯

However, there are plenty of other R-orders in $M_n(K)$, which may or may not be relevant in the study of the quotient singularity $\mathbb{C}^d/G$.

Example I.3
If $f, g \in R - \{0\}$, then the free R-submodule of rank n^2 of the cyclic K-algebra $\Sigma = (f, g)_n$ of example *I.1*

$$A = \sum_{i,j=0}^{n-1} Rx^i y^j$$

is an R-order. But there is really no need to go for this "canonical" example. Someone more twisted may take I and J any two nonzero ideals of R, and consider

$$A_{IJ} = \sum_{i,j=0}^{n-1} I^i J^j x^i y^j$$

which is also an R-order in Σ, far from being a projective R-module unless I and J are invertible R-ideals.

For example, in $M_n(K)$ we can take the "obvious" R-order $M_n(R)$ but one might also take the subring

$$\begin{bmatrix} R & I \\ J & R \end{bmatrix}$$

which is an R-order if I and J are nonzero ideals of R.   ⬜

From a geometric viewpoint, our goal is to construct lots of affine PGL_n-varieties M such that the algebraic quotient M/PGL_n is isomorphic to X and, moreover, such that there is a Zariski open subset $U \subset X$

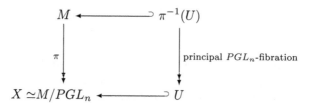

for which the quotient map is a principal PGL_n-fibration, that is, all fibers $\pi^{-1}(u) \simeq PGL_n$ for $u \in U$. For the connection between such varieties M and orders A in central simple algebras think of M as the affine variety of n-dimensional representations $\texttt{rep}_n\ A$ and of U as the Zariski open subset of all simple n-dimensional representations.

Naturally, one can only expect the R-order A (or the corresponding PGL_n-variety M) to be useful in the study of resolutions of X if A is *smooth* in some appropriate noncommutative sense. There are many characterizations of *commutative* smooth domains R:

- R is *regular*, that is, has finite global dimension

- R is *smooth*, that is, X is a smooth variety

and generalizing either of them to the noncommutative world leads to quite different concepts. We will call an R-order A a central simple K-algebra Σ:

- **Serre-smooth** if A has finite global dimension together with some extra features such as Auslander regularity or Cohen-Macaulay property, see for example [80].

- **Cayley-smooth** if the corresponding PGL_n-affine variety M is a smooth variety as we will clarify later.

For applications of Serre-smooth orders to desingularizations we refer to the paper [102]. We will concentrate on the properties of Cayley-smooth orders instead. Still, it is worth pointing out the strengths and weaknesses of both definitions.

Serre-smooth orders are excellent if you want to control homological properties, for example, if you want to study the derived categories of their modules. At this moment there is no local characterization of Serre-smooth orders if $\dim X \geq 3$. Cayley-smooth orders are excellent if you want to have smooth moduli spaces of semistable representations. As we will see later, in each dimension there are only a finite number of local types of Cayley-smooth orders and these will be classified in this book. The downside of this is that Cayley-smooth orders are less versatile than Serre-smooth orders. In general though, both theories are quite different.

Example I.4
The skew group algebra $\mathbb{C}[x_1, \ldots, x_d] \# G$ is always a Serre-smooth order but it is virtually never a Cayley-smooth order. $\square$

Example I.5
Let X be the variety of matrix-invariants, that is

$$X = M_n(\mathbb{C}) \oplus M_n(\mathbb{C})/PGL_n$$

where PGL_n acts on pairs of $n \times n$ matrices by simultaneous conjugation. The *trace ring* of two generic $n \times n$ matrices A is the subalgebra of $M_n(\mathbb{C}[M_n(\mathbb{C}) \oplus M_n(\mathbb{C})])$ generated over $\mathbb{C}[X]$ by the two *generic matrices*

$$X = \begin{bmatrix} x_{11} & \cdots & x_{1n} \\ \vdots & & \vdots \\ x_{n1} & \cdots & x_{nn} \end{bmatrix} \quad \text{and} \quad Y = \begin{bmatrix} y_{11} & \cdots & y_{1n} \\ \vdots & & \vdots \\ y_{n1} & \cdots & y_{nn} \end{bmatrix}$$

Then, A is an R-order in a division algebra of dimension n^2 over K, called the *generic division algebra*. Moreover, A is a Cayley-smooth order but is Serre-smooth only when $n = 2$, see [78]. $\square$

Descent theory allows construction of elaborate examples out of trivial ones by bringing in topology and enables one to classify objects that are only *locally* (but not necessarily globally) trivial. For applications to orders there are two topologies to consider : the well-known Zariski topology and the perhaps lesser-known étale topology. Let us try to give a formal definition of Zariski and étale *covers* aimed at ring theorists. Much more detail on étale topology will be given in section 3.1.

A *Zariski cover* of X is a finite product of localizations at elements of R

$$S_z = \prod_{i=1}^{k} R_{f_i} \qquad \text{such that} \qquad (f_1, \ldots, f_k) = R$$

and is therefore a faithfully flat extension of R. Geometrically, the ring-morphism $R \longrightarrow S_z$ defines a cover of $X = \text{spec } R$ by k disjoint sheets $\text{spec } S_z = \sqcup_i \text{spec } R_{f_i}$, each corresponding to a Zariski open subset of X, the complement of $\mathbb{V}(f_i)$, and the condition is that these closed subsets $\mathbb{V}(f_i)$ do not have a point in common. That is, we have the picture of figure I.2.

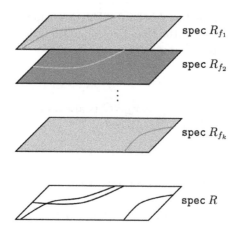

FIGURE I.2:   A Zariski cover of $X = \text{spec } R$.

Zariski covers form a *Grothendieck topology*, that is, two Zariski covers $S_z^1 = \prod_{i=1}^{k} R_{f_i}$ and $S_z^2 = \prod_{j=1}^{l} R_{g_j}$ have a common refinement

$$S_z = S_z^1 \otimes_R S_z^2 = \prod_{i=1}^{k} \prod_{j=1}^{l} R_{f_i g_j}$$

For a given Zariski cover $S_z = \prod_{i=1}^{k} R_{f_i}$ a corresponding *étale cover* is a product

$$S_e = \prod_{i=1}^{k} \frac{R_{f_i}[x(i)_1, \ldots, x(i)_{k_i}]}{(g(i)_1, \ldots, g(i)_{k_i})} \quad \text{with} \quad \begin{bmatrix} \frac{\partial g(i)_1}{\partial x(i)_1} & \cdots & \frac{\partial g(i)_1}{\partial x(i)_{k_i}} \\ \vdots & & \vdots \\ \frac{\partial g(i)_{k_i}}{\partial x(i)_1} & \cdots & \frac{\partial g(i)_{k_i}}{\partial x(i)_{k_i}} \end{bmatrix}$$

a unit in the i-th component of S_e. In fact, for applications to orders it is usually enough to consider *special etale extensions*

$$S_e = \prod_{i=1}^{k} \frac{R_{f_i}[x]}{(x^{k_i} - a_i)} \quad \text{where} \quad a_i \text{ is a unit in } R_{f_i}$$

Geometrically, an étale cover determines for every Zariski sheet **spec** R_{f_i} a *locally isomorphic* (for the analytic topology) multicovering and the number of sheets may vary with i (depending on the degrees of the polynomials $g(i)_j \in R_{f_i}[x(i)_1, \ldots, x(i)_{k_i}]$. That is, the mental picture corresponding to an étale cover is given in figure I.3.

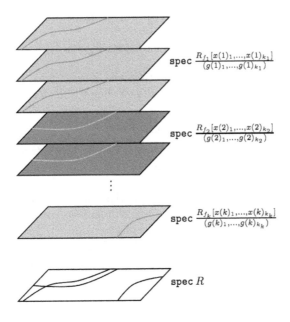

$$\text{spec } \frac{R_{f_1}[x(1)_1,\ldots,x(1)_{k_1}]}{(g(1)_1,\ldots,g(1)_{k_1})}$$

$$\text{spec } \frac{R_{f_2}[x(2)_1,\ldots,x(2)_{k_2}]}{(g(2)_1,\ldots,g(2)_{k_2})}$$

$$\vdots$$

$$\text{spec } \frac{R_{f_k}[x(k)_1,\ldots,x(k)_{k_k}]}{(g(k)_1,\ldots,g(k)_{k_k})}$$

$$\text{spec } R$$

FIGURE I.3:   An étale cover of $X = \text{spec } R$.

Again, étale covers form a Zariski topology as the common refinement $S_e^1 \otimes_R$ S_e^2 of two étale covers is again étale because its components are of the form

$$\frac{R_{f_i g_j}[x(i)_1,\ldots,x(i)_{k_i},y(j)_1,\ldots,y(j)_{l_j}]}{(g(i)_1,\ldots,g(i)_{k_i},h(j)_1,\ldots,h(j)_{l_j})}$$

and the Jacobian-matrix condition for each of these components is again satisfied. Because of the local isomorphism property many ring theoretical local properties (such as smoothness, normality, etc.) are preserved under étale covers.

For a fixed R-order B in some central simple K-algebra Σ, then a *Zariski twisted form* A of B is an R-algebra such that

$$A \otimes_R S_z \simeq B \otimes_R S_z$$

for some Zariski cover S_z of R. If $P \in X$ is a point with corresponding maximal ideal $\mathfrak{m}$, then $P \in \text{spec } R_{f_i}$ for some of the components of S_z and as $A_{f_i} \simeq B_{f_i}$ we have for the local rings at P

$$A_\mathfrak{m} \simeq B_\mathfrak{m}$$

that is, the Zariski local information of any Zariski-twisted form of B is that of B itself.

Likewise, an *étale twisted form* A of B is an R-algebra such that

$$A \otimes_R S_e \simeq B \otimes_R S_e$$

for some étale cover S_e of R. This time the Zariski local information of A and B may be different at a point $P \in X$ but we do have that the $\mathfrak{m}$-adic completions of A and B

$$\hat{A}_\mathfrak{m} \simeq \hat{B}_\mathfrak{m}$$

are isomorphic as $\hat{R}_\mathfrak{m}$-algebras. Thus, the Zariski local structure of A determines the localization $A_\mathfrak{m}$, the étale local structure determines the completion $\hat{A}_\mathfrak{m}$.

Descent theory allows us to classify Zariski- or étale twisted forms of an R-order B by means of the corresponding cohomology groups of the automorphism schemes. For more details on this please read [57], [82] or section 3.1. If one applies descent to the most trivial of all R-orders, the full matrix algebra $M_n(R)$, one arrives at *Azumaya algebras*. A Zariski twisted form of $M_n(R)$ is an R-algebra A such that

$$A \otimes_R S_z \simeq M_n(S_z) = \prod_{i=1}^{k} M_n(R_{f_i})$$

Conversely, you can construct such twisted forms by *gluing together* the matrix rings $M_n(R_{f_i})$. The easiest way to do this is to glue $M_n(R_{f_i})$ with $M_n(R_{f_j})$ over $R_{f_i f_j}$ via the natural embeddings

$$R_{f_i} \lhook\joinrel\longrightarrow R_{f_i f_j} \longleftarrow\joinrel\rhook R_{f_j}$$

Not surprisingly, we obtain in this way $M_n(R)$ back. However there are more clever ways to perform the gluing by bringing in the noncommutativity of matrix-rings. We can glue

$$M_n(R_{f_i}) \lhook\joinrel\longrightarrow M_n(R_{f_i f_j}) \overset{g_{ij} \cdot g_{ij}^{-1}}{\underset{\simeq}{\longrightarrow}} M_n(R_{f_i f_j}) \longleftarrow\joinrel\rhook M_n(R_{f_j})$$

over their intersection via conjugation with an invertible matrix g_{ij} in $GL_n(R_{f_i f_j})$. If the elements g_{ij} for $1 \leq i, j \leq k$ satisfy the *cocycle condition* (meaning that the different possible gluings are compatible over their common localization $R_{f_i f_j f_l}$), we obtain a sheaf of noncommutative algebras $\mathcal{A}$ over $X = \mathrm{spec}\, R$ such that its global sections are not necessarily $M_n(R)$.

PROPOSITION I.1

Any Zariski twisted form of $M_n(R)$ is isomorphic to $End_R(P)$ where P is a projective R-module of rank n. Two such twisted forms are isomorphic as R-algebras

$$End_R(P) \simeq End_R(Q) \quad \text{iff} \quad P \simeq Q \otimes I$$

for some invertible R-ideal I.

PROOF   We have an exact sequence of group schemes

$$1 \longrightarrow \mathbb{G}_m \longrightarrow \mathrm{GL}_n \longrightarrow \mathrm{PGL}_n \longrightarrow 1$$

(here, $\mathbb{G}_m$ is the sheaf of units) and taking Zariski cohomology groups over X we have a sequence

$$1 \longrightarrow H^1_{Zar}(X, \mathbb{G}_m) \longrightarrow H^1_{Zar}(X, \mathrm{GL}_n) \longrightarrow H^1_{Zar}(X, \mathrm{PGL}_n)$$

where the first term is isomorphic to the Picard group $Pic(R)$ and the second term classifies projective R-modules of rank n upto isomorphism. The final term classifies the Zariski twisted forms of $M_n(R)$ as the automorphism group of $M_n(R)$ is PGL_n.   ▯

Example I.6
Let I and J be two invertible ideals of R, then

$$End_R(I \oplus J) \simeq \begin{bmatrix} R & I^{-1}J \\ IJ^{-1} & R \end{bmatrix} \subset M_2(K)$$

and if $IJ^{-1} = (r)$ then $I \oplus J \simeq (Rr \oplus R) \otimes J$ and indeed we have an isomorphism

$$\begin{bmatrix} 1 & 0 \\ 0 & r^{-1} \end{bmatrix} \begin{bmatrix} R & I^{-1}J \\ IJ^{-1} & R \end{bmatrix} \begin{bmatrix} 1 & 0 \\ 0 & r \end{bmatrix} = \begin{bmatrix} R & R \\ R & R \end{bmatrix}$$

▯

The situation becomes a lot more interesting when we replace the Zariski topology by the étale topology.

DEFINITION I.1   *An n-Azumaya algebra over R is an étale twisted form A of $M_n(R)$. If A is also a Zariski twisted form we call A a* trivial *Azumaya algebra.*

LEMMA I.1
If A is an n-Azumaya algebra over R, then:

1. *The center $Z(A) = R$ and A is a projective R-module of rank n^2.*

2. *All simple A-representations have dimension n and for every maximal ideal $\mathfrak{m}$ of R we have*
$$A/\mathfrak{m}A \simeq M_n(\mathbb{C})$$

PROOF   For (2) take $M \cap R = \mathfrak{m}$ where M is the kernel of a simple representation $A \longrightarrow M_k(\mathbb{C})$, then as $\hat{A}_{\mathfrak{m}} \simeq M_n(\hat{R}_{\mathfrak{m}})$ it follows that

$$A/\mathfrak{m}A \simeq M_n(\mathbb{C})$$

and hence that $k = n$ and $M = A\mathfrak{m}$.   □

It is clear from the definition that when A is an n-Azumaya algebra and A' is an m-Azumaya algebra over R, $A \otimes_R A'$ is an mn-Azumaya and also that

$$A \otimes_R A^{op} \simeq End_R(A)$$

where A^{op} is the *opposite* algebra (that is, equipped with the reverse multiplication rule). These facts allow us to define the *Brauer group* $\mathrm{Br}\,R$ to be the set of equivalence classes $[A]$ of Azumaya algebras over R where

$$[A] = [A'] \quad \text{iff} \quad A \otimes_R A' \simeq End_R(P)$$

for some projective R-module P and where multiplication is induced from the rule

$$[A].[A'] = [A \otimes_R A']$$

One can extend the definition of the Brauer group from affine varieties to arbitrary schemes and A. Grothendieck has shown that the Brauer group of a projective smooth variety is a birational invariant, see [40]. Moreover, he conjectured a cohomological description of the Brauer group $\mathrm{Br}\,R$, which was subsequently proved by O. Gabber in [34].

THEOREM I.1
The Brauer group is an étale cohomology group

$$\mathrm{Br}\,R \simeq H^2_{et}(X, \mathbb{G}_m)_{torsion}$$

where $\mathbb{G}_m$ is the unit sheaf and where the subscript denotes that we take only torsion elements. If R is regular, then $H^2_{et}(X, \mathbb{G}_m)$ is torsion so we can forget the subscript.

This result should be viewed as the ring theory analogon of the *crossed product theorem* for central simple algebras over fields. Observe that in Gabber's result there is no sign of singularities in the description of the Brauer group. In fact, with respect to the desingularization project, Azumaya algebras are only as good as their centers.

PROPOSITION I.2
If A is an n-Azumaya algebra over R, then

1. *A is Serre-smooth iff R is commutative regular.*

2. *A is Cayley-smooth iff R is commutative regular.*

PROOF (1) follows from faithfully flat descent and (2) from lemma I.1, which asserts that the PGL_n-affine variety corresponding to A is a principal PGL_n-fibration in the étale topology, which shows that both n-Azumaya algebras and principal PGL_n-fibrations are classified by the étale cohomology group $H^1_{et}(X, PGL_n)$. More details are given in chapter 3. ▯

In the correspondence between R-orders and PGL_n-varieties, Azumaya algebras correspond to *principal PGL_n-fibrations* over X and with respect to desingularizations, Azumaya algebras are of little use. So let us bring in *ramification* in order to construct orders that may be more useful.

Example I.7
Consider the R-order in $M_2(K)$

$$A = \begin{bmatrix} R & R \\ I & R \end{bmatrix}$$

where I is some ideal of R and let $P \in X$ be a point with corresponding maximal ideal $\mathfrak{m}$. For I not contained in $\mathfrak{m}$ we have $A_\mathfrak{m} \simeq M_2(R_\mathfrak{m})$ whence A is an Azumaya algebra in P. For $I \subset \mathfrak{m}$ we have

$$A_\mathfrak{m} \simeq \begin{bmatrix} R_\mathfrak{m} & R_\mathfrak{m} \\ I_\mathfrak{m} & R_\mathfrak{m} \end{bmatrix} \neq M_2(R_\mathfrak{m})$$

whence A is not Azumaya in P. ▯

DEFINITION I.2 *The* ramification locus *of an R-order A is the Zariski closed subset of X consisting of those points P such that for the corresponding maximal ideal $\mathfrak{m}$*

$$A/\mathfrak{m}A \not\simeq M_n(\mathbb{C})$$

That is, ram *A is the locus of X where A is not an Azumaya algebra. Its complement* azu *A is called the* Azumaya locus *of A, which is always a Zariski open subset of X.*

DEFINITION I.3 *An R-order A is said to be a* reflexive n-Azumaya algebra *iff*

1. ram *A has codimension at least two in X, and*

2. *A is a reflexive R-module*

that is, $A \simeq Hom_R(Hom_R(A, R), R) = A^{**}$.

The origin of the terminology is that when A is a reflexive n-Azumaya algebra we have that $A_{\mathfrak{p}}$ is n-Azumaya for every height one prime ideal $\mathfrak{p}$ of R and that $A = \cap_{\mathfrak{p}} A_{\mathfrak{p}}$ where the intersection is taken over all height one primes.

For example, in example I.7 if I is a divisorial ideal of R, then A is not reflexive Azumaya as $A_{\mathfrak{p}}$ is not Azumaya for $\mathfrak{p}$ a height one prime containing I and if I has at least height two, then A is often not a reflexive Azumaya algebra because A is not reflexive as an R-module. For example take

$$A = \begin{bmatrix} \mathbb{C}[x,y] & \mathbb{C}[x,y] \\ (x,y) & \mathbb{C}[x,y] \end{bmatrix}$$

then the reflexive closure of A is $A^{**} = M_2(\mathbb{C}[x,y])$.

Sometimes though, we get reflexivity of A for free, for example when A is a Cohen-Macaulay R-module. An other important fact to remember is that for A a reflexive Azumaya, A is Azumaya if and only if A is projective as an R-module.

Example I.8
Let $A = \mathbb{C}[x_1, \ldots, x_d]\#G$, then A is a reflexive Azumaya algebra whenever G acts freely away from the origin and $d \geq 2$. Moreover, A is never an Azumaya algebra as its ramification locus is the isolated singularity.   ⬚

In analogy with the Brauer group one can define the *reflexive Brauer group* $\beta(R)$ whose elements are the equivalence classes $[A]$ for A a reflexive Azumaya algebra over R with equivalence relation

$$[A] = [A'] \qquad \text{iff} \qquad (A \otimes_R A')^{**} \simeq End_R(M)$$

where M is a reflexive R-module and with multiplication induced by the rule

$$[A].[A'] = [(A \otimes_R A')^{**}]$$

In [66] it was shown that the reflexive Brauer group does have a cohomological description similar to Gabber's result above.

PROPOSITION I.3
The reflexive Brauer group is an étale cohomology group

$$\beta(R) \simeq H^2_{et}(X_{sm}, \mathbb{G}_m)$$

where X_{sm} is the smooth locus of X.

This time we see that the singularities of X do appear in the description so perhaps reflexive Azumaya algebras are a class of orders more suitable for our project. This is even more evident if we impose noncommutative smoothness conditions on A.

PROPOSITION I.4

Let A be a reflexive Azumaya algebra over R, then:

1. *if A is Serre-smooth, then* ram $A = X_{sing}$, *and*

2. *if A is Cayley-smooth, then X_{sing} is contained in* ram A.

PROOF   (1) was proved in [68] the essential point being that if A is Serre-smooth then A is a Cohen-Macaulay R-module whence it must be projective over a Cayley-smooth point of X but then it is not just an reflexive Azumaya but actually an Azumaya algebra in that point. The second statement can be further refined as we will see later.   ▯

Many classes of well-studied algebras are reflexive Azumaya algebras.

- Trace rings $\mathbb{T}_{m,n}$ of m generic $n \times n$ matrices (unless $(m,n) = (2,2)$), see [65].

- Quantum enveloping algebras $U_q(\mathfrak{g})$ of semisimple Lie algebras at roots of unity, see for example [16].

- Quantum function algebras $O_q(G)$ for semisimple Lie groups at roots of unity, see for example [17].

- Symplectic reflection algebras $A_{t,c}$, see [18].

Now that we have a large supply of orders, it is time to clarify the connection with PGL_n-equivariant geometry. We will introduce a class of noncommutative algebras, the so-called *Cayley-Hamilton algebras*, which are the level n generalization of the category of commutative algebras and which contain all R-orders.

A *trace map tr* is a $\mathbb{C}$-linear function $A \longrightarrow A$ satisfying for all $a, b \in A$

$$tr(tr(a)b) = tr(a)tr(b) \qquad tr(ab) = tr(ba) \qquad \text{and} \qquad tr(a)b = btr(a)$$

so in particular, the image $tr(A)$ is contained in the center of A. If $M \in M_n(R)$ where R is a commutative $\mathbb{C}$-algebra, then its characteristic polynomial

$$\chi_M = det(t1_n - M) = t^n + a_1 t^{n-1} + a_2 t^{n-2} + \ldots + a_n$$

has coefficients a_i which are polynomials with rational coefficients in traces of powers of M

$$a_i = f_i(tr(M), tr(M^2), \ldots, tr(M^{n-1}))$$

Hence, if we have an algebra A with a trace map tr we can define a *formal characteristic polynomial* of degree n for every $a \in A$ by taking

$$\chi_a = t^n + f_1(tr(a), \ldots, tr(a^{n-1}))t^{n-1} + \ldots + f_n(tr(a), \ldots, tr(a^{n-1}))$$

which allows us to define the category **alg@n** of Cayley-Hamilton algebras of degree n.

DEFINITION I.4   *An object A in* **alg@n** *is a Cayley-Hamilton algebra of degree n, that is, a $\mathbb{C}$-algebra with trace map* $tr : A \longrightarrow A$ *satisfying*

$$tr(1) = n \qquad and \qquad \forall a \in A : \chi_a(a) = 0$$

Morphisms $A \longrightarrow B$ in **alg@n** *are trace preserving $\mathbb{C}$-algebra morphisms.*

Example I.9

Azumaya algebras, reflexive Azumaya algebras and more generally every R-order A in a central simple K-algebra of dimension n^2 is a Cayley-Hamilton algebra of degree n. For, consider the inclusions

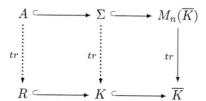

Here, $tr : M_n(\overline{K}) \longrightarrow \overline{K}$ is the usual trace map. By Galois descent this induces a trace map, the so-called *reduced trace*, $tr : \Sigma \longrightarrow K$. Finally, because R is integrally closed in K and A is a finitely generated R-module it follows that $tr(a) \in R$ for every element $a \in A$. ⬜

If A is a finitely generated object in **alg@n**, we can define an affine PGL_n-scheme, $\mathbf{trep}_n \, A$, classifying all trace preserving n-dimensional representations $A \xrightarrow{\phi} M_n(\mathbb{C})$ of A. The action of PGL_n on $\mathbf{trep}_n \, A$ is induced by conjugation in the target space, that is, $g.\phi$ is the trace preserving algebra map

$$A \xrightarrow{\phi} M_n(\mathbb{C}) \xrightarrow{g.-.g^{-1}} M_n(\mathbb{C})$$

Orbits under this action correspond precisely to isomorphism classes of representations. The scheme $\mathbf{trep}_n \, A$ is a closed subscheme of $\mathbf{rep}_n \, A$ the more familiar PGL_n-affine scheme of all n-dimensional representations of A. In general, both schemes may be different.

Example I.10

Let A be the quantum plane at -1, that is

$$A = \frac{\mathbb{C}\langle x, y \rangle}{(xy + yx)}$$

then A is an order with center $R = \mathbb{C}[x^2, y^2]$ in the quaternion algebra $(x, y)_2 = K1 \oplus Ku \oplus Kv \oplus Kuv$ over $K = \mathbb{C}(x, y)$ where $u^2 = x.v^2 = y$ and $uv = -vu$. Observe that $tr(x) = tr(y) = 0$ as the embedding $A \hookrightarrow (x, y)_2 \hookrightarrow M_2(\mathbb{C}[u, y])$ is given by

$$x \mapsto \begin{bmatrix} u & 0 \\ 0 & -u \end{bmatrix} \quad \text{and} \quad y \mapsto \begin{bmatrix} 0 & 1 \\ y & 0 \end{bmatrix}$$

Therefore, a trace preserving algebra map $A \longrightarrow M_2(\mathbb{C})$ is fully determined by the images of x and y, which are trace zero 2×2 matrices

$$\phi(x) = \begin{bmatrix} a & b \\ c & -a \end{bmatrix} \quad \text{and} \quad \phi(y) = \begin{bmatrix} d & e \\ f & -d \end{bmatrix} \quad \text{satisfying} \quad bf + ce = 0$$

That is, $\mathbf{trep}_2\, A$ is the hypersurface $\mathbb{V}(bf + ce) \subset \mathbb{A}^6$, which has a unique isolated singularity at the origin. However, $\mathbf{rep}_2\, A$ contains more points, for example

$$\phi(x) = \begin{bmatrix} a & 0 \\ 0 & b \end{bmatrix} \quad \text{and} \quad \phi(y) = \begin{bmatrix} 0 & 0 \\ 0 & 0 \end{bmatrix}$$

is a point in $\mathbf{rep}_2\, A - \mathbf{trep}_2\, A$ whenever $b \neq -a$. $\quad\quad$ ⧠

A functorial description of $\mathbf{trep}_n\, A$ is given by the following universal property due to C. Procesi [87], which will be proved in chapter 2.

THEOREM I.2

Let A be a $\mathbb{C}$-algebra with trace map tr_A, then there is a trace preserving algebra morphism

$$j_A : A \longrightarrow M_n(\mathbb{C}[\mathbf{trep}_n\, A])$$

satisfying the following universal property. If C is a commutative $\mathbb{C}$-algebra and there is a trace preserving algebra map $A \xrightarrow{\psi} M_n(C)$ (with the usual trace on $M_n(C)$), then there is a unique algebra morphism $\mathbb{C}[\mathbf{trep}_n\, A] \xrightarrow{\phi} C$ such that the diagram

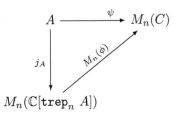

is commutative. Moreover, A is an object in alg@n *if and only if j_A is a monomorphism.*

The PGL_n-action on trep$_n$ A induces an action of PGL_n by automorphisms on $\mathbb{C}[\text{trep}_n\ A]$. On the other hand, PGL_n acts by conjugation on $M_n(\mathbb{C})$ so we have a combined action on $M_n(\mathbb{C}[\text{trep}_n\ A]) = M_n(\mathbb{C}) \otimes \mathbb{C}[\text{trep}_n\ A]$ and it follows from the universal property that the image of j_A is contained in the ring of PGL_n-invariants

$$A \xrightarrow{\ j_A\ } M_n(\mathbb{C}[\text{trep}_n\ A])^{PGL_n}$$

which is an inclusion if A is a Cayley-Hamilton algebra. In fact, C. Procesi proved in [87] the following important result that allows reconstruction of orders and their centers from PGL_n-equivariant geometry. This result will be proved in chapter 2.

THEOREM I.3
The functor
$$\text{trep}_n : \text{alg@n} \longrightarrow \text{PGL(n)-affine}$$

has a left *inverse*

$$\text{A}_- : \text{PGL(n)-affine} \longrightarrow \text{alg@n}$$

defined by $\text{A}_Y = M_n(\mathbb{C}[Y])^{PGL_n}$. *In particular, we have for any A in* alg@n

$$A = M_n(\mathbb{C}[\text{trep}_n\ A])^{PGL_n} \qquad \text{and} \qquad tr(A) = \mathbb{C}[\text{trep}_n\ A]^{PGL_n}$$

That is the central subalgebra $tr(A)$ is the coordinate ring of the algebraic quotient variety

$$\text{trep}_n\ A/PGL_n = \text{triss}_n\ A$$

classifying isomorphism classes of trace preserving semisimple n-dimensional representations of A.

The category alg@n is to noncommutative geometry@n what comm, the category of all commutative algebras is to commutative algebraic geometry. In fact, alg@1 $\simeq$ comm by taking as trace maps the identity on every commutative algebra. Further we have a natural commutative diagram of functors

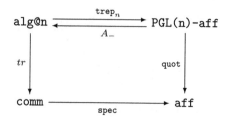

where the bottom map is the antiequivalence between affine algebras and affine schemes and the top map is the correspondence between Cayley-Hamilton algebras and affine PGL_n-schemes, which is *not* an equivalence of categories.

Example I.11
Conjugacy classes of nilpotent matrices in $M_n(\mathbb{C})$ correspond bijective to partitions $\lambda = (\lambda_1 \geq \lambda_2 \geq \ldots)$ of n (the λ_i determine the sizes of the Jordan blocks). It follows from the Gerstenhaber-Hesselink theorem that the closures of such orbits

$$\overline{\mathcal{O}_\lambda} = \cup_{\mu \leq \lambda} \mathcal{O}_\mu$$

where $\leq$ is the dominance order relation. Each $\overline{\mathcal{O}_\lambda}$ is an affine PGL_n-variety and the corresponding algebra is

$$\mathsf{A}_{\overline{\mathcal{O}_\lambda}} = \mathbb{C}[x]/(x^{\lambda_1})$$

whence many orbit closures (all of which are affine PGL_n-varieties) correspond to the same algebra. More details are given in section 2.7. ❏

Among the many characterizations of commutative smooth (that is, regular) algebras is the following, due to A. Grothendieck.

THEOREM I.4
A commutative $\mathbb{C}$-algebra A is smooth if and only if it satisfies the following lifting property: if (B, I) is a test-object such that B is a commutative algebra and I is a nilpotent ideal of B, then for any algebra map ϕ, there exists a lifted algebra morphism $\tilde{\phi}$

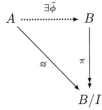

As the category `comm` of all commutative $\mathbb{C}$-algebras is just `alg@1` it makes sense to define Cayley-smooth Cayley-Hamilton algebras by the same lifting property. This was done first by W. Schelter [91] in the category of algebras satisfying all polynomial identities of $n \times n$ matrices and later by C. Procesi [87] in `alg@n`. Cayley-smooth algebras and their representation theory will be the main topic of this book.

DEFINITION I.5   *A Cayley-smooth algebra A is an object in `alg@n` satisfying the following lifting property. If (B, I) is a test-object in `alg@n`, that is, B is an object in `alg@n`, I is a nilpotent ideal in B such that B/I*

is an object in alg@n *and such that the natural map* $B \xrightarrow{\pi} B/I$ *is trace preserving, then every trace preserving algebra map* ϕ *has a lift* $\tilde{\phi}$

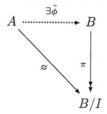

making the diagram commutative. If A *is in addition an order, we say that* A *is a* Cayley-smooth order.

In the next section we will give a large class of Cayley-smooth orders, but it should be stressed that there is no connection between this notion of non-commutative smoothness and the more homological notion of Serre-smooth orders (except in dimension one when all notions coincide). Under the correspondence between alg@n and PGL(n)-aff, Cayley-smooth Cayley-Hamilton algebras correspond to smooth PGL_n-varieties.

THEOREM I.5
An object A *in* alg@n *is Cayley-smooth if and only if the corresponding affine* PGL_n*-scheme* $\text{trep}_n A$ *is smooth (and hence, in particular, reduced).*

PROOF   (One implication) Assume A is Cayley-smooth, then to prove that $\text{trep}_n A$ is smooth we have to prove that $\mathbb{C}[\text{trep}_n A]$ satisfies Grothendieck's lifting property. So let (B, I) be a test-object in comm and take an algebra morphism $\phi: \mathbb{C}[\text{trep}_n A] \longrightarrow B/I$. Consider the following diagram

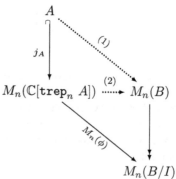

the morphism (1) follows from Cayley-smoothness of A applied to the morphism $M_n(\phi) \circ j_A$. From the universal property of the map j_A it follows that there is a morphism (2), which is of the form $M_n(\psi)$ for some algebra

morphism $\psi : \mathbb{C}[\mathsf{trep}_n\, A] \longrightarrow B$. This ψ is the required lift. The inverse
implication will be proved in section 4.1. ⬚

Example I.12
Trace rings $\mathbb{T}_{m,n}$ are the free algebras generated by m elements in alg@n and
as such trivially satisfy the lifting property whence are Cayley-smooth orders.
Alternatively, because

$$\mathsf{trep}_n\, \mathbb{T}_{m,n} \simeq M_n(\mathbb{C}) \oplus \ldots \oplus M_n(\mathbb{C}) = \mathbb{C}^{mn^2}$$

is a smooth PGL_n-variety, $\mathbb{T}_{m,n}$ is Cayley-smooth by the previous result. ⬚

Example I.13
Consider again the quantum plane at -1

$$A = \frac{\mathbb{C}\langle x, y\rangle}{(xy + yx)}$$

then we have seen that $\mathsf{trep}_2\, A = \mathbb{V}(bf + ce) \subset \mathbb{A}^6$ has a unique isolated
singularity at the origin. Hence, A is not a Cayley-smooth order. ⬚

I.2 Noncommutative geometry

We will associate to $A \in$ alg@n a *noncommutative variety* max A and argue that this gives a noncommutative manifold when A is a Cayley-smooth
order. In particular, we will show that for fixed n and central dimension d
there are a finite number of étale types of such orders. This fact is the noncommutative analogon of the classical result that a commutative manifold is
locally diffeomorphic to affine space or, in ring theory terms, that the m-adic
completion of a smooth algebra C of dimension d has just one étale type :
$\hat{C}_{\mathfrak{m}} \simeq \mathbb{C}[[x_1, \ldots, x_d]]$. There is one new feature which noncommutative geometry has to offer compared to commutative geometry : distinct points can
lie infinitesimally close to each other. As desingularization is the process of
separating bad tangents, this fact should be useful somehow in our project.

Recall that if X is an affine commutative variety with coordinate ring R,
then to each point $P \in X$ corresponds a maximal ideal $\mathfrak{m}_P \lhd R$ and a one-
dimensional simple representation

$$S_P = \frac{R}{\mathfrak{m}_P}$$

A basic tool in the study of Hilbert schemes is that finite closed subschemes of X can be decomposed according to their support. In algebraic terms this means that there are no extensions between different points, that if $P \neq Q$ then

$$Ext_R^1(S_P, S_Q) = 0 \qquad \text{whereas} \qquad Ext_R^1(S_P, S_P) = T_P\, X$$

That is, all infinitesimal information of X near P is contained in the self-extensions of S_P and separate points do not contribute. This is no longer the case for noncommutative algebras.

Example I.14

Take the path algebra A of the *quiver* $\circ\!\longrightarrow\!\circ$, that is

$$A \simeq \begin{bmatrix} \mathbb{C} & \mathbb{C} \\ 0 & \mathbb{C} \end{bmatrix}$$

Then A has two maximal ideals and two corresponding one-dimensional simple representations

$$S_1 = \begin{bmatrix} \mathbb{C} \\ 0 \end{bmatrix} = \begin{bmatrix} \mathbb{C} & \mathbb{C} \\ 0 & \mathbb{C} \end{bmatrix} \Big/ \begin{bmatrix} 0 & \mathbb{C} \\ 0 & \mathbb{C} \end{bmatrix} \qquad \text{and} \qquad S_2 = \begin{bmatrix} 0 \\ \mathbb{C} \end{bmatrix} = \begin{bmatrix} \mathbb{C} & \mathbb{C} \\ 0 & \mathbb{C} \end{bmatrix} \Big/ \begin{bmatrix} \mathbb{C} & \mathbb{C} \\ 0 & 0 \end{bmatrix}$$

Then, there is a nonsplit exact sequence with middle term the second column of A

$$0 \longrightarrow S_1 = \begin{bmatrix} \mathbb{C} \\ 0 \end{bmatrix} \longrightarrow M = \begin{bmatrix} \mathbb{C} \\ \mathbb{C} \end{bmatrix} \longrightarrow S_2 = \begin{bmatrix} 0 \\ \mathbb{C} \end{bmatrix} \longrightarrow 0$$

Whence $Ext_A^1(S_2, S_1) \neq 0$ whereas $Ext_A^1(S_1, S_2) = 0$. It is no accident that these two facts are encoded into the quiver. ▯

DEFINITION I.6   *For A an algebra in* alg@n, *define its* maximal ideal spectrum max A *to be the set of all maximal two-sided ideals M of A equipped with the* noncommutative Zariski topology, *that is, a typical open set of* max A *is of the form*

$$\mathbb{X}(I) = \{ M \in \max A \mid I \not\subset M \}$$

Recall that for every $M \in$ max A the quotient

$$\frac{A}{M} \simeq M_k(\mathbb{C}) \qquad \text{for some } k \leq n$$

that is, M determines a unique k-dimensional simple representation S_M of A.

Every maximal ideal M of A intersects the center R in a maximal ideal $\mathfrak{m}_P = M \cap R$ so, in the case of an R-order A a continuous map

$$\max A \xrightarrow{\;c\;} X \quad \text{defined by} \quad M \mapsto P \quad \text{where } M \cap R = \mathfrak{m}_P$$

Ring theorists have studied the fibers $c^{-1}(P)$ of this map in the seventies and eighties in connection with localization theory. The oldest description is the *Bergman-Small* theorem, see for example [8]

THEOREM I.6 Bergman-Small
If $c^{-1}(P) = \{M_1, \ldots, M_k\}$ then there are natural numbers $e_i \in \mathbb{N}_+$ such that

$$n = \sum_{i=1}^{k} e_i d_i \qquad \text{where } d_i = dim_{\mathbb{C}} S_{M_i}$$

In particular, $c^{-1}(P)$ is finite for all P.

Here is a modern proof of this result based on the results of this book. Because X is the algebraic quotient $\mathbf{trep}_n A/GL_n$, points of X correspond to *closed* GL_n-orbits in $\mathbf{rep}_n A$. By a result of M. Artin [2] (see section 2.4) closed orbits are precisely the isomorphism classes of *semisimple n-dimensional representations*, and therefore we denote the quotient variety

$$X = \mathbf{trep}_n A/GL_n = \mathbf{triss}_n A$$

A point P determines a semisimple n-dimensional A-representation

$$M_P = S_1^{\oplus e_1} \oplus \ldots \oplus S_k^{\oplus e_k}$$

with the S_i the distinct simple components, say of dimension $d_i = dim_{\mathbb{C}} S_i$ and occurring in M_P with multiplicity $e_i \geq 1$. This gives $n = \sum e_i d_i$ and clearly the annihilator of S_i is a maximal ideal M_i of A lying over $\mathfrak{m}_P$.

Another interpretation of $c^{-1}(P)$ follows from the work of A. V. Jategaonkar and B. Müller. Define a *link diagram* on the points of $\mathbf{max}\ A$ by the rule

$$M \rightsquigarrow M' \qquad \Leftrightarrow \qquad Ext_A^1(S_M, S_{M'}) \neq 0$$

In fancier language, $M \rightsquigarrow M'$ if and only if M and M' lie infinitesimally close in $\mathbf{max}\ A$. In fact, the definition of the link diagram in [47, Ch.5] or [39, Ch.11] is slightly different but amounts to the same thing.

THEOREM I.7 Jategaonkar-Müller
The connected components of the link diagram on $\mathbf{max}\ A$ are all finite and are in one-to-one correspondence with $P \in X$. That is, if

$$\{M_1, \ldots, M_k\} = c^{-1}(P) \subset \mathbf{max}\ A$$

then this set is a connected component of the link diagram.

In $\mathbf{max}\ A$ there is a Zariski open set of *Azumaya points*, that is those $M \in \mathbf{max}\ A$ such that $A/M \simeq M_n(\mathbb{C})$. It follows that each of these maximal ideals is a singleton connected component of the link diagram. So on this open set there is a one-to-one correspondence between points of X and maximal ideals of A so we can say that $\max A$ and X are *birational*. However, over the ramification locus there may be several maximal ideals of A lying over the same central maximal ideal and these points should be thought of as lying infinitesimally close to each other.

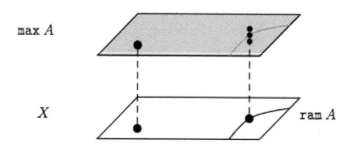

One might hope that the cluster of infinitesimally nearby points of $\max A$ lying over a central singularity $P \in X$ can be used to separate tangent information in P rather than having to resort to a blowing-up process to achieve this.

Because an R-order A in a central simple K-algebra Σ of dimension n^2 is a finite R-module, we can associate with A the sheaf $\mathcal{O}_A$ of noncommutative $\mathcal{O}_X$-algebras using central localization. That is, the section over a basic affine open piece $\mathbb{X}(f) \subset X$ are

$$\Gamma(\mathbb{X}(f), \mathcal{O}_A) = A_f = A \otimes_R R_f$$

which is readily checked to be a sheaf with global sections $\Gamma(X, \mathcal{O}_A) = A$. As we will investigate Cayley-smooth orders via their (central) étale structure, (that is, information about $\hat{A}_{\mathfrak{m}_P}$), we will only need the structure sheaf $\mathcal{O}_A$ over X in this book. However, in the 1970s F. Van Oystaeyen [103] and A. Verschoren [105] introduced genuine noncommutative structure sheaves associated to an R-order A. It is not my intention to promote nostalgia here but perhaps these noncommutative structure sheaves $\mathcal{O}_A^{nc}$ on $\max A$ deserve renewed investigation.

DEFINITION I.7   *$\mathcal{O}_A^{nc}$ is defined by taking as the sections over the typical open set $\mathbb{X}(I)$ (for I a two-sided ideal of A) in $\max A$*

$$\Gamma(\mathbb{X}(I), \mathcal{O}_A^{nc}) = \{\delta \in \Sigma \mid \exists l \in \mathbb{N} \; : \; I^l \delta \subset A \}$$

By [103] this defines a sheaf of noncommutative algebras over $\max A$ with global sections $\Gamma(\max A, \mathcal{O}_A^{nc}) = A$. The stalk of this sheaf at a point $M \in \max A$ is the symmetric localization

$$\mathcal{O}_{A,M}^{nc} = Q_{A-M}(A) = \{\delta \in \Sigma \mid I\delta \subset A \text{ for some ideal } I \not\subset P \}$$

Example I.15

Let $X = \mathbb{A}^1$, that is, $R = \mathbb{C}[x]$ and consider the order

$$A = \begin{bmatrix} R & R \\ \mathfrak{m} & R \end{bmatrix}$$

where $\mathfrak{m} = (x) \lhd R$. A is an hereditary order so is both a Serre-smooth order and a Cayley-smooth order. The ramification locus of A is $P_0 = \mathbb{V}(x)$ so over any $P_0 \neq P \in \mathbb{A}^1$ there is a unique maximal ideal of A lying over $\mathfrak{m}_P$ and the corresponding quotient is $M_2(\mathbb{C})$. However, over $\mathfrak{m}$ there are two maximal ideals of A

$$M_1 = \begin{bmatrix} \mathfrak{m} & R \\ \mathfrak{m} & R \end{bmatrix} \quad \text{and} \quad M_2 = \begin{bmatrix} R & R \\ \mathfrak{m} & \mathfrak{m} \end{bmatrix}$$

Both M_1 and M_2 determine a one-dimensional simple representation of A, so the Bergman-Small number are $e_1 = e_2 = 1$ and $d_1 = d_2 = 1$. That is, we have the following picture

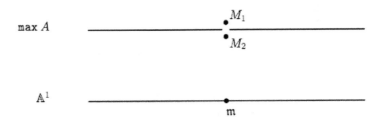

There is one nonsingleton connected component in the link diagram of A, namely

with the vertices corresponding to $\{M_1, M_2\}$. The stalk of $\mathcal{O}_A$ at the central point P_0 is clearly

$$\mathcal{O}_{A,P_0} = \begin{bmatrix} R_\mathfrak{m} & R_\mathfrak{m} \\ \mathfrak{m}_\mathfrak{m} & R_\mathfrak{m} \end{bmatrix}$$

On the other hand the stalks of the noncommutative structure sheaf $\mathcal{O}_A^{nc}$ in M_1 resp. M_2 can be computed to be

$$\mathcal{O}_{A,M_1}^{nc} = \begin{bmatrix} R_\mathfrak{m} & R_\mathfrak{m} \\ R_\mathfrak{m} & R_\mathfrak{m} \end{bmatrix} \quad \text{and} \quad \mathcal{O}_{A,M_2}^{nc} = \begin{bmatrix} R_\mathfrak{m} & x^{-1}R_\mathfrak{m} \\ xR_\mathfrak{m} & R_\mathfrak{m} \end{bmatrix}$$

and hence both stalks are Azumaya algebras. Observe that we recover the central stalk $\mathcal{O}_{A,P_0}$ as the intersection of these two rings in $M_2(K)$. Hence, somewhat surprisingly, the noncommutative structure sheaf of the hereditary non-Azumaya algebra A is a sheaf of Azumaya algebras over $\mathtt{max}\ A$. $\square$

Consider the continuous map for the Zariski topology

$$\mathtt{max}\ A \xrightarrow{\ c\ } X$$

and let for a central point $P \in X$ the fiber be $\{M_1, \ldots, M_k\}$ where the M_i are maximal ideals of A with corresponding simple d_i-dimensional representation S_i. We have introduced the *Bergman-Small data*, that is

$$\alpha = (e_1, \ldots, e_k) \text{ and } \beta = (d_1, \ldots, d_k) \in \mathbb{N}_+^k \quad \text{satisfying} \quad \alpha.\beta = \sum_{i=1}^{k} e_i d_i = n$$

(recall that e_i is the multiplicity of S_i in the semisimple n-dimensional representation corresponding to P). Moreover, we have the *Jategaonkar-Müller data*, which is a directed connected graph on the vertices $\{v_1, \ldots, v_k\}$ (corresponding to the M_i) with an arrow

$$v_i \rightsquigarrow v_j \qquad \text{iff} \qquad Ext_A^1(S_i, S_j) \neq 0$$

We will associate a combinatorial object to this local data. To begin, introduce a quiver setting (Q, α) where Q is a *quiver* (that is, a directed graph) on the vertices $\{v_1, \ldots, v_k\}$ with the number of arrows from v_i to v_j equal to the dimension of $Ext_A^1(S_i, S_j)$

$$\#\ (\ v_i \longrightarrow v_j\ )\ =\ dim_\mathbb{C}\ Ext_A^1(S_i, S_j)$$

and where $\alpha = (e_1, \ldots, e_k)$ is the *dimension vector* of the multiplicities e_i.

Recall that the representation space $\mathbf{rep}_\alpha Q$ of a quiver-setting is $\oplus_a M_{e_i \times e_j}(\mathbb{C})$ where the sum is taken over all arrows $a : v_i \longrightarrow v_j$ of Q. On this space there is a natural action by the group

$$GL(\alpha) = GL_{e_1} \times \ldots \times GL_{e_k}$$

by base-change in the vertex-spaces $V_i = \mathbb{C}^{e_i}$. The ring theoretic relevance of the quiver-setting (Q, α) is that

$$\mathbf{rep}_\alpha\ Q \simeq Ext_A^1(M_P, M_P) \qquad \text{as } GL(\alpha)\text{-modules}$$

where M_P is the semisimple n-dimensional A-module corresponding to P

$$M_P = S_1^{\oplus e_1} \oplus \ldots \oplus S_k^{\oplus e_k}$$

and because $GL(\alpha)$ is the automorphism group of M_P there is an induced action on $Ext_A^1(M_P, M_P)$.

Because M_P is n-dimensional, an element $\psi \in Ext_A^1(M_P, M_P)$ defines an algebra morphism

$$A \xrightarrow{\rho} M_n(\mathbb{C}[\epsilon])$$

where $\mathbb{C}[\epsilon] = \mathbb{C}[x]/(x^2)$ is the ring of *dual numbers*. As we are working in the category alg@n we need the stronger assumption that ρ is trace preserving. For this reason we have to consider the $GL(\alpha)$-subspace

$$Ext_A^{tr}(M_P, M_P) \subset Ext_A^1(M_P, M_P)$$

of *trace preserving extensions*. As traces only use blocks on the diagonal (corresponding to loops in Q) and as any subspace $M_{e_i}(\mathbb{C})$ of $\mathbf{rep}_\alpha Q$ decomposes as a $GL(\alpha)$-module in simple representations

$$M_{e_i}(\mathbb{C}) = M_{e_i}^0(\mathbb{C}) \oplus \mathbb{C}$$

where $M_{e_i}^0(\mathbb{C})$ is the subspace of trace zero matrices, we see that

$$\mathbf{rep}_\alpha Q^\bullet \simeq Ext_A^{tr}(M_P, M_P) \qquad \text{as } GL(\alpha)\text{-modules}$$

where $Q^\bullet$ is a *marked quiver* that has the same number of arrows between distinct vertices as Q has, but may have fewer loops and some of these loops may acquire a *marking* meaning that their corresponding component in $\mathbf{rep}_\alpha Q^\bullet$ is $M_{e_i}^0(\mathbb{C})$ instead of $M_{e_i}(\mathbb{C})$.

Summarizing, if the local structure of the noncommutative variety max A near the fiber $c^{-1}(P)$ of a central point $P \in X$ is determined by the Bergman-Small data

$$\alpha = (e_1, \ldots, e_k) \qquad \text{and} \qquad \beta = (d_1, \ldots, d_k)$$

and by the Jategoankar-Müller data, which is encoded in the marked quiver $Q^\bullet$ on k-vertices, then we associate to P the combinatorial data

$$\mathbf{type}(P) = (Q^\bullet, \alpha, \beta)$$

We call $(Q^\bullet, \alpha)$ the *marked quiver setting* associated to A in $P \in X$. The dimension vector $\beta = (d_1, \ldots, d_k)$ will be called the *Morita setting* associated to A in P.

Example I.16

If A is an Azumaya algebra over R. Then for every maximal ideal $\mathfrak{m}$ corresponding to a point $P \in X$ we have that

$$A/\mathfrak{m}A = M_n(\mathbb{C})$$

so there is a unique maximal ideal $M = \mathfrak{m}A$ lying over $\mathfrak{m}$ whence the Jategaonkar-Müller data are $\alpha = (1)$ and $\beta = (n)$. If $S_P = R/\mathfrak{m}$ is the simple representation of R we have

$$Ext_A^1(M_P, M_P) \simeq Ext_R^1(S_P, S_P) = T_P X$$

and as all the extensions come from the center, the corresponding algebra representations $A \longrightarrow M_n(\mathbb{C}[\epsilon])$ are automatically trace preserving. That is, the marked quiver-setting associated to A in P is

where the number of loops is equal to the dimension of the tangent space $T_P X$ in P at X and the Morita-setting associated to A in P is (n). ▯

Example I.17

Consider the order of example I.15, which is generated as a $\mathbb{C}$-algebra by the elements

$$a = \begin{bmatrix} 1 & 0 \\ 0 & 0 \end{bmatrix} \quad b = \begin{bmatrix} 0 & 1 \\ 0 & 0 \end{bmatrix} \quad c = \begin{bmatrix} 0 & 0 \\ x & 0 \end{bmatrix} \quad d = \begin{bmatrix} 0 & 0 \\ 0 & 1 \end{bmatrix}$$

and the 2-dimensional semisimple representation M_{P_0} determined by $\mathfrak{m}$ is given by the algebra morphism $A \longrightarrow M_2(\mathbb{C})$ sending a and d to themselves and b and c to the zero matrix. A calculation shows that

$$Ext^1_A(M_{P_0}, M_{P_0}) = \mathbf{rep}_\alpha Q \qquad \text{for} \qquad (Q, \alpha) = \text{①} \overset{u}{\underset{v}{\rightleftarrows}} \text{①}$$

and as the correspondence with algebra maps to $M_2(\mathbb{C}[\epsilon])$ is given by

$$a \mapsto \begin{bmatrix} 1 & 0 \\ 0 & 0 \end{bmatrix} \quad b \mapsto \begin{bmatrix} 0 & \epsilon v \\ 0 & 0 \end{bmatrix} \quad c \mapsto \begin{bmatrix} 0 & 0 \\ \epsilon u & 0 \end{bmatrix} \quad d \mapsto \begin{bmatrix} 0 & 0 \\ 0 & 1 \end{bmatrix}$$

each of these maps is trace preserving so the marked quiver setting is (Q, α) and the Morita-setting is $(1, 1)$. ▯

Because the combinatorial data $\mathbf{type}(P) = (Q^\bullet, \alpha, \beta)$ encodes the infinitesimal information of the cluster of maximal ideals of A lying over the central point $P \in X$, $(\mathbf{rep}_\alpha Q^\bullet, \beta)$ should be viewed as analogous to the usual tangent space $T_P X$. If $P \in X$ is a singular point, then the tangent space is too large so we have to impose additional relations to describe the variety X in a neighborhood of P, but if P is a smooth point we can recover the local structure of X from $T_P X$. In the noncommutative case we might expect a similar phenomenon: in general the data $(\mathbf{rep}_\alpha Q^\bullet, \beta)$ will be too big to describe $\hat{A}_{\mathfrak{m}P}$ unless A is a Cayley-smooth order in P in which case we can recover $\hat{A}_{\mathfrak{m}P}$.

We begin by defining some algebras that can be described combinatorially from $(Q^\bullet, \alpha, \beta)$. For every arrow $a : v_i \longrightarrow v_j$ define a *generic rectangular matrix* of size $e_j \times e_i$

$$X_a = \begin{bmatrix} x_{11}(a) & \cdots\cdots & x_{1e_i}(a) \\ \vdots & & \vdots \\ x_{e_j1}(a) & \cdots\cdots & x_{e_je_i}(a) \end{bmatrix}$$

In the case when a is a marked loop, make this a generic trace zero matrix, that is, let

$$x_{e_i e_i}(a) = -x_{11}(a) - x_{22}(a) - \ldots - x_{e_i-1 e_i-1}(a)$$

Then, the coordinate ring $\mathbb{C}[\mathbf{rep}_\alpha \, Q^\bullet]$ is the polynomial ring in the entries of all X_a. For an oriented path p in the marked quiver $Q^\bullet$ with starting vertex v_i and terminating vertex v_j

$$v_i \xdashrightarrow{p} v_j \quad = \quad v_i \xrightarrow{a_1} v_{i_1} \xrightarrow{a_2} \ldots \xrightarrow{a_{l-1}} v_{i_l} \xrightarrow{a_l} v_j$$

we can form the square $e_j \times e_i$ matrix

$$X_p = X_{a_l} X_{a_{l-1}} \ldots X_{a_2} X_{a_1}$$

which has all its entries polynomials in $\mathbb{C}[\mathbf{rep}_\alpha \, Q^\bullet]$. In particular, if the path is an oriented *cycle* c in $Q^\bullet$ starting and ending in v_i then X_c is a square $e_i \times e_i$ matrix and we can take its trace $tr(X_c) \in \mathbb{C}[\mathbf{rep}_\alpha \, Q^\bullet]$ which is a polynomial invariant under the action of $GL(\alpha)$ on $\mathbf{rep}_\alpha \, Q^\bullet$. In fact, we will prove in section 4.3 that these *traces along oriented cycles* generate the invariant ring

$$R_{Q^\bullet}^\alpha = \mathbb{C}[\mathbf{rep}_\alpha \, Q^\bullet]^{GL(\alpha)} \subset \mathbb{C}[\mathbf{rep}_\alpha \, Q^\bullet]$$

Next we bring in the Morita-setting $\beta = (d_1, \ldots, d_k)$ and define a block-matrix ring

$$A_{Q^\bullet}^{\alpha,\beta} = \begin{bmatrix} M_{d_1 \times d_1}(P_{11}) & \ldots & M_{d_1 \times d_k}(P_{1k}) \\ \vdots & & \vdots \\ M_{d_k \times d_1}(P_{k1}) & \ldots & M_{d_k \times d_k}(P_{kk}) \end{bmatrix} \subset M_n(\mathbb{C}[\mathbf{rep}_\alpha \, Q^\bullet])$$

where P_{ij} is the $R_{Q^\bullet}^\alpha$-submodule of $M_{e_j \times e_i}(\mathbb{C}[\mathbf{rep}_\alpha \, Q^\bullet])$ generated by all X_p where p is an oriented path in $Q^\bullet$ starting in v_i and ending in v_k. Observe that for triples $(Q^\bullet, \alpha, \beta_1)$ and $(Q^\bullet, \alpha, \beta_2)$ we have that

$$A_{Q^\bullet}^{\alpha,\beta_1} \qquad \text{is Morita-equivalent to} \qquad A_{Q^\bullet}^{\alpha,\beta_2}$$

whence the name Morita-setting for β. Recall that the *Euler-form* of the underlying quiver Q of $Q^\bullet$ (that is, forgetting the markings of some loops) is the bilinear form χ_Q on $\mathbb{Z}^k$ such that $\chi_Q(e_i, e_j)$ is equal to δ_{ij} minus the number of arrows from v_i to v_j. The next result will be proved in section 5.2.

THEOREM I.8

For a triple $(Q^\bullet, \alpha, \beta)$ with $\alpha.\beta = n$ we have

1. $A_{Q^\bullet}^{\alpha,\beta}$ *is an R_Q^α-order in* $\mathtt{alg@n}$ *if and only if α is the dimension vector of a simple representation of $Q^\bullet$, that is, for all vertex-dimensions δ_i we have*

$$\chi_Q(\alpha, \delta_i) \leq 0 \qquad \text{and} \qquad \chi_Q(\delta_i, \alpha) \leq 0$$

unless $Q^\bullet$ is an oriented cycle of type $\tilde{A}_{k-1}$ then α must be $(1, \ldots, 1)$.

2. If this condition is satisfied, the dimension of the center $R_{Q^\bullet}^\alpha$ is equal to

$$\dim R_{Q^\bullet}^\alpha = 1 - \chi_Q(\alpha,\alpha) - \#\{\text{marked loops in } Q^\bullet\}$$

These combinatorial algebras determine the étale local structure of Cayley-smooth orders as was proved in [74] (or see section 5.2). The principal technical ingredient in the proof is the *Luna slice theorem*, see, for example, [99] [81] or section 3.8.

THEOREM I.9

Let A be a Cayley-smooth order over R in $\mathbf{alg@n}$ and let $P \in X$ with corresponding maximal ideal $\mathfrak{m}$. If the marked quiver setting and the Morita-setting associated to A in P is given by the triple $(Q^\bullet, \alpha, \beta)$, then there is a Zariski open subset $\mathbb{X}(f_i)$ containing P and an étale extension S of both R_{f_i} and the algebra $R_{Q^\bullet}^\alpha$ such that we have the following diagram

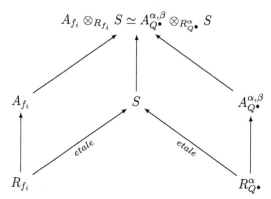

In particular, we have

$$\hat{R}_{\mathfrak{m}} \simeq \hat{R}_{Q^\bullet}^\alpha \qquad \text{and} \qquad \hat{A}_{\mathfrak{m}} \simeq \hat{A}_{Q^\bullet}^{\alpha,\beta}$$

where the completions at the right-hand sides are with respect to the maximal (graded) ideal of $R_{Q^\bullet}^\alpha$ corresponding to the zero representation.

Example I.18

From example I.16 we recall that the triple $(Q^\bullet, \alpha, \beta)$ associated to an Azumaya algebra in a point $P \in X$ is given by

$$\begin{array}{ccc} \text{①} & \text{and} & \beta = (n) \end{array}$$

where the number of arrows is equal to $dim_{\mathbb{C}} T_P X$. In case P is a Cayley-smooth point of X this number is equal to $d = \dim X$. Observe that $GL(\alpha) = \mathbb{C}^*$ acts trivially on $\mathbf{rep}_\alpha Q^\bullet = \mathbb{C}^d$ in this case. Therefore we have that

$$R_{Q^\bullet}^\alpha \simeq \mathbb{C}[x_1,\dots,x_d] \quad \text{and} \quad A_{Q^\bullet}^{\alpha,\beta} = M_n(\mathbb{C}[x_1,\dots,x_d])$$

Because A is a Cayley-smooth order in such points we get that

$$\hat{A}_{\mathfrak{m}_{\mathfrak{P}}} \simeq M_n(\mathbb{C}[[x_1,\ldots,x_d]])$$

consistent with our étale local knowledge of Azumaya algebras.   ❏

Because $\alpha.\beta = n$, the number of vertices of $Q^{\bullet}$ is bounded by n and as

$$d = 1 - \chi_Q(\alpha,\alpha) - \#\{\text{marked loops}\}$$

the number of arrows and (marked) loops is also bounded. This means that for a particular dimension d of the central variety X there are only a finite number of étale local types of Cayley-smooth orders in alg@n. This might be seen as a noncommutative version of the fact that there is just one étale type of a Cayley-smooth variety in dimension d namely, $\mathbb{C}[[x_1,\ldots,x_d]]$. At this moment a similar result for Serre-smooth orders seems to be far out of reach.

The reduction steps below were discovered by R. Bocklandt in his Ph.D. thesis (see also [10] or section 5.7) in which he classified quiver settings having a Serre-smooth ring of invariants. These steps were slightly extended in [14] (or section 5.8) in order to classify central singularities of Cayley-smooth orders. All reductions are made locally around a vertex in the marked quiver. There are three types of allowed moves.

Vertex removal: Assume we have a marked quiver setting $(Q^{\bullet},\alpha)$ and a vertex v such that the local structure of $(Q^{\bullet},\alpha)$ near v is indicated by the picture on the left below, that is, inside the vertices we have written the components of the dimension vector and the subscripts of an arrow indicate how many such arrows there are in $Q^{\bullet}$ between the indicated vertices. Define the new marked quiver setting $(Q_R^{\bullet},\alpha_R)$ obtained by the operation R_V^v, which removes the vertex v and composes all arrows through v, the dimensions of the other vertices are unchanged

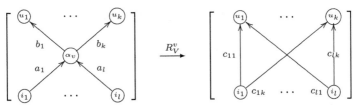

where $c_{ij} = a_i b_j$ (observe that some of the incoming and outgoing vertices may be the same so that one obtains loops in the corresponding vertex). This reduction can be made provided

$$\alpha_v \geq \sum_{j=1}^{l} a_j i_j \qquad \text{or} \qquad \alpha_v \geq \sum_{j=1}^{k} b_j u_j$$

(observe that if we started off from a marked quiver setting $(Q^{\bullet},\alpha)$ coming from an order, then these inequalities must actually be equalities).

Small loop removal: If v is a vertex with vertex-dimension $\alpha_v = 1$ and having $k \geq 1$ loops. Let $(Q_R^\bullet, \alpha_R)$ be the marked quiver setting obtained by the loop removal operation R_l^v

$$\left[\quad \underset{1}{\overset{k}{\bigcirc}} \quad \right] \xrightarrow{R_l^v} \left[\quad \underset{1}{\overset{k-1}{\bigcirc}} \quad \right]$$

removing one loop in v and keeping the same dimension vector.

Loop removal: If the local situation in v is such that there is exactly one (marked) loop in v, the dimension vector in v is $k \geq 2$ and there is exactly one arrow leaving v and this to a vertex with dimension vector 1, then one is allowed to make the reduction R_L^v indicated below

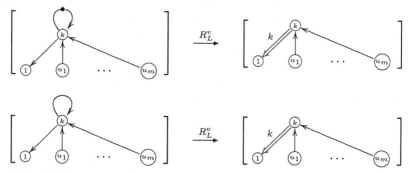

Similarly, if there is one (marked) loop in v and $\alpha_v = k \geq 2$ and there is only one arrow arriving at v coming from a vertex of dimension vector 1, then one is allowed to make the reduction R_L^v

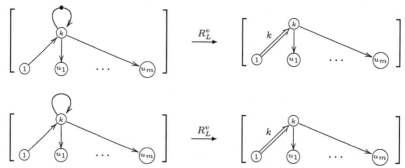

The relevance of these reducation steps on marked quiver settings is that if

$$(Q_1^\bullet, \alpha_1) \rightsquigarrow (Q_2^\bullet, \alpha_2)$$

is a sequence of legal moves then

$$R_{Q_1^\bullet}^{\alpha_1} \simeq R_{Q_2^\bullet}^{\alpha_2}[y_1, \ldots, y_z]$$

where z is the sum of all loops removed in R_l^v reductions plus the sum of α_v for each reduction step R_L^v involving a genuine loop and the sum of $\alpha_v - 1$ for each reduction step R_L^v involving a marked loop. In section 5.3 we will prove that every marked quiver setting $(Q^\bullet, \alpha)$ can be reduced *uniquely* to a *reduced* or *zero-setting* $Z(Q^\bullet, \alpha) = (Q_0^\bullet, \alpha_0)$ that is a setting from which no further reductions can be made. Therefore it is sufficient to classify these zero-settings if we want to classify all central singularities of a Cayley-smooth order for a given central dimension d.

To start, can we decide when P is a smooth point of X? If A is an Azumaya algebra in P, we know that A can only be Cayley-smooth if X is smooth in P. For Cayley-smooth orders the situation is more delicate but we have a complete solution in terms of the reduction steps, as will be proved in section 5.7.

THEOREM I.10
If A is a Cayley-smooth R-order and $(Q^\bullet, \alpha, \beta)$ is the combinatorial data associated to A in $P \in X$. Then, P is a smooth point of X if and only if the unique zero-setting

$$Z(Q^\bullet, \alpha) \in \{ \; \}$$

The Azumaya points are such that $Z(Q^\bullet, \alpha) = \textcircled{1}$ hence the singular locus of X is contained in the ramification locus $\mathbf{ram}\ A$ but may be strictly smaller.

To classify the central singularities of Cayley-smooth orders we may reduce to zero-settings $(Q^\bullet, \alpha) = Z(Q^\bullet, \alpha)$. For such a setting we have for all vertices v_i the inequalities

$$\chi_Q(\alpha, \delta_i) < 0 \qquad \text{and} \qquad \chi_Q(\delta_i, \alpha) < 0$$

and the dimension of the central variety can be computed from the Euler-form χ_Q. This gives us an estimate of $d = \dim X$, which is very efficient to classify the singularities in low dimensions.

THEOREM I.11
Let $(Q^\bullet, \alpha) = Z(Q^\bullet, \alpha)$ be a zero-setting on $k \geq 2$ vertices. Then

$$\dim X \geq 1 + \sum_{\substack{\textcircled{a}}}^{a \geq 1} a + \sum^{a > 1} (2a - 1) + \sum^{a > 1} (2a) + \sum^{a > 1} (a^2 + a - 2) +$$

$$\sum^{a > 1} (a^2 + a - 1) + \sum^{a > 1} (a^2 + a) + \ldots + \sum^{a > 1} ((k + l - 1)a^2 + a - k) + \ldots$$

In this sum the contribution of a vertex v with $\alpha_v = a$ is determined by the number of (marked) loops in v. By the reduction steps (marked) loops only occur at vertices where $\alpha_v > 1$.

Example I.19 dimension 2

When $\dim X = 2$ no zero-settings on at least two vertices satisfy the inequality of theorem 5.14, so the only zero-position possible to be obtained from a marked quiver-setting $(Q^\bullet, \alpha)$ in dimension two is

$$Z(Q^\bullet, \alpha) = \textcircled{\scriptsize 1}$$

and therefore the central two-dimensional variety X of a Cayley-smooth order is smooth.   ☐

Example I.20 dimension 3

If $(Q^\bullet, \alpha)$ is a zero-setting for dimension ≤ 3 then $Q^\bullet$ can have at most two vertices. If there is just one vertex it must have dimension 1 (reducing again to $\textcircled{\scriptsize 1}$ whence smooth) or must be

$$Z(Q^\bullet, \alpha) = \qquad \text{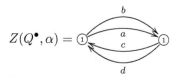}$$

which is again a smooth setting. If there are two vertices both must have dimension 1 and both must have at least two incoming and two outgoing arrows (for otherwise we could perform an additional vertex-removal reduction). As there are no loops possible in these vertices for zero-settings, it follows from the formula $d = 1 - \chi_Q(\alpha, \alpha)$ that the only possibility is

$$Z(Q^\bullet, \alpha) = \textcircled{\scriptsize 1} \underset{\substack{\\ d}}{\overset{\substack{b\\ }}{\rightrightarrows\!\!\!\rightleftarrows}} \textcircled{\scriptsize 1}$$

The ring of polynomial invariants $R^\alpha_{Q^\bullet}$ is generated by traces along oriented cycles in $Q^\bullet$ so in this case it is generated by the invariants

$$x = ac, \quad y = ad, \quad u = bc \quad \text{and} \quad v = bd$$

and there is one relation between these generators, so

$$R^\alpha_{Q^\bullet} \simeq \frac{\mathbb{C}[x, y, u, v]}{(xy - uv)}$$

Therefore, the only étale type of central singularity in dimension three is the *conifold singularity.*   ☐

Example I.21 dimension 4

If $(Q^\bullet, \alpha)$ is a zero-setting for dimension 4 then $Q^\bullet$ can have at most three vertices. If there is just one, its dimension must be 1 (smooth setting) or 2 in which case the only new type is

$$Z(Q^\bullet, \alpha) = \quad$$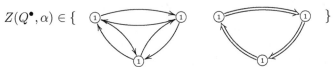

which is again a smooth setting.

If there are two vertices, both must have dimension 1 and have at least two incoming and outgoing arrows as in the previous example. The only new type that occurs is

$$Z(Q^\bullet, \alpha) = $$

for which one calculates as before the ring of invariants to be

$$R_{Q^\bullet}^\alpha = \frac{\mathbb{C}[a, b, c, d, e, f]}{(ae - bd, af - cd, bf - ce)}$$

If there are three vertices all must have dimension 1 and each vertex must have at least two incoming and two outgoing vertices. There are just two such possibilities in dimension 4

$$Z(Q^\bullet, \alpha) \in \{ \qquad \qquad \qquad \}$$

The corresponding rings of polynomial invariants are

$$R_{Q^\bullet}^\alpha = \frac{\mathbb{C}[x_1, x_2, x_3, x_4, x_5]}{(x_4 x_5 - x_1 x_2 x_3)} \qquad \text{resp.} \qquad R_{Q^\bullet}^\alpha = \frac{\mathbb{C}[x_1, x_2, x_3, x_4, y_1, y_2, y_3, y_4]}{R_2}$$

where R_2 is the ideal generated by all 2×2 minors of the matrix

$$\begin{bmatrix} x_1 & x_2 & x_3 & x_4 \\ y_1 & y_2 & y_3 & y_4 \end{bmatrix}$$

Hence, in low dimensions (and we will extend the above calculation to dimension 5 in section 5.8) there is a full classification of the central singularities $\hat{R}_m$ of a Cayley-smooth order in **alg@n**. There is even a classification of all isolated simgularities that can arise in arbitrary dimension.

THEOREM I.12

Let A be a Cayley-smooth order over R and let $(Q^\bullet, \alpha, \beta)$ be the combinatorial data associated to a A in a point $P \in X$. Then, P is an isolated singularity *if and only if* $Z(Q^\bullet, \alpha) = T(k_1, \ldots, k_l)$ *where*

$$T(k_1, \ldots, k_l) =$$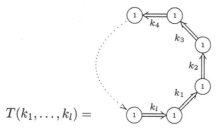

with $d = \dim X = \sum_i k_i - l + 1$. *Moreover, two such singularities, corresponding to* $T(k_1, \ldots, k_l)$ *and* $T(k'_1, \ldots, k'_{l'})$, *are isomorphic if and only if*

$$l = l' \quad \text{and} \quad k'_i = k_{\sigma(i)}$$

for some permutation $\sigma \in S_l$.

I.3   Noncommutative desingularizations

In view of the last theorem, the coordinate ring $R = \mathbb{C}[\mathbb{C}^d/G]$ of a quotient singularity can never be the center of a Cayley-smooth order A. However, there are nice orders $A \in \mathtt{alg@n}$ that are quotients $A \simeq A_{Q^\bullet}^{\alpha,\beta}/I$ of the Cayley-smooth order $A_{Q^\bullet}^{\alpha,\beta}$ modulo an ideal I of relations and having $\mathbb{C}[\mathbb{C}^d/G]$ as their center.

Example I.22 Kleinian singularities

For a Kleinian singularity, that is, a quotient singularity $\mathbb{C}^2/G$ with $G \subset SL_2(\mathbb{C})$ there is an extended Dynkin diagram D associated.

Let Q be the *double quiver* of D, that is to each arrow $\circ \xrightarrow{x} \circ$ in D we adjoin an arrow $\circ \xleftarrow{x^*} \circ$ in Q in the opposite direction and let α be the unique minimal dimension vector such that $\chi_D(\alpha, \alpha) = 0$. Further, consider the *moment element*

$$m = \sum_{x \in D} [x, x^*]$$

in the order A_Q^α then

$$A = \frac{A_Q^\alpha}{(m)}$$

is an order with center $R = \mathbb{C}[\mathbb{C}^2/G]$ which is isomorphic to the skew-group algebra $\mathbb{C}[x, y] \# G$. Moreover, A is Morita equivalent to the *preprojective algebra* which is the quotient of the path algebra of Q by the ideal generated by the moment element

$$\Pi_0 = \mathbb{C}Q/(\sum [x, x^*])$$

For more details we refer to the lecture notes by W. Crawley-Boevey [27] and section 5.6. ☐

Example I.23

Consider a quotient singularity $X = \mathbb{C}^d/G$ with $G \subset SL_d(\mathbb{C})$ and Q be the *McKay quiver* of G acting on $V = \mathbb{C}^d$. That is, the vertices $\{v_1, \ldots, v_k\}$ of Q are in one-to-one correspondence with the irreducible representations $\{R_1, \ldots, R_k\}$ of G such that $R_1 = \mathbb{C}_{triv}$ is the trivial representation. Decompose the tensorproduct in irreducibles

$$V \otimes_{\mathbb{C}} R_j = R_1^{\oplus j_1} \oplus \ldots \oplus R_k^{\oplus j_k}$$

then the number of arrows in Q from v_i to v_j

$$\#(v_i \longrightarrow v_j) = j_i$$

is the multiplicity of R_i in $V \otimes R_j$. Let $\alpha = (e_1, \ldots, e_k)$ be the dimension vector where $e_i = dim_{\mathbb{C}} R_i$. The relevance of this quiver-setting is that

$$\mathbf{rep}_\alpha Q = Hom_G(R, R \otimes V)$$

where R is the *regular representation*, see, for example, [23]. Consider $Y \subset \mathbf{rep}_\alpha Q$ the affine subvariety of all α-dimensional representations of Q for which the corresponding G-equivariant map $B \in Hom_G(R, V \otimes R)$ satisfies

$$B \wedge B = 0 \in Hom_G(R, \wedge^2 V \otimes R)$$

Y is called the *variety of commuting matrices* and its defining relations can be expressed as linear equations between paths in Q evaluated in $\mathbf{rep}_\alpha Q$, say $(l_1, \ldots, l_z)$. Then

$$A = \frac{A_Q^\alpha}{(l_1, \ldots, l_z)}$$

is an order with center $R = \mathbb{C}[\mathbb{C}^d/G]$. In fact, A is the skew group algebra

$$A = \mathbb{C}[x_1, \ldots, x_d] \# G$$

☐

Example I.24

Consider the natural action of $\mathbb{Z}_3$ on $\mathbb{C}^2$ via its embedding in $SL_2(\mathbb{C})$ sending the generator to the matrix

$$\begin{bmatrix} \rho & 0 \\ 0 & \rho^{-1} \end{bmatrix}$$

where ρ is a primitive 3rd root of unity. $\mathbb{Z}_3$ has three one-dimensional simples $R_1 = \mathbb{C}_{triv}, R_2 = \mathbb{C}_\rho$ and $R_2 = \mathbb{C}_{\rho^2}$. As $V = \mathbb{C}^2 = R_2 \oplus R_3$ it follows that the McKay quiver setting (Q, α) is

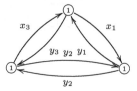

Consider the matrices

$$X = \begin{bmatrix} 0 & 0 & x_3 \\ x_1 & 0 & 0 \\ 0 & x_2 & 0 \end{bmatrix} \quad \text{and} \quad Y = \begin{bmatrix} 0 & y_1 & 0 \\ 0 & 0 & y_2 \\ y_3 & 0 & 0 \end{bmatrix}$$

then the variety of commuting matrices is determined by the matrix-entries of $[X, Y]$ that is

$$I = (x_3y_3 - y_1x_1, x_1y_1 - y_2x_2, x_2y_2 - y_3x_3)$$

so the skew-group algebra is the quotient of the Cayley-smooth order A_Q^α (which incidentally is one of our zero-settings for dimension 4)

$$\mathbb{C}[x, y]\#\mathbb{Z}_3 \simeq \frac{A_Q^\alpha}{(x_3y_3 - y_1x_1, x_1y_1 - y_2x_2, x_2y_2 - y_3x_3)}$$

Taking $y_i = x_i^*$ this coincides with the description via preprojective algebras as the moment element is

$$m = \sum_{i=1}^{3} [x_i, x_i^*] = (x_3y_3 - y_1x_1)e_1 + (x_1y_1 - y_2x_2)e_2 + (x_2y_2 - y_3x_3)e_3$$

where the e_i are the vertex-idempotents.   ◻

From now on we will restrict to quotient algebras $A = A_{Q^\bullet}^\alpha/I$ satisfying the following conditions :

- $\alpha = (e_1, \ldots, e_k)$ is the dimension vector of a simple representation of A, and

- the center $R = Z(A)$ is an integrally closed domain.

These requirements imply that A is an order over R in $\mathbf{alg@n}$ where n is the total dimension of the simple representation, that is $|\alpha| = \sum_i e_i$.

For such an $A = A_{Q^\bullet}^\alpha/I$ define the affine variety of α-dimensional representations

$$\mathbf{rep}_\alpha A = \{V \in \mathbf{rep}_\alpha Q^\bullet \mid r(V) = 0 \ \forall r \in I\}$$

The action of $GL(\alpha) = \prod_i GL_{e_i}$ by base change on $\mathbf{rep}_\alpha\, Q^\bullet$ induces an action on $\mathbf{rep}_\alpha\, A$. Usually, $\mathbf{rep}_\alpha\, A$ will have singularities but it may be smooth on the Zariski open subset of θ-semistable representations, which we will now define.

A *character* of $GL(\alpha)$ is determined by an integral k-tuple $\theta = (t_1, \ldots, t_k) \in \mathbb{Z}^k$

$$\chi_\theta \;:\; GL(\alpha) \longrightarrow \mathbb{C}^* \qquad (g_1, \ldots, g_k) \mapsto det(g_1)^{t_1} \ldots det(g_k)^{t_k}$$

Characters define *stability structures* on A-representations but as the acting group on $\mathbf{rep}_\alpha\, A$ is really $PGL(\alpha) = GL(\alpha)/\mathbb{C}^*(1_{e_1}, \ldots, 1_{e_k})$ we only consider characters θ satisfying $\theta.\alpha = \sum_i t_i e_i = 0$. If $V \in \mathbf{rep}_\alpha\, A$ and $V' \subset V$ is an A-subrepresentation we denote the dimension vector of V' by $\mathbf{dim}V'$.

DEFINITION I.8 *For θ satisfying $\theta.\alpha = 0$, a representation $V \in \mathbf{rep}_\alpha\, A$ is said to be*

- θ-semistable *if and only if for every proper A-subrepresentation $0 \neq V' \subset V$ we have $\theta.\mathbf{dim}V' \geq 0$.*

- θ-stable *if and only if for every proper A-subrepresentation $0 \neq V' \subset V$ we have $\theta.\mathbf{dim}V' > 0$.*

For any setting $\theta.\alpha = 0$ we have the following inclusions of Zariski open $GL(\alpha)$-stable subsets of $\mathbf{rep}_\alpha\, A$

$$\mathbf{rep}_\alpha^{simple}\, A \subset \mathbf{rep}_\alpha^{\theta-stable}\, A \subset \mathbf{rep}_\alpha^{\theta-semist}\, A \subset \mathbf{rep}_\alpha\, A$$

but one should note that some of these open subsets may actually be empty!

Recall that a point of the algebraic quotient variety $\mathbf{iss}_\alpha\, A = \mathbf{rep}_\alpha A/GL(\alpha)$ represents the orbit of an α-dimensional semisimple representation V and such representations can be separated by the values $f(V)$ where f is a polynomial invariant on $\mathbf{rep}_\alpha\, A$. For θ-stable and θ-semistable representations there are similar results and morally one should view θ-stable representations as corresponding to simple representations whereas θ-semistables are arbitrary representations. For this reason we will only be able to classify direct sums of θ-stable representations by certain algebraic varieties, which are called *moduli spaces* of semistables representations. More details will be given in section 4.8.

The notion corresponding to a polynomial invariant in this more general setting is that of a *polynomial semi-invariant*. A polynomial function $f \in \mathbb{C}[\mathbf{rep}_\alpha\, A]$ is said to be a θ-semi-invariant of *weight* l if for all $g \in GL(\alpha)$ we have

$$g.f = \chi_\theta(g)^l f$$

where χ_θ is the character of $GL(\alpha)$ corresponding to θ. A representation $V \in \mathbf{rep}_\alpha\, A$ is θ-semistable if and only if there is a θ-semi-invariant f of some weight l such that $f(V) \neq 0$. Clearly, θ-semi-invariants of weight zero are just polynomial invariants and the multiplication of θ-semi-invariants of weight l resp. l' has weight $l + l'$. Hence, the ring of all θ-semi-invariants

$$\mathbb{C}[\text{rep}_\alpha \ A]^{GL(\alpha),\theta} = \oplus_{l=0}^\infty \{f \in \mathbb{C}[\text{rep}_\alpha \ A] \ | \forall g \in GL(\alpha) \ : \ g.f = \chi_\theta(g)^l f \ \}$$

is a graded algebra with part of degree zero $\mathbb{C}[\text{iss}_\alpha \ A]$. But then we have a *projective morphism*

$$\text{proj } \mathbb{C}[\text{rep}_\alpha \ A]^{GL(\alpha),\theta} \xrightarrow{\pi} \text{iss}_\alpha \ A$$

such that all fibers of π are projective varieties. The main properties of π can be deduced from [54] or section 4.8.

THEOREM I.13
Points in $\text{proj } \mathbb{C}[\text{rep}_\alpha \ A]^{GL(\alpha),\theta}$ *are in one-to-one correspondence with isomorphism classes of direct sums of θ-stable representations of total dimension α. If α is such that there are α-dimensional simple A-representations, then π is a birational map.*

DEFINITION I.9
We call $\text{proj } \mathbb{C}[\text{rep}_\alpha \ A]^{GL(\alpha),\theta}$ *the moduli space of θ-semistable representations of A and denote it with* $\text{moduli}_\alpha^\theta \ A$.

Example I.25
In the case of Kleinian singularities, example I.22, if we take θ to be a generic character such that $\theta.\alpha = 0$, then the projective map

$$\text{moduli}_\alpha^\theta \ A \longrightarrow X = \mathbb{C}^2/G$$

is a minimal resolution of singularities. Note that the map is birational as α is the dimension vector of a simple representation of $A = \Pi_0$, see [27]. □

Example I.26
For general quotient singularities, see example I.23, assume that the first vertex in the McKay quiver corresponds to the trivial representation. Take a character $\theta \in \mathbb{Z}^k$ such that $t_1 < 0$ and all $t_i > 0$ for $i \geq 2$, for example take

$$\theta = (-\sum_{i=2}^k \dim R_i, 1, \ldots, 1)$$

Then, the corresponding moduli space is isomorphic to

$$\text{moduli}_\alpha^\theta \ A \simeq G - \text{Hilb } \mathbb{C}^d$$

the *G-equivariant Hilbert scheme*, which classifies all #G-codimensional ideals $I \lhd \mathbb{C}[x_1, \ldots, x_d]$ where

$$\frac{\mathbb{C}[x_1, \ldots, x_d]}{I} \simeq \mathbb{C}G$$

as G-modules, hence, in particular I must be stable under the action of G. It is well known that the natural map

$$G - \texttt{Hilb } \mathbb{C}^d \longrightarrow X = \mathbb{C}^d/G$$

is a minimal resolution if $d = 2$ and if $d = 3$ it is often a crepant resolution, for example whenever G is Abelian. In non-Abelian cases it may have remaining singularities though which often are of conifold type. See [23] for more details.

⬜

Example I.27
In the $\mathbb{C}^2/\mathbb{Z}_3$-example one can take $\theta = (-2, 1, 1)$. The following representations

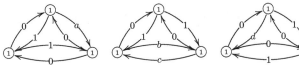

are all nilpotent and are θ-stable. In fact if $bc = 0$ they are representants of the exceptional fiber of the desingularization

$$\texttt{moduli}^\theta_\alpha \, A \longrightarrow \texttt{iss}_\alpha \, A = \mathbb{C}^2/\mathbb{Z}_3$$

⬜

THEOREM I.14
Let $A = A^\alpha_{Q\bullet}/(R)$ be an R-order in **alg@n**. *Assume that there exists a stability structure $\theta \in \mathbb{Z}^k$ such that the Zariski open subset* $\texttt{rep}^{\theta-semist}_\alpha A$ *of all θ-semistable α-dimensional representations of A is a smooth variety. Then there exists a sheaf $\mathcal{A}$ of Cayley-smooth orders over* $\texttt{moduli}^\theta_\alpha A$ *such that the diagram below is commutative*

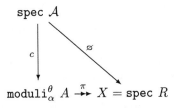

Here, **spec** $\mathcal{A}$ *is a noncommutative variety obtained by gluing affine noncommutative varieties* **spec** A_i *together and c is the map that intersects locally a maximal ideal with the center. As $\mathcal{A}$ is a sheaf of Cayley-smooth orders,*

ϕ *can be viewed as a noncommutative desingularization of* X. *The map* π *itself is a partial resolution of* X *and we have full control over the remaining singularities in* $\texttt{moduli}_\alpha^\theta$ A, *that is, all remaining singularities are of the form described in the previous section. Moreover, if* θ *is such that all* θ-*semistable* A-*representations are actually* θ-*stable, then* $\mathcal{A}$ *is a sheaf of Azumaya algebras over* $\texttt{moduli}_\alpha^\theta$ A *and in this case* π *is a commutative desingularization of* X. *If, in addition, also* $gcd(\alpha) = 1$, *then* $\mathcal{A} \simeq End\ \mathcal{P}$ *for some vectorbundle of rank* n *over* $\texttt{moduli}_\alpha^\theta$ A.

Example I.28

In the case of Kleinian singularities, example I.22, there exists a suitable stability structure θ such that $\texttt{rep}_\alpha^{\theta-semist}\ \Pi_0$ is smooth. Consider the *moment map*

$$\texttt{rep}_\alpha\ Q \xrightarrow{\ \mu\ } \texttt{lie}\ GL(\alpha) = M_\alpha(\mathbb{C}) = M_{e_1}(\mathbb{C}) \oplus \ldots \oplus M_{e_k}(\mathbb{C})$$

defined by sending $V = (V_a, V_{a^*})$ to

$$\left(\sum_{\underset{a}{\circ \to ①}} V_a V_{a^*} - \sum_{\underset{a}{① \to \circ}} V_{a^*} V_a, \ldots, \sum_{\underset{a}{\circ \to ⓚ}} V_a V_{a^*} - \sum_{\underset{a}{ⓚ \to \circ}} V_{a^*} V_a \right)$$

The differential $d\mu$ can be verified to be surjective in any representation $V \in \texttt{rep}_\alpha\ Q$ that has a stabilizer subgroup $\mathbb{C}^*(1_{e_1}, \ldots, 1_{e_k})$ (a so-called *Schur representation*) see, for example, [26, lemma 6.5]. Further, any θ-stable representation is Schurian. Moreover, for a generic stability structure $\theta \in \mathbb{Z}^k$ we have that every θ-semistable α-dimensional representation is θ-stable as the $gcd(\alpha) = 1$. Combining these facts it follows that $\mu^{-1}(0) = \texttt{rep}_\alpha\ \Pi_0$ is smooth in all θ-stable representations. ▯

Example I.29

Another case where smoothness of $\texttt{rep}_\alpha^{\theta-semist}\ A$ is evident is when $A = A_Q^\alpha$. is a Cayley-smooth order as then $\texttt{rep}_\alpha\ A$ itself is smooth. This observation can be used to resolve the remaining singularities in the partial resolution. If $gcd(\alpha) = 1$ then for a sufficiently general θ all θ-semistable representations are actually θ-stable whence the quotient map

$$\texttt{rep}_\alpha^{\theta-semist}\ A \longrightarrow\!\!\!\!\rightarrow \texttt{moduli}_\alpha^\theta\ A$$

is a principal $PGL(\alpha)$-fibration and as the total space is smooth, so is $\texttt{moduli}_\alpha^\theta\ A$. Therefore, the projective map

$$\texttt{moduli}_\alpha^\theta\ A \xrightarrow{\ \pi\ } \texttt{iss}_\alpha\ A$$

is a resolution of singularities in this case. However, if $l = gcd(\alpha)$, then $\texttt{moduli}_\alpha^\theta\ A$ will usually contain singularities that are as bad as the quotient

variety singularity of tuples of $l \times l$ matrices under simultaneous conjugation.
▯

The bulk of the proof of the theorem follows from the results of the last section. Because A has $\mathbb{C}^k$ as the subalgebra generated by the vertex idempotents, the trace map on A determines the trace on $\mathbb{C}^k$ and hence, the dimension vector α. Moreover, we will see that

$$\mathtt{trep}_n \, A = GL_n \times^{GL(\alpha)} \mathtt{rep}_\alpha \, A$$

and hence is a *principal fiber bundle*. As is the case for any principal fiber bundle, this gives a natural one-to-one correspondence between

- GL_n-orbits in $\mathtt{trep}_n \, A$, and

- $GL(\alpha)$-orbits in $\mathtt{rep}_\alpha \, A$.

Moreover the corresponding quotient varieties $\mathtt{triss}_n \, A = \mathtt{trep}_n \, A/GL_n$ and $\mathtt{iss}_\alpha \, A = \mathtt{rep}_\alpha \, A/GL(\alpha)$ are isomorphic so we can apply all our GL_n-results of the previous section to this setting. Further, we claim that we can cover the moduli space

$$\mathtt{moduli}_\alpha^\theta \, A = \bigcup_D X_D$$

where X_D is an affine open subset such that under the canonical quotient map

$$\mathtt{rep}_\alpha^{\theta-semist} \, A \xrightarrow{\ \pi\ } \mathtt{moduli}_\alpha^\theta \, A$$

we have that

$$\pi^{-1}(X_D) = \mathtt{rep}_\alpha \, A_D$$

for some $\mathbb{C}[X_D]$-order A_D in $\mathtt{alg@n}$. If in addition $\mathtt{rep}_\alpha^{\theta-semist} \, A$ is a smooth variety, each of the $\mathtt{rep}_\alpha \, A_D$ are smooth affine $GL(\alpha)$-varieties whence the orders A_D are all Cayley-smooth and the result will follow from the foregoing sections.

Because $\mathtt{moduli}_\alpha^\theta \, A = \mathtt{proj} \, \mathbb{C}[\mathtt{rep}_\alpha \, A]^{GL(\alpha),\theta}$ we need control on the generators of all θ-semi-invariants. Such a generating set was found by Aidan Schofield and Michel Van den Bergh in [95] and will be proved in section 7.2. Reorder the vertices in $Q^\bullet$ such that the entries of θ are separated in three strings

$$\theta = (\underbrace{t_1, \ldots, t_i}_{>0}, \underbrace{t_{i+1}, \ldots, t_j}_{=0}, \underbrace{t_{j+1}, \ldots, t_k}_{<0})$$

and let θ be such that $\theta.\alpha = 0$. Fix a weight $l \in \mathbb{N}_+$ and take arbitrary natural numbers $\{l_{i+1}, \ldots, l_j\}$. Consider a rectangular matrix $L = (L_{r,c})$ where each entry of the block $L_{r,c}$ is a linear combination of oriented paths in the marked quiver $Q^\bullet$ with starting vertex v_c and ending vertex v_r and the sizes of the blocks $L_{r,c}$ are such that L is a square matrix on $lt_1 + \ldots + lt_i + l_{i+1} + \ldots + l_j$

rows and $l_{i+1} + \ldots + l_j - lt_{j+1} - \ldots - lt_k$ columns. So we can evaluate all entries of L at all representations $V \in \mathbf{rep}_\alpha A$, consider $D(V) = \det L(V)$ and verify that D is a $GL(\alpha)$-semi-invariant polynomial on $\mathbf{rep}_\alpha A$ of weight χ_θ^l. The result of [95] asserts that these *determinantal semi-invariants* are algebra generators of the graded algebra

$$\mathbb{C}[\mathbf{rep}_\alpha A]^{GL(\alpha),\theta}$$

Because a representation $V \in \mathbf{rep}_\alpha A$ is θ-semistable if and only if some semi-invariant of weight χ_θ^l for some l is nonzero on it. This proves the following.

THEOREM I.15

The Zariski open subset of θ-semistable α-dimensional A-representations can be covered by affine $GL(\alpha)$-stable open subsets

$$\mathbf{rep}_\alpha^{\theta-semist} A = \bigsqcup_D \{V \mid D(V) = \det L(V) \neq 0\}$$

and hence the moduli space can also be covered by affine open subsets

$$\mathtt{moduli}_\alpha^\theta A = \bigcup_D X_D$$

where $X_D = \{[V] \in \mathtt{moduli}_\alpha^\theta A \mid D(V) = \det L(V) \neq 0\}$.

Example I.30

In the $\mathbb{C}^2/\mathbb{Z}_3$ example, the θ-semistable representations

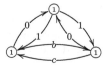

with $\theta = (-2,1,1)$ all lie in the affine open subset X_D where L is a matrix of the form

$$L = \begin{bmatrix} x_1 & 0 \\ * & y_3 \end{bmatrix}$$

where $*$ is any path in Q starting in x_1 and ending in x_3.   ▯

Analogous to the rectangular matrix L we define a rectangular matrix N with $lt_1 + \ldots + lt_i + l_{i+1} + \ldots + l_j$ columns and $l_{i+1} + \ldots + l_j - lt_{j+1} - \ldots - lt_k$ rows filled with new variables and define an *extended marked quiver* $Q_D^\bullet$ where we adjoin for each entry in $N_{r,c}$ an additional arrow from v_c to v_r and denote it with the corresponding variable from N. Let I_1 (resp. I_2) be the set of relations in $\mathbb{C}Q_D^\bullet$ determined from the matrix-equations that $L.N$ and $N.L$

are made up of identity matrix vertex-blocks. Define a new noncommutative order

$$A_D = \frac{A^{\alpha}_{Q^{\bullet}_D}}{(I, I_1, I_2)}$$

then A_D is a $\mathbb{C}[X_D]$-order in $\texttt{alg@n}$.

Example I.31

In the setting of example I.30 with $* = y_3$, the extended quiver-setting (Q_D, α) is

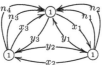

Hence, with

$$L = \begin{bmatrix} x_1 & 0 \\ y_3 & y_3 \end{bmatrix} \qquad N = \begin{bmatrix} n_1 & n_3 \\ n_2 & n_4 \end{bmatrix}$$

the defining equations of the order A_D become

$$\begin{cases} I = (x_3 y_3 - y_1 x_1, x_1 y_1 - y_2 x_2, x_2 y_2 - y_3 x_3) \\ I_1 = (n_1 x_1 + n_3 y_3 - v_1, n_3 y_3, n_2 x_1 + n_4 y_3, n_4 y_3 - v_1) \\ I_2 = (x_1 n_1 - v_2, x_1 n_3, y_3 n_1 + y_2 n_2, y_3 n_3 + y_3 n_4 - v_3) \end{cases}$$

$\square$

This construction may seem a bit mysterious at first but what we are really doing is to construct the *universal localization* as in, for example [92], (or see section 7.3) associated to the map between projective A-modules determined by L, but this time not in the category $\texttt{alg}$ of all algebras but in $\texttt{alg@}\alpha$. We have the situation

$$\textbf{rep}^{\theta-semist} \ A \longleftarrow \pi^{-1}(X_D) \simeq \textbf{rep}_{\alpha} \ A_D$$

$$\pi \downarrow \qquad\qquad\qquad\qquad\qquad \downarrow$$

$$\textbf{moduli}^{\theta}_{\alpha} \ A \longleftarrow X_D$$

and theorem I.14 follows from the next result.

THEOREM I.16

The following statements are equivalent:

1. $V \in \textbf{rep}^{\theta-semist}_{\alpha} \ A$ lies in $\pi^{-1}(X_D)$, and

2. *There is a unique extension $\tilde{V}$ of V such that $\tilde{V} \in \mathbf{rep}_\alpha A_D$.*

PROOF $1 \Rightarrow 2$: Because $L(V)$ is invertible we can take $N(V)$ to be its inverse and decompose it into blocks corresponding to the new arrows in $Q_D^\bullet$. This then defines the unique extension $\tilde{V} \in \mathbf{rep}_\alpha Q_D^\bullet$ of V. As $\tilde{V}$ satisfies R (because V does) and I_1 and I_2 (because $N(V) = L(V)^{-1}$) we have that $\tilde{V} \in \mathbf{rep}_\alpha A_D$.

$2 \Rightarrow 1$: Restrict $\tilde{V}$ to the arrows of Q to get a $V \in \mathbf{rep}_\alpha Q$. As $\tilde{V}$ (and hence V) satisfies R, $V \in \mathbf{rep}_\alpha A$. Moreover, V is such that $L(V)$ is invertible (this follows because $\tilde{V}$ satisfies I_1 and I_2). Hence, $D(V) \neq 0$ and because D is a θ-semi-invariant it follows that V is an α-dimensional θ-semistable representation of A.   ☐

Example I.32
In the setting of example I.30 with $* = y_3$ we have that the uniquely determined extension of the A-representation

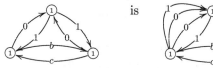

Observe that this extension is a simple A_D-representation for every $b, c \in \mathbb{C}$.
☐

There remains one more thing to clarify: how are the different A_D's glued together to form a sheaf $\mathcal{A}$ of noncommutative algebras over $\mathbf{moduli}_\alpha^\theta A$ and how can we construct the noncommutative variety $\mathbf{spec}\ \mathcal{A}$? The solution to both problems follows from the universal property of A_D. Let A_{D_1} (resp. A_{D_2}) be the algebra constructed from a rectangular matrix L_1 (resp. L_2), then we can construct the direct sum map $L = L_1 \oplus L_2$ for which the corresponding semi-invariant $D = D_1 D_2$. As $A \longrightarrow A_D$ makes the projective module morphisms associated to L_1 and L_2 an isomorphism we have uniquely determined maps in $\mathbf{alg@}\alpha$

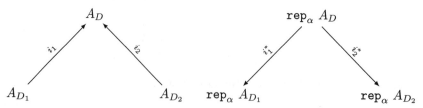

As $\mathbf{rep}_\alpha A_D = \pi^{-1}(X_D)$ (and similarly for D_i) we have that i_j^* are embeddings as are the i_j. This way we can glue the sections $\Gamma(X_{D_1}, \mathcal{A}) = A_{D_1}$ with $\Gamma(X_{D_2}, \mathcal{A}) = A_{D_2}$ over their intersection $X_D = X_{D_1} \cap X_{D_2}$ via the inclusions i_j. Hence we get a coherent sheaf of noncommutative algebras $\mathcal{A}$

over $\mathtt{moduli}_{\alpha}^{\theta}$ A. Observe that many of the orders A_D are isomorphic. In example I.30 all matrices L with fixed diagonal entries x_1 and y_3 but with varying $*$-entry have isomorphic orders A_D (use the universal property). In a similar way we would like to glue $\mathtt{max}$ A_{D_1} with $\mathtt{max}$ A_{D_2} over $\mathtt{max}$ A_D using the algebra maps i_j to form a noncommutative variety $\mathtt{spec}$ $\mathcal{A}$. However, the construction of $\mathtt{max}$ A and the noncommutative structure sheaf is *not* functorial in general.

Example I.33

Consider the inclusion map map in $\mathtt{alg@2}$

$$A = \begin{bmatrix} R & R \\ I & R \end{bmatrix} \longhookrightarrow \begin{bmatrix} R & R \\ R & R \end{bmatrix} = A'$$

then all two-sided maximal ideals of A' are of the form $M_2(\mathfrak{m})$ where $\mathfrak{m}$ is a maximal ideal of R. If $I \subset \mathfrak{m}$ then the intersection

$$\begin{bmatrix} \mathfrak{m} & \mathfrak{m} \\ \mathfrak{m} & \mathfrak{m} \end{bmatrix} \cap \begin{bmatrix} R & R \\ I & R \end{bmatrix} = \begin{bmatrix} \mathfrak{m} & \mathfrak{m} \\ I & \mathfrak{m} \end{bmatrix}$$

which is *not* a maximal ideal of A as

$$\begin{bmatrix} \mathfrak{m} & R \\ I & R \end{bmatrix} \begin{bmatrix} R & R \\ I & \mathfrak{m} \end{bmatrix} = \begin{bmatrix} \mathfrak{m} & \mathfrak{m} \\ I & \mathfrak{m} \end{bmatrix}$$

and so there is no natural map $\mathtt{max}\, A' \longrightarrow \mathtt{max}\, A$, let alone a continuous one.
▯

In [86] it was proved that if $A \xrightarrow{\ f\ } B$ is an *extension* (that is, an algebra map such that $B = f(A)Z_B(A)$ where $Z_B(A) = \{b \in B \mid bf(a) = f(a)b \ \forall a \in A\}$) then the map

$$\mathtt{spec}\ B \longrightarrow \mathtt{spec}\ R \qquad\qquad P \longrightarrow f^{-1}(P)$$

is well-defined and continuous for the Zariski topology. In our situation, the maps $i_j : A_{D_j} \longrightarrow A_D$ are even *central extensions*, that is

$$A_D = A_{D_j} Z(A_D)$$

which follows again from the universal property by localizing A_{D_j} at the central element D. Hence, we can define a genuine noncommutative variety $\mathtt{spec}$ $\mathcal{A}$ with central scheme $\mathtt{moduli}_{\alpha}^{\theta}$ A, finishing the proof of the noncommutative desingularization approach.

About the Author

Lieven Le Bruyn was born in Brecht, Belgium, on May 30, 1958. Before joining the University of Antwerp as Professor in Mathematics in 2000, he served as Research Director at the Belgian Science Foundation. Following graduation in 1980, he received his Ph.D. in 1983 and his Habilitation in 1985, all from the University of Antwerp. He is a laureate of the Belgian Academy of Sciences and winner of the Louis Empain Prize for Mathematics. He works and lives in Antwerp and blogs at http://www.neverendingbooks.org.

About the Author

Chapter 1

Cayley-Hamilton Algebras

In this chapter we will define the category $\mathbf{alg@n}$ of Cayley-Hamilton algebras of degree n. These are affine $\mathbb{C}$-algebras A equipped with a trace map tr_A such that all trace identities holding in $n \times n$ matrices also hold in A. Hence, we have to study trace identities and, closely related to them, necklace relations. This requires the description of the *generic algebras*

$$\int_n \mathbb{C}\langle x_1, \ldots, x_m \rangle = \mathbb{T}_n^m \qquad \text{and} \qquad \oint_n \mathbb{C}\langle x_1, \ldots, x_m \rangle = \mathbb{N}_n^m$$

called the *trace algebra of m generic $n \times n$ matrices*, respectively the *necklace algebra of m generic $n \times n$ matrices*. For every $A \in \mathbf{alg@n}$ there are epimorphisms $\mathbb{T}_n^m \longrightarrow\!\!\!\!\!\rightarrow A$ and $\mathbb{N}_n^m \longrightarrow\!\!\!\!\!\rightarrow tr_A(A)$ for some m.

In chapter 2 we will reconstruct the Cayley-Hamilton algebra A (and its central subalgebra $tr_A(A)$) as the ring of GL_n-equivariant polynomial functions (resp. invariant polynomials) on the representation scheme $\mathbf{rep}_n\ A$. Using the Reynolds operator in geometric invariant theory, it suffices to prove these results for the generic algebras mentioned above. An n-dimensional representation of the free algebra $\mathbb{C}\langle x_1, \ldots, x_m \rangle$ is determined by the images of the generators x_i in $M_n(\mathbb{C})$, whence

$$\mathbf{rep}_n\ \mathbb{C}\langle x_1, \ldots, x_m \rangle \simeq \underbrace{M_n(\mathbb{C}) \oplus \ldots \oplus M_n(\mathbb{C})}_{m}$$

and the GL_n-action on it is *simultaneous conjugation*. For this reason we have to understand the fundamental results on the invariant theory of m-tuples on $n \times n$ matrices, due to Claudio Procesi [85].

1.1 Conjugacy classes of matrices

In this section we recall the standard results in the case when $m = 1$, that is, the study of conjugacy classes of $n \times n$ matrices. Clearly, the conjugacy classes are determined by matrices in Jordan normal form. Though this gives a complete set-theoretic solution to the orbit problem in this case, there cannot be an orbit variety due to the existence of nonclosed orbits. Hence, the geometric study of the conjugacy classes splits up into a *quotient problem* (the polynomial invariants determine an affine variety whose points correspond to the closed orbits) and a *nullcone problem* (the study of the orbits having a given closed orbit in their closures). In this section we will solve the first part in full detail, the second part will be solved in section 2.7. A recurrent theme of this book will be to generalize this two-part approach to the orbit-space problem to other representation varieties.

We denote by M_n the space of all $n \times n$ matrices $M_n(\mathbb{C})$ and by GL_n the general linear group $GL_n(\mathbb{C})$. A matrix $A \in M_n$ determines by left multiplication a linear operator on the n-dimensional vector space $V_n = \mathbb{C}^n$ of *column vectors* . If $g \in GL_n$ is the matrix describing the *base change* from the canonical basis of V_n to a new basis, then the linear operator expressed in this new basis is represented by the matrix gAg^{-1}. For a given matrix A we want to find a suitable basis such that the *conjugated matrix* gAg^{-1} has a simple form.

Consider the linear action of GL_n on the n^2-dimensional vector space M_n

$$GL_n \times M_n \longrightarrow M_n \qquad (g, A) \mapsto g.A = gAg^{-1}$$

The *orbit* $\mathcal{O}(A) = \{gAg^{-1} \mid g \in GL_n \}$ of A under this action is called the *conjugacy class* of A. We look for a particularly nice representative in a given conjugacy class. The answer to this problem is, of course, given by the *Jordan normal form* of the matrix.

With e_{ij} we denote the matrix whose unique nonzero entry is 1 at entry (i, j). Recall that the group GL_n is generated by the following three classes of matrices :

- the *permutation* matrices $p_{ij} = \mathbb{1}_n + e_{ij} + e_{ji} - e_{ii} - e_{jj}$ for all $i \neq j$,

- the *addition* matrices $a_{ij}(\lambda) = \mathbb{1}_n + \lambda e_{ij}$ for all $i \neq j$ and $0 \neq \lambda$, and

- the *multiplication* matrices $m_i(\lambda) = \mathbb{1}_n + (\lambda - 1)e_{ii}$ for all i and $0 \neq \lambda$.

Conjugation by these matrices determine the three types of *Jordan moves* on $n \times n$ matrices, as depicted below, where the altered rows and columns are indicated

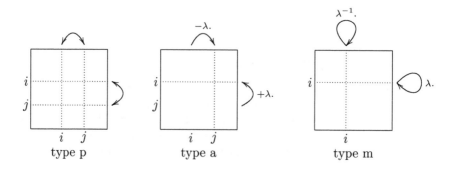

type p   type a   type m

Therefore, it suffices to consider sequences of these moves on a given $n \times n$ matrix $A \in M_n$. The *characteristic polynomial* of A is defined to be the polynomial of degree n in the variable t

$$\chi_A(t) = det(t \mathbb{1}_n - A) \in \mathbb{C}[t]$$

As $\mathbb{C}$ is algebraically closed, $\chi_A(t)$ decomposes as a product of linear terms

$$\prod_{i=1}^{e}(t - \lambda_i)^{d_i}$$

Here, the $\{\lambda_1, \ldots, \lambda_e\}$ are called the *eigenvalues* of the matrix A. Observe that λ_i is an eigenvalue of A if and only if there is a nonzero *eigenvector* $v \in V_n = \mathbb{C}^n$ with eigenvalue λ_i, that is, $A.v = \lambda_i v$. In particular, the *rank* r_i of the matrix $A_i = \lambda_i \mathbb{1}_n - A$ satisfies $n - d_i \leq r_i < n$. A nice inductive procedure using only Jordan moves is given in [35] and proves the *Jordan-Weierstrass theorem* .

THEOREM 1.1 Jordan-Weierstrass

Let $A \in M_n$ with characteristic polynomial $\chi_A(t) = \prod_{i=1}^{e}(t - \lambda_i)^{d_i}$. Then, A determines unique partitions

$$p_i = (a_{i1}, a_{i2}, \ldots, a_{im_i}) \quad of \quad d_i$$

associated to the eigenvalues λ_i of A such that A is conjugated to a unique (up to permutation of the blocks) block-diagonal matrix

$$J_{(p_1, \ldots, p_e)} = \begin{bmatrix} B_1 & 0 & \ldots & 0 \\ 0 & B_2 & & 0 \\ \vdots & & \ddots & \vdots \\ 0 & 0 & \ldots & B_m \end{bmatrix}$$

with $m = m_1 + \ldots + m_e$ and exactly one block B_l of the form $J_{a_{ij}}(\lambda_i)$ for all $1 \leq i \leq e$ and $1 \leq j \leq m_i$ where

$$J_{a_{ij}}(\lambda_i) = \begin{bmatrix} \lambda_i & 1 & & \\ & \lambda_i & \ddots & \\ & & \ddots & 1 \\ & & & \lambda_i \end{bmatrix} \in M_{a_{ij}}(\mathbb{C})$$

Let us prove uniqueness of the partitions p_i of d_i corresponding to the eigenvalue λ_i of A. Assume A is conjugated to another Jordan block matrix

$J_{(q_1,\ldots,q_e)}$, necessarily with partitions $q_i = (b_{i1}, \ldots, b_{im'_i})$ of d_i. To begin, observe that for a Jordan block of size k we have that

$$rk\ J_k(0)^l = k - l \quad \text{for all } l \leq k \text{ and if } \mu \neq 0 \text{ then} \quad rk\ J_k(\mu)^l = k$$

for all l. As $J_{(p_1,\ldots,p_e)}$ is conjugated to $J_{(q_1,\ldots,q_e)}$ we have for all $\lambda \in \mathbb{C}$ and all l

$$rk\ (\lambda \mathbb{1}_n - J_{(p_1,\ldots,p_e)})^l = rk\ (\lambda \mathbb{1}_n - J_{(q_1,\ldots,q_e)})^l$$

Take $\lambda = \lambda_i$ then only the Jordan blocks with eigenvalue λ_i are important in the calculation and one obtains for the ranks

$$n - \sum_{h=1}^{l} \#\{j \mid a_{ij} \geq h\} \quad \text{respectively} \quad n - \sum_{h=1}^{l} \#\{j \mid b_{ij} \geq h\}$$

For any partition $p = (c_1, \ldots, c_u)$ and any natural number h we see that the number $z = \#\{j \mid c_j \geq h\}$

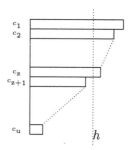

is the number of blocks in the h-th row of the dual partition p^*, which is defined to be the partition obtained by interchanging rows and columns in the *Young diagram* of p (see section 1.5 for the definition). Therefore, the above rank equality implies that $p_i^* = q_i^*$ and hence that $p_i = q_i$. As we can repeat this argument for the other eigenvalues we have the required uniqueness.

Hence, the Jordan normal form shows that the classification of GL_n-orbits in M_n consists of two parts: a discrete part choosing

- a partition $p = (d_1, d_2, \ldots, d_e)$ of n, and for each d_i

- a partition $p_i = (a_{i1}, a_{i2}, \ldots, a_{im_i})$ of d_i

determining the sizes of the Jordan blocks and a continuous part choosing

- an e-tuple of distinct complex numbers $(\lambda_1, \lambda_2, \ldots, \lambda_e)$

fixing the eigenvalues. Moreover, this e-tuple $(\lambda_1, \ldots, \lambda_e)$ is determined only up to permutations of the subgroup of all permutations π in the symmetric group S_e such that $p_i = p_{\pi(i)}$ for all $1 \leq i \leq e$.

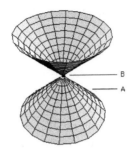

FIGURE 1.1: Orbit closure for 2×2 matrices.

Whereas this gives a satisfactory set-theoretical description of the orbits we cannot put an Hausdorff topology on this set due to the existence of non-closed orbits in M_n. For example, if $n = 2$, consider the matrices

$$A = \begin{bmatrix} \lambda & 1 \\ 0 & \lambda \end{bmatrix} \quad \text{and} \quad B = \begin{bmatrix} \lambda & 0 \\ 0 & \lambda \end{bmatrix}$$

which are in different normal form so correspond to distinct orbits. For any $\epsilon \neq 0$ we have that

$$\begin{bmatrix} \epsilon & 0 \\ 0 & 1 \end{bmatrix} \cdot \begin{bmatrix} \lambda & 1 \\ 0 & \lambda \end{bmatrix} \cdot \begin{bmatrix} \epsilon^{-1} & 0 \\ 0 & 1 \end{bmatrix} = \begin{bmatrix} \lambda & \epsilon \\ 0 & \lambda \end{bmatrix}$$

belongs to the orbit of A. Hence if $\epsilon \longrightarrow 0$, we see that B lies in the closure of $\mathcal{O}(A)$. As any matrix in $\mathcal{O}(A)$ has trace 2λ, the orbit is contained in the 3-dimensional subspace

$$\begin{bmatrix} \lambda + x & y \\ z & \lambda - x \end{bmatrix} \hookrightarrow M_2$$

In this space, the orbit-closure $\overline{\mathcal{O}(A)}$ is the set of points satisfying $x^2 + yz = 0$ (the determinant has to be λ^2), which is a cone having the origin as its top: The orbit $\mathcal{O}(B)$ is the top of the cone and the orbit $\mathcal{O}(A)$ is the complement, see figure 1.1.

Still, for general n we can try to find the best separated topological quotient space for the action of GL_n on M_n. We will prove that this space coincide with the quotient variety determined by the invariant polynomial functions.

If two matrices are conjugated $A \sim B$, then A and B have the same *unordered* n-tuple of eigenvalues $\{\lambda_1, \ldots, \lambda_n\}$ (occurring with multiplicities). Hence any symmetric function in the λ_i will have the same values in A as in B. In particular this is the case for the *elementary symmetric functions* σ_l

$$\sigma_l(\lambda_1, \ldots, \lambda_l) = \sum_{i_1 < i_2 < \ldots < i_l} \lambda_{i_1} \lambda_{i_2} \ldots \lambda_{i_l}.$$

Observe that for every $A \in M_n$ with eigenvalues $\{\lambda_1, \ldots, \lambda_n\}$ we have

$$\prod_{j=1}^{n}(t - \lambda_j) = \chi_A(t) = det(t\mathbb{1}_n - A) = t^n + \sum_{i=1}^{n}(-1)^i \sigma_i(A)t^{n-i}$$

Developing the determinant $det(t\mathbb{1}_n - A)$ we see that each of the coefficients $\sigma_i(A)$ is in fact a *polynomial function* in the entries of A. Therefore, $\sigma_i(A)$ is a complex valued continous function on M_n. The above equality also implies that the functions $\sigma_i : M_n \longrightarrow \mathbb{C}$ are constant along orbits. We now construct the continuous map

$$M_n \xrightarrow{\pi} \mathbb{C}^n$$

sending a matrix $A \in M_n$ to the point $(\sigma_1(A), \ldots, \sigma_n(A))$ in $\mathbb{C}^n$. Clearly, if $A \sim B$ then they map to the same point in $\mathbb{C}^n$. We claim that π is surjective. Take any point $(a_1, \ldots, a_n) \in \mathbb{C}^n$ and consider the matrix $A \in M_n$

$$A = \begin{bmatrix} 0 & & & & a_n \\ -1 & 0 & & & a_{n-1} \\ & \ddots & \ddots & & \vdots \\ & & -1 & 0 & a_2 \\ & & & -1 & a_1 \end{bmatrix} \tag{1.1}$$

then we will show that $\pi(A) = (a_1, \ldots, a_n)$, that is

$$det(t\mathbb{1}_n - A) = t^n - a_1 t^{n-1} + a_2 t^{n-2} - \ldots + (-1)^n a_n$$

Indeed, developing the determinant of $t\mathbb{1}_n - A$ along the first column we obtain

$$\begin{vmatrix} t & 0 & 0 & 0 & -a_n \\ 1 & t & 0 & 0 & -a_{n-1} \\ 0 & 1 & t & 0 & -a_{n-2} \\ & \ddots & \ddots & \vdots & \vdots \\ 0 & 0 & 1 & t & -a_2 \\ 0 & 0 & & 1 & t-a_1 \end{vmatrix} - \begin{vmatrix} t & 0 & 0 & \cdots & 0 & -a_n \\ 1 & t & 0 & 0 & -a_{n-1} \\ 0 & 1 & t & 0 & -a_{n-2} \\ & \ddots & \ddots & \vdots & \vdots \\ 0 & 0 & 1 & t & -a_2 \\ 0 & 0 & & 1 & t-a_1 \end{vmatrix}$$

Here, the second determinant is equal to $(-1)^{n-1}a_n$ and by induction on n the first determinant is equal to $t.(t^{n-1} - a_1 t^{n-2} + \ldots + (-1)^{n-1}a_{n-1})$, proving the claim.

Next, we will determine which $n \times n$ matrices can be conjugated to a matrix in the canonical form A as above. We call a matrix $B \in M_n$ *cyclic* if there is a (column) vector $v \in \mathbb{C}^n$ such that $\mathbb{C}^n$ is spanned by the vectors $\{v, B.v, B^2.v, \ldots, B^{n-1}.v\}$. Let $g \in GL_n$ be the basechange transforming the standard basis to the ordered basis

$$(v, -B.v, B^2.v, -B^3.v, \ldots, (-1)^{n-1}B^{n-1}.v)$$

In this new basis, the linear map determined by B (or equivalently, $g.B.g^{-1}$) is equal to the matrix in canonical form

$$
\begin{bmatrix}
0 & & & & b_n \\
-1 & 0 & & & b_{n-1} \\
& \ddots & \ddots & & \vdots \\
& & -1 & 0 & b_2 \\
& & & -1 & b_1
\end{bmatrix}
$$

where $B^n.v$ has coordinates $(b_n, \ldots, b_2, b_1)$ in the new basis. Conversely, any matrix in this form is a cyclic matrix.

We claim that the set of all cyclic matrices in M_n is a *dense* open subset. To see this take $v = (x_1, \ldots, x_n)^\tau \in \mathbb{C}^n$ and compute the determinant of the $n \times n$ matrix

This gives a polynomial of total degree n in the x_i with all its coefficients polynomial functions c_j in the entries b_{kl} of B. Now, B is a cyclic matrix if and only if at least one of these coefficients is nonzero. That is, the set of non-cyclic matrices is exactly the intersection of the finitely many *hypersurfaces*

$$
V_j = \{ B = (b_{kl})_{k,l} \in M_n \mid c_j(b_{11}, b_{12}, \ldots, b_{nn}) = 0 \}
$$

in the vector space M_n.

THEOREM 1.2
The best continuous approximation to the orbit space is given by the surjection

$$
M_n \xrightarrow{\ \pi\ } \mathbb{C}^n
$$

mapping a matrix $A \in M_n(\mathbb{C})$ to the n-tuple $(\sigma_1(A), \ldots, \sigma_n(A))$

Let $f : M_n \longrightarrow \mathbb{C}$ be a continuous function which is constant along conjugacy classes. We will show that f factors through π, that is, f is really a continuous function in the $\sigma_i(A)$. Consider the diagram

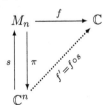

where s is the *section* of π (that is, $\pi \circ s = id_{\mathbb{C}^n}$) determined by sending a point $(a_1, \ldots, a_n)$ to the cyclic matrix in canonical form A as in equation (1.1). Clearly, s is continuous, hence so is $f' = f \circ s$. The approximation property follows if we prove that $f = f' \circ \pi$. By continuity, it suffices to check equality on the dense open set of cyclic matrices in M_n.

There it is a consequence of the following three facts we have proved before (1) : any cyclic matrix lies in the same orbit as one in standard form, (2) : s is a section of π and (3): f is constant along orbits.

Example 1.1 Orbits in M_2

A 2×2 matrix A can be conjugated to an upper triangular matrix with diagonal entries the eigenvalues λ_1, λ_2 of A. As the trace and determinant of both matrices are equal we have

$$\sigma_1(A) = tr(A) \text{ and } \sigma_2(A) = det(A)$$

The best approximation to the orbitspace is therefore given by the surjective map

$$M_2 \xrightarrow{\ \pi\ } \mathbb{C}^2 \qquad \begin{bmatrix} a & b \\ c & d \end{bmatrix} \mapsto (a + d, ad - bc)$$

The matrix A has two equal eigenvalues if and only if the discriminant of the characteristic polynomial $t^2 - \sigma_1(A)t + \sigma_2(A)$ is zero, that is when $\sigma_1(A)^2 - 4\sigma_2(A) = 0$. This condition determines a closed curve C in $\mathbb{C}^2$ where

$$C = \{(x, y) \in \mathbb{C}^2 \mid x^2 - 4y = 0\}.$$

Observe that C is a smooth 1-dimensional submanifold of $\mathbb{C}^2$. We will describe the *fibers* (that is, the inverse images of points) of the surjective map π.

If $p = (x, y) \in \mathbb{C}^2 - C$, then $\pi^{-1}(p)$ consists of precisely one orbit (which is then necessarily closed in M_2) namely, that of the diagonal matrix

$$\begin{bmatrix} \lambda_1 & 0 \\ 0 & \lambda_2 \end{bmatrix} \quad \text{where} \quad \lambda_{1,2} = \frac{-x \pm \sqrt{x^2 - 4y}}{2}$$

If $p = (x, y) \in C$ then $\pi^{-1}(p)$ consists of two orbits

$$\mathcal{O}\begin{bmatrix} \lambda & 1 \\ 0 & \lambda \end{bmatrix} \quad \text{and} \quad \mathcal{O}\begin{bmatrix} \lambda & 0 \\ 0 & \lambda \end{bmatrix}$$

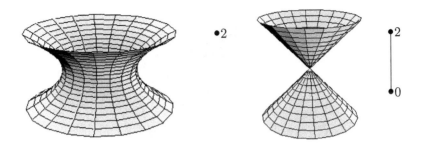

FIGURE 1.2: Orbit closures of 2×2 matrices.

where $\lambda = \frac{1}{2}x$. We have seen that the second orbit lies in the closure of the first. Observe that the second orbit reduces to one point in M_2 and hence is closed. Hence, also $\pi^{-1}(p)$ contains a unique closed orbit.

To describe the fibers of π as closed subsets of M_2 it is convenient to write any matrix A as a linear combination

$$A = u(A) \begin{bmatrix} \frac{1}{2} & 0 \\ 0 & \frac{1}{2} \end{bmatrix} + v(A) \begin{bmatrix} \frac{1}{2} & 0 \\ 0 & -\frac{1}{2} \end{bmatrix} + w(A) \begin{bmatrix} 0 & 1 \\ 0 & 0 \end{bmatrix} + z(A) \begin{bmatrix} 0 & 0 \\ 1 & 0 \end{bmatrix}$$

Expressed in the coordinate functions u, v, w and z the fibers $\pi^{-1}(p)$ of a point $p = (x, y) \in \mathbb{C}^2$ are the common zeroes of

$$\begin{cases} u & = x \\ v^2 + 4wz & = x^2 - 4y \end{cases}$$

The first equation determines a three-dimensional affine subspace of M_2 in which the second equation determines a *quadric*. If $p \notin C$ this quadric is non-degenerate and thus $\pi^{-1}(p)$ is a smooth 2-dimensional submanifold of M_2. If $p \in C$, the quadric is a cone with top lying in the point $\frac{x}{2}\mathbb{1}_2$. Under the GL_2-action, the unique singular point of the cone must be clearly fixed giving us the closed orbit of dimension 0 corresponding to the diagonal matrix. The other orbit is the complement of the top and hence is a smooth 2-dimensional (nonclosed) submanifold of M_2. The graphs in figure 1.2 represent the orbit-closures and the dimensions of the orbits. □

Example 1.2 Orbits in M_3

We will describe the fibers of the surjective map $M_3 \xrightarrow{\pi} \mathbb{C}^3$. If a 3×3 matrix has multiple eigenvalues then the *discriminant* $d = (\lambda_1 - \lambda_2)^2(\lambda_2 - \lambda_3)^2(\lambda_3 - \lambda_1)^2$ is zero. Clearly, d is a symmetric polynomial and hence can be expressed in terms of σ_1, σ_2 and σ_3. More precisely

$$d = 4\sigma_1^3\sigma_3 + 4\sigma_2^3 + 27\sigma_3^2 - \sigma_1^2\sigma_2^2 - 18\sigma_1\sigma_2\sigma_3$$

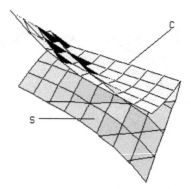

FIGURE 1.3: Representation strata for 3×3 matrices.

The set of points in $\mathbb{C}^3$ where d vanishes is a surface S with *singularities*. These singularities are the common zeroes of the $\frac{\partial d}{\partial \sigma_i}$ for $1 \leq i \leq 3$. One computes that these singularities form a *twisted cubic* curve C in $\mathbb{C}^3$, that is

$$C = \{(3c, 3c^2, c^3) \mid c \in \mathbb{C}\}$$

The description of the fibers $\pi^{-1}(p)$ for $p = (x, y, z) \in \mathbb{C}^3$ is as follows. When $p \notin S$, then $\pi^{-1}(p)$ consists of a unique orbit (which is therefore closed in M_3), the conjugacy class of a matrix with paired distinct eigenvalues. If $p \in S - C$, then $\pi^{-1}(p)$ consists of the orbits of

$$A_1 = \begin{bmatrix} \lambda & 1 & 0 \\ 0 & \lambda & 0 \\ 0 & 0 & \mu \end{bmatrix} \quad \text{and} \quad A_2 = \begin{bmatrix} \lambda & 0 & 0 \\ 0 & \lambda & 0 \\ 0 & 0 & \mu \end{bmatrix}$$

Finally, if $p \in C$, then the matrices in the fiber $\pi^{-1}(p)$ have a single eigenvalue $\lambda = \frac{1}{3}x$ and the fiber consists of the orbits of the matrices

$$B_1 = \begin{bmatrix} \lambda & 1 & 0 \\ 0 & \lambda & 1 \\ 0 & 0 & \lambda \end{bmatrix} \quad B_2 = \begin{bmatrix} \lambda & 1 & 0 \\ 0 & \lambda & 0 \\ 0 & 0 & \lambda \end{bmatrix} \quad B_3 = \begin{bmatrix} \lambda & 0 & 0 \\ 0 & \lambda & 0 \\ 0 & 0 & \lambda \end{bmatrix}$$

We observe that the *strata* with distinct fiber behavior (that is, $\mathbb{C}^3 - S$, $S - C$ and C) are all submanifolds of $\mathbb{C}^3$, see figure 1.3.

The dimension of an orbit $\mathcal{O}(A)$ in M_n is computed as follows. Let C_A be the subspace of all matrices in M_n commuting with A. Then, the *stabilizer* subgroup of A is a dense open subset of C_A whence the dimension of $\mathcal{O}(A)$ is equal to $n^2 - \dim C_A$.

Performing these calculations for the matrices given above, we obtain the following graphs representing orbit-closures and the dimensions of orbits

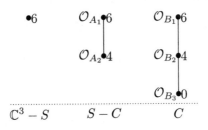

Returning to M_n, the set of cyclic matrices is a Zariski open subset of M_n. For, consider the generic matrix of coordinate functions and generic column vector

$$X = \begin{bmatrix} x_{11} & \cdots & x_{1n} \\ \vdots & & \vdots \\ x_{n1} & \cdots & x_{nn} \end{bmatrix} \quad \text{and} \quad v = \begin{bmatrix} v_1 \\ \vdots \\ v_n \end{bmatrix}$$

and form the square matrix

$$\begin{bmatrix} v & X.v & X^2.v & \cdots & X^{n-1}.v \end{bmatrix} \in M_n(\mathbb{C}[x_{11}, x_{12}, \ldots, x_{nn}, v_1, \ldots, v_n])$$

Then its determinant can be written as $\sum_{l=1}^{z} p_l(x_{ij})q_l(v_k)$ where the q_l are polynomials in the v_k and the p_l polynomials in the x_{ij}. Let $A \in M_n$ be such that at least one of the $p_l(A) \neq 0$, then the polynomial $d = \sum_l p_l(A)q_l(v_k) \in \mathbb{C}[v_1, \ldots, v_k]$ is nonzero. But then there is a $c = (c_1, \ldots, c_n) \in \mathbb{C}^n$ such that $d(c) \neq 0$ and hence c^τ is a cyclic vector for A. The converse implication is obvious.

THEOREM 1.3
Let $f : M_n \longrightarrow \mathbb{C}$ is a regular (that is, polynomial) function on M_n which is constant along conjugacy classes, then

$$f \in \mathbb{C}[\sigma_1(X), \ldots, \sigma_n(X)]$$

PROOF   Consider again the diagram

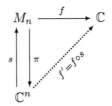

The function $f' = f \circ s$ is a regular function on $\mathbb{C}^n$ whence is a polynomial in the coordinate functions of $\mathbb{C}^m$ (which are the $\sigma_i(X)$), so

$$f' \in \mathbb{C}[\sigma_1(X), \ldots, \sigma_n(X)] \longhookrightarrow \mathbb{C}[M_n].$$

Moreover, f and f' are equal on a Zariski open (dense) subset of M_n whence they are equal as polynomials in $\mathbb{C}[M_n]$.   ☐

The ring of polynomial functions on M_n which are constant along conjugacy classes can also be viewed as a ring of invariants. The group GL_n acts as algebra automorphisms on the polynomial ring $\mathbb{C}[M_n]$. The automorphism ϕ_g determined by $g \in GL_n$ sends the variable x_{ij} to the (i,j)-entry of the matrix $g^{-1}.X.g$, which is a linear form in $\mathbb{C}[M_n]$. This action is determined by the property that for all $g \in GL_n$, $A \in A$ and $f \in \mathbb{C}[M_n]$ we have that

$$\phi_g(f)(A) = f(g.A.g^{-1})$$

The *ring of polynomial invariants* is the algebra of polynomials left invariant under this action

$$\mathbb{C}[M_n]^{GL_n} = \{ f \in \mathbb{C}[M_n] \mid \phi_g(f) = f \text{ for all } g \in GL_n \}$$

and hence is the ring of polynomial functions on M_n that are constant along orbits. The foregoing theorem determines the ring of polynomials invariants

$$\mathbb{C}[M_n]^{GL_n} = \mathbb{C}[\sigma_1(X), \ldots, \sigma_n(X)]$$

We will give an equivalent description of this ring below.

Consider the variables $\lambda_1, \ldots, \lambda_n$ and consider the polynomial

$$f_n(t) = \prod_{i=1}^{n}(t - \lambda_i) = t^n + \sum_{i=1}^{n}(-1)^i \sigma_i t^{n-i}$$

then σ_i is the i-th elementary symmetric polynomial in the λ_j. We know that these polynomials are algebraically independent and generate the *ring of symmetric polynomials* in the λ_j, that is,

$$\mathbb{C}[\sigma_1, \ldots, \sigma_n] = \mathbb{C}[\lambda_1, \ldots, \lambda_n]^{S_n}$$

where S_n is the *symmetric group* on n letters acting by automorphisms on the polynomial ring $\mathbb{C}[\lambda_1, \ldots, \lambda_n]$ via $\pi(\lambda_i) = \lambda_{\pi(i)}$ and the algebra of polynomials, which are fixed under these automorphisms are precisely the symmetric polynomials in the λ_j.

Consider the symmetric Newton functions $s_i = \lambda_1^i + \ldots + \lambda_n^i$, then we claim that this is another generating set of symmetric polynomials, that is

$$\mathbb{C}[\sigma_1, \ldots, \sigma_n] = \mathbb{C}[s_1, \ldots, s_n]$$

To prove this it suffices to express each σ_i as a polynomial in the s_j. More precisely, we claim that the following identities hold for all $1 \leq j \leq n$

$$s_j - \sigma_1 s_{j-1} + \sigma_2 s_{j-2} - \ldots + (-1)^{j-1} \sigma_{j-1} s_1 + (-1)^j \sigma_j . j = 0 \qquad (1.2)$$

For $j = n$ this identity holds because we have

$$0 = \sum_{i=1}^{n} f_n(\lambda_i) = s_n + \sum_{i=1}^{n} (-1)^i \sigma_i s_{n-i}$$

if we take $s_0 = n$. Assume now $j < n$ then the left hand side of equation 1.2 is a symmetric function in the λ_i of degree $\leq j$ and is therefore a polynomial $p(\sigma_1, \ldots, \sigma_j)$ in the first j elementary symmetric polynomials. Let ϕ be the algebra epimorphism

$$\mathbb{C}[\lambda_1, \ldots, \lambda_n] \xrightarrow{\phi} \mathbb{C}[\lambda_1, \ldots, \lambda_j]$$

defined by mapping $\lambda_{j+1}, \ldots, \lambda_j$ to zero. Clearly, $\phi(\sigma_i)$ is the i-th elementary symmetric polynomial in $\{\lambda_1, \ldots, \lambda_j\}$ and $\phi(s_i) = \lambda_1^i + \ldots + \lambda_j^i$. Repeating the above $j = n$ argument (replacing n by j) we have

$$0 = \sum_{i=1}^{j} f_j(\lambda_i) = \phi(s_j) + \sum_{i=1}^{j} (-1)^i \phi(\sigma_i) \phi(s_{n-i})$$

(this time with $s_0 = j$). But then, $p(\phi(\sigma_1), \ldots, \phi(\sigma_j)) = 0$ and as the $\phi(\sigma_k)$ for $1 \leq k \leq j$ are algebraically independent we must have that p is the zero polynomial finishing the proof of the claimed identity.

If $\lambda_1, \ldots, \lambda_n$ are the eigenvalues of an $n \times n$ matrix A, then A can be conjugated to an upper triangular matrix B with diagonal entries $(\lambda_1, \ldots, \lambda_1)$. Hence, the *trace* $tr(A) = tr(B) = \lambda_1 + \ldots + \lambda_n = s_1$. In general, A^i can be conjugated to B^i which is an upper triangular matrix with diagonal entries $(\lambda_1^i, \ldots, \lambda_n^i)$ and hence the traces of A^i and B^i are equal to $\lambda_1^i + \ldots + \lambda_n^i = s_i$. Concluding, we have

THEOREM 1.4
Consider the action of conjugation by GL_n on M_n. Let X be the generic matrix of coordinate functions on M_n

$$X = \begin{bmatrix} x_{11} & \ldots & x_{nn} \\ \vdots & & \vdots \\ x_{n1} & \ldots & x_{nn} \end{bmatrix}$$

Then, the ring of polynomial invariants is generated by the traces of powers of X, that is

$$\mathbb{C}[M_n]^{GL_n} = \mathbb{C}[tr(X), tr(X^2), \ldots, tr(X^n)]$$

PROOF The result follows from theorem 1.3 and the fact that

$$\mathbb{C}[\sigma_1(X), \ldots, \sigma_n(X)] = \mathbb{C}[tr(X), \ldots, tr(X^n)]$$

⬜

1.2 Simultaneous conjugacy classes

As mentioned in the introduction, we need to extend what we have done for conjugacy classes of matrices to simultaneous conjugacy classes of *m-tuples of matrices* . Consider the mn^2-dimensional complex vector space

$$M_n^m = \underbrace{M_n \oplus \ldots \oplus M_n}_{m}$$

of m-tuples $(A_1, \ldots, A_m)$ of $n \times n$-matrices $A_i \in M_n$. On this space we let the group GL_n act by simultaneous conjugation, that is

$$g.(A_1, \ldots, A_m) = (g.A_1.g^{-1}, \ldots, g.A_m.g^{-1})$$

for all $g \in GL_n$ and all m-tuples $(A_1, \ldots, A_m)$. Unfortunately, there is no substitute for the Jordan normalform result in this more general setting.

Still, for small m and n one can work out the GL_n-orbits by brute force methods. In this section we will give the details for the first nontrivial case, that of couples of 2×2 matrices. These explicit calculations will already exhibit some of the general features we will prove later. For example, that all subvarieties of the quotient variety determined by points of the same representation type are smooth and that the fiber structure depends only on the representation type.

Example 1.3 Orbits in $M_2^2 = M_2 \oplus M_2$
We can try to mimic the geometric approach to the conjugacy class problem, that is, we will try to approximate the orbitspace via polynomial functions on M_2^2 that are constant along orbits. For $(A, B) \in M_2^2 = M_2 \oplus M_2$ clearly the polynomial functions we have encountered before $tr(A), det(A)$ and $tr(B), det(B)$ are constant along orbits. However, there are more: for example $tr(AB)$. In the next section, we will show that these five functions generate all polynomials functions that are constant along orbits. Here, we will show that the map $M_2^2 = M_2 \oplus M_2 \xrightarrow{\ \pi\ } \mathbb{C}^5$ defined by

$$(A, B) \mapsto (tr(A), det(A), tr(B), det(B), tr(AB))$$

is surjective such that each fiber contains precisely one closed orbit. In the next chapter, we will see that this property characterizes the best polynomial approximation to the (nonexistent) orbit space.

First, we will show surjectivity of π, that is, for every $(x_1, \ldots, x_5) \in \mathbb{C}^5$ we will construct a couple of 2×2 matrices (A, B) (or rather its orbit) such that $\pi(A, B) = (x_1, \ldots, x_5)$. Consider the open set where $x_1^2 \neq 4x_2$. We have seen that this property characterizes those $A \in M_2$ such that A has distinct eigenvalues and hence diagonalizable. Hence, we can take a representative of the orbit $\mathcal{O}(A, B)$ to be a couple

$$(\begin{bmatrix} \lambda & 0 \\ 0 & \mu \end{bmatrix} , \begin{bmatrix} c_1 & c_2 \\ c_3 & c_4 \end{bmatrix})$$

with $\lambda \neq \mu$. We need a solution to the set of equations

$$\begin{cases} x_3 = c_1 + c_4 \\ x_4 = c_1 c_4 - c_2 c_3 \\ x_5 = \lambda c_1 + \mu c_4 \end{cases}$$

Because $\lambda \neq \mu$ the first and last equation uniquely determine c_1, c_4 and substitution in the second gives us $c_2 c_3$. Analogously, points of $\mathbb{C}^5$ lying in the open set $x_3^2 \neq x_4$ lie in the image of π. Finally, for a point in the complement of these open sets, that is, when $x_1^2 = x_2$ and $x_3^2 = 4x_4$ we can consider a couple (A, B)

$$(\begin{bmatrix} \lambda & 1 \\ 0 & \lambda \end{bmatrix} , \begin{bmatrix} \mu & 0 \\ c & \mu \end{bmatrix})$$

where $\lambda = \frac{1}{2}x_1$ and $\mu = \frac{1}{2}x_3$. Observe that the remaining equation $x_5 = tr(AB) = 2\lambda\mu + c$ has a solution in c.

Now, we will describe the fibers of π. Assume (A, B) is such that A and B have a common eigenvector v. Simultaneous conjugation with a $g \in GL_n$ expressing a basechange from the standard basis to $\{v, w\}$ for some w shows that the orbit $\mathcal{O}(A, B)$ contains a couple of upper-triangular matrices. We want to describe the image of these matrices under π. Take an upper triangular representative in $\mathcal{O}(A, B)$

$$(\begin{bmatrix} a_1 & a_2 \\ 0 & a_3 \end{bmatrix} , \begin{bmatrix} b_1 & b_2 \\ 0 & b_3 \end{bmatrix}).$$

with π-image $(x_1, \ldots, x_5)$. The coordinates x_1, x_2 determine the eigenvalues a_1, a_3 of A only as an unordered set (similarly, x_3, x_4 only determine the set of eigenvalues $\{b_1, b_3\}$ of B). Hence, $tr(AB)$ is one of the following two expressions

$$a_1 b_1 + a_3 b_3 \quad \text{or} \quad a_1 b_3 + a_3 b_1$$

and therefore satisfies the equation

$$(tr(AB) - a_1 b_1 - a_3 b_3)(tr(AB) - a_1 b_3 - a_3 b_1) = 0.$$

Recall that $x_1 = a_1 + a_3, x_2 = a_1 a_3, x_3 = b_1 + b_3, x_4 = b_1 b_3$ and $x_5 = tr(AB)$ we can express this equation as

$$x_5^2 - x_1 x_3 x_5 + x_1^2 x_4 + x_3^2 x_2 - 4x_2 x_4 = 0$$

This determines an hypersurface $H \longhookrightarrow \mathbb{C}^5$. If we view the left-hand side as a polynomial f in the coordinate functions of $\mathbb{C}^5$ we see that H is a four dimensional subset of $\mathbb{C}^5$ with singularities the common zeroes of the partial derivatives

$$\frac{\partial f}{\partial x_i} \text{ for } 1 \le i \le 5$$

These singularities for the 2-dimensional submanifold S of points of the form $(2a, a^2, 2b, b^2, 2ab)$. We now claim that the smooth submanifolds $\mathbb{C}^5 - H$, $H - S$ and S of $\mathbb{C}^5$ describe the different types of fiber behavior. In chapter 6 we will see that the subsets of points with different fiber behavior (actually, of different representation type) are manifolds for m-tuples of $n \times n$ matrices.

If $p \notin H$ we claim that $\pi^{-1}(p)$ is a unique orbit, which is therefore closed in M_2^2. Let $(A, B) \in \pi^{-1}$ and assume first that $x_1^2 \ne 4x_2$ then there is a representative in $\mathcal{O}(A, B)$ of the form

$$(\begin{bmatrix} \lambda & 0 \\ 0 & \mu \end{bmatrix} , \begin{bmatrix} c_1 & c_2 \\ c_3 & c_4 \end{bmatrix})$$

with $\lambda \ne \mu$. Moreover, $c_2 c_3 \ne 0$ (for otherwise A and B would have a common eigenvector whence $p \in H$) hence we may assume that $c_2 = 1$ (eventually after simultaneous conjugation with a suitable diagonal matrix $diag(t, t^{-1})$). The value of λ, μ is determined by x_1, x_2. Moreover, c_1, c_3, c_4 are also completely determined by the system of equations

$$\begin{cases} x_3 & = c_1 + c_4 \\ x_4 & = c_1 c_4 - c_3 \\ x_5 & = \lambda c_1 + \mu c_4 \end{cases}$$

and hence the point $p = (x_1, \ldots, x_5)$ completely determines the orbit $\mathcal{O}(A, B)$. Remains to consider the case when $x_1^2 = 4x_2$ (that is, when A has a single eigenvalue). Consider the couple $(uA + vB, B)$ for $u, v \in \mathbb{C}^*$. To begin, $uA + vB$ and B do not have a common eigenvalue. Moreover, $p = \pi(A, B)$ determines $\pi(uA + vB, B)$ as

$$\begin{cases} tr(uA + vB) & = utr(A) + vtr(B) \\ det(uA + vB) & = u^2 det(A) + v^2 det(B) + uv(tr(A)tr(B) - tr(AB)) \\ tr((uA + vB)B) & = utr(AB) + v(tr(B)^2 - 2det(B)) \end{cases}$$

Assume that for all $u, v \in \mathbb{C}^*$ we have the equality $tr(uA + vB)^2 = 4det(uA + vB)$ then comparing coefficients of this equation expressed as a polynomial in u and v we obtain the conditions $x_1^2 = 4x_2$, $x_3^2 = 4x_4$ and $2x_5 = x_1 x_3$ whence $p \in S \longhookrightarrow H$, a contradiction. So, fix u, v such that $uA + vB$ has distinct eigenvalues. By the above argument $\mathcal{O}(uA + vB, B)$ is the unique orbit lying over $\pi(uA + vB, B)$, but then $\mathcal{O}(A, B)$ must be the unique orbit lying over p.

Let $p \in H - S$ and $(A, B) \in \pi^{-1}(p)$, then A and B are simultaneous upper triangularizable, with eigenvalues a_1, a_2 respectively b_1, b_2. Either $a_1 \neq a_2$ or $b_1 \neq b_2$ for otherwise $p \in S$. Assume $a_1 \neq a_2$, then there is a representative in the orbit $\mathcal{O}(A, B)$ of the form

$$
\left(\begin{bmatrix} a_i & 0 \\ 0 & a_j \end{bmatrix} , \begin{bmatrix} b_k & b \\ 0 & b_l \end{bmatrix} \right)
$$

for $\{i, j\} = \{1, 2\} = \{k, l\}$. If $b \neq 0$ we can conjugate with a suitable diagonal matrix to get $b = 1$ hence we get at most 9 possible orbits. Checking all possibilities we see that only three of them are distinct, those corresponding to the couples

$$
\left(\begin{bmatrix} a_1 & 0 \\ 0 & a_2 \end{bmatrix} , \begin{bmatrix} b_1 & 1 \\ 0 & b_2 \end{bmatrix} \right) \quad \left(\begin{bmatrix} a_1 & 0 \\ 0 & a_2 \end{bmatrix} , \begin{bmatrix} b_1 & 0 \\ 0 & b_2 \end{bmatrix} \right) \quad \left(\begin{bmatrix} a_2 & 0 \\ 0 & a_1 \end{bmatrix} , \begin{bmatrix} b_1 & 1 \\ 0 & b_2 \end{bmatrix} \right)
$$

Clearly, the first and last orbit have the middle one lying in its closure. Observe that the case assuming that $b_1 \neq b_2$ is handled similarly. Hence, if $p \in H - S$ then $\pi^{-1}(p)$ consists of three orbits, two of dimension three whose closures intersect in a (closed) orbit of dimension two.

Finally, consider the case when $p \in S$ and $(A, B) \in \pi^{-1}(p)$. Then, both A and B have a single eigenvalue and the orbit $\mathcal{O}(A, B)$ has a representative of the form

$$
\left(\begin{bmatrix} a & x \\ 0 & a \end{bmatrix} , \begin{bmatrix} b & y \\ 0 & b \end{bmatrix} \right)
$$

for certain $x, y \in \mathbb{C}$. If either x or y are nonzero, then the subgroup of GL_2 fixing this matrix consists of the matrices of the form

$$
Stab \begin{bmatrix} c & 1 \\ 0 & c \end{bmatrix} = \{ \begin{bmatrix} u & v \\ 0 & u \end{bmatrix} \mid u \in \mathbb{C}^*, v \in \mathbb{C} \}
$$

but these matrices also fix the second component. Therefore, if either x or y is nonzero, the orbit is fully determined by $[x : y] \in \mathbb{P}^1$. That is, for $p \in S$, the fiber $\pi^{-1}(p)$ consists of an infinite family of orbits of dimension 2 parameterized by the points of the projective line $\mathbb{P}^1$ together with the orbit of

$$
\left(\begin{bmatrix} a & 0 \\ 0 & a \end{bmatrix} , \begin{bmatrix} b & 0 \\ 0 & b \end{bmatrix} \right)
$$

which consists of one point (hence is closed in M_2^2) and lies in the closure of each of the 2-dimensional orbits.

Concluding, we see that each fiber $\pi^{-1}(p)$ contains a unique closed orbit (that of minimal dimension). The orbit closure and dimension diagrams have

the following shapes

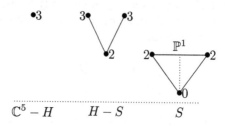

$$\mathbb{C}^5 - H \qquad H - S \qquad S$$

⬜

The reader is invited to try to extend this to the case of three 2×2 matrices (relatively easy) or to two 3×3 matrices (substantially harder). By the end of this book you will have learned enough techniques to solve the general case, *at least in principle.* As this problem is the archetypical example of a *wild representation problem* it is customary to view it as "hopeless." Hence, sooner or later we will hit the wall, but what this book will show you is that you can push the wall a bit further than was generally expected.

1.3   Matrix invariants and necklaces

In this section we will determine the ring of all polynomial maps

$$M_n^m = \underbrace{M_n \oplus \ldots \oplus M_n}_{m} \xrightarrow{\ f\ } \mathbb{C}$$

which are constant along orbits under the action of GL_n on M_n^m by simultaneous conjugation. The strategy we will use is classical in invariant theory.

- First, we will determine the *multilinear* maps which are constant along orbits, equivalently, the *linear* maps

$$M_n^{\otimes m} = \underbrace{M_n \otimes \ldots \otimes M_n}_{m} \longrightarrow \mathbb{C}$$

which are constant along GL_n-orbits where GL_n acts by the diagonal action, that is

$$g.(A_1 \otimes \ldots \otimes A_m) = gA_1g^{-1} \otimes \ldots \otimes gA_mg^{-1}.$$

- Afterward, we will be able to obtain from them all polynomial invariant maps by using *polarization* and *restitution* operations.

First, we will translate our problem into one studied in classical invariant theory of GL_n.

Let $V_n \simeq \mathbb{C}^n$ be the n-dimensional vector space of column vectors on which GL_n acts naturally by left multiplication

$$V_n = \begin{bmatrix} \mathbb{C} \\ \mathbb{C} \\ \vdots \\ \mathbb{C} \end{bmatrix} \quad \text{with action} \quad g.\begin{bmatrix} \nu_1 \\ \nu_2 \\ \vdots \\ \nu_n \end{bmatrix}$$

In order to define an action on the dual space $V_n^* = Hom(V_n, \mathbb{C}) \simeq \mathbb{C}^n$ of *covectors* (or, row vectors) we have to use the *contragradient* action

$$V_n^* = \begin{bmatrix} \mathbb{C} \ \mathbb{C} \ldots \mathbb{C} \end{bmatrix} \quad \text{with action} \quad \begin{bmatrix} \phi_1 \ \phi_2 \ \ldots \ \phi_n \end{bmatrix}.g^{-1}$$

Observe that we have an *evaluation* map $V_n^* \times V_n \longrightarrow \mathbb{C}$ which is given by the scalar product $f(v)$ for all $f \in V_n^*$ and $v \in V_n$

$$\begin{bmatrix} \phi_1 \ \phi_2 \ \ldots \ \phi_n \end{bmatrix}.\begin{bmatrix} \nu_1 \\ \nu_2 \\ \vdots \\ \nu_n \end{bmatrix} = \phi_1\nu_1 + \phi_2\nu_2 + \ldots + \phi_n\nu_n$$

which is invariant under the diagonal action of GL_n on $V_n^* \times V_n$. Further, we have the natural identification

$$M_n = V_n \otimes V_n^* = \begin{bmatrix} \mathbb{C} \\ \mathbb{C} \\ \vdots \\ \mathbb{C} \end{bmatrix} \otimes \begin{bmatrix} \mathbb{C} \ \mathbb{C} \ldots \mathbb{C} \end{bmatrix}$$

Under this identification, a *pure tensor* $v \otimes f$ corresponds to the rank one matrix (or rank one endomorphism of V_n) defined by

$$v \otimes f : V_n \longrightarrow V_n \quad \text{with} \quad w \mapsto f(w)v$$

and we observe that the rank one matrices span M_n. The diagonal action of GL_n on $V_n \otimes V_n^*$ is then determined by its action on the pure tensors where it is equal to

$$g.\begin{bmatrix} \nu_1 \\ \nu_2 \\ \ldots \\ \nu_n \end{bmatrix} \otimes \begin{bmatrix} \phi_1 \ \phi_2 \ \ldots \ \phi_n \end{bmatrix}.g^{-1}$$

and therefore coincides with the action of conjugation on M_n. Now, let us consider the identification

$$(V_n^{*\otimes m} \otimes V_n^{\otimes m})^* \simeq End(V_n^{\otimes m})$$

obtained from the nondegenerate pairing

$$End(V_n^{\otimes m}) \times (V_n^{*\otimes m} \otimes V_n^{\otimes m}) \longrightarrow \mathbb{C}$$

given by the formula

$$\langle \lambda, f_1 \otimes \ldots \otimes f_m \otimes v_1 \otimes \ldots \otimes v_m \rangle = f_1 \otimes \ldots \otimes f_m(\lambda(v_1 \otimes \ldots \otimes v_m))$$

GL_n acts diagonally on $V_n^{\otimes m}$ and hence again by conjugation on $End(V_n^{\otimes m})$ after embedding $GL_n \hookrightarrow GL(V_n^{\otimes m}) = GL_{mn}$. Thus, the above identifications are isomorphism as vector spaces with GL_n-action. But then, the space of GL_n-invariant linear maps

$$V_n^{*\otimes m} \otimes V_n^{\otimes m} \longrightarrow \mathbb{C}$$

can be identified with the space $End_{GL_n}(V_n^{\otimes m})$ of GL_n-linear endomorphisms of $V_n^{\otimes m}$. We will now give a different presentation of this vector space relating it to the symmetric group.

Apart from the diagonal action of GL_n on $V_n^{\otimes m}$ given by

$$g.(v_1 \otimes \ldots \otimes v_m) = g.v_1 \otimes \ldots \otimes g.v_m$$

we have an action of the symmetric group S_m on m letters on $V_n^{\otimes m}$ given by

$$\sigma.(v_1 \otimes \ldots \otimes v_m) = v_{\sigma(1)} \otimes \ldots \otimes v_{\sigma(m)}$$

These two actions commute with each other and give embeddings of GL_n and S_m in $End(V_n^{\otimes m})$. The subspace of $V_n^{\otimes m}$ spanned by the image of GL_n will be denoted by $\langle GL_n \rangle$. Similarly, with $\langle S_m \rangle$ we denote the subspace spanned by the image of S_m.

THEOREM 1.5
With notations as above we have

1. $\langle GL_n \rangle = End_{S_m}(V_n^{\otimes m})$

2. $\langle S_m \rangle = End_{GL_n}(V_n^{\otimes m})$.

PROOF   (1): Under the identification $End(V_n^{\otimes m}) = End(V_n)^{\otimes m}$ an element $g \in GL_n$ is mapped to the symmetric tensor $g \otimes \ldots \otimes g$. On the other hand, the image of $End_{S_m}(V_n^{\otimes m})$ in $End(V_n)^{\otimes m}$ is the subspace of all symmetric tensors in $End(V)^{\otimes m}$. We can give a basis of this subspace as follows. Let $\{e_1, \ldots, e_{n^2}\}$ be a basis of $End(V_n)$, then the vectors $e_{i_1} \otimes \ldots \otimes e_{i_m}$ form a basis of $End(V_n)^{\otimes m}$ that is stable under the S_m-action. Further, any S_m-orbit contains a unique representative of the form

$$e_1^{\otimes h_1} \otimes \ldots \otimes e_{n^2}^{\otimes h_{n^2}}$$

with $h_1 + \ldots + h_{n^2} = m$. If we denote by $r(h_1, \ldots, h_{n^2})$ the sum of all elements in the corresponding S_m-orbit then these vectors are a basis of the symmetric tensors in $End(V_n)^{\otimes m}$.

The claim follows if we can show that every linear map λ on the symmetric tensors, which is zero on all $g \otimes \ldots \otimes g$ with $g \in GL_n$ is the zero map. Write $e = \sum x_i e_i$, then

$$\lambda(e \otimes \ldots \otimes e) = \sum x_1^{h_1} \ldots x_{n^2}^{h_{n^2}} \lambda(r(h_1, \ldots, h_{n^2}))$$

is a polynomial function on $End(V_n)$. As GL_n is a Zariski open subset of $End(V)$ on which by assumption this polynomial vanishes, it must be the zero polynomial. Therefore, $\lambda(r(h_1, \ldots, h_{n^2})) = 0$ for all $(h_1, \ldots, h_{n^2})$ finishing the proof.

(2) : Recall that the groupalgebra $\mathbb{C}S_m$ of S_m is a *semisimple algebra* . Any epimorphic image of a semisimple algebra is semisimple. Therefore, $\langle S_m \rangle$ is a semisimple subalgebra of the matrix algebra $End(V_n^{\otimes m}) \simeq M_{nm}$. By the *double centralizer theorem* (see, for example [84]), it is therefore equal to the centralizer of $End_{S_m}(V_m^{\otimes m})$. By the first part, it is the centralizer of $\langle GL_n \rangle$ in $End(V_n^{\otimes m})$ and therefore equal to $End_{GL_n}(V_n^{\otimes m})$. $\square$

Because $End_{GL_n}(V_n^{\otimes m}) = \langle S_m \rangle$, every GL_n-endomorphism of $V_n^{\otimes m}$ can be written as a linear combination of the morphisms λ_σ describing the action of $\sigma \in S_m$ on $V_n^{\otimes m}$. Our next job is to trace back these morphisms λ_σ through the canonical identifications until we can express them in terms of matrices.

To start let us compute the linear invariant

$$\mu_\sigma : V_n^{*\otimes m} \otimes V_n^{\otimes m} \longrightarrow \mathbb{C}$$

corresponding to λ_σ under the identification $(V_n^{*\otimes m} \otimes V_n^{\otimes m})^* \simeq End(V_n^{\otimes m})$. By the identification we know that $\mu_\sigma(f_1 \otimes \ldots f_m \otimes v_1 \otimes \ldots \otimes v_m)$ is equal to

$$\langle \lambda_\sigma, f_1 \otimes \ldots f_m \otimes v_1 \otimes \ldots \otimes v_m \rangle = f_1 \otimes \ldots \otimes f_m(v_{\sigma(1)} \otimes \ldots v_{\sigma(m)})$$
$$= \prod_i f_i(v_{\sigma(i)})$$

That is, we have proved the following.

PROPOSITION 1.1
Any multilinear GL_n-invariant map

$$\gamma : V_n^{*\otimes m} \otimes V_n^{\otimes m} \longrightarrow \mathbb{C}$$

is a linear combination of the invariants

$$\mu_\sigma(f_1 \otimes \ldots f_m \otimes v_1 \otimes \ldots \otimes v_m) = \prod_i f_i(v_{\sigma(i)})$$

for $\sigma \in S_m$.

Using the identification $M_n(\mathbb{C}) = V_n \otimes V_n^*$ a multilinear GL_n-invariant map

$$(V_n^* \otimes V_n)^{\otimes m} = V_n^{*\otimes m} \otimes V_n^{\otimes m} \longrightarrow \mathbb{C}$$

corresponds to a multilinear GL_n-invariant map

$$M_n(\mathbb{C}) \otimes \ldots \otimes M_n(\mathbb{C}) \longrightarrow \mathbb{C}$$

We will now give a description of the generating maps μ_σ in terms of matrices. Under the identification, matrix multiplication is induced by composition on rank one endomorphisms and here the rule is given by

$$v \otimes f.v' \otimes f' = f(v')v \otimes f'$$

$$\begin{bmatrix} \nu_1 \\ \vdots \\ \nu_n \end{bmatrix} \otimes \begin{bmatrix} \phi_1 & \cdots & \phi_n \end{bmatrix} \cdot \begin{bmatrix} \nu_1' \\ \vdots \\ \nu_n' \end{bmatrix} \otimes \begin{bmatrix} \phi_1' & \cdots & \phi_n' \end{bmatrix} = \begin{bmatrix} \nu_1 \\ \vdots \\ \nu_n \end{bmatrix} f(v') \otimes \begin{bmatrix} \phi_1' & \cdots & \phi_n' \end{bmatrix}$$

Moreover, the trace map on M_n is induced by that on rank one endomorphisms where it is given by the rule

$$tr(v \otimes f) = f(v)$$

$$tr(\begin{bmatrix} \nu_1 \\ \vdots \\ \nu_n \end{bmatrix} \otimes \begin{bmatrix} \phi_1 & \cdots & \phi_n \end{bmatrix}) = tr(\begin{bmatrix} \nu_1\phi_1 & \cdots & \nu_1\phi_n \\ \vdots & \ddots & \vdots \\ \nu_n\phi_1 & \cdots & \nu_n\phi_n \end{bmatrix}) = \sum_i \nu_i\phi_i = f(v)$$

With these rules we can now give a matrix-interpretation of the GL_n-invariant maps μ_σ.

PROPOSITION 1.2
Let $\sigma = (i_1 i_2 \ldots i_\alpha)(j_1 j_2 \ldots j_\beta) \ldots (z_1 z_2 \ldots z_\zeta)$ *be a decomposition of* $\sigma \in S_m$ *into cycles (including those of length one). Then, under the above identification we have*

$$\mu_\sigma(A_1 \otimes \ldots \otimes A_m) = tr(A_{i_1} A_{i_2} \ldots A_{i_\alpha}) tr(A_{j_1} A_{j_2} \ldots A_{j_\beta}) \ldots tr(A_{z_1} A_{z_2} \ldots A_{z_\zeta})$$

PROOF   Both sides are multilinear, hence it suffices to verify the equality for rank one matrices. Write $A_i = v_i \otimes f_i$, then we have that

$$\mu_\sigma(A_1 \otimes \ldots \otimes A_m) = \mu_\sigma(v_1 \otimes \ldots v_m \otimes f_1 \otimes \ldots \otimes f_m)$$
$$= \prod_i f_i(v_{\sigma(i)})$$

Consider the subproduct

$$f_{i_1}(v_{i_2}) f_{i_2}(v_{i_3}) \ldots f_{i_{\alpha-1}}(v_{i_\alpha}) = S$$

Now, look at the matrixproduct

$$v_{i_1} \otimes f_{i_1}.v_{i_2} \otimes f_{i_2}. \;\cdots\; .v_{i_\alpha} \otimes f_{i_\alpha}$$

which is by the product rule equal to

$$f_{i_1}(v_{i_2})f_{i_2}(v_{i_3})\ldots f_{i_{\alpha-1}}(v_{i_\alpha})v_{i_1} \otimes f_{i_\alpha}$$

Hence, by the trace rule we have that

$$tr(A_{i_1}A_{i_2}\ldots A_{i_\alpha}) = \prod_{j=1}^{\alpha} f_{i_j}(v_{\sigma(i_j)}) = S$$

▯

Having found a description of the multilinear invariant polynomial maps

$$M_n^m = \underbrace{M_n \oplus \ldots \oplus M_n}_{m} \longrightarrow \mathbb{C}$$

we will now describe all polynomial maps that are constant along orbits by polarization. The coordinate algebra $\mathbb{C}[M_n^m]$ is the polynomial ring in mn^2 variables $x_{ij}(k)$ where $1 \leq k \leq m$ and $1 \leq i,j \leq n$. Consider the m generic $n \times n$ matrices

$$\boxed{k} = X_k = \begin{bmatrix} x_{11}(k) & \ldots & x_{1n}(k) \\ \vdots & & \vdots \\ x_{n1}(k) & \ldots & x_{nn}(k) \end{bmatrix} \in M_n(\mathbb{C}[M_n^m])$$

The action of GL_n on polynomial maps $f \in \mathbb{C}[M_n^m]$ is fully determined by the action on the coordinate functions $x_{ij}(k)$. As in the case of one $n \times n$ matrix we see that this action is given by

$$g.x_{ij}(k) = (g^{-1}.X_k.g)_{ij}$$

We see that this action preserves the subspaces spanned by the entries of any of the generic matrices. Hence, we can define a gradation on $\mathbb{C}[M_n^m]$ by $deg(x_{ij}(k)) = (0,\ldots,0,1,0,\ldots,0)$ (with 1 at place k) and decompose

$$\mathbb{C}[M_n^m] = \bigoplus_{(d_1,\ldots,d_m)\in\mathbb{N}^m} \mathbb{C}[M_n^m]_{(d_1,\ldots,d_m)}$$

where $\mathbb{C}[M_n^m]_{(d_1,\ldots,d_m)}$ is the subspace of all multihomogeneous forms f in the $x_{ij}(k)$ of degree $(d_1,\ldots,d_m)$, that is, in each monomial term of f there are exactly d_k factors coming from the entries of the generic matrix X_k for all $1 \leq k \leq m$. The action of GL_n stabilizes each of these subspaces, that is

if $f \in \mathbb{C}[M_n^m]_{(d_1,\ldots,d_m)}$ then $g.f \in \mathbb{C}[M_n^m]_{(d_1,\ldots,d_m)}$ for all $g \in GL_n$

In particular, if f determines a polynomial map on M_n^m which is constant along orbits, that is, if f belongs to the ring of invariants $\mathbb{C}[M_n^m]^{GL_n}$ then each of its multihomogeneous components is also an invariant and therefore it suffices to determine all multihomogeneous invariants.

Let $f \in \mathbb{C}[M_n^m]_{(d_1,\ldots,d_m)}$ and take for each $1 \leq k \leq m$ d_k new variables $t_1(k),\ldots,t_{d_k}(k)$. Expand

$$f(t_1(1)A_1(1) + \ldots + t_{d_1}A_{d_1}(1),\ldots,t_1(m)A_1(m) + \ldots + t_{d_m}(m)A_{d_m}(m))$$

as a polynomial in the variables $t_i(k)$, then we get an expression

$$\sum t_1(1)^{s_1(1)}\ldots t_{d_1}^{s_{d_1}(1)}\ldots t_1(m)^{s_1(m)}\ldots t_{d_m}(m)^{s_{d_m}(m)}.$$
$$f_{(s_1(1),\ldots,s_{d_1}(1),\ldots,s_1(m),\ldots,s_{d_m}(m))}(A_1(1),\ldots,A_{d_1}(1),\ldots,A_1(m),\ldots,A_{d_m}(m))$$

such that for all $1 \leq k \leq m$ we have $\sum_{i=1}^{d_k} s_i(k) = d_k$. Moreover, each of the $f_{(s_1(1),\ldots,s_{d_1}(1),\ldots,s_1(m),\ldots,s_{d_m}(m))}$ is a multihomogeneous polynomial function on

$$\underbrace{M_n \oplus \ldots \oplus M_n}_{d_1} \underbrace{\oplus M_n \oplus \ldots \oplus M_n}_{d_2} \oplus \ldots \oplus \underbrace{M_n \oplus \ldots \oplus M_n}_{d_m}$$

of multidegree $(s_1(1),\ldots,s_{d_1}(1),\ldots,s_1(m),\ldots,s_{d_m}(m))$. Observe that if f is an invariant polynomial function on M_n^m, then each of these multihomogeneous functions is an invariant polynomial function on M_n^D where $D = d_1 + \ldots + d_m$.

In particular, we consider the multi*linear* function

$$f_{1,\ldots,1} : M_n^D = M_n^{d_1} \oplus \ldots \oplus M_n^{d_m} \longrightarrow \mathbb{C}$$

which we call the *polarization* of the polynomial f and denote with $Pol(f)$. Observe that $Pol(f)$ in symmetric in each of the entries belonging to a block $M_n^{d_k}$ for every $1 \leq k \leq m$. If f is invariant under GL_n, then so is the multilinear function $Pol(f)$ and we know the form of all such functions by the results given before (replacing M_n^m by M_n^D).

Finally, we want to recover f back from its polarization. We claim to have the equality

$$Pol(f)(\underbrace{A_1,\ldots,A_1}_{d_1},\ldots,\underbrace{A_m,\ldots,A_m}_{d_m}) = d_1!\ldots d_m!f(A_1,\ldots,A_m)$$

and hence we recover f. This process is called *restitution* . The claim follows from the observation that

$$f(t_1(1)A_1 + \ldots + t_{d_1}(1)A_1,\ldots,t_1(m)A_m + \ldots + t_{d_m}(m)A_m) =$$
$$f((t_1(1) + \ldots + t_{d_1}(1))A_1,\ldots,(t_1(m) + \ldots + t_{d_m}(m))A_m) =$$
$$(t_1(1) + \ldots + t_{d_1}(1))^{d_1}\ldots(t_1(m) + \ldots + t_{d_m}(m))^{d_m}f(A_1,\ldots,A_m)$$

and the definition of $Pol(f)$. Hence we have proved that any multi-homogeneous invariant polynomial function f on M_n^m of multidegree $(d_1, \ldots, d_m)$ can be obtained by restitution of a multilinear invariant function

$$Pol(f) : M_n^D = M_n^{d_1} \oplus \ldots \oplus M_n^{d_m} \longrightarrow \mathbb{C}$$

If we combine this fact with our description of all multilinear invariant functions on $M_n \oplus \ldots \oplus M_n$ we finally obtain the following.

THEOREM 1.6 First fundamental theorem of matrix invariants
Any polynomial function $M_n^m \xrightarrow{f} \mathbb{C}$ *that is constant along orbits under the action of* GL_n *by simultaneous conjugation is a polynomial in the invariants*

$$tr(X_{i_1} \ldots X_{i_l})$$

where $X_{i_1} \ldots X_{i_l}$ *run over all possible noncommutative polynomials in the generic matrices* $\{X_1, \ldots, X_m\}$.

We will call the algebra $\mathbb{C}[M_n^m]$ generated by these invariants the *necklace algebra* $\mathbb{N}_n^m = \mathbb{C}[M_n^m]^{GL_n}$. The terminology is justified by the observation that the generators

$$tr(X_{i_1} X_{i_2} \ldots X_{i_l})$$

are only determined up to cyclic permutation of the factors X_j. They correspond to a *necklace word* w

where each i-colored bead $\boxed{i}$ corresponds to a generic matrix X_i. To obtain an invariant, these bead-matrices are cyclically multiplied to obtain an $n \times n$ matrix with coefficients in $M_n(\mathbb{C}[M_n^m])$. The trace of this matrix is called $tr(w)$ and theorem 1.6 asserts that these elements generate the ring of polynomial invariants.

1.4 The trace algebra

In this section we will prove that there is a bound on the length of the necklace words w necessary for the $tr(w)$ to generate $\mathbb{N}_n^m$. Later, after we have determined the relations between these necklaces $tr(w)$, we will be able to improve this bound.

First, we will characterize all GL_n-*equivariant maps* from M_n^m to M_n, that is all polynomial maps $M_n^m \xrightarrow{f} M_n$, such that for all $g \in GL_n$ the diagram below is commutative

$$
\begin{array}{ccc}
M_n^m & \xrightarrow{\ f\ } & M_n \\
{\scriptstyle g.g^{-1}}\downarrow & & \downarrow{\scriptstyle g.g^{-1}} \\
M_n^m & \xrightarrow{\ f\ } & M_n
\end{array}
$$

With pointwise addition and multiplication in the target algebra M_n, these polynomial maps form a noncommutative algebra $\mathbb{T}_n^m$ called the *trace algebra*. Obviously, the trace algebra is a subalgebra of the algebra of *all* polynomial maps from M_n^m to M_n, that is

$$
\mathbb{T}_n^m \lhook\joinrel\longrightarrow M_n(\mathbb{C}[M_n^m])
$$

Clearly, using the diagonal embedding of $\mathbb{C}$ in M_n any invariant polynomial on M_n^m determines a GL_n-equivariant map. Equivalently, using the diagonal embedding of $\mathbb{C}[M_n^m]$ in $M_n(\mathbb{C}[M_n^m])$ we can embed the necklace algebra

$$
\mathbb{N}_n^m = \mathbb{C}[M_n^m]^{GL_n} \lhook\joinrel\longrightarrow \mathbb{T}_n^m
$$

Another source of GL_n-equivariant maps are the *coordinate maps*

$$
X_i : M_n^m = M_n \oplus \ldots \oplus M_n^m \longrightarrow M_n \qquad (A_1, \ldots, A_m) \mapsto A_i
$$

Observe that the coordinate map X_i is represented by the generic matrix $\boxed{i} = X_i$ in $M_n(\mathbb{C}[M_n^m])$.

PROPOSITION 1.3

As an algebra over the necklace algebra $\mathbb{N}_n^m$, the trace algebra $\mathbb{T}_n^m$ is generated by the elements $\{X_1, \ldots, X_m\}$.

PROOF   Consider a GL_n-equivariant map $M_n^m \xrightarrow{f} M_n$ and associate to it the polynomial map

$$
M_n^{m+1} = M_n^m \oplus M_n \xrightarrow{\ tr(fX_{m+1})\ } \mathbb{C}
$$

defined by sending $(A_1, \ldots, A_m, A_{m+1})$ to $tr(f(A_1, \ldots, A_m).A_{m+1})$. For all $g \in GL_n$ we have that $f(g.A_1.g^{-1}, \ldots, g.A_m.g^{-1})$ is equal to

$g.f(A_1,\ldots,A_m).g^{-1}$ and hence

$$tr(f(g.A_1.g^{-1},\ldots,g.A_m.g^{-1}).g.A_{m+1}.g^{-1}) =$$
$$tr(g.f(A_1,\ldots,A_m).g^{-1}.g.A_{m+1}.g^{-1}) =$$
$$tr(g.f(A_1,\ldots,A_m).A_{m+1}.g^{-1}) = tr(f(A_1,\ldots,A_m).A_{m+1})$$

so $tr(fX_{m+1})$ is an invariant polynomial function on M_n^{m+1}, which is *linear* in X_{m+1}. By theorem 1.6 we can therefore write

$$tr(fX_{m+1}) = \sum_{\underbrace{}_{\in \mathbb{N}_n^m}} g_{i_1\ldots i_l}\, tr(X_{i_1}\ldots X_{i_l}X_{m+1})$$

Here, we used the necklace property allowing to permute cyclically the trace terms in which X_{m+1} occurs such that X_{m+1} occurs as the last factor. But then, $tr(fX_{m+1}) = tr(gX_{m+1})$ where

$$g = \sum g_{i_1\ldots i_l} X_{i_1}\ldots X_{i_l}.$$

Finally, using the *nondegeneracy* of the trace map on M_n (that is, if $A,B \in M_n$ such that $tr(AC) = tr(BC)$ for all $C \in M_n$, then $A = B$) it follows that $f = g$.
□

If we give each of the generic matrices X_i degree one, we see that the trace algebra $\mathbb{T}_n^m$ is a *connected positively graded algebra*

$$\mathbb{T}_n^m = T_0 \oplus T_1 \oplus T_2 \oplus \ldots \qquad \text{with } T_0 = \mathbb{C}$$

Our aim is to bound the length of the monomials in the X_i necessary to generate $\mathbb{T}_n^m$ as a module over the necklace algebra $\mathbb{N}_n^m$. Before we can do this we need to make a small detour in one of the more exotic realms of noncommutative algebra: the *Nagata-Higman problem* .

THEOREM 1.7 Nagata-Higman
Let R be an associative algebra without a unit element. *Assume there is a fixed natural number n such that $x^n = 0$ for all $x \in R$. Then, $R^{2^n-1} = 0$, that is*

$$x_1.x_2.\ldots.x_{2^n-1} = 0$$

for all $x_j \in R$.

PROOF We use induction on n, the case $n = 1$ being obvious. Consider for all $x,y \in R$

$$f(x,y) = yx^{n-1} + xyx^{n-2} + x^2yx^{n-3} + \ldots + x^{n-2}yx + x^{n-1}y.$$

Because for all $c \in \mathbb{C}$ we must have that

$$0 = (y + cx)^n = x^n c^n + f(x, y)c^{n-1} + \ldots + y^n$$

it follows that all the coefficients of the c^i with $1 \le i < n$ must be zero, in particular $f(x, y) = 0$. But then we have for all $x, y, z \in R$ that

$$
\begin{aligned}
0 &= f(x, z)y^{n-1} + f(x, zy)y^{n-2} + f(x, zy^2)y^{n-3} + \ldots + f(x, zy^{n-1}) \\
&= nx^{n-1}zy^{n-1} + zf(y, x^{n-1}) + xzf(y, x^{n-2}) + \\
&\quad x^2 zf(y, x^{n-3}) + \ldots + x^{n-2}zf(y, x)
\end{aligned}
$$

and therefore $x^{n-1}zy^{n-1} = 0$. Let $I \lhd R$ be the two-sided ideal of R generated by all elements x^{n-1}, then we have that $I.R.I = 0$. In the quotient algebra $\overline{R} = R/I$ every element $\overline{x}$ satisfies $\overline{x}^{n-1} = 0$.

By induction we may assume that $\overline{R}^{2^{n-1}-1} = 0$, or equivalently that $R^{2^{n-1}-1}$ is contained in I. But then

$$R^{2^n - 1} = R^{2(2^{n-1}-1)+1} = R^{2^{n-1}-1}.R.R^{2^{n-1}-1} \hookrightarrow I.R.I = 0$$

finishing the proof. $\qquad\qquad\qquad\qquad\qquad\qquad\qquad\qquad\qquad\qquad\qquad$ □

PROPOSITION 1.4

The trace algebra $\mathbb{T}_n^m$ is spanned as a module over the necklace algebra $\mathbb{N}_n^m$ by all monomials in the generic matrices

$$X_{i_1} X_{i_2} \ldots X_{i_l}$$

of degree $l \le 2^n - 1$.

PROOF   By the diagonal embedding of $\mathbb{N}_n^m$ in $M_n(\mathbb{C}[M_n^m])$ it is clear that $\mathbb{N}_n^m$ commutes with any of the X_i. Let $\mathbb{T}_+$ and $\mathbb{N}_+$ be the strict positive degrees of $\mathbb{T}_n^m$ and $\mathbb{N}_n^m$ and form the graded associative algebra (without unit element)

$$R = \mathbb{T}_+/\mathbb{N}_+.\mathbb{T}_+$$

Observe that any element $t \in \mathbb{T}_+$ satisfies an equation of the form

$$t^n + c_1 t^{n-1} + c_2 t^{n-2} + \ldots + c_n = 0$$

with all of the $c_i \in \mathbb{N}_+$. Indeed we have seen that all the coefficients of the characteristic polynomial of a matrix can be expressed as polynomials in the traces of powers of the matrix. But then, for any $x \in R$ we have that $x^n = 0$.

By the Nagata-Higman theorem we know that $R^{2^n-1} = (R_1)^{2^n-1} = 0$. Let $\mathbb{T}'$ be the graded $\mathbb{N}_n^m$-submodule of $\mathbb{T}_n^m$ spanned by all monomials in the generic matrices X_i of degree at most $2^n - 1$, then the above can be reformulated as

$$\mathbb{T}_n^m = \mathbb{T}' + \mathbb{N}_+.\mathbb{T}_n^m$$

We claim that $\mathbb{T}_m^n = \mathbb{T}'$. Otherwise there is a homogeneous $t \in \mathbb{T}_n^m$ of *minimal degree* d not contained in $\mathbb{T}'$ but still we have a description

$$t = t' + c_1.t_1 + \ldots + c_s.t_s$$

with t' and all c_i, t_i homogeneous elements. As $deg(t_i) < d$, $t_i \in \mathbb{T}'$ for all i but then is $t \in \mathbb{T}'$ a contradiction. □

Finally we are in a position to bound the length of the necklaces generating $\mathbb{N}_n^m$ as an algebra.

THEOREM 1.8
The necklace algebra $\mathbb{N}_n^m$ is generated by all necklaces $tr(w)$ where w is a necklace word in the bead-matrices $\{X_1, \ldots, X_m\}$ of length $l \leq 2^n$.

PROOF Let $\mathbb{T}'$ be the $\mathbb{C}$-subalgebra of $\mathbb{T}_n^m$ generated by the generic matrices X_i. Then, $tr(\mathbb{T}'_+)$ generates the ideal $\mathbb{N}_+$. Let $\mathbb{S}$ be the set of all monomials in the X_i of degree at most $2^n - 1$. By the foregoing proposition we know that $\mathbb{T}' \hookrightarrow \mathbb{N}_n^m.\mathbb{S}$. The trace map

$$tr : \mathbb{T}_n^m \longrightarrow \mathbb{N}_n^m$$

is $\mathbb{N}_n^m$-linear and therefore, because $\mathbb{T}'_+ \subset \mathbb{T}'.(\mathbb{C}X_1 + \ldots + \mathbb{C}X_m)$ we have

$$tr(\mathbb{T}'_+) \subset tr(\mathbb{N}_n^m.\mathbb{S}.(\mathbb{C}X_1 + \ldots + \mathbb{C}X_m)) \subset \mathbb{N}_n^m.tr(\mathbb{S}')$$

where $\mathbb{S}'$ is the set of monomials in the X_i of degree at most 2^n. If $\mathbb{N}'$ is the $\mathbb{C}$-subalgebra of $\mathbb{N}_n^m$ generated by all $tr(S')$, then we have $tr(\mathbb{T}'_+) \subset \mathbb{N}_n^m.\mathbb{N}'_+$. But then, we have

$$\mathbb{N}_+ = \mathbb{N}_n^m tr(\mathbb{T}_+) \subset \mathbb{N}_n^m \mathbb{N}'_+ \quad \text{and thus} \quad \mathbb{N}_n^m = \mathbb{N}' + \mathbb{N}_n^m \mathbb{N}'_+$$

from which it follows that $\mathbb{N}_n^m = \mathbb{N}'$ by a similar argument as in the foregoing proof. □

Example 1.4 The algebras $\mathbb{N}_2^2$ and $\mathbb{T}_2^2$
When working with 2×2 matrices, the following identities are often helpful

$$0 = A^2 - tr(A)A + det(A)$$
$$A.B + B.A = tr(AB) - tr(A)tr(B) + tr(A)B + tr(B)A$$

for all $A, B \in M_2$. Let $\mathbb{N}'$ be the subalgebra of $\mathbb{N}_2^2$ generated by $tr(X_1), tr(X_2)$, $det(X_1), det(X_2)$ and $tr(X_1X_2)$. Using the two formulas above and $\mathbb{N}_2^2$-linearity of the trace on $\mathbb{T}_2^2$ we see that the trace of any monomial in X_1

and X_2 of degree $d \geq 3$ can be expressed in elements of $\mathbb{N}'$ and traces of monomials of degree $\leq d - 1$. Hence, we have

$$\mathbb{N}_2^2 = \mathbb{C}[tr(X_1), tr(X_2), det(X_1), det(X_2), tr(X_1 X_2)].$$

Observe that there can be no algebraic relations between these generators as we have seen that the induced map $\pi : M_2^2 \longrightarrow \mathbb{C}^5$ is surjective. Another consequence of the above identities is that over $\mathbb{N}_2^2$ any monomial in the X_1, X_2 of degree $d \geq 3$ can be expressed as a linear combination of $1, X_1, X_2$ and $X_1 X_2$ and so these elements generate $\mathbb{T}_2^2$ as a $\mathbb{N}_2^2$-module. In fact, they are a basis of $\mathbb{T}_2^2$ over $\mathbb{N}_2^2$. Assume otherwise, there would be a relation, say

$$X_1 X_2 = \alpha I_2 + \beta X_1 + \gamma X_2$$

with $\alpha, \beta, \gamma \in \mathbb{C}(tr(X_1), tr(X_2), det(X_1), det(X_2), tr(X_1 X_2))$. Then this relation has to hold for all matrix couples $(A, B) \in M_2^2$ and we obtain a contradiction if we take the couple

$$A = \begin{bmatrix} 0 & 1 \\ 0 & 0 \end{bmatrix} \quad B = \begin{bmatrix} 0 & 0 \\ 1 & 0 \end{bmatrix} \quad \text{whence} \quad AB = \begin{bmatrix} 1 & 0 \\ 0 & 0 \end{bmatrix}.$$

Concluding, we have the following description of $\mathbb{N}_2^2$ and $\mathbb{T}_2^2$ as a subalgebra of $\mathbb{C}[M_2^2]$ respectively, $M_2(\mathbb{C}[M_2^2])$

$$\begin{cases} \mathbb{N}_2^2 = & \mathbb{C}[tr(X_1), tr(X_2), det(X_1), det(X_2), tr(X_1 X_2)] \\ \mathbb{T}_2^2 = & \mathbb{N}_2^2 . I_2 \oplus \mathbb{N}_2^2 . X_1 \oplus \mathbb{N}_2^2 . X_2 \oplus \mathbb{N}_2^2 . X_1 X_2 \end{cases}$$

Observe that we might have taken the generators $tr(X_i^2)$ rather than $det(X_i)$ because $det(X_i) = \frac{1}{2}(tr(X_i)^2 - tr(X_i)^2)$ as follows from taking the trace of characteristic polynomial of X_i. $\square$

1.5   The symmetric group

Let S_d be the symmetric group of all permutations on d letters. The group algebra $\mathbb{C} \, S_d$ is a semisimple algebra. In particular, any simple S_d-representation is isomorphic to a minimal left ideal of $\mathbb{C} \, S_d$, which is generated by an *idempotent*. We will now determine these idempotents.

To start, conjugacy classes in S_d correspond naturally to *partitions* $\lambda = (\lambda_1, \ldots, \lambda_k)$ of d, that is, decompositions in natural numbers

$$d = \lambda_1 + \ldots + \lambda_k \quad \text{with} \quad \lambda_1 \geq \lambda_2 \geq \ldots \geq \lambda_k \geq 1$$

The correspondence associates to a partition $\lambda = (\lambda_1, \ldots, \lambda_k)$ the conjugacy class of a permutation consisting of disjoint cycles of lengths $\lambda_1, \ldots, \lambda_k$. It is traditional to assign to a partition $\lambda = (\lambda_1, \ldots, \lambda_k)$ a *Young diagram* with λ_i boxes in the i-th row, the rows of boxes lined up to the left. The *dual partition* $\lambda^* = (\lambda_1^*, \ldots, \lambda_r^*)$ to λ is defined by interchanging rows and columns in the Young diagram of λ .

For example, to the partition $\lambda = (3, 2, 1, 1)$ of 7 we assign the Young diagram

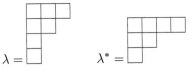

with dual partition $\lambda^* = (4, 2, 1)$. A *Young tableau* is a numbering of the boxes of a Young diagram by the integers $\{1, 2, \ldots, d\}$. For example, two distinct Young tableaux of type λ are

1	2	3
4	5	
6		
7		

1	3	5
2	4	
6		
7		

Now, fix a Young tableau T of type λ and define subgroups of S_d by

$$P_\lambda = \{\sigma \in S_d \mid \sigma \text{ preserves each row } \}$$

$$Q_\lambda = \{\sigma \in S_d \mid \sigma \text{ preserves each column } \}$$

For example, for the second Young tableaux given above we have that

$$\begin{cases} P_\lambda &= S_{\{1,3,5\}} \times S_{\{2,4\}} \times \{(6)\} \times \{(7)\} \\ Q_\lambda &= S_{\{1,2,6,7\}} \times S_{\{3,4\}} \times \{(5)\} \end{cases}$$

Observe that different Young tableaux for the same λ define different subgroups and different elements to be defined below. Still, the simple representations we will construct from them turn out to be isomorphic.

Using these subgroups, we define the following elements in the groupalgebra $\mathbb{C}S_d$

$$a_\lambda = \sum_{\sigma \in P_\lambda} e_\sigma \quad , \quad b_\lambda = \sum_{\sigma \in Q_\lambda} sgn(\sigma) e_\sigma \quad \text{and} \quad c_\lambda = a_\lambda . b_\sigma$$

The element c_λ is called a *Young symmetrizer* . The next result gives an explicit one-to-one correspondence between the simple representations of $\mathbb{C}S_d$ and the conjugacy classes in S_d (or, equivalently, Young diagrams).

THEOREM 1.9
For every partition λ of d the left ideal $\mathbb{C}S_d.c_\lambda = V_\lambda$ is a simple S_d-representations and, conversely, any simple S_d-representation is isomorphic to V_λ for a unique partition λ.

PROOF (sketch) Observe that $P_\lambda \cap Q_\lambda = \{e\}$ (any permutation preserving rows as well as columns preserves all boxes) and so any element of S_d can be written in at most one way as a product $p.q$ with $p \in P_\lambda$ and $q \in Q_\lambda$. In particular, the Young symmetrizer can be written as $c_\lambda = \sum \pm e_\sigma$ with $\sigma = p.q$ for unique p and q and the coefficient $\pm 1 = sgn(q)$. From this it follows that for all $p \in P_\lambda$ and $q \in Q_\lambda$ we have

$$p.a_\lambda = a_\lambda.p = a_\lambda \quad , \quad sgn(q)q.b_\lambda = b_\lambda.sgn(q)q = b_\lambda \quad , \quad p.c_\lambda.sgn(q)q = c_\lambda$$

Moreover, we claim that c_λ is the unique element in $\mathbb{C}S_d$ (up to a scalar factor) satisfying the last property. This requires a few preparations.

Assume $\sigma \notin P_\lambda.Q_\lambda$ and consider the tableaux $T' = \sigma T$, that is, replacing the label i of each box in T by $\sigma(i)$. We claim that there are two distinct numbers that belong to the same row in T and to the same column in T'. If this were not the case, then all the distinct numbers in the first row of T appear in different columns of T'. But then we can find an element q_1' in the subgroup $\sigma.Q_\lambda.\sigma^{-1}$ preserving the columns of T' to take all these elements to the first row of T'. But then, there is an element $p_1 \in T_\lambda$ such that $p_1 T$ and $q_1' T'$ have the same first row. We can proceed to the second row and so on and obtain elements $p \in P_\lambda$ and $q' \in \sigma.Q_\lambda.\sigma^{-1}$ such that the tableaux pT and $q'T'$ are equal. Hence, $pT = q'\sigma T$ entailing that $p = q'\sigma$. Further, $q' = \sigma.q.\sigma^{-1}$ but then $p = q'\sigma = \sigma q$ whence $\sigma = p.q^{-1} \in P_\lambda.Q_\lambda$, a contradiction. Therefore, to $\sigma \notin P_\lambda.Q_\lambda$ we can assign a *transposition* $\tau = (ij)$ (replacing the two distinct numbers belonging to the same row in T and to the same column in T') for which $p = \tau \in P_\lambda$ and $q = \sigma^{-1}.\tau.\sigma \in Q_\lambda$.

After these preliminaries, assume that $c' = \sum a_\sigma e_\sigma$ is an element such that

$$p.c'.sgn(q)q = c' \quad \text{for all} \quad p \in P_\lambda, q \in Q_\lambda$$

We claim that $a_\sigma = 0$ whenever $\sigma \notin P_\lambda.Q_\lambda$. Take the transposition τ found above and $p = \tau$, $q = \sigma^{-1}.\tau.\sigma$, then $p.\sigma.q = \tau.\sigma.\sigma^{-1}.\tau.\sigma = \sigma$. However, the coefficient of σ in c' is a_σ and that of $p.c'.q$ is $-a_\sigma$ proving the claim. That is

$$c' = \sum_{p,q} a_{pq} e_{p.q}$$

but then by the property of c' we must have that $a_{pq} = sgn(q)a_e$ whence $c' = a_e c_\lambda$ finishing the proof of the claimed uniqueness of the element c_λ.

As a consequence we have for all elements $x \in \mathbb{C}S_d$ that $c_\lambda.x.c_\lambda = \alpha_x c_\lambda$ for some scalar $\alpha_x \in \mathbb{C}$ and in particular that $c_\lambda^2 = n_\lambda c_\lambda$, for

$$p.(c_\lambda.x.c_\lambda).sgn(q)q = p.a_\lambda.b_\lambda.x.a_\lambda.b_\lambda.sgn(q)q$$
$$= a_\lambda.b_\lambda.x.a_\lambda.b_\lambda = c_\lambda.x.c_\lambda$$

and the statement follows from the uniqueness result for c_λ.

Define $V_\lambda = \mathbb{C}S_d.c_\lambda$ then we have $c_\lambda.V_\lambda \subset \mathbb{C}c_\lambda$. We claim that V_λ is a simple S_d-representation. Let $W \subset V_\lambda$ be a simple subrepresentation, then being a left ideal of $\mathbb{C}S_d$, we can write $W = \mathbb{C}S_d.x$ with $x^2 = x$ (note that W is a direct summand). Assume that $c_\lambda.W = 0$, then $W.W \subset \mathbb{C}S_d.c_\lambda.W = 0$ implying that $x = 0$ whence $W = 0$, a contradiction. Hence, $c_\lambda.W = \mathbb{C}c_\lambda \subset W$, but then

$$V_\lambda = \mathbb{C}S_d.c_\lambda \subset W \quad \text{whence} V_\lambda = W$$

is simple. It remains to be shown that for different partitions, the corresponding simple representations cannot be isomorphic.

We put a *lexicographic* ordering on the partitions by the rule that

$$\lambda > \mu \quad \text{if the first nonvanishing } \lambda_i - \mu_i \text{ is positive}$$

We claim that if $\lambda > \mu$ then $a_\lambda.\mathbb{C}S_d.b_\mu = 0$. It suffices to check that $a_\lambda.\sigma.b_\mu = 0$ for $\sigma \in S_d$. As $\sigma.b_\mu.\sigma^{-1}$ is the b-element constructed from the tableau $b.T'$ where T' is the tableaux fixed for μ, it is sufficient to check that $a_\lambda.b_\mu = 0$. As $\lambda > \mu$ there are distinct numbers i and j belonging to the same row in T and to the same column in T'. If not, the distinct numbers in any fixed row of T must belong to different columns of T', but this can only happen for all rows if $\mu \geq \lambda$. So consider $\tau = (ij)$, which belongs to P_λ and to Q_μ, whence $a_\lambda.\tau = a_\lambda$ and $\tau.b_\mu = -b_\mu$. But then

$$a_\lambda.b_\mu = a_\lambda.\tau, \tau, b_\mu = -a_\lambda.b_\mu$$

proving the claim.

If $\lambda \neq \mu$ we claim that V_λ is not isomorphic to V_μ. Assume that $\lambda > \mu$ and ϕ a $\mathbb{C}S_d$-isomorphism with $\phi(V_\lambda) = V_\mu$, then

$$\phi(c_\lambda V_\lambda) = c_\lambda\phi(V_\lambda) = c_\lambda V_\mu = c_\lambda\mathbb{C}S_d c_\mu = 0$$

Hence, $c_\lambda V_\lambda = \mathbb{C}c_\lambda \neq 0$ lies in the kernel of an isomorphism that is clearly absurd.

Summarizing, we have constructed to distinct partitions of d, λ and μ non-isomorphic simple $\mathbb{C}S_d$-representations V_λ and V_μ. As we know that there are as many isomorphism classes of simples as there are conjugacy classes in S_d (or partitions), the V_λ form a complete set of isomorphism classes of simple S_d-representations, finishing the proof of the theorem. $\quad\square$

1.6   Necklace relations

In this section we will prove that all the relations holding among the elements of the necklace algebra $\mathbb{N}_n^m$ are formal consequences of the Cayley-Hamilton theorem. First, we will have to set up some notation to clarify what we mean by this.

For technical reasons it is sometimes convenient to have an infinite supply
of noncommutative variables $\{x_1, x_2, \ldots, x_i, \ldots\}$. Two monomials of the same
degree d in these variables

$$M = x_{i_1} x_{i_2} \ldots x_{i_d} \quad \text{and} \quad M' = x_{j_1} x_{j_2} \ldots x_{j_d}$$

are said to be *equivalent* if M' is obtained from M by a cyclic permutation,
that is, there is a k such that $i_1 = j_k$ and all $i_a = j_b$ with $b = k + a - 1 \bmod d$.
That is, if they determine the same necklace word

with each of the beads one of the noncommuting variables $\boxed{i} = x_i$. To each
equivalence class we assign a formal variable that we denote by

$$t(x_{i_1} x_{i_2} \ldots x_{i_d}).$$

The *formal necklace algebra* $\mathbb{N}^\infty$ is then the polynomial algebra on all these
(infinitely many) letters. Similarly, we define the *formal trace algebra* $\mathbb{T}^\infty$ to
be the algebra

$$\mathbb{T}^\infty = \mathbb{N}^\infty \otimes_\mathbb{C} \mathbb{C}\langle x_1, x_2, \ldots, x_i, \ldots \rangle$$

that is, the free associative algebra on the noncommuting variables x_i with
coefficients in the polynomial algebra $\mathbb{N}^\infty$.

Crucial for our purposes is the existence of an $\mathbb{N}^\infty$-linear *formal trace map*

$$t : \mathbb{T}^\infty \longrightarrow \mathbb{N}^\infty$$

defined by the formula

$$t\left(\sum a_{i_1 \ldots i_k} x_{i_1} \ldots x_{i_k}\right) = \sum a_{i_1 \ldots i_k} t(x_{i_1} \ldots x_{i_k})$$

where $a_{i_1 \ldots i_k} \in \mathbb{N}^\infty$.

In an analogous manner we will define infinite versions of the necklace and trace algebras. Let M_n^∞ be the space of all ordered sequences
$(A_1, A_2, \ldots, A_i, \ldots)$ with $A_i \in M_n$ and all but finitely many of the A_i are
the zero matrix. Again, GL_n acts on M_n^∞ by simultaneous conjugation and
we denote the *infinite necklace algebra* $\mathbb{N}_n^\infty$ to be the algebra of polynomial
functions f

$$M_n^\infty \xrightarrow{\ f\ } \mathbb{C}$$

which are constant along orbits. Clearly, $\mathbb{N}_n^\infty$ is generated as $\mathbb{C}$-algebra by the
invariants $tr(M)$ where M runs over all monomials in the coordinate generic

matrices $X_k = (x_{ij}(k))_{i,j}$ belonging to the k-th factor of M_n^∞. Similarly, the *infinite trace algebra* $\mathbb{T}_n^\infty$ is the algebra of GL_n-equivariant polynomial maps

$$M_n^\infty \longrightarrow M_n.$$

Clearly, $\mathbb{T}_n^\infty$ is the $\mathbb{C}$-algebra generated by $\mathbb{N}_n^\infty$ and the generic matrices X_k for $1 \leq k < \infty$. Observe that $\mathbb{T}_n^\infty$ is a subalgebra of the matrix ring

$$\mathbb{T}_n^\infty \hookrightarrow M_n(\mathbb{C}[M_n^\infty])$$

and as such has a trace map tr defined on it and from our knowledge of the generators of $\mathbb{N}_n^\infty$ we know that $tr(\mathbb{T}_n^\infty) = \mathbb{N}_n^\infty$.

Now, there are natural algebra epimorphisms

$$\mathbb{T}^\infty \xrightarrow{\ \tau\ } \mathbb{T}_n^\infty \quad \text{and} \quad \mathbb{N}^\infty \xrightarrow{\ \nu\ } \mathbb{N}_n^\infty$$

defined by $\tau(t(x_{i_1} \ldots x_{i_k})) = \nu(t(x_{i_1} \ldots x_{i_k})) = tr(X_{i_1} \ldots X_{i_k})$ and $\tau(x_i) = X_i$. That is, ν and τ are compatible with the trace maps

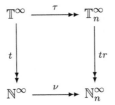

We are interested in describing the *necklace relations*, that is, the kernel of ν. In the next section we will describe the *trace relations* that are the kernel of τ. Note that we obtain the relations holding among the necklaces in $\mathbb{N}_n^m$ by setting all $x_i = 0$ with $i > m$ and all $t(x_{i_1} \ldots x_{i_k}) = 0$ containing a variable with $i_j > m$.

In the description a map $T : \mathbb{C}S_d \longrightarrow \mathbb{N}^\infty$ will be important. Let S_d be the symmetric group of permutations on $\{1, \ldots, d\}$ and let

$$\sigma = (i_1 i_1 \ldots i_\alpha)(j_1 j_2 \ldots j_\beta) \ldots (z_1 z_2 \ldots z_\zeta)$$

be a decomposition of $\sigma \in S_d$ into cycles including those of length one. The map T assigns to σ a formal necklace $T_\sigma(x_1, \ldots, x_d)$ defined by

$$T_\sigma(x_1, \ldots, x_d) = t(x_{i_1} x_{i_2} \ldots x_{i_\alpha}) t(x_{j_1} x_{j_2} \ldots x_{j_\beta}) \ldots t(x_{z_1} x_{z_2} \ldots x_{z_\zeta})$$

Let $V = V_n$ be again the n-dimensional vector space of column vectors, then S_d acts naturally on $V^{\otimes d}$ via

$$\sigma.(v_1 \otimes \ldots \otimes v_d) = v_{\sigma(1)} \otimes \ldots \otimes v_{\sigma(d)}$$

hence determines a linear map $\lambda_\sigma \in End(V^{\otimes d})$. Recall from section 3 that under the natural identifications

$$(M_n^{\otimes d})^* \simeq (V^{*\otimes d} \otimes V^{\otimes d})^* \simeq End(V^{\otimes d})$$

the map λ_σ defines the multilinear map

$$\mu_\sigma : \underbrace{M_n \otimes \ldots \otimes M_n}_{d} \longrightarrow \mathbb{C}$$

defined by (using the cycle decomposition of σ as before)

$$\mu_\sigma(A_1 \otimes \ldots \otimes A_d) = tr(A_{i_1}A_{i_2}\ldots A_{i_\alpha})tr(A_{j_1}A_{j_2}\ldots A_{j_\beta})\ldots tr(A_{z_1}A_{z_2}\ldots A_{z_\zeta})$$

Therefore, a linear combination $\sum a_\sigma T_\sigma(x_1,\ldots,x_d)$ is a necklace relation (that is, belongs to $Ker\ \nu$) if and only if the multilinear map $\sum a_\sigma \mu_\sigma$: $M_n^{\otimes d} \longrightarrow \mathbb{C}$ is zero. This, in turn, is equivalent to the endomorphism $\sum a_\sigma \lambda_\sigma \in End(V^{\otimes m})$, induced by the action of the element $\sum a_\sigma e_\sigma \in \mathbb{C}S_d$ on $V^{\otimes d}$, being zero. In order to answer the latter problem we have to understand the action of a Young symmetrizer $c_\lambda \in \mathbb{C}S_d$ on $V^{\otimes d}$.

Let $\lambda = (\lambda_1, \lambda_2, \ldots, \lambda_k)$ be a partition of d and equip the corresponding Young diagram with the standard tableau (that is, order first the boxes in the first row from left to right, then the second row from left to right and so on) as shown

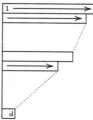

The subgroup P_λ of S_d which preserves each row then becomes

$$P_\lambda = S_{\lambda_1} \times S_{\lambda_2} \times \ldots \times S_{\lambda_k} \longrightarrow S_d$$

As $a_\lambda = \sum_{p \in P_\lambda} e_p$ we see that the image of the action of a_λ on $V^{\otimes d}$ is the subspace

$$Im(a_\lambda) = Sym^{\lambda_1}\ V \otimes Sym^{\lambda_2}\ V \otimes \ldots \otimes Sym^{\lambda_k}\ V \longrightarrow V^{\otimes d}$$

Here, $Sym^i\ V$ denotes the subspace of symmetric tensors in $V^{\otimes i}$.

Similarly, equip the Young diagram of λ with the tableau by ordering first the boxes in the first column from top to bottom, then those of the second column from top to bottom and so on as shown

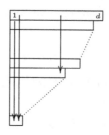

Equivalently, give the Young diagram corresponding to the dual partition of λ

$$\lambda^* = (\mu_1, \mu_2, \ldots, \mu_l)$$

the standard tableau. Then, the subgroup Q_λ of S_d, which preserves each row of λ (or equivalently, each column of λ^*) is

$$Q_\lambda = S_{\mu_1} \times S_{\mu_2} \times \ldots \times S_{\mu_l} \hookrightarrow S_d$$

As $b_\lambda = \sum_{q \in Q_\lambda} sgn(q) e_q$ we see that the image of b_λ on $V^{\otimes d}$ is the subspace

$$Im(b_\lambda) = \overset{\mu_1}{\bigwedge} V \otimes \overset{\mu_2}{\bigwedge} V \otimes \ldots \otimes \overset{\mu_l}{\bigwedge} V \hookrightarrow V^{\otimes d} \quad .$$

Here, $\bigwedge^i V$ is the subspace of all antisymmetric tensors in $V^{\otimes i}$. Note that $\bigwedge^i V = 0$ whenever i is greater than the dimension $dim\ V = n$. That is, the image of the action of b_λ on $V^{\otimes d}$ is zero whenever the dual partition λ^* contains a row of length $\geq n+1$, or equivalently, whenever λ has $\geq n+1$ rows. Because the Young symmetrizer $c_\lambda = a_\lambda . b_\lambda \in \mathbb{C}\ S_d$ we have proved the first result on necklace relations.

THEOREM 1.10 Second fundamental theorem of matrix invariants

A formal necklace

$$\sum_{\sigma \in S_d} a_\sigma T_\sigma(x_1, \ldots, x_d)$$

is a necklace relation (for $n \times n$ matrices) if and only if the element

$$\sum a_\sigma e_\sigma \in \mathbb{C} S_d$$

belongs to the ideal of $\mathbb{C} S_d$ spanned by the Young symmetrizers c_λ relative to partitions $\lambda = (\lambda_1, \ldots, \lambda_k)$

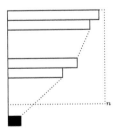

with a least $n+1$ rows, that is, $k \geq n+1$.

Example 1.5

(Fundamental necklace and trace relation.) Consider the partition $\lambda = (1, 1, \ldots, 1)$ of $n + 1$, with corresponding Young tableau

$$
\begin{array}{|c|}
\hline
1 \\
\hline
2 \\
\hline
\vdots \\
\hline
n+1 \\
\hline
\end{array}
$$

Then, $P_\lambda = \{e\}$, $Q_\lambda = S_{n+1}$ and we have the Young symmetrizer

$$
a_\lambda = 1 \qquad b_\lambda = c_\lambda = \sum_{\sigma \in S_{n+1}} sgn(\sigma) e_\sigma
$$

The corresponding element is called the *fundamental necklace relation*

$$
\mathtt{fund}_n(x_1, \ldots, x_{n+1}) = \sum_{\sigma \in S_{n+1}} sgn(\sigma) T_\sigma(x_1, \ldots, x_{n+1})
$$

Clearly, $\mathtt{fund}_n(x_1, \ldots, x_{n+1})$ is multilinear of degree $n + 1$ in the variables $\{x_1, \ldots, x_{n+1}\}$. Conversely, any multilinear necklace relation of degree $n + 1$ must be a scalar multiple of $\mathtt{fund}_n(x_1, \ldots, x_{n+1})$. This follows from the proposition as the ideal described there is for $d = n+1$ just the scalar multiples of $\sum_{\sigma \in S_{n+1}} sgn(\sigma) e_\sigma$.

Because $\mathtt{fund}_n(x_1, \ldots, x_{n+1})$ is multilinear in the variables x_i we can use the cyclic permutation property of the formal trace t to write it in the form

$$
\mathtt{fund}_n(x_1, \ldots, x_{n+1}) = t(\mathtt{cha}_n(x_1, \ldots, x_n) x_{n+1}) \mathrm{with} \mathtt{cha}_n(x_1, \ldots, x_n) \in \mathbb{T}^\infty
$$

Observe that $\mathtt{cha}_n(x_1, \ldots, x_n)$ is multilinear in the variables x_i. Moreover, by the nondegeneracy of the trace map tr and the fact that $\mathtt{fund}_n(x_1, \ldots, x_{n+1})$ is a necklace relation, it follows that $\mathtt{cha}_n(x_1, \ldots, x_n)$ is a trace relation. Again, any multilinear trace relation of degree n in the variables $\{x_1, \ldots, x_n\}$ is a scalar multiple of $\mathtt{cha}_n(x_1, \ldots, x_n)$. This follows from the corresponding uniqueness result for $\mathtt{fund}_n(x_1, \ldots, x_{n+1})$.

We can give an explicit expression of this *fundamental trace relation*

$$
\mathtt{cha}_n(x_1, \ldots, x_n) =
$$

$$
\sum_{k=0}^{n} (-1)^k \sum_{i_1 \neq i_2 \neq \ldots \neq i_k} x_{i_1} x_{i_2} \ldots x_{i_k} \sum_{\sigma \in S_J} sgn(\sigma) T_\sigma(x_{j_1}, \ldots, x_{j_{n-k}})
$$

where $J = \{1, \ldots, n\} - \{i_1, \ldots, i_k\}$. In a moment we will see that $\mathtt{cha}_n(x_1, \ldots, x_n)$ and hence also $\mathtt{fund}_n(x_1, \ldots, x_{n+1})$ is obtained by polarization of the Cayley-Hamilton identity for $n \times n$ matrices. □

We will explain what we mean by the Cayley-Hamilton polynomial for an element of $\mathbb{T}^\infty$. Recall that when $X \in M_n(A)$ is a matrix with coefficients in a commutative $\mathbb{C}$-algebra A its *characteristic polynomial* is defined to be

$$\chi_X(t) = det(t\mathbb{1}_n - X) \in A[t]$$

and by the *Cayley-Hamilton theorem* we have the basic relation that $\chi_X(X) = 0$. We have seen that the coefficients of the characteristic polynomial can be expressed as polynomial functions in the $tr(X^i)$ for $1 \le i \le n$.

For example if $n = 2$, then the characteristic polynomial can we written as

$$\chi_X(t) = t^2 - tr(X)t + \frac{1}{2}(tr(X)^2 - tr(X^2))$$

For general n the method for finding these polynomial functions is based on the formal recursive algorithm expressing elementary symmetric functions in term of *Newton functions* , usually expressed by the formulae

$$f(t) = \prod_{i=1}^{n}(t - \lambda_i),$$

$$\frac{f'(t)}{f(t)} = \frac{d \log f(t)}{dt} = \sum_{i=1}^{n} \frac{1}{t - \lambda_i} = \sum_{k=0}^{\infty} \frac{1}{t^{k+1}}(\sum_{i=1}^{n} \lambda_i^k)$$

Note, if λ_i are the eigenvalues of $X \in M_n$, then $f(t) = \chi_X(t)$ and $\sum_{i=1}^{n} \lambda_i^k = tr(X^k)$. Therefore, one can use the formulae to express $f(t)$ in terms of the elements $\sum_{i=1}^{n} \lambda_i^k$. To get the required expression for the characteristic polynomial of X one only has to substitute $\sum_{i=1}^{n} \lambda_i^k$ with $tr(X^k)$.

This allows us to construct a *formal Cayley-Hamilton polynomial* $\chi_x(x) \in \mathbb{T}^\infty$ of an element $x \in \mathbb{T}^\infty$ by replacing in the above characteristic polynomial the term $tr(X^k)$ with $t(x^k)$ and t^l with x^l. If x is one of the variables x_i then $\chi_x(x)$ is an element of $\mathbb{T}^\infty$ homogeneous of degree n. Moreover, by the Cayley-Hamilton theorem it follows immediately that $\chi_x(x)$ is a trace relation. Hence, if we fully polarize $\chi_x(x)$ (say, using the variables $\{x_1, \ldots, x_n\}$) we obtain a multilinear trace relation of degree n. By the argument given in the example above we know that this element must be a scalar multiple of $\mathbf{cha}_n(x_1, \ldots, x_n)$. In fact, one can see that this scale factor must be $(-1)^n$ as the leading term of the multilinearization is $\sum_{\sigma \in S_n} x_{\sigma(1)} \cdots x_{\sigma(n)}$ and compare this with the explicit form of $\mathbf{cha}_n(x_1, \ldots, x_n)$.

Example 1.6

Consider the case $n = 2$. The formal Cayley-Hamilton polynomial of an element $x \in \mathbb{T}^\infty$ is

$$\chi_x(x) = x^2 - t(x)x + \frac{1}{2}(t(x)^2 - t(x^2))$$

Polarization with respect to the variables x_1 and x_2 gives the expression

$$x_1 x_2 + x_2 x_1 - t(x_1)x_2 - t(x_2)x_1 + t(x_1)t(x_2) - t(x_1 x_2)$$

which is $\mathtt{cha}_2(x_1, x_2)$. Indeed, multiplying it on the right with x_3 and applying the formal trace t to it we obtain

$$t(x_1 x_2 x_3) + t(x_2 x_1 x_3) - t(x_1)t(x_2 x_3) - t(x_2)t(x_1 x_3)$$
$$+ t(x_1)t(x_2)t(x_3) - t(x_1 x_2)t(x_3)$$
$$= \quad T_{(123)}(x_1, x_2, x_3) + T_{(213)}(x_1, x_2, x_3) - T_{(1)(23)}(x_1, x_2, x_3)$$
$$- T_{(2)(13)}(x_1, x_2, x_3) + T_{(1)(2)(3)}(x_1, x_2, x_3) - T_{(12)(3)}(x_1, x_2, x_3)$$
$$= \textstyle\sum_{\sigma \in S_3} T_\sigma(x_1, x_2, x_3) = \mathtt{fund}_2(x_1, x_2, x_3)$$

as required.   ☐

THEOREM 1.11
The necklace relations $Ker\ \nu$ is the ideal of $\mathbb{N}^\infty$ generated by all the elements

$$\mathtt{fund}_n(m_1, \ldots, m_{n+1})$$

where the m_i run over all monomials in the variables $\{x_1, x_2, \ldots, x_i, \ldots\}$

PROOF   Take a homogeneous necklace relation $f \in Ker\ \nu$ of degree d and polarize it to get a multilinear element $f' \in \mathbb{N}^\infty$. Clearly, f' is also a necklace relation and if we can show that f' belongs to the described ideal, then so does f as the process of restitution maps this ideal into itself.

Therefore, we may assume that f is multilinear of degree d. A priori f may depend on more than d variables x_k, but we can separate f as a sum of multilinear polynomials f_i each depending on precisely d variables such that for $i \neq j$ f_i and f_j do not depend on the same variables. Setting some of the variables equal to zero, we see that each of the f_i is again a necklace relation.

Thus, we may assume that f is a multilinear necklace identity of degree d depending on the variables $\{x_1, \ldots, x_d\}$. But then we know from theorem 1.10 that we can write

$$f = \sum_{\tau \in S_d} a_\tau T_\tau(x_1, \ldots, x_d)$$

where $\sum a_\tau e_\tau \in \mathbb{C}S_d$ belongs to the ideal spanned by the Young symmetrizers of Young diagrams λ having at least $n + 1$ rows.

We claim that this ideal is generated by the Young symmetrizer of the partition $(1, \ldots, 1)$ of $n + 1$ under the natural embedding of S_{n+1} into S_d. Let λ be a Young diagram having $k \geq n + 1$ boxes and let c_λ be a Young symmetrizer with respect to a tableau where the boxes in the first column are labeled by the numbers $I = \{i_1, \ldots, i_k\}$ and let S_I be the obvious subgroup of S_d. As $Q_\lambda = S_I \times Q'$ we see that $b_\lambda = (\sum_{\sigma \in S_I} sgn(\sigma)e_\sigma).b'$ with $b' \in \mathbb{C}Q'$.

Hence, c_λ belongs to the two-sided ideal generated by $c_I = \sum_{\sigma \in S_I} sgn(\sigma)e_\sigma$ but this is also the two-sided ideal generated by $c_k = \sum_{\sigma \in S_k} sgn(\sigma)e_\sigma$ as one verifies by conjugation with a partition sending I to $\{1, \ldots, k\}$. Moreover, by induction one shows that the two-sided ideal generated by c_k belongs to the two-sided ideal generated by $c_d = \sum_{\sigma \in S_d} sgn(\sigma)e_\sigma$, finishing the proof of the claim.

From this claim, we can write

$$\sum_{\tau \in S_d} a_\tau e_\tau = \sum_{\tau_i, \tau_j \in S_d} a_{ij} e_{\tau_i} . \left(\sum_{\sigma \in S_{n+1}} sgn(\sigma)e_\sigma \right). e_{\tau_j}$$

and therefore it suffices to analyze the form of the necklace identity associated to an element of the form

$$e_\tau . \left(\sum_{\sigma \in S_{n+1}} sgn(\sigma)e_\sigma \right). e_{\tau'} \quad \text{with} \quad \tau, \tau' \in S_d$$

Now, if a group element $\sum_{\mu \in S_d} b_\mu e_\mu$ corresponds to the formal necklace polynomial $g(x_1, \ldots, x_d)$, then the element $e_\tau . (\sum_{\mu \in S_d} b_\mu e_\mu). e_{\tau^{-1}}$ corresponds to the formal necklace polynomial $g(x_{\tau(1)}, \ldots, x_{\tau(d)})$.

Therefore, we may replace the element $e_\tau . (\sum_{\sigma \in S_{n+1}} sgn(\sigma)e_\sigma). e_{\tau'}$ by the element

$$\left(\sum_{\sigma \in S_{n+1}} sgn(\sigma)e_\sigma \right). e_\eta \quad \text{with} \quad \eta = \tau'.\tau \in S_d$$

We claim that we can write $\eta = \sigma'.\theta$ with $\sigma' \in S_{n+1}$ and $\theta \in S_d$ such that each cycle of θ contains at most one of the elements from $\{1, 2, \ldots, n+1\}$. Indeed assume that η contains a cycle containing more than one element from $\{1, \ldots, n+1\}$, say 1 and 2, that is

$$\eta = (1i_1i_2 \ldots i_r 2j_1j_2 \ldots j_s)(k_1 \ldots k_\alpha) \ldots (z_1 \ldots z_\zeta)$$

then we can express the product $(12).\eta$ in cycles as

$$(1i_1i_2 \ldots i_r)(2j_1j_2 \ldots j_s)(k_1 \ldots k_\alpha) \ldots (z_1 \ldots z_\zeta)$$

Continuing in this manner we reduce the number of elements from $\{1, \ldots, n+1\}$ in every cycle to at most one.

But then as $\sigma' \in S_{n+1}$ we have seen that $(\sum sgn(\sigma)e_\sigma). e_{\sigma'} = sgn(\sigma')(\sum sgn(\sigma)e_\sigma)$ and consequently

$$\left(\sum_{\sigma \in S_{n+1}} sgn(\sigma)e_\sigma \right). e_\eta = \pm \left(\sum_{\sigma \in S_{n+1}} sgn(\sigma)e_\sigma \right). e_\theta$$

where each cycle of θ contains at most one of $\{1, \ldots, n+1\}$. Let us write

$$\theta = (1i_1 \ldots i_\alpha)(2j_1 \ldots j_\beta) \ldots (n+1s_1 \ldots s_\kappa)(t_1 \ldots t_\lambda) \ldots (z_1 \ldots z_\zeta)$$

Now, let $\sigma \in S_{n+1}$ then the cycle decomposition of $\sigma.\theta$ is obtained as follows. Substitute in each cycle of σ the element 1 formally by the string $1i_1 \ldots i_\alpha$, the element 2 by the string $2j_1 \ldots j_\beta$, and so on until the element $n+1$ by the string $n+1s_1 \ldots s_\kappa$ and finally adjoin the cycles of θ in which no elements from $\{1, \ldots, n+1\}$ appear.

Finally, we can write out the formal necklace element corresponding to the element $(\sum_{\sigma \in S_{n+1}} sgn(\sigma)e_\sigma).e_\theta$ as

$$\mathbf{fund}_n(x_1 x_{i_1} \ldots x_{i_\alpha}, x_2 x_{j_1} \ldots x_{j_\beta}, \ldots, x_{n+1} x_{s_1} \ldots x_{s_\kappa})$$
$$t(x_{t_1} \ldots x_{t_\lambda}) \ldots t(x_{z_1} \ldots x_{z_\zeta})$$

finishing the proof of the theorem.   □

1.7 Trace relations

We will again use the nondegeneracy of the trace map to deduce the trace relations. That is, we will describe the kernel of the epimorphism

$$\tau \ : \ \int \mathbb{C}\langle x_1, x_2, \ldots \rangle = \mathbb{T}^\infty \longrightarrow \mathbb{T}^\infty_n = \int_n \mathbb{C}\langle x_1, x_2, \ldots \rangle$$

from the description of the necklace relations.

THEOREM 1.12
The trace relations $Ker\ \tau$ is the two-sided ideal of the formal trace algebra $\mathbb{T}^\infty$ generated by all elements

$$\mathbf{fund}_n(m_1, \ldots, m_{n+1}) \quad \text{and} \quad \mathbf{cha}_n(m_1, \ldots, m_n)$$

where the m_i run over all monomials in the variables $\{x_1, x_2, \ldots, x_i, \ldots\}$.

PROOF   Consider a trace relation $\mathbf{h}(x_1, \ldots, x_d) \in Ker\ \tau$. Then, we have a necklace relation of the form

$$t(\mathbf{h}(x_1, \ldots, x_d)x_{d+1}) \in Ker\ \nu$$

By theorem 1.11 we know that this element must be of the form

$$\sum n_{i_1 \ldots i_{n+1}} \mathbf{fund}_n(m_{i_1}, \ldots, m_{i_{n+1}})$$

where the m_i are monomials, the $n_{i_1 \ldots i_{n+1}} \in \mathbb{N}^\infty$ and the expression must be linear in the variable x_{d+1}. That is, x_{d+1} appears linearly in each of the terms

$$n\mathbf{fund}_n(m_1, \ldots, m_{n+1})$$

so appears linearly in n or in precisely one of the monomials m_i. If x_{d+1} appears linearly in n we can write

$$n = t(n'.x_{d+1}) \quad \text{where} \quad n' \in \mathbb{T}^\infty$$

If x_{d+1} appears linearly in one of the monomials m_i we may assume that it does so in m_{n+1}, permuting the monomials if necessary. That is, we may assume $m_{n+1} = m'_{n+1}.x_{d+1}.m"_{n+1}$ with m, m' monomials. But then, we can write

$$n\mathbf{fund}_n(m_1,\ldots,m_{n+1}) = nt(\mathbf{cha}_n(m_1,\ldots,m_n).m'_{n+1}.x_{d+1}.m"_{n+1})$$
$$= t(n.m"_{n+1}.\mathbf{cha}_n(m_1,\ldots,m_n).m'_{n+1}.x_{d+1})$$

using $\mathbb{N}^\infty$-linearity and the cyclic permutation property of the formal trace t. But then, separating the two cases, one can write the total expression

$$t(\mathbf{h}(x_1,\ldots,x_d)x_{d+1}) = t([\sum_i n'_{i_1\ldots i_{n+1}} \mathbf{fund}_n(m_{i_1},\ldots,m_{i_{n+1}})$$

$$+ \sum_j n_{j_1\ldots j_{n+1}}.m"_{j_{n+1}}.\mathbf{cha}_n(m_{j_1},\ldots,m_{j_n}).m'_{j_{n+1}}]x_{d+1})$$

Finally, observe that two formal trace elements $\mathbf{h}(x_1,\ldots,x_d)$ and $\mathbf{k}(x_1,\ldots,x_d)$ are equal if the formal necklaces

$$t(\mathbf{h}(x_1,\ldots,x_d)x_{d+1}) = t(\mathbf{k}(x_1,\ldots,x_d)x_{d+1})$$

are equal, finishing the proof.   ⬜

We will give another description of the necklace relations $Ker\ \tau$, which is better suited for the categorical interpretation of $\mathbb{T}^\infty_n$ to be given in the next chapter. Consider formal trace elements $m_1, m_2, \ldots, m_i, \ldots$ with $m_j \in \mathbb{T}^\infty$. The *formal substitution*

$$f \mapsto f(m_1, m_2, \ldots, m_i, \ldots)$$

is the uniquely determined algebra endomorphism of $\mathbb{T}^\infty$, which maps the variable x_i to m_i and is compatible with the formal trace t. That is, the substitution sends a monomial $x_{i_1} x_{i_2} \ldots x_{i_k}$ to the element $g_{i_1} g_{i_2} \ldots g_{i_k}$ and an element $t(x_{i_1} x_{i_2} \ldots x_{i_k})$ to the element $t(g_{i_1} g_{i_2} \ldots g_{i_k})$. A *substitution invariant ideal* of $\mathbb{T}^\infty$ is a two-sided ideal of $\mathbb{T}^\infty$ that is closed under all possible substitutions as well as under the formal trace t. For any subset of elements $E \subset \mathbb{T}^\infty$ there is a minimal substitution invariant ideal containing E. This is the ideal generated by all elements obtained from E by making all possible substitutions and taking all their formal traces. We will refer to this ideal as the *substitution invariant ideal generated by E*.

Recall the definition of the formal Cayley-Hamilton polynomial $\chi_x(x)$ of an element $x \in \mathbb{T}^\infty$ given in the previous section.

THEOREM 1.13

The trace relations $Ker\ \tau$ is the substitution invariant ideal of $\mathbb{T}^\infty$ generated by the formal Cayley-Hamilton polynomials

$$\chi_x(x) \quad \text{for all} \quad x \in \mathbb{T}^\infty$$

PROOF The result follows from theorem 1.12 and the definition of a substitution invariant ideal once we can show that the full polarization of $\chi_x(x)$, which we have seen is $\mathtt{cha}_n(x_1,\ldots,x_n)$, lies in the substitution invariant ideal generated by the $\chi_x(x)$.

This is true since we may replace the process of polarization with the process of multilinearization, whose first step is to replace, for instance

$$\chi_x(x) \quad \text{by} \quad \chi_{x+y}(x+y) - \chi_x(x) - \chi_y(y)$$

The final result of multilinearization is the same as of full polarization and the claim follows as multilinearizing a polynomial in a substitution invariant ideal, we remain in the same ideal. $\quad\square$

We will use our knowledge on the necklace and trace relations to improve the bound of $2^n - 1$ in the *Nagata-Higman problem* to n^2. Recall that this problem asks for a number $N(n)$ with the property that if R is an associative $\mathbb{C}$-algebra without unit such that $r^n = 0$ for all $r \in R$, then we must have for all $r_i \in R$ the identity

$$r_1 r_2 \ldots r_{N(n)} = 0 \quad \text{in} \quad R$$

We start by reformulating the problem. Consider the positive part $\mathbb{F}_+$ of the free $\mathbb{C}$-algebra generated by the variables $\{x_1, x_2, \ldots, x_i, \ldots\}$

$$\mathbb{F}_+ = \mathbb{C}\langle x_1, x_2, \ldots, x_i, \ldots\rangle_+$$

which is an associative $\mathbb{C}$-algebra without unit. Let $T(n)$ be the two-sided ideal of $\mathbb{F}_+$ generated by all n-powers f^n with $f \in \mathbb{F}_+$. Note that the ideal $T(n)$ is invariant under all substitutions of $\mathbb{F}_+$. The Nagata-Higman problem then asks for a number $N(n)$ such that the product

$$x_1 x_2 \ldots x_{N(n)} \in T(n)$$

We will now give an alternative description of the quotient algebra $\mathbb{F}_+/T(n)$. Let $\mathbb{N}_+$ be the positive part of the infinite necklace algebra $\mathbb{N}_n^\infty$ and $\mathbb{T}_+$ the positive part of the infinite trace algebra $\mathbb{T}_n^\infty$. Consider the quotient associative $\mathbb{C}$-algebra without unit

$$\overline{\mathbb{T}_+} = \mathbb{T}_+/(\mathbb{N}_+\mathbb{T}_n^\infty).$$

Observe the following facts about $\overline{\mathbb{T}_+}$: as a $\mathbb{C}$-algebra it is generated by the variables $X_1, X_2, \ldots$ as all the other algebra generators of the form

$t(x_{i_1} \ldots x_{i_r})$ of $\mathbb{T}^\infty$ are mapped to zero in $\overline{\mathbb{T}_+}$. Further, from the form of the Cayley-Hamilton polynomial it follows that every $t \in \overline{\mathbb{T}_+}$ satisfies $t^n = 0$. That is, we have an algebra epimorphism

$$\mathbb{F}_+/T(n) \longrightarrow\!\!\!\!\!\rightarrow \overline{\mathbb{T}_+}$$

and we claim that it is also injective. To see this, observe that the quotient $\mathbb{T}^\infty/N_+^\infty \mathbb{T}^\infty$ is just the free $\mathbb{C}$-algebra on the variables $\{x_1, x_2, \ldots\}$. To obtain $\overline{\mathbb{T}_+}$ we have to factor out the ideal of trace relations. Now, a formal Cayley-Hamilton polynomial $\chi_x(x)$ is mapped to x^n in $\mathbb{T}^\infty/N_+^\infty \mathbb{T}^\infty$. That is, to obtain $\overline{\mathbb{T}_+}$ we factor out the substitution invariant ideal (observe that t is zero here) generated by the elements x^n, but this is just the definition of $\mathbb{F}_+/T(n)$.

Therefore, a reformulation of the Nagata-Higman problem is to find a number $N = N(n)$ such that the product of the first N generic matrices

$$X_1 X_2 \ldots X_N \in N_+^\infty \mathbb{T}_n^\infty \quad \text{or, equivalently that} \quad tr(X_1 X_2 \ldots X_N X_{N+1})$$

can be expressed as a linear combination of products of traces of lower degree. Using the description of the necklace relations given in theorem 1.10 we can reformulate this condition in terms of the group algebra $\mathbb{C}S_{N+1}$. Let us introduce the following subspaces of the group algebra as follows

- A will be the subspace spanned by all $N+1$ cycles in S_{N+1}

- B will be the subspace spanned by all elements except $N+1$ cycles

- $L(n)$ will be the ideal of $\mathbb{C}S_{N+1}$ spanned by the Young symmetrizers associated to partitions

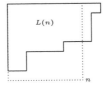

with $\leq n$ rows, and

- $M(n)$ will be the ideal of $\mathbb{C}S_{N+1}$ spanned by the Young symmetrizers associated to partitions

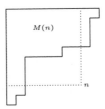

having more than n rows.

With these notations, we can reformulate the above condition as

$$(12 \ldots N \, N + 1) \in B + M(n) \quad \text{and consequently} \quad \mathbb{C}S_{N+1} = B + M(n)$$

Define an inner product on the group algebra $\mathbb{C}S_{N+1}$ such that the group elements form an orthonormal basis, then A and B are orthogonal complements and also $L(n)$ and $M(n)$ are orthogonal complements. But then, taking orthogonal complements the condition can be rephrased as

$$(B + M(n))^{\perp} = A \cap L(n) = 0$$

Finally, let us define an automorphism τ on $\mathbb{C}S_{N+1}$ induced by sending e_σ to $sgn(\sigma)e_\sigma$. Clearly, τ is just multiplication by $(-1)^N$ on A and therefore the above condition is equivalent to

$$A \cap L(n) \cap \tau L(n) = 0$$

Moreover, for any Young tableau λ we have that $\tau(a_\lambda) = b_{\lambda^*}$ and $\tau(b_\lambda) = a_{\lambda^*}$. Hence, the automorphism τ sends the Young symmetrizer associated to a partition to the Young symmetrizer of the dual partition. This gives the following characterization

- $\tau L(n)$ is the ideal of $\mathbb{C}S_{N+1}$ spanned by the Young symmetrizers associated to partitions

  with $\leq n$ columns.

Now, specialize to the case $N = n^2$. Clearly, any Young diagram having $n^2 + 1$ boxes must have either more than n columns or more than n rows

and consequently we indeed have for $N = n^2$ that

$$A \cap L(n) \cap \tau L(n) = 0$$

finishing the proof of the promised refinement of the Nagata-Higman bound

THEOREM 1.14

Let R be an associative $\mathbb{C}$-algebra without unit element. Assume that $r^n = 0$ for all $r \in R$. Then, for all $r_i \in R$ we have

$$r_1 r_2 \ldots r_{n^2} = 0$$

THEOREM 1.15

The necklace algebra $\mathbb{N}_n^m$ is generated as a $\mathbb{C}$-algebra by all elements of the form

$$tr(X_{i_1} X_{i_2} \ldots X_{i_l})$$

with $l \leq n^2 + 1$. The trace algebra $\mathbb{T}_n^m$ is spanned as a module over the necklace algebra $\mathbb{N}_n^m$ by all monomials in the generic matrices

$$X_{i_1} X_{i_2} \ldots X_{i_l}$$

of degree $l \leq n^2$.

1.8 Cayley-Hamilton algebras

In this section we define the category **alg@n** of Cayley-Hamilton algebras of degree n.

DEFINITION 1.1 *A trace map on an (affine) $\mathbb{C}$-algebra A is a $\mathbb{C}$-linear map*

$$tr : A \longrightarrow A$$

satisfying the following three properties for all $a, b \in A$:

1. *$tr(a)b = btr(a)$,*

2. *$tr(ab) = tr(ba)$ and*

3. *$tr(tr(a)b) = tr(a)tr(b)$.*

Note that it follows from the first property that the image $tr(A)$ of the trace map is contained in the *center* of A. Consider two algebras A and B equipped with a trace map, which we will denote by tr_A, respectively, tr_B. A *trace morphism* $\phi : A \longrightarrow B$ will be a $\mathbb{C}$-algebra morphism that is compatible with the trace maps, that is, the following diagram commutes

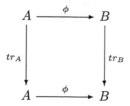

This definition turns algebras with a trace map into a category, denoted by **alg@**. We will say that an algebra A with trace map tr is *trace generated* by a subset of elements $I \subset A$ if the $\mathbb{C}$-algebra generated by B and $tr(B)$ is equal to A where B is the $\mathbb{C}$-subalgebra generated by the elements of I. Note that A does not have to be generated as a $\mathbb{C}$-algebra by the elements from I.

Observe that for $\mathbb{T}^\infty$ the formal trace $t : \mathbb{T}^\infty \longrightarrow \mathbb{N}^\infty \hookrightarrow \mathbb{T}^\infty$ is a trace map. Property (1) follows because $\mathbb{N}^\infty$ commutes with all elements of $\mathbb{T}^\infty$, property (2) is the cyclic permutation property for t and property (3) is the fact that t is a $\mathbb{N}^\infty$-linear map. The formal trace algebra $\mathbb{T}^\infty$ is trace generated by the variables $\{x_1, x_2, \ldots, x_i, \ldots\}$ but *not* as a $\mathbb{C}$-algebra.

Actually, $\mathbb{T}^\infty$ is the *free algebra* in the generators $\{x_1, x_2, \ldots, x_i, \ldots\}$ in the category of algebras with a trace map, **alg@**. That is, if A is an algebra with trace tr which is trace generated by $\{a_1, a_2, \ldots\}$, then there is a trace preserving algebra epimorphism

$$\mathbb{T}^\infty \xrightarrow{\;\pi\;} A$$

For example, define $\pi(x_i) = a_i$ and $\pi(t(x_{i_1} \ldots x_{i_l})) = tr(\pi(x_{i_1}) \ldots \pi(x_{i_l}))$. Also, the formal trace algebra $\mathbb{T}^m$, that is, the subalgebra of $\mathbb{T}^\infty$ trace generated by $\{x_1, \ldots, x_m\}$, is the free algebra in the category of algebras with trace that are trace generated by at most m elements.

Given a trace map tr on A, we can define for any $a \in A$ a *formal Cayley-Hamilton polynomial of degree n* . Indeed, express

$$f(t) = \prod_{i=1}^{n} (t - \lambda_i)$$

as a polynomial in t with coefficients polynomial functions in the Newton functions $\sum_{i=1}^{n} \lambda_i^k$. Replacing the Newton function $\sum \lambda_i^k$ by $tr(a^k)$ we obtain the Cayley-Hamilton polynomial of degree n

$$\chi_a^{(n)}(t) \in A[t]$$

DEFINITION 1.2   *An (affine) $\mathbb{C}$-algebra A with trace map $tr : A \longrightarrow A$ is said to be a* Cayley-Hamilton algebra of degree n *if the following two properties are satisfied:*

1. $tr(1) = n$, and

2. For all $a \in A$ we have $\chi_a^{(n)}(a) = 0$ in A.

alg@n *is the category of Cayley-Hamilton algebras of degree n with trace preserving morphisms.*

Observe that if R is a commutative $\mathbb{C}$-algebra, then $M_n(R)$ is a Cayley-Hamilton algebra of degree n. The corresponding trace map is the composition of the usual trace with the inclusion of $R \longrightarrow M_n(R)$ via scalar matrices. As a consequence, the infinite trace algebra $\mathbb{T}_n^\infty$ has a trace map induced by the natural inclusion

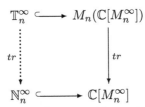

which has image $tr(\mathbb{T}_n^\infty)$ the infinite necklace algebra $\mathbb{N}_n^\infty$. Clearly, being a trace-preserving inclusion, $\mathbb{T}_n^\infty$ is a Cayley-Hamilton algebra of degree n. With this definition, we have the following categorical description of the trace algebra $\mathbb{T}_n^\infty$.

THEOREM 1.16
The trace algebra $\mathbb{T}_n^\infty$ is the free algebra in the generic matrix generators $\{X_1, X_2, \ldots, X_i, \ldots\}$ in the category of Cayley-Hamilton algebras of degree n.

For any m, the trace algebra $\mathbb{T}_n^m$ is the free algebra in the generic matrix generators $\{X_1, \ldots, X_m\}$ in the category **alg@n** *of Cayley-Hamilton algebras of degree n which are trace generated by at most m elements.*

PROOF Let F_n be the free algebra in the generators $\{y_1, y_2, \ldots\}$ in the category **alg@n**, then by freeness of $\mathbb{T}^\infty$ there is a trace preserving algebra epimorphism

$$\mathbb{T}^\infty \xrightarrow{\pi} F_n \quad \text{with} \quad \pi(x_i) = y_i$$

By the universal property of F_n, the ideal $Ker\ \pi$ is the minimal ideal I of $\mathbb{T}^\infty$ such that $\mathbb{T}^\infty/I$ is Cayley-Hamilton of degree n.

We claim that $Ker\ \pi$ is substitution invariant. Consider a substitution endomorphism ϕ of $\mathbb{T}^\infty$ and consider the diagram

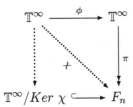

then $Ker\ \chi$ is an ideal closed under traces such that $\mathbb{T}^{\infty}/Ker\ \chi$ is a Cayley-Hamilton algebra of degree n (being a subalgebra of F_n). But then $Ker\ \pi \subset Ker\ \chi$ (by minimality of $Ker\ \pi$) and therefore χ factors over F_n, that is, the substitution endomorphism ϕ descends to an endomorphism $\overline{\phi} : F_n \longrightarrow F_n$ meaning that $Ker\ \pi$ is left invariant under ϕ, proving the claim. Further, any formal Cayley-Hamilton polynomial $\chi_x^{(n)}(x)$ of degree n of $x \in \mathbb{T}^{\infty}$ maps to zero under π. By substitution invariance it follows that the ideal of trace relations $Ker\ \tau \subset Ker\ \pi$. We have seen that $\mathbb{T}^{\infty}/Ker\ \tau = \mathbb{T}_n^{\infty}$ is the infinite trace algebra, which is a Cayley-Hamilton algebra of degree n. Thus, by minimality of $Ker\ \pi$ we must have $Ker\ \tau = Ker\ \pi$ and hence $F_n \simeq \mathbb{T}_n^{\infty}$. The second assertion follows immediately. $\qquad\qquad\qquad\qquad\qquad\qquad\qquad\qquad$ $\square$

Let A be a Cayley-Hamilton algebra of degree n that is trace generated by the elements $\{a_1,\ldots,a_m\}$. We have a trace preserving algebra epimorphism p_A defined by $p(X_i) = a_i$

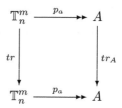

and hence a *presentation* $A \simeq \mathbb{T}_n^m/T_A$ where $T_A \lhd \mathbb{T}_n^m$ is the *ideal of trace relations* holding among the generators a_i. We recall that $\mathbb{T}_n^m$ is the ring of GL_n-equivariant polynomial maps $M_n^m \xrightarrow{f} M_n$, that is

$$M_n(\mathbb{C}[M_n^m])^{GL_n} = \mathbb{T}_n^m$$

where the action of GL_n is the diagonal action on $M_n(\mathbb{C}[M_n^m]) = M_n \otimes \mathbb{C}[M_n^m]$.

Observe that if R is a commutative algebra, then any two-sided ideal $I \lhd M_n(R)$ is of the form $M_n(J)$ for an ideal $J \lhd R$. Indeed, the subsets J_{ij} of (i,j) entries of elements of I is an ideal of R as can be seen by multiplication with scalar matrices. Moreover, by multiplying on both sides with *permutation matrices* one verifies that $J_{ij} = J_{kl}$ for all i,j,k,l proving the claim.

Applying this to the induced ideal $M_n(\mathbb{C}[M_n^m])\ T_A\ M_n(\mathbb{C}[M_n^m]) \lhd M_n(\mathbb{C}[M_n^m])$ we find an ideal $N_A \lhd \mathbb{C}[M_n^m]$ such that

$$M_n(\mathbb{C}[M_n^m])\ T_A\ M_n(\mathbb{C}[M_n^m]) = M_n(N_A)$$

Observe that both the induced ideal and N_A are stable under the respective GL_n-actions.

Assume that V and W are two (not necessarily finite dimensional) $\mathbb{C}$-vector spaces with a locally finite GL_n-action (that is, every finite dimensional subspace is contained in a finite dimensional GL_n-stable subspace) and that $V \xrightarrow{f} W$ is a linear map commuting with the GL_n-action. In section 2.5

we will see that we can decompose V and W uniquely in direct sums of simple representations and in their isotypical components (that is, collecting all factors isomorphic to a given simple GL_n-representation) and prove that $V_{(0)} = V^{GL_n}$, respectively, $W_{(0)} = W^{GL_n}$ where (0) denotes the trivial GL_n-representation. We obtain a commutative diagram

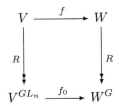

where R is the *Reynolds operator*, that is, the canonical projection to the isotypical component of the trivial representation. Clearly, the Reynolds operator commutes with the GL_n-action. Moreover, using complete decomposability we see that f_0 is surjective (resp. injective) if f is surjective (resp. injective).

Because N_A is a GL_n-stable ideal of $\mathbb{C}[M_n^m]$ we can apply the above in the situation

$$M_n(\mathbb{C}[M_n^m]) \xrightarrow{\pi} M_n(\mathbb{C}[M_n^m]/N_A)$$

$$\downarrow R \qquad\qquad\qquad\qquad \downarrow R$$

$$\mathbb{T}_n^m \xrightarrow{\pi_0} M_n(\mathbb{C}[M_n^m]/N_A)^{GL_n}$$

and the bottom map factorizes through $A = \mathbb{T}_n^m/T_A$ giving a surjection

$$A \longrightarrow M_n(\mathbb{C}[M_n^m]/N_A)^{GL_n}$$

In order to verify that this map is injective (and hence an isomorphism) it suffices to check that

$$M_n(\mathbb{C}[M_n^m]) \, T_A \, M_n(\mathbb{C}[M_n^m]) \cap \mathbb{T}_n^m = T_A$$

Using the functor property of the Reynolds operator with respect to multiplication in $M_n(\mathbb{C}[M_n^\infty])$ with an element $x \in \mathbb{T}_n^m$ or with respect to the trace map (both commuting with the GL_n-action) we deduce the following relations :

- For all $x \in \mathbb{T}_n^m$ and all $z \in M_n(\mathbb{C}[M_n^\infty])$ we have $R(xz) = xR(z)$ and $R(zx) = R(z)x$.

- For all $z \in M_n(\mathbb{C}[M_n^\infty])$ we have $R(tr(z)) = tr(R(z))$.

Assume that $z = \sum_i t_i x_i n_i \in M_n(\mathbb{C}[M_n^m]) \, T_A \, M_n(\mathbb{C}[M_n^m]) \cap \mathbb{T}_n^m$ with $m_i, n_i \in M_n(\mathbb{C}[M_n^m])$ and $t_i \in T_A$. Now, consider $X_{m+1} \in \mathbb{T}_n^\infty$. Using the cyclic property of traces we have

$$tr(zX_{m+1}) = \sum_i tr(m_i t_i n_i X_{m+1}) = \sum_i tr(n_i X_{m+1} m_i t_i)$$

and if we apply the Reynolds operator to it we obtain the equality

$$tr(zX_{m+1}) = tr(\sum_i R(n_i X_{m+1} m_i) t_i)$$

For any i, the term $R(n_i X_{m+1} m_i)$ is invariant so belongs to $\mathbb{T}_n^{m+1}$ and is linear in X_{m+1}. Knowing the generating elements of $\mathbb{T}_n^{m+1}$ we can write

$$R(n_i X_{m+1} m_i) = \sum_j s_{ij} X_{m+1} t_{ij} + \sum_k tr(u_{ik} X_{m+1}) v_{ik}$$

with all of the elements s_{ij}, t_{ij}, u_{ik} and v_{ik} in $\mathbb{T}_n^m$. Substituting this information and again using the cyclic property of traces we obtain

$$tr(zX_{m+1}) = tr((\sum_{i,j,k} s_{ij} t_{ij} t_i + tr(v_{ik} t_i)) X_{m+1})$$

and by the nondegeneracy of the trace map we again deduce from this the equality

$$z = \sum_{i,j,k} s_{ij} t_{ij} t_i + tr(v_{ik} t_i)$$

Because $t_i \in T_A$ and T_A is stable under taking traces we deduce from this that $z \in T_A$ as required.

Because $A = M_n(\mathbb{C}[M_n^m]/N_A)^{GL_n}$ we can apply the functor property of the Reynolds operator to the setting

$$
\begin{array}{ccc}
M_n(\mathbb{C}[M_n^m]/N_A) & \xrightarrow{\;\;tr\;\;} & \mathbb{C}[M_n]/N_A \\[2mm]
{\scriptstyle R}\Big\downarrow & & \Big\downarrow{\scriptstyle R} \\[2mm]
A & \xrightarrow[\;\;tr_A\;\;]{} & (\mathbb{C}[M_n]/N_A)^{GL_n}
\end{array}
$$

Concluding we also have the equality

$$tr_A(A) = (\mathbb{C}[M_n^m]/J_A)^{GL_n}$$

Summarizing, we have proved the following invariant theoretic reconstruction result for Cayley-Hamilton algebras.

THEOREM 1.17
Let A be a Cayley-Hamilton algebra of degree n, with trace map tr_A, which is trace generated by at most m elements. Then , there is a canonical ideal

$N_A \lhd \mathbb{C}[M_n^m]$ *from which we can reconstruct the algebras A and $tr_A(A)$ as invariant algebras*

$$A = M_n(\mathbb{C}[M_n^m]/N_A)^{GL_n} \quad \text{and} \quad tr_A(A) = (\mathbb{C}[M_n^m]/N_A)^{GL_n}$$

A direct consequence of the above proof is the following *universal property* of the embedding

$$A \overset{i_A}{\hookrightarrow} M_n(\mathbb{C}[M_n^m]/N_A)$$

Let R be a commutative $\mathbb{C}$-algebra, then $M_n(R)$ with the usual trace is a Cayley-Hamilton algebra of degree n. If $f : A \longrightarrow M_n(R)$ is a trace preserving morphism, we claim that there exists a natural algebra morphism $\overline{f} : \mathbb{C}[M_n^m]/N_A \longrightarrow R$ such that the diagram

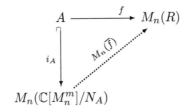

where $M_n(\overline{f})$ is the algebra morphism defined entrywise. To see this, consider the composed trace preserving morphism $\phi : \mathbb{T}_n^m \longrightarrow A \overset{f}{\longrightarrow} M_n(R)$. Its image is fully determined by the images of the trace generators X_k of $\mathbb{T}_n^m$ which are, say, $m_k = (m_{ij}(k))_{i,j}$. But then we have an algebra morphism $\mathbb{C}[M_n^m] \overset{g}{\longrightarrow} R$ defined by sending the variable $x_{ij}(k)$ to $m_{ij}(k)$. Clearly, $T_A \subset Ker\ \phi$ and after inducing to $M_n(\mathbb{C}[M_n^m])$ it follows that $N_A \subset Ker\ g$ proving that g factors through $\mathbb{C}[M_n^m]/J_A \longrightarrow R$. This morphism has the required universal property.

References

The first fundamental theorem of matrix invariants, theorem 1.6, is due independently to G. B. Gurevich [41], C. Procesi [85] and K. S. Siberskii [98]. The second fundamental theorem of matrix invariants, theorem 1.10 is due independently to C. Procesi [85] and Y. P. Razmyslov [88]. Our treatment follows the paper [85] of C. Procesi, supplemented with material taken from the lecture notes of H.P. Kraft [64] and E. Formanek [32]. The invariant theoretic reconstruction result, theorem 1.17, is due to C. Procesi [87].

Chapter 2

Reconstructing Algebras

We will associate to an affine $\mathbb{C}$-algebra A its affine scheme of n-dimensional representations $\mathtt{rep}_n\ A$. There is a base change action by GL_n on this scheme and its orbits are exactly the isomorphism classes of n-dimensional representations. We will prove the *Hilbert criterium* which describes the nullcone via one-parameter subgroups and apply it to prove Michael Artin's result that the closed orbits in $\mathtt{rep}_n\ A$ correspond to *semisimple* representations.

We recall the basic results on algebraic quotient varieties in geometric invariant theory and apply them to prove Procesi's reconstruction result. If $A \in \mathtt{alg@n}$, then we can recover A as

$$A \simeq \Uparrow^n [\mathtt{trep}_n\ A]$$

the ring of GL_n-equivariant polynomial maps from the trace preserving representation scheme $\mathtt{trep}_n\ A$ to $M_n(\mathbb{C})$. However, the functors

$$\mathtt{alg@n} \underset{\Uparrow^n}{\overset{\mathtt{trep}_n}{\rightleftarrows}} \mathtt{GL(n)-affine}$$

do not determine an antiequivalence of categories (as they do in commutative algebraic geometry, which is the special case $n = 1$). We will illustrate this by calculating the rings of equivariant maps of orbit-closures of nilpotent matrices. These orbit-closures are described by the *Gerstenhaber-Hesselink* theorem. Later, we will be able to extend this result and study the nullcones of more general representation varieties.

2.1 Representation schemes

For a noncommutative affine algebra A with generating set $\{a_1, \ldots, a_m\}$, there is an epimorphism

$$\mathbb{C}\langle x_1, \ldots, x_m \rangle \xrightarrow{\phi} A$$

defined by $\phi(x_i) = a_i$. That is, a presentation of A as

$$A \simeq \mathbb{C}\langle x_1, \ldots, x_m \rangle / I_A$$

where I_A is the *two-sided ideal of relations* holding among the a_i. For example, if $A = \mathbb{C}[x_1, \ldots, x_m]$, then I_A is the two-sided ideal of $\mathbb{C}\langle x_1, \ldots, x_m \rangle$ generated by the elements $x_i x_j - x_j x_i$ for all $1 \leq i, j \leq m$.

An *n-dimensional representation* of A is an algebra morphism

$$A \xrightarrow{\ \psi\ } M_n$$

from A to the algebra of $n \times n$ matrices over $\mathbb{C}$. If A is generated by $\{a_1, \ldots, a_m\}$, then ψ is fully determined by the point

$$(\psi(a_1), \psi(a_2), \ldots, \psi(a_m)) \in M_n^m = \underbrace{M_n \oplus \ldots \oplus M_n}_{m}$$

We claim that $rep_n(A)$, the set of all n-dimensional representations of A, forms a Zariski closed subset of M_n^m. To begin, observe that

$$rep_n(\mathbb{C}\langle x_1, \ldots, x_m \rangle) = M_n^m$$

as any m-tuple of $n \times n$ matrices $(A_1, \ldots, A_m) \in M_n^m$ determines an algebra morphism $\mathbb{C}\langle x_1, \ldots, x_m \rangle \xrightarrow{\ \psi\ } M_n$ by taking $\psi(x_i) = A_i$.

Given a presentation $A = \mathbb{C}\langle x_1, \ldots, x_m \rangle / I_A$ an m-tuple $(A_1, \ldots, A_m) \in M_n^m$ determines an n-dimensional representation of A if (and only if) for every noncommutative polynomial $r(x_1, \ldots, x_m) \in I_A \lhd \mathbb{C}\langle x_1, \ldots, x_m \rangle$ we have that

$$r(A_1, \ldots, A_m) = \begin{bmatrix} 0 \ldots 0 \\ \vdots \quad \vdots \\ 0 \ldots 0 \end{bmatrix} \in M_n$$

Hence, consider the ideal $I_A(n)$ of $\mathbb{C}[M_n^m] = \mathbb{C}[x_{ij}(k) \mid 1 \leq i, j \leq n, 1 \leq k \leq m]$ generated by all the entries of the matrices in $M_n(\mathbb{C}[M_n^m])$ of the form

$$r(X_1, \ldots, X_m) \quad \text{for all } r(x_1, \ldots, x_m) \in I_A$$

We see that the *reduced representation variety $rep_n A$* is the set of simultaneous zeroes of the ideal $I_A(n)$, that is

$$rep_n A = \mathbb{V}(I_A(n)) \hookrightarrow M_n^m$$

proving the claim. Here, $\mathbb{V}$ denotes the *closed* set in the *Zariski topology* determined by an ideal. The complement of $\mathbb{V}(I)$ we will denote with $\mathbb{X}(I))$. Observe that, even when A is not finitely presented, the ideal $I_A(n)$ is finitely generated as an ideal of the commutative (Noetherian) polynomial algebra $\mathbb{C}[M_n^m]$.

Example 2.1

It may happen that $rep_n A = \emptyset$. For example, consider the *Weyl algebra*

$$A_1(\mathbb{C}) = \mathbb{C}\langle x, y \rangle / (xy - yx - 1)$$

If a couple of $n \times n$-matrices $(A, B) \in rep_n \ A_1(\mathbb{C})$ then we must have

$$A.B - B.A = \mathbb{1}_n \in M_n$$

However, taking traces on both sides gives a contradiction as $tr(AB) = tr(BA)$ and $tr(\mathbb{1}_n) = n \neq 0$. ▯

Often, the ideal $I_A(n)$ contains more information than the closed subset $rep_n(A) = \mathbb{V}(I_A(n))$ which, using the *Hilbert Nullstellensatz*, only determines the radical ideal of $I_A(n)$. This fact forces us to consider the *representation variety* (or scheme) $\mathbf{rep}_n \ A$.

In the foregoing chapter we studied the action of GL_n by simultaneous conjugation on M_n^m. We claim that $rep_n \ A \hookrightarrow M_n^m$ is *stable* under this action, that is, if $(A_1, \ldots, A_m) \in rep_n \ A$, then also $(gA_1g^{-1}, \ldots, gA_mg^{-1}) \in rep_n \ A$. This is clear by composing the n-dimensional representation ψ of A determined by $(A_1, \ldots, A_m)$ with the algebra automorphism of M_n given by conjugation with $g \in GL_n$

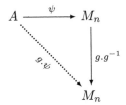

Therefore, $rep_n \ A$ is a GL_n-*variety*. We will give an interpretation of the orbits under this action.

Recall that a *left A-module* M is a vector space on which elements of A act on the left as linear operators satisfying the conditions

$$1.m = m \quad \text{and} \quad a.(b.m) = (ab).m$$

for all $a, b \in A$ and all $m \in M$. An A-module *morphism* $M \xrightarrow{f} N$ between two left A-modules is a linear map such that $f(a.m) = a.f(m)$ for all $a \in A$ and all $m \in M$. An A-module *automorphism* is an A-module morphism $M \xrightarrow{f} N$ such that there is an A-module morphism $N \xrightarrow{g} M$ such that $f \circ g = id_M$ and $g \circ f = id_N$.

Assume the A-module M has dimension n, then after fixing a basis we can identify M with $\mathbb{C}^n$ (column vectors). For any $a \in A$ we can represent the linear action of a on M by an $n \times n$ matrix $\psi(a) \in M_n$. The condition that $a.(b.m) = (ab).m$ for all $m \in M$ asserts that $\psi(ab) = \psi(a)\psi(b)$ for all $a, b \in A$, that is, ψ is an algebra morphism $A \xrightarrow{\psi} M_n$ and hence M determines an n-dimensional representation of A. Conversely, an n-dimensional representation $A \xrightarrow{\psi} M_n$ determines an A-module structure on $\mathbb{C}^n$ by the rule

$$a.v = \psi(a)v \quad \text{for all} \quad v \in \mathbb{C}^n$$

Hence, there is a one-to-one correspondence between the n-dimensional representations of A and the A-module structures on $\mathbb{C}^n$. If two n-dimensional A-module structures M and N on $\mathbb{C}^n$ are isomorphic (determined by a linear invertible map $g \in GL_n$) then for all $a \in A$ we have the commutative diagram

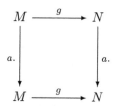

Hence, if the action of a on M is represented by the matrix A, then the action of a on M is represented by the matrix $g.A.g^{-1}$. Therefore, two A-module structures on $\mathbb{C}^n$ are isomorphic if and only if the points of $rep_n\, A$ corresponding to them lie in the same GL_n-orbit. Concluding, studying n-dimensional A-modules up to isomorphism is the same as studying the GL_n-orbits in the reduced representation variety $rep_n\, A$.

If the defining ideal $I_A(n)$ is a radical ideal, the above suffices. In general, the *scheme* structure of the representation variety $\mathtt{rep}_n\, A$ will be important. By definition, the scheme $\mathtt{rep}_n\, A$ is the functor assigning to any (affine) commutative $\mathbb{C}$-algebra R, the set

$$\mathtt{rep}_n\, A(R) = Alg_{\mathbb{C}}(\mathbb{C}[M_n^m]/I_A(n), R)$$

of $\mathbb{C}$-algebra morphisms $\frac{\mathbb{C}[M_n^m]}{I_A(n)} \xrightarrow{\psi} R$. Such a map ψ is determined by the image $\psi(x_{ij}(k)) = r_{ij}(k) \in R$. That is, $\psi \in \mathtt{rep}_n\, A(R)$ determines an m-tuple of $n \times n$ matrices with coefficients in R

$$(r_1, \ldots, r_m) \in \underbrace{M_n(R) \oplus \ldots \oplus M_n(R)}_{m} \quad \text{where} \quad r_k = \begin{bmatrix} r_{11}(k) & \ldots & r_{1n}(k) \\ \vdots & & \vdots \\ r_{n1}(k) & \ldots & r_{nn}(k) \end{bmatrix}$$

Clearly, for any $r(x_1, \ldots, x_m) \in I_A$ we must have that $r(r_1, \ldots, r_m)$ is the zero matrix in $M_n(R)$. That is, ψ determines uniquely an R-algebra morphism

$$\psi : R \otimes_{\mathbb{C}} A \longrightarrow M_n(R) \quad \text{by mapping} \quad x_k \mapsto r_k$$

Alternatively, we can identify the set $\mathtt{rep}_n(R)$ with the set of left $R \otimes A$-module structures on the free R-module $R^{\oplus n}$ of rank n.

2.2 Some algebraic geometry

Throughout this book we assume that the reader has some familiarity with algebraic geometry, as contained in the first two chapters of the textbook [43].

In this section we restrict to the dimension formulas and the relation between Zariski and analytic closures. We will illustrate these results by examples from representation varieties. We will consider only the reduced varieties in this section.

A morphism $X \xrightarrow{\phi} Y$ between two affine *irreducible varieties* (that is, the coordinate rings $\mathbb{C}[X]$ and $\mathbb{C}[Y]$ are domains) is said to be *dominant* if the image $\phi(X)$ is Zariski dense in Y. On the level of the coordinate algebras dominance is equivalent to $\phi^* : \mathbb{C}[Y] \longrightarrow \mathbb{C}[X]$ being injective and hence inducing a fieldextension $\phi^* : \mathbb{C}(Y) \longhookrightarrow \mathbb{C}(X)$ between the function fields. Indeed, for $f \in \mathbb{C}[Y]$ the function $\phi^*(f)$ is by definition the composition

$$X \xrightarrow{\phi} Y \xrightarrow{f} \mathbb{C}$$

and therefore $\phi^*(f) = 0$ iff $f(\phi(X)) = 0$ iff $f(\overline{\phi(X)}) = 0$.

A morphism $X \xrightarrow{\phi} Y$ between two affine varieties is said to be *finite* if under the algebra morphism ϕ^* the coordinate algebra $\mathbb{C}[X]$ is a finite $\mathbb{C}[Y]$-module. An important property of finite morphisms is that they are *closed*, that is the image of a closed subset is closed. Indeed, we can replace without loss of generality Y by the closed subset $\overline{\phi(X)} = \mathbb{V}_Y(Ker \; \phi^*)$ and hence assume that ϕ^* is an inclusion $\mathbb{C}[Y] \longhookrightarrow \mathbb{C}[X]$. The claim then follows from the fact that in a finite extension there exists for any maximal ideal $N \triangleleft \mathbb{C}[Y]$ a maximal ideal $M \triangleleft \mathbb{C}[X]$ such that $M \cap \mathbb{C}[Y] = \mathbb{C}[X]$.

Example 2.2
Let X be an irreducible affine variety of dimension d. By the *Noether normalization* result $\mathbb{C}[X]$ is a finite module over a polynomial subalgebra $\mathbb{C}[f_1, \ldots, f_d]$. But then, the finite inclusion $\mathbb{C}[f_1, \ldots, f_d] \longhookrightarrow \mathbb{C}[X]$ determines a finite surjective morphism

$$X \xrightarrow{\phi} \mathbb{C}^d$$

◻

An important source of finite morphisms is given by integral extensions. Recall that, if $R \longhookrightarrow S$ is an inclusion of domains we call S *integral* over R if every $s \in S$ satisfies an equation

$$s^n = \sum_{i=0}^{n-1} r_i s^i \quad \text{with} \quad r_i \in R.$$

A *normal* domain R has the property that any element of its field of fractions, which is integral over R, belongs already to R. If $X \xrightarrow{\phi} Y$ is a dominant morphism between two irreducible affine varieties, then ϕ is finite if and only if $\mathbb{C}[X]$ in integral over $\mathbb{C}[Y]$ for the embedding coming from ϕ^*.

PROPOSITION 2.1

Let $X \xrightarrow{\phi} Y$ be a dominant morphism between irreducible affine varieties. Then, for any $x \in X$ and any irreducible component C of the fiber $\phi^{-1}(\phi(z))$ we have

$$dim\ C \geq dim\ X - dim\ Y.$$

Moreover, there is a nonempty open subset U of Y contained in the image $\phi(X)$ such that for all $u \in U$ we have

$$dim\ \phi^{-1}(u) = dim\ X - dim\ Y$$

PROOF   Let $d = dim\ X - dim\ Y$ and apply the Noether normalization result to the affine $\mathbb{C}(Y)$-algebra $\mathbb{C}(Y)\mathbb{C}[X]$. Then, we can find a function $g \in \mathbb{C}[Y]$ and algebraic independent functions $f_1, \ldots, f_d \in \mathbb{C}[X]_g$ (g clears away any denominators that occur after applying the normalization result) such that $\mathbb{C}[X]_g$ is *integral* over $\mathbb{C}[Y]_g[f_1, \ldots, f_d]$. That is, we have the commutative diagram

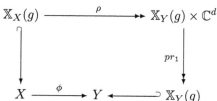

where we know that ρ is finite and surjective. But then we have that the open subset $\mathbb{X}_Y(g)$ lies in the image of ϕ and in $\mathbb{X}_Y(g)$ all fibers of ϕ have dimension d. For the first part of the statement we have to recall the statement of *Krull's Hauptideal result*: if X is an irreducible affine variety and $g_1, \ldots, g_r \in \mathbb{C}[X]$ with $(g_1, \ldots, g_r) \neq \mathbb{C}[X]$, then any component C of $\mathbb{V}_X(g_1, \ldots, g_r)$ satisfies the inequality

$$dim\ C \geq dim\ X - r$$

If $dim\ Y = r$ apply this result to the g_i determining the morphism

$$X \xrightarrow{\phi} Y \twoheadrightarrow \mathbb{C}^r$$

where the latter morphism is the one from example 2.2.   ⬚

In fact, a stronger result holds. *Chevalley's theorem* asserts the following.

THEOREM 2.1

Let $X \xrightarrow{\phi} Y$ be a morphism between affine varieties, the function

$$X \longrightarrow \mathbb{N}   \text{defined by}   x \mapsto dim_x\ \phi^{-1}(\phi(x))$$

is upper-semicontinuous. That is, for all $n \in \mathbb{N}$, the set

$$\{x \in X \mid dim_x\ \phi^{-1}(\phi(x)) \leq n\}$$

is Zariski open in X.

PROOF Let $Z(\phi, n)$ be the set $\{x \in X \mid dim_x\ \phi^{-1}(\phi(x)) \geq n\}$. We will prove that $Z(\phi, n)$ is closed by induction on the dimension of X. We first make some reductions. We may assume that X is irreducible. For, let $X = \cup_i X_i$ be the decomposition of X into irreducible components, then $Z(\phi, n) = \cup Z(\phi \mid X_i, n)$. Next, we may assume that $Y = \overline{\phi(X)}$ whence Y is also irreducible and ϕ is a dominant map. Now, we are in the setting of proposition 2.1. Therefore, if $n \leq dim\ X - dim\ Y$ we have $Z(\phi, n) = X$ by that proposition, so it is closed. If $n > dim\ X - dim\ Y$ consider the open set U in Y of proposition 2.1. Then, $Z(\phi, n) = Z(\phi \mid (X - \phi^{-1}(U)), n)$. the dimension of the closed subvariety $X - \phi^{-1}(U)$ is strictly smaller that $dim\ X$ hence by induction we may assume that $Z(\phi \mid (X - \phi^{-1}(U)), n)$ is closed in $X - \phi^{-1}(U)$ whence closed in X. □

An immediate consequence of the foregoing proposition is that for any morphism $X \xrightarrow{\phi} Y$ between affine varieties, the image $\phi(X)$ contains an open dense subset of $\overline{\phi(Z)}$ (reduce to irreducible components and apply the proposition).

Example 2.3
Let A be an affine $\mathbb{C}$-algebra and $M \in rep_n\ A$. We claim that the orbit

$$\mathcal{O}(M) = GL_n.M \quad \text{is Zariski open in its closure} \quad \overline{\mathcal{O}(M)}$$

Consider the *"orbit-map"* $GL_n \xrightarrow{\phi} rep_n\ A$ defined by $g \mapsto g.M$. Then, by the above remark $\overline{\mathcal{O}(M)} = \overline{\phi(GL_n)}$ contains a Zariski open subset U of $\overline{\mathcal{O}(M)}$ contained in the image of ϕ which is $\mathcal{O}(M)$. But then

$$\mathcal{O}(M) = GL_n.M = \cup_{g \in GL_n} g.U$$

is also open in $\overline{\mathcal{O}(M)}$. Next, we claim that $\overline{\mathcal{O}(M)}$ contains a closed orbit. Indeed, assume $\mathcal{O}(M)$ is not closed, then the complement $C_M = \overline{\mathcal{O}(M)} - \mathcal{O}(M)$ is a proper Zariski closed subset whence $dim\ C < dim\ \overline{\mathcal{O}(M)}$. But, C is the union of GL_n-orbits $\mathcal{O}(M_i)$ with $dim\ \overline{\mathcal{O}(M_i)} < dim\ \overline{\mathcal{O}(M)}$. Repeating the argument with the M_i and induction on the dimension we will obtain a closed orbit in $\overline{\mathcal{O}(M)}$. □

Next, we want to relate the Zariski closure with the $\mathbb{C}$-closure (that is, closure in the usual complex or analytic topology). Whereas they are usually not equal (for example, the unit circle in $\mathbb{C}^1$), we will show that they coincide for the important class of *constructible* subsets. A subset Z of an affine variety X is said to be *locally closed* if Z is open in its Zariski closure $\overline{Z}$. A subset Z is said to be *constructible* if Z is the union of finitely many locally closed subsets. Clearly, finite unions, finite intersections and complements of constructible subsets are again constructible. The importance of constructible sets for algebraic geometry is clear from the following result.

PROPOSITION 2.2
Let $X \xrightarrow{\phi} Y$ be a morphism between affine varieties. If Z is a constructible subset of X, then $\phi(Z)$ is a constructible subset of Y.

PROOF   Because every open subset of X is a finite union of special open sets, which are themselves affine varieties, it suffices to show that $\phi(X)$ is constructible. We will use induction on $dim\ \overline{\phi(X)}$. There exists an open subset $U \subset \overline{\phi(X)}$ that is contained in $\phi(X)$. Consider the closed complement $W = \overline{\phi(X)} - U$ and its inverse image $X' = \phi^{-1}(W)$. Then, X' is an affine variety and by induction we may assume that $\phi(X')$ is constructible. But then, $\phi(X) = U \cup \phi(X')$ is also constructible.   □

Example 2.4
Let A be an affine $\mathbb{C}$-algebra. The subset $ind_n\ A \hookrightarrow rep_n\ A$ of the *indecomposable* n-dimensional A-modules is constructible. Indeed, define for any pair k, l such that $k + l = n$ the morphism

$$GL_n \times rep_k\ A \times rep_l\ A \longrightarrow rep_n\ A$$

by sending a triple (g, M, N) to $g.(M \oplus N)$. By the foregoing result the image of this map is constructible. The decomposable n-dimensional A-modules belong to one of these finitely many sets whence are constructible, but then so is its complement which in $ind_n\ A$.   □

Apart from being closed, finite morphisms often satisfy the *going-down property* . That is, consider a finite and surjective morphism

$$X \xrightarrow{\phi} Y$$

where X is irreducible and Y is normal (that is, $\mathbb{C}[Y]$ is a normal domain). Let $Y' \hookrightarrow Y$ an irreducible Zariski closed subvariety and $x \in X$ with image $\phi(x) = y' \in Y'$. Then, the going-down property asserts the existence of an irreducible Zariski closed subvariety $X' \hookrightarrow X$ such that $x \in X'$ and $\phi(X') = Y'$. In particular, the morphism $X' \xrightarrow{\phi} Y'$ is again finite and surjective and in particular $dim\ X' = dim\ Y'$.

LEMMA 2.1
Let $x \in X$ an irreducible affine variety and U a Zariski open subset. Then, there is an irreducible curve $C \hookrightarrow X$ through x and intersecting U.

PROOF   If $d = dim\ X$ consider the finite surjective morphism $X \xrightarrow{\phi} \mathbb{C}^d$ of example 2.2. Let $y \in \mathbb{C}^d - \phi(X - U)$ and consider the line L through y and $\phi(x)$. By the going-down property there is an irreducible curve $C \hookrightarrow X$ containing x such that $\phi(C) = L$ and by construction $C \cap U \neq \emptyset$. ☐

PROPOSITION 2.3

Let $X \xrightarrow{\phi} Y$ be a dominant morphism between irreducible affine varieties any $y \in Y$. Then, there is an irreducible curve $C \hookrightarrow X$ such that $y \in \overline{\phi(C)}$.

PROOF   Consider an open dense subset $U \hookrightarrow Y$ contained in the image $\phi(X)$. By the lemma there is a curve $C' \hookrightarrow Y$ containing y and such that $C' \cap U \neq \emptyset$. Then, again applying the lemma to an irreducible component of $\phi^{-1}(C')$ not contained in a fiber, we obtain an irreducible curve $C \hookrightarrow X$ with $\overline{\phi(C)} = \overline{C'}$. ☐

Any affine variety $X \hookrightarrow \mathbb{C}^k$ can also be equipped with the induced $\mathbb{C}$-topology (or *analytic topology*) from $\mathbb{C}^k$, which is much finer than the *Zariski topology* . Usually there is no relation between the closure $\overline{Z}^\mathbb{C}$ of a subset $Z \hookrightarrow X$ in the $\mathbb{C}$-topology and the Zariski closure $\overline{Z}$.

LEMMA 2.2

Let $U \subset \mathbb{C}^k$ containing a subset V that is Zariski open and dense in $\overline{U}$. Then,

$$\overline{U}^\mathbb{C} = \overline{U}$$

PROOF   By reducing to irreducible components, we may assume that $\overline{U}$ is irreducible. Assume first that $dim\ \overline{U} = 1$, that is, $\overline{U}$ is an irreducible curve in $\mathbb{C}^k$. Let U_s be the subset of points where $\overline{U}$ is a complex manifold, then $\overline{U} - U_s$ is finite and by the *implicit function theorem* in analysis every $u \in U_s$ has a $\mathbb{C}$-open neighborhood which is $\mathbb{C}$-homeomorphic to the complex line $\mathbb{C}^1$, whence the result holds in this case.

If $\overline{U}$ is general and $x \in \overline{U}$ we can take by the lemma above an irreducible curve $C \hookrightarrow \overline{U}$ containing z and such that $C \cap V \neq \emptyset$. Then, $C \cap V$ is Zariski open and dense in C and by the curve argument above $x \in \overline{(C \cap V)}^\mathbb{C} \subset \overline{U}^\mathbb{C}$. We can do this for any $x \in \overline{U}$ finishing the proof. ☐

Consider the embedding of an affine variety $X \hookrightarrow \mathbb{C}^k$, proposition 2.2 and the fact that any constructible set Z contains a subset U which is open and dense in $\overline{Z}$ we deduce from the lemma at once the next result.

PROPOSITION 2.4
If Z is a constructible subset of an affine variety X, then

$$\overline{Z}^{\mathbb{C}} = \overline{Z}$$

Example 2.5
Let A be an affine $\mathbb{C}$-algebra and $M \in rep_n A$. We have proved in example 2.3 that the orbit $\mathcal{O}(M) = GL_n.M$ is Zariski open in its closure $\overline{\mathcal{O}(M)}$. Therefore, the orbit $\mathcal{O}(M)$ is a constructible subset of $rep_n A$. By the proposition above, the Zariski closure $\overline{\mathcal{O}(M)}$ of the orbit coincides with the closure of $\mathcal{O}(M)$ in the $\mathbb{C}$-topology.   ▯

2.3   The Hilbert criterium

A *one-parameter subgroup* of a linear algebraic group G is a morphism

$$\lambda : \mathbb{C}^* \longrightarrow G$$

of affine algebraic groups. That is, λ is both a group morphism and a morphism of affine varieties. The set of all one-parameter subgroup of G will be denoted by $Y(G)$.

If G is commutative algebraic group, then $Y(G)$ is an Abelian group with additive notation

$$\lambda_1 + \lambda_2 : \mathbb{C}^* \longrightarrow G \quad \text{with } (\lambda_1 + \lambda_2)(t) = \lambda_1(t).\lambda_2(t)$$

Recall that an n-dimensional torus is an affine algebraic group isomorphic to

$$\underbrace{\mathbb{C}^* \times \ldots \times \mathbb{C}^*}_{n} = T_n$$

the closed subgroup of invertible diagonal matrices in GL_n.

LEMMA 2.3
$Y(T_n) \simeq \mathbb{Z}^n$. *The correspondence is given by assigning to* $(r_1, \ldots, r_n) \in \mathbb{Z}^n$ *the one-parameter subgroup*

$$\lambda : \mathbb{C}^* \longrightarrow T_n \quad \text{given by} \quad t \mapsto (t^{r_1}, \ldots, t^{r_n})$$

PROOF   For any two affine algebraic groups G and H there is a canonical bijection $Y(G \times H) = Y(G) \times Y(H)$ so it suffices to verify that $Y(\mathbb{C}^*) \simeq \mathbb{Z}$

with any $\lambda : \mathbb{C}^* \longrightarrow \mathbb{C}^*$ given by $t \mapsto t^r$ for some $r \in \mathbb{Z}$. This is obvious as λ induces the algebra morphism

$$\mathbb{C}[\mathbb{C}^*] = \mathbb{C}[x, x^{-1}] \xrightarrow{\lambda^*} \mathbb{C}[x, x^{-1}] = \mathbb{C}[\mathbb{C}^*]$$

which is fully determined by the image of x which must be an invertible element. Now, any invertible element in $\mathbb{C}[x, x^{-1}]$ is homogeneous of the form cx^r for some $r \in \mathbb{Z}$ and $c \in \mathbb{C}^*$. The corresponding morphism maps t to ct^r, which is only a group morphism if it maps the identity element 1 to 1 so $c = 1$, finishing the proof. ∎

PROPOSITION 2.5

Any one-parameter subgroup $\lambda : \mathbb{C}^* \longrightarrow GL_n$ *is of the form*

$$t \mapsto g^{-1}. \begin{bmatrix} t^{r_1} & & 0 \\ & \ddots & \\ 0 & & t^{r_n} \end{bmatrix} .g$$

for some $g \in GL_n$ *and some* n-tuple $(r_1, \ldots, r_n) \in \mathbb{Z}^n$.

PROOF Let H be the image under λ of the subgroup μ_∞ of roots of unity in $\mathbb{C}^*$. We claim that there is a base change matrix $g \in GL_n$ such that

$$g.H.g^{-1} \hookrightarrow \begin{bmatrix} \mathbb{C}^* & & 0 \\ & \ddots & \\ 0 & & \mathbb{C}^* \end{bmatrix}$$

Assume $h \in H$ not a scalar matrix, then h has a proper eigenspace decomposition $V \oplus W = \mathbb{C}^n$. We use that $h^l = \mathbb{1}_n$ and hence all its Jordan blocks must have size one as for any $\lambda \neq 0$ we have

$$\begin{bmatrix} \lambda & 1 & & 0 \\ & \ddots & \ddots & \\ & & \ddots & 1 \\ 0 & & & \lambda \end{bmatrix}^l = \begin{bmatrix} \lambda^l & l\lambda^{l-1} & & * \\ & \ddots & \ddots & \\ & & \ddots & l\lambda^{l-1} \\ & & & \lambda^l \end{bmatrix} \neq \mathbb{1}_n$$

Because H is commutative, both V and W are stable under H. By induction on n we may assume that the images of H in $GL(V)$ and $GL(W)$ are diagonalizable, but then the same holds in GL_n.

As μ_∞ is infinite, it is Zariski dense in $\mathbb{C}^*$ and because the diagonal matrices are Zariski closed in GL_n we have

$$g.\lambda(\mathbb{C}^*).g^{-1} = g.\overline{H}.g^{-1} \hookrightarrow T_n$$

and the result follows from the lemma above   ⬚

Let V be a general GL_n-representation considered as an affine space with GL_n-action, let X be a GL_n-stable closed subvariety and consider a point $x \in X$. A one-parameter subgroup $\mathbb{C}^* \xrightarrow{\lambda} GL_n$ determines a morphism

$$\mathbb{C}^* \xrightarrow{\lambda_x} X \quad \text{defined by} \quad t \mapsto \lambda(t).x$$

Observe that the image $\lambda_x(\mathbb{C}^*)$ lies in the orbit $GL_n.x$ of x. Assume there is a continuous extension of this map to the whole of $\mathbb{C}$. We claim that this extension must then be a morphism. If not, the induced algebra morphism

$$\mathbb{C}[X] \xrightarrow{\lambda_x^*} \mathbb{C}[t, t^{-1}]$$

does not have its image in $\mathbb{C}[t]$, so for some $f \in \mathbb{C}[Z]$ we have that

$$\lambda_x^*(f) = \frac{a_0 + a_1 t + \ldots + a_z t^z}{t^s} \quad \text{with} \quad a_0 \neq 0 \text{ and } s > 0$$

But then $\lambda_x^*(f)(t) \longrightarrow \pm\infty$ when t goes to zero, that is, λ_x^* cannot have a continuous extension, a contradiction.

So, *if* a continuous extension exists there is morphism $\lambda_x : \mathbb{C} \longrightarrow X$. Then, $\lambda_x(0) = y$ and we denote this by

$$\lim_{t \to 0} \lambda(t).x = y$$

Clearly, the point $y \in X$ must belong to the orbitclosure $\overline{GL_n.x}$ in the Zariski topology (or in the $\mathbb{C}$-topology as orbits are constructible). Conversely, one might ask whether if $y \in \overline{GL_n.x}$ we can always approach y via a one-parameter subgroup. The *Hilbert criterium* gives situations when this is indeed possible.

The only ideals of the formal power series $\mathbb{C}[[t]]$ are principal and generated by t^r for some $r \in \mathbb{N}_+$. With $\mathbb{C}((t))$ we will denote the field of fractions of the domain $\mathbb{C}((t))$.

LEMMA 2.4
Let V be a GL_n-representation, $v \in V$ and a point $w \in V$ lying in the orbitclosure $\overline{GL_n.v}$. Then, there exists a matrix g with coefficients in the field $\mathbb{C}((t))$ and $\det(g) \neq 0$ such that

$$(g.v)_{t=0} \quad \text{is well defined and is equal to} \quad w$$

PROOF   Note that $g.v$ is a vector with coordinates in the field $\mathbb{C}((t))$. If all coordinates belong to $\mathbb{C}[[t]]$ we can set $t = 0$ in this vector and obtain a vector in V. It is this vector that we denote with $(g.v)_{t=0}$.

Consider the orbit map $\mu : GL_n \longrightarrow V$ defined by $g' \mapsto g'.v$. As $w \in \overline{GL_n.v}$ we have seen that there is an irreducible curve $C \hookrightarrow GL_n$ such that $w \in \overline{\mu(C)}$. We obtain a diagram of $\mathbb{C}$-algebras

$$
\begin{array}{ccccc}
\mathbb{C}[GL_n] & \longrightarrow & \mathbb{C}[C] & \hookleftarrow & \mathbb{C}(C) \\
\uparrow & & \uparrow & & \uparrow \\
\mu^* \big\uparrow & & \mu^* \big\uparrow & & \big\uparrow \\
\mathbb{C}[V] & \longrightarrow & \mathbb{C}[\overline{\mu(C)}] & \hookleftarrow & \mathbb{C}[C']
\end{array}
$$

Here, $\mathbb{C}[C]$ is defined to be the integral closure of $\mathbb{C}[\overline{\mu(C)}]$ in the function field $\mathbb{C}(C)$ of C. Two things are important to note here : $C' \longrightarrow \overline{\mu(C)}$ is finite, so surjective and take $c \in C'$ be a point lying over $w \in \overline{\mu(C)}$. Further, C' having an integrally closed coordinate ring is a complex manifold. Hence, by the implicit function theorem polynomial functions on C can be expressed in a neighborhood of c as power series in one variable, giving an embedding $\mathbb{C}[C'] \hookrightarrow \mathbb{C}[[t]]$ with $(t) \cap \mathbb{C}[C'] = M_c$. This inclusion extends to one on the level of their fields of fractions. That is, we have a diagram of $\mathbb{C}$-algebra morphisms

$$
\begin{array}{ccccccc}
\mathbb{C}[GL_n] & \longrightarrow & \mathbb{C}(C) & = & \mathbb{C}(C') & \hookleftarrow & \mathbb{C}((t)) \\
\uparrow & & \uparrow & & \uparrow & & \uparrow \\
\mu^* \big\uparrow & & \big\uparrow & & \big\uparrow & & \big\uparrow \\
\mathbb{C}[V] & \longrightarrow & \mathbb{C}[\overline{\mu(C)}] & \hookleftarrow & \mathbb{C}[C'] & \hookleftarrow & \mathbb{C}[[t]]
\end{array}
$$

The upper composition defines an invertible matrix $g(t)$ with coefficients in $\mathbb{C}((t))$, its (i,j)-entry being the image of the coordinate function $x_{ij} \in \mathbb{C}[GL_n]$. Moreover, the inverse image of the maximal ideal $(t) \triangleleft \mathbb{C}[[t]]$ under the lower composition gives the maximal ideal $M_w \triangleleft \mathbb{C}[V]$. This proves the claim. $\qquad\square$

LEMMA 2.5
Let g be an $n \times n$ matrix with coefficients in $\mathbb{C}((t))$ and $\det g \neq 0$. Then there exist $u_1, u_2 \in GL_n(\mathbb{C}[[t]])$ such that

$$
g = u_1 . \begin{bmatrix} t^{r_1} & & 0 \\ & \ddots & \\ 0 & & t^{r_n} \end{bmatrix} . u_2
$$

with $r_i \in \mathbb{Z}$ and $r_1 \leq r_2 \leq \ldots \leq r_n$.

PROOF By multiplying g with a suitable power of t we may assume that $g = (g_{ij}(t))_{i,j} \in M_n(\mathbb{C}[[t]])$. If $f(t) = \sum_{i=0}^{\infty} f_i t^i \in \mathbb{C}[[t]]$ define $v(f(t))$ to be the minimal i such that $a_i \neq 0$. Let (i_0, j_0) be an entry where $v(g_{ij}(t))$ attains a minimum, say, r_1. That is, for all (i, j) we have $g_{ij}(t) = t^{r_1} t^r f(t)$ with $r \geq 0$ and $f(t)$ an invertible element of $\mathbb{C}[[t]]$.

By suitable row and column interchanges we can take the entry (i_0, j_0) to the $(1, 1)$-position. Then, multiplying with a unit we can replace it by t^{r_1} and by elementary row and column operations all the remaining entries in the first row and column can be made zero. That is, we have invertible matrices $a_1, a_2 \in GL_n(\mathbb{C}[[t]])$ such that

$$g = a_1 . \begin{bmatrix} t^{r_1} & 0 \dots 0 \\ \hline 0 & \\ \vdots & g_1 \\ 0 & \end{bmatrix} . a_2$$

Repeating the same idea on the submatrix g_1 and continuing gives the result.
□

We can now state and prove the *Hilbert criterium*, which allows us to study orbit-closures by one-parameter subgroups.

THEOREM 2.2
Let V be a GL_n-representation and $X \hookrightarrow V$ a closed GL_n-stable subvariety. Let $\mathcal{O}(x) = GL_n.x$ be the orbit of a point $x \in X$. Let $Y \hookrightarrow \overline{\mathcal{O}(x)}$ be a closed GL_n-stable subset. Then, there exists a one-parameter subgroup $\lambda : \mathbb{C}^* \longrightarrow GL_n$ such that

$$\lim_{t \to 0} \lambda(t).x \in Y$$

PROOF   It suffices to prove the result for $X = V$. By lemma 2.4 there is an invertible matrix $g \in M_n(\mathbb{C}((t)))$ such that

$$(g.x)_{t=0} = y \in Y$$

By lemma 2.5 we can find $u_1, u_2 \in GL_n(\mathbb{C}[[t]])$ such that

$$g = u_1.\lambda'(t).u_2 \quad \text{with} \quad \lambda'(t) = \begin{bmatrix} t^{r_1} & & 0 \\ & \ddots & \\ 0 & & t^{r_n} \end{bmatrix}$$

a one-parameter subgroup. There exist $x_i \in V$ such that $u_2.x = \sum_{i=0}^{\infty} z_i t^i$ in particular $u_2(0).x = x_0$. But then

$$(\lambda'(t).u_2.x)_{t=0} = \sum_{i=0}^{\infty} (\lambda'(t).x_i t^i)_{t=0}$$

$$= (\lambda'(t).x_0)_{t=0} + (\lambda'(t).x_1 t)_{t=0} + \dots$$

But one easily verifies (using a basis of eigenvectors of $\lambda'(t)$) that

$$\lim_{s \to 0} \lambda'^{-1}(s).(\lambda'(t)x_i t^i)_{t=0} = \begin{cases} (\lambda'(t).x_0)_{t=0} & \text{if } i = 0, \\ 0 & \text{if } i \neq 0 \end{cases}$$

As $(\lambda'(t).u_2.x)_{t=0} \in Y$ and Y is a closed GL_n-stable subset, we also have that

$$\lim_{s \to 0} \lambda'^{-1}(s).(\lambda'(t).u_2.x)_{t=0} \in Y \quad \text{that is,} \quad (\lambda'(t).x_0)_{t=0} \in Y$$

But then, we have for the one-parameter subgroup $\lambda(t) = u_2(0)^{-1}.\lambda'(t).u_2(0)$ that

$$\lim_{t \to 0} \lambda(t).x \in Y$$

finishing the proof. $\square$

An important special case occurs when $x \in V$ belongs to the *nullcone*, that is, when the orbit closure $\overline{\mathcal{O}(x)}$ contains the fixed point $0 \in V$. The original Hilbert criterium asserts the following.

PROPOSITION 2.6
Let V be a GL_n-representation and $x \in V$ in the nullcone. Then, there is a one-parameter subgroup $\mathbb{C}^* \xrightarrow{\lambda} GL_n$ such that

$$\lim_{t \to 0} \lambda(t).x = 0$$

In the statement of theorem 2.2 it is important that Y is closed. In particular, it does *not* follow that any orbit $\mathcal{O}(y) \hookrightarrow \overline{\mathcal{O}(x)}$ can be reached via one-parameter subgroups, see example 2.7 below.

2.4 Semisimple modules

In this section we will characterize the closed GL_n-orbits in the representation variety $rep_n A$ for an affine $\mathbb{C}$-algebra A. We have seen that any point $\psi \in rep_n A$ (that is any n-dimensional representation $A \xrightarrow{\psi} M_n$) determines an n-dimensional A-module, which we will denote with M_ψ.

A *finite filtration* F on an n-dimensional module M is a sequence of A-submodules

$$F \quad : \quad 0 = M_{t+1} \subset M_t \subset \ldots \subset M_1 \subset M_0 = M$$

The *associated graded* A-module is the n-dimensional module

$$gr_F M = \oplus_{i=0}^t M_i/M_{i+1}$$

We have the following ring theoretical interpretation of the action of one-parameter subgroups of GL_n on the representation variety rep_n A.

LEMMA 2.6
Let $\psi, \rho \in rep_n$ A. *Equivalent are*

1. *There is a one-parameter subgroup* $\mathbb{C}^* \xrightarrow{\lambda} GL_n$ *such that*

$$\lim_{t \to 0} \lambda(t).\psi = \rho$$

2. *There is a finite filtration* F *on the* A-*module* M_ψ *such that*

$$gr_F \ M_\psi \simeq M_\rho$$

   as A-*modules.*

PROOF $(1) \Rightarrow (2)$: If V is any GL_n-representation and $\mathbb{C}^* \xrightarrow{\lambda} GL_n$ a one-parameter subgroup, we have an induced *weight space decomposition* of V

$$V = \oplus_i V_{\lambda,i} \quad \text{where} \quad V_{\lambda,i} = \{v \in V \mid \lambda(t).v = t^i v, \forall t \in \mathbb{C}^*\}$$

In particular, we apply this to the underlying vector space of M_ψ, which is $V = \mathbb{C}^n$ (column vectors) on which GL_n acts by left multiplication. We define

$$M_j = \oplus_{i>j} V_{\lambda,i}$$

and claim that this defines a finite filtration on M_ψ with associated graded A-module M_ρ. For any $a \in A$ (it suffices to vary a over the generators of A) we can consider the linear maps

$$\phi_{ij}(a) : V_{\lambda,i} \lhook\joinrel\longrightarrow V = M_\psi \xrightarrow{a.} M_\psi = V \longrightarrow\!\!\!\!\!\rightarrow V_{\lambda,j}$$

(that is, we express the action of a in a block matrix Φ_a with respect to the decomposition of V). Then, the action of a on the module corresponding to $\lambda(t).\psi$ is given by the matrix $\Phi'_a = \lambda(t).\Phi_a.\lambda(t)^{-1}$ with corresponding blocks

$$
\begin{array}{ccc}
V_{\lambda,i} & \xrightarrow{\phi_{ij}(a)} & V_{\lambda,j} \\
\Big\uparrow{\lambda(t)^{-1}} & & \Big\downarrow{\lambda(t)} \\
V_{\lambda,i} & \xrightarrow[\phi'_{ij}(a)]{} & V_{\lambda,j}
\end{array}
$$

that is, $\phi'_{ij}(a) = t^{j-i}\phi_{ij}(a)$. Therefore, if $lim_{t \to 0}\lambda(t).\psi$ exists we must have that

$$\phi_{ij}(a) = 0 \quad \text{for all} \quad j < i$$

But then, the action by a sends any $M_k = \oplus_{i>k} V_{\lambda,i}$ to itself, that is, the M_k are A-submodules of M_ψ. Moreover, for $j > i$ we have

$$\lim_{t\to 0} \phi'_{ij}(a) = \lim_{t\to 0} t^{j-i}\phi_{ij}(a) = 0$$

Consequently, the action of a on ρ is given by the diagonal block matrix with blocks $\phi_{ii}(a)$, but this is precisely the action of a on $V_i = M_{i-1}/M_i$, that is, ρ corresponds to the associated graded module.

(2) $\Rightarrow$ (1): Given a finite filtration on M_ψ

$$F: \quad 0 = M_{t+1} \subset \ldots \subset M_1 \subset M_0 = M_\psi$$

we have to find a one-parameter subgroup $\mathbb{C}^* \xrightarrow{\lambda} GL_n$, which induces the filtration F as in the first part of the proof. Clearly, there exist subspaces V_i for $0 \le i \le t$ such that

$$V = \oplus_{i=0}^t V_i \quad \text{and} \quad M_j = \oplus_{j=i}^t V_i$$

Then we take λ to be defined by $\lambda(t) = t^i Id_{V_i}$ for all i and it verifies the claims. □

Example 2.6

Let M_ψ be the 2-dimensional $\mathbb{C}[x]$-module determined by the Jordan block and consider the canonical basevectors

$$\begin{bmatrix} \lambda & 1 \\ 0 & \lambda \end{bmatrix} \quad e_1 = \begin{bmatrix} 1 \\ 0 \end{bmatrix} \quad e_2 = \begin{bmatrix} 0 \\ 1 \end{bmatrix}$$

Then, $\mathbb{C}e_1$ is a $\mathbb{C}[x]$-submodule of M_ψ and we have a filtration

$$0 = M_2 \subset \mathbb{C}e_1 = M_1 \subset \mathbb{C}e_1 \oplus \mathbb{C}e_2 = M_0 = M_\psi$$

Using the conventions of the second part of the above proof we then have

$$V_1 = \mathbb{C}e_1 \quad V_2 = \mathbb{C}e_2 \quad \text{hence} \quad \lambda(t) = \begin{bmatrix} t & 0 \\ 0 & 1 \end{bmatrix}$$

Indeed, we then obtain that

$$\begin{bmatrix} t & 0 \\ 0 & 1 \end{bmatrix} \cdot \begin{bmatrix} \lambda & 1 \\ 0 & \lambda \end{bmatrix} \cdot \begin{bmatrix} t^{-1} & 0 \\ 0 & 1 \end{bmatrix} = \begin{bmatrix} \lambda & t \\ 0 & \lambda \end{bmatrix}$$

and the limit $t \longrightarrow 0$ exists and is the associated graded module $gr_F\, M_\psi = M_\rho$ determined by the diagonal matrix. □

Consider two modules $M_\psi, M_\psi \in rep_n\, A$. Assume that $\mathcal{O}(M_\rho) \lhook\joinrel\longrightarrow \overline{\mathcal{O}(M_\psi)}$ and that we can reach the orbit of M_ρ via a one-parameter subgroup. Then, lemma 2.6 asserts that M_ρ must be *decomposable*

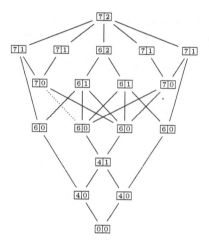

FIGURE 2.1:   Kraft's diamond describing the nullcone of M_3^2.

as it is the associated graded of a nontrivial filtration on M_ψ. This gives us a criterion to construct examples showing that the closedness assumption in the formulation of Hilbert's criterium is essential.

Example 2.7 Nullcone of $M_3^2 = M_3 \oplus M_3$

In chapter 6 we will describe a method to determine the nullcones of m-tuples of $n \times n$ matrices. The special case of two 3×3 matrices has been worked out by H.P. Kraft in [62, p. 202]. The orbits are depicted in figure 2.1 In this picture, each node corresponds to a torus. The right-hand number is the dimension of the torus and the left-hand number is the dimension of the orbit represented by a point in the torus. The solid or dashed lines indicate orbit closures. For example, the dashed line corresponds to the following two points in $M_3^2 = M_3 \oplus M_3$

$$\psi = (\begin{bmatrix} 0\,0\,1 \\ 0\,0\,1 \\ 0\,0\,0 \end{bmatrix}, \begin{bmatrix} 0\,1\,0 \\ 0\,0\,0 \\ 0\,0\,0 \end{bmatrix}) \qquad \rho = (\begin{bmatrix} 0\,0\,1 \\ 0\,0\,0 \\ 0\,0\,0 \end{bmatrix}, \begin{bmatrix} 0\,1\,0 \\ 0\,0\,0 \\ 0\,0\,0 \end{bmatrix})$$

We claim that M_ρ is an indecomposable 3-dimensional module of $\mathbb{C}\langle x, y \rangle$. Indeed, the only subspace of the column vectors $\mathbb{C}^3$ left invariant under both x and y is equal to

$$\begin{bmatrix} \mathbb{C} \\ 0 \\ 0 \end{bmatrix}$$

hence M_ρ cannot have a direct sum decomposition of two or more modules. Next, we claim that $\mathcal{O}(M_\rho) \hookrightarrow \overline{\mathcal{O}(M_\psi)}$. Indeed, simultaneous conjugating

ψ with the invertible matrix

$$\begin{bmatrix} 1 & \epsilon^{-1}-1 & 0 \\ 0 & 1 & 0 \\ 0 & 0 & \epsilon^{-1} \end{bmatrix} \quad \text{we obtain the couple} \quad (\begin{bmatrix} 0 & 0 & 1 \\ 0 & 0 & \epsilon \\ 0 & 0 & 0 \end{bmatrix}, \begin{bmatrix} 0 & 1 & 0 \\ 0 & 0 & 0 \\ 0 & 0 & 0 \end{bmatrix})$$

and letting $\epsilon \longrightarrow 0$ we see that the limiting point is ρ. ▯

The *Jordan-Hölder theorem* , see, for example [84, 2.6], asserts that any finite dimensional A-module M has a *composition series* , that is, M has a finite filtration

$$F \quad : \quad 0 = M_{t+1} \subset M_t \subset \ldots \subset M_1 \subset M_0 = M$$

such that the successive quotients $S_i = M_i/M_{i+1}$ are all *simple* A-modules for $0 \le i \le t$. Moreover, these *composition factors* S and their *multiplicities* are independent of the chosen composition series, that is, the set $\{S_0, \ldots, S_t\}$ is the same for every composition series. In particular, the associated graded module for a composition series is determined only up to isomorphism and is the semisimple n-dimensional module

$$gr\ M = \oplus_{i=0}^{t} S_i$$

THEOREM 2.3
Let A be an affine $\mathbb{C}$-algebra and $M \in rep_n A$.

1. *The orbit $\mathcal{O}(M)$ is closed in $rep_n A$ if and only if M is an n-dimensional semisimple A-module.*

2. *The orbit closure $\overline{\mathcal{O}(M)}$ contains exactly one closed orbit, corresponding to the direct sum of the composition factors of M.*

3. *The points of the quotient variety of $rep_n A$ under GL_n classify the isomorphism classes of n-dimensional semisimple A-modules. We will denote the quotient variety by $iss_n A$.*

PROOF (1): Assume that the orbit $\mathcal{O}(M)$ is Zariski closed. Let $gr\ M$ be the associated graded module for a composition series of M. From lemma 2.6 we know that $\mathcal{O}(gr\ M)$ is contained in $\overline{\mathcal{O}(M)} = \mathcal{O}(M)$. But then $gr\ M \simeq M$ whence M is semisimple.

Conversely, assume M is semisimple. We know that the orbit closure $\overline{\mathcal{O}(M)}$ contains a closed orbit, say $\mathcal{O}(N)$. By the Hilbert criterium we have a one-parameter subgroup $\mathbb{C}^* \xrightarrow{\lambda} GL_n$ such that

$$\lim_{t \to 0} \lambda(t).M = N' \simeq N.$$

By lemma 2.6 this means that there is a finite filtration F on M with associated graded module $gr_F\ M \simeq N$. For the semisimple module M the only possible finite filtrations are such that each of the submodules is a direct sum of simple components, so $gr_F\ M \simeq M$, whence $M \simeq N$ and hence the orbit $\mathcal{O}(M)$ is closed.

(2): Remains only to prove uniqueness of the closed orbit in $\overline{\mathcal{O}(M)}$. This either follows from the Jordan-Hölder theorem or, alternatively, from the separation property of the quotient map to be proved in the next section.

(3): We will prove in the next section that the points of a quotient variety parameterize the closed orbits.   □

Example 2.8

Recall the description of the orbits in $M_2^2 = M_2 \oplus M_2$ given in the previous chapter

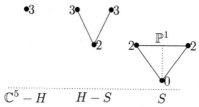

$$\mathbb{C}^5 - H \qquad H - S \qquad S$$

and each fiber contains a unique closed orbit. Over a point in $H - S$ this orbit corresponds to the matrix couple

$$\left(\begin{bmatrix} a_1 & 0 \\ 0 & a_2 \end{bmatrix}, \begin{bmatrix} b_1 & 0 \\ 0 & b_2 \end{bmatrix} \right)$$

which is indeed a semisimple module of $\mathbb{C}\langle x, y \rangle$ (the direct sum of two 1-dimensional simple representations determined by $x \mapsto a_i$ and $y \mapsto b_i$). In case $a_1 = a_2$ and $b_1 = b_2$ these two simples coincide and the semisimple module having this factor with multiplicity two is the unique closed orbit in the fiber of a point in S.   □

Example 2.9

Assume A is a finite dimensional $\mathbb{C}$-algebra. Then, there are only a finite number, say, k, of nonisomorphic n-dimensional semisimple A-modules. Hence $iss_n\ A$ is a finite number of k points, whence $rep_n\ A$ is the disjoint union of k connected components, each consisting of all n-dimensional A-modules with the same composition factors. Connectivity follows from the fact that the orbit of the sum of the composition factors lies in the closure of each orbit.   □

Example 2.10

Let A be an affine commutative algebra with presentation $A =$

$\mathbb{C}[x_1, \ldots, x_m]/I_A$ and let X be the affine variety $\mathbb{V}(I_A)$. Observe that any simple A-module is one-dimensional hence corresponds to a point in X. (Indeed, for any algebra A a simple k-dimensional module determines an epimorphism $A \longrightarrow M_k(\mathbb{C})$ and $M_k(\mathbb{C})$ is only commutative if $k = 1$). Applying the Jordan-Hölder theorem we see that

$$iss_n \ A \simeq X^{(n)} = \underbrace{X \times \ldots \times X}_{n} / S_n$$

the n-th symmetric product of X. $\quad\square$

2.5 Some invariant theory

The results in this section hold for arbitrary reductive algebraic groups. Because we will only work with GL_n (or later with products $GL(\alpha) = GL_{a_1} \times \ldots \times GL_{a_k}$) we include a proof in this case. Our first aim is to prove that GL_n is a *reductive group* , that is, all GL_n-representations are completely reducible. Consider the *unitary group*

$$U_n = \{A \in GL_n \mid A.A^* = 1_n\}$$

where A^* is the *Hermitian transpose* of A. Clearly, U_n is a compact Lie group. Any compact Lie group has a so-called *Haar measure*, which allows one to integrate continuous complex valued functions over the group in an invariant way. That is, there is a linear function assigning to every continuous function $f : U_n \longrightarrow \mathbb{C}$ its integral

$$f \mapsto \int_{U_n} f(g)dg \in \mathbb{C}$$

which is normalized such that $\int_{U_n} dg = 1$ and is left and right invariant, which means that for all $u \in U_n$ we have the equalities

$$\int_{U_n} f(gu)dg = \int_{U_n} f(g)dg = \int_{U_n} f(ug)dg$$

This integral replaces the classical idea in representation theory of averaging functions over a finite group.

PROPOSITION 2.7
Every U_n-representation is completely reducible.

PROOF Take a finite dimensional complex vector space V with a U_n-action and assume that W is a subspace of V left invariant under this action.

Extending a basis of W to V we get a linear map $V \xrightarrow{\phi} W$ which is the identity on W. For any $v \in V$ we have a continuous map

$$U_n \longrightarrow W \qquad g \mapsto g.\phi(g^{-1}.v)$$

(use that W is left invariant) and hence we can integrate it over U_n (integrate the coordinate functions). Hence we can define a map $\phi_0 : V \longrightarrow W$ by

$$\phi_0(v) = \int_{U_n} g.\phi(g^{-1}.v)dg$$

Clearly, ϕ_0 is linear and is the identity on W. Moreover,

$$\phi_0(u.v) = \int_{U_n} g.\phi(g^{-1}u.v)dg = u. \int_{U_n} u^{-1}g.\phi(g^{-1}u.v)dg$$
$$\overset{*}{=} u. \int_{U_n} g\phi(g^{-1}.v)dg = u.\phi_0(v)$$

where the starred equality uses the invariance of the Haar measure. Hence, $V = W \oplus Ker\ \phi_0$ is a decomposition as U_n-representations. Continuing whenever one of the components has a nontrivial subrepresentation we arrive at a decomposition of V into simple U_n-representations. $\square$

We claim that for any n, U_n is Zariski dense in GL_n. Let D_n be the group of all diagonal matrices in GL_n. The *Cartan decomposition* for GL_n asserts that

$$GL_n = U_n.D_n.U_n$$

For, take $g \in GL_n$ then $g.g^*$ is an Hermitian matrix and hence diagonalizable by unitary matrices. So, there is a $u \in U_n$ such that

$$u^{-1}.g.g^*.u = \begin{bmatrix} \alpha_1 & & \\ & \ddots & \\ & & \alpha_n \end{bmatrix} = \underbrace{s^{-1}.g.s}_{p} . \underbrace{s^{-1}.g^*.s}_{p^*}$$

Then, each $\alpha_i > 0 \in \mathbb{R}$ as $\alpha_i = \sum_{j=1}^n \| p_{ij} \|^2$. Let $\beta_i = \sqrt{\alpha_i}$ and let d the diagonal matrix $diag(\beta_1, \ldots, \beta_n)$. Clearly

$$g = u.d.(d^{-1}.u^{-1}.g) \quad \text{and we claim} \quad v = d^{-1}.u^{-1}.g \in U_n$$

Indeed, we have

$$v.v^* = (d^{-1}.u^{-1}.g).(g^*.u.d^{-1}) = d^{-1}.(u^{-1}.g.g^*.u).d^{-1}$$
$$= d^{-1}.d^2.d^{-1} = \mathbb{1}_n$$

proving the Cartan decomposition. Now, $D_n = \mathbb{C}^* \times \ldots \times \mathbb{C}^*$ and $D_n \cap U_n = U_1 \times \ldots \times U_1$ and because $U_1 = \mu$ is Zariski dense (being infinite) in $D_1 = \mathbb{C}^*$, we have that D_n is contained in the Zariski closure of U_n. By the Cartan decomposition we then have that the Zariski closure of U_n is GL_n.

THEOREM 2.4

GL_n is a reductive group. That is, all GL_n-representations are completely reducible.

PROOF Let V be a GL_n-representation having a subrepresentation W. In particular, V and W are U_n-representations, so by the foregoing proposition we have a decomposition $V = W \oplus W'$ as U_n-representations. Consider the subgroup

$$N = N_{GL_n}(W') = \{g \in GL_n \mid g.W' \subset W'\}$$

then N is a Zariski closed subgroup of GL_n containing U_n. As the Zariski closure of U_n is GL_n we have $N = GL_n$ and hence that W' is a representation of GL_n. Continuing gives a decomposition of V in simple GL_n-representations.
$\square$

Let $S = S_{GL_n}$ be the set of isomorphism classes of simple GL_n-representations. If W is a simple GL_n-representation belonging to the isomorphism class $s \in S$, we say that W is *of type s* and denote this by $W \in s$. Let X be a complex vector space (not necessarily finite dimensional) with a linear action of GL_n. We say that the action is *locally finite* on X if, for any finite dimensional subspace Y of X, there exists a finite dimensional subspace $Y \subset Y' \subset X$ which is a GL_n-representation. The *isotypical component* of X of type $s \in S$ is defined to be the subspace

$$X_{(s)} = \sum \{W \mid W \subset X, W \in s\}$$

If V is a GL_n-representation, we have seen that V is completely reducible. Then, $V = \oplus V_{(s)}$ and every isotypical component $V_{(s)} \simeq W^{\oplus e_s}$ for $W \in s$ and some number e_s. Clearly, $e_s \neq 0$ for only finitely many classes $s \in S$. We call the decomposition $V = \oplus_{s \in S} V_{(s)}$ the *isotypical decomposition* of V and we say that the simple representation $W \in s$ occurs *with multiplicity e_s* in V.

If V' is another GL_n-representation and if $V \xrightarrow{\phi} V'$ is a morphism of GL_n-representations (that is, a linear map commuting with the action), then for any $s \in S$ we have that $\phi(V_{(s)}) \subset V'_{(s)}$. If the action of GL_n on X is locally finite, we can reduce to finite dimensional GL_n-subrepresentation and obtain a decomposition

$$X = \oplus_{s \in S} X_{(s)}$$

which is again called the *isotypical decomposition* of X.

Let V be a GL_n-representation of some dimension m. Then, we can view V as an affine space $\mathbb{C}^m$ and we have an induced action of GL_n on the polynomial functions $f \in \mathbb{C}[V]$ by the rule

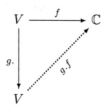

that is, $(g.f)(v) = f(g^{-1}.v)$ for all $g \in GL_n$ and all $v \in V$. If $\mathbb{C}[V] = \mathbb{C}[x_1, \ldots, x_m]$ is graded by giving all the x_i degree one, then each of the homogeneous components of $\mathbb{C}[V]$ is a finite dimensional GL_n-representation. Hence, the action of GL_n on $\mathbb{C}[V]$ is locally finite. Indeed, let $\{y_1, \ldots, y_l\}$ be a basis of a finite dimensional subspace $Y \subset \mathbb{C}[V]$ and let d be the maximum of the $deg(y_i)$. Then $Y' = \oplus_{i=0}^d \mathbb{C}[V]_i$ is a GL_n-representation containing Y.

Therefore, we have an isotypical decomposition $\mathbb{C}[V] = \oplus_{s \in S} \mathbb{C}[V]_{(s)}$. In particular, if $0 \in S$ denotes the isomorphism class of the trivial GL_n-representation ($\mathbb{C}_{triv} = \mathbb{C}x$ with $g.x = x$ for every $g \in GL_n$) then we have

$$\mathbb{C}[V]_{(0)} = \{f \in \mathbb{C}[V] \mid g.f = f, \forall g \in GL_n\} = \mathbb{C}[V]^{GL_n}$$

the ring of *polynomial invariants* , that is, of polynomial functions which are constant along orbits in V.

LEMMA 2.7
Let V be a GL_n-representation.

1. *Let $I \lhd \mathbb{C}[V]$ be a GL_n-stable ideal, that is, $g.I \subset I$ for all $g \in GL_n$, then*

$$(\mathbb{C}[V]/I)^{GL_n} \simeq \mathbb{C}[V]^{GL_n}/(I \cap \mathbb{C}[V]^{GL_n}).$$

2. *Let $J \lhd \mathbb{C}[V]^{GL_n}$ be an ideal, then we have a lying-over property*

$$J = J\mathbb{C}[V] \cap \mathbb{C}[V]^{GL_n}.$$

Hence, $\mathbb{C}[V]^{GL_n}$ is Noetherian, that is, every increasing chain of ideals stabilizes.

3. *Let I_j be a family of GL_n-stable ideals of $\mathbb{C}[V]$, then*

$$\left(\sum_j I_j\right) \cap \mathbb{C}[V]^{GL_n} = \sum_j (I_j \cap \mathbb{C}[V]^{GL_n}).$$

PROOF   (1): As I has the induced GL_n-action, which is locally finite, we have the isotypical decomposition $I = \oplus I_{(s)}$ and clearly $I_{(s)} = \mathbb{C}[V]_{(s)} \cap I$. But then also, taking quotients we have

$$\oplus_s (\mathbb{C}[V]/I)_{(s)} = \mathbb{C}[V]/I = \oplus_s \mathbb{C}[V]_{(s)}/I_{(s)}$$

Therefore, $(\mathbb{C}[V]/I)_{(s)} = \mathbb{C}[V]_{(s)}/I_{(s)}$ and taking the special case $s = 0$ is the statement.

(2): For any $f \in \mathbb{C}[V]^{GL_n}$ left-multiplication by f in $\mathbb{C}[V]$ commutes with the GL_n-action, whence $f.\mathbb{C}[V]_{(s)} \subset \mathbb{C}[V]_{(s)}$. That is, $\mathbb{C}[V]_{(s)}$ is a $\mathbb{C}[V]^{GL_n}$-module. But then, as $J \subset \mathbb{C}[V]^{GL_n}$ we have

$$\oplus_s (J\mathbb{C}[V])_{(s)} = J\mathbb{C}[V] = \oplus_s J\mathbb{C}[V]_{(s)}$$

Therefore, $(J\mathbb{C}[V])_{(s)} = J\mathbb{C}[V]_{(s)}$ and again taking the special value $s = 0$ we obtain $J\mathbb{C}[V] \cap \mathbb{C}[V]^{GL_n} = (J\mathbb{C}[V])_{(0)} = J$. The Noetherian statement follows from the fact that $\mathbb{C}[V]$ is Noetherian (the Hilbert basis theorem).

(3): For any j we have the decomposition $I_j = \oplus_s (I_j)_{(s)}$. But then, we have

$$\oplus_s (\sum_j I_j)_{(s)} = \sum_j I_j = \sum_j \oplus_s (I_j)_{(s)} = \oplus_s \sum_j (I_j)_{(s)}$$

Therefore, $(\sum_j I_j)_{(s)} = \sum_j (I_j)_{(s)}$ and taking $s = 0$ gives the required statement. □

THEOREM 2.5
Let V be a GL_n-representation. Then, the ring of polynomial invariants $\mathbb{C}[V]^{GL_n}$ is an affine $\mathbb{C}$-algebra.

PROOF Because the action of GL_n on $\mathbb{C}[V]$ preserves the gradation, the ring of invariants is also graded

$$\mathbb{C}[V]^{GL_n} = R = \mathbb{C} \oplus R_1 \oplus R_2 \oplus \ldots$$

From lemma 2.7(2) we know that $\mathbb{C}[V]^{GL_n}$ is Noetherian and hence the ideal $R_+ = R_1 \oplus R_2 \oplus \ldots$ is finitely generated $R_+ = Rf_1 + \ldots + Rf_l$ by homogeneous elements $f_1, \ldots, f_l$. We claim that as a $\mathbb{C}$-algebra $\mathbb{C}[V]^{GL_n}$ is generated by the f_i. Indeed, we have $R_+ = \sum_{i=1}^{l} \mathbb{C}f_i + R_+^2$ and then also

$$R_+^2 = \sum_{i,j=1}^{l} \mathbb{C}f_i f_j + R_+^3$$

and iterating this procedure we obtain for all powers m that

$$R_+^m = \sum_{\sum m_i = m} \mathbb{C}f_1^{m_1} \ldots f_l^{m_l} + R_+^{m+1}$$

Now, consider the subalgebra $\mathbb{C}[f_1, \ldots, f_l]$ of $R = \mathbb{C}[V]^{GL_n}$, then we obtain for any power $d > 0$ that

$$\mathbb{C}[V]^{GL_n} = \mathbb{C}[f_1, \ldots, f_l] + R_+^d$$

For any i we then have for the homogeneous components of degree i

$$R_i = \mathbb{C}[f_1, \ldots, f_l]_i + (R_+^d)_i$$

Now, if $d > i$ we have that $(R_+^d)_i = 0$ and hence that $R_i = \mathbb{C}[f_1, \ldots, f_l]_i$. As this holds for all i we proved the claim. $\quad\square$

Choose generating invariants $f_1, \ldots, f_l$ of $\mathbb{C}[V]^{GL_n}$, consider the morphism

$$V \xrightarrow{\phi} \mathbb{C}^l \quad \text{defined by} \quad v \mapsto (f_1(v), \ldots, f_l(v))$$

and define W to be the Zariski closure $\overline{\phi(V)}$ in $\mathbb{C}^l$. Then, we have a diagram

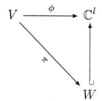

and an isomorphism $\mathbb{C}[W] \xrightarrow{\pi^*} \mathbb{C}[V]^{GL_n}$. More general, let X be a closed GL_n-stable subvariety of V, then $X = \mathbb{V}_V(I)$ for some GL_n-stable ideal I of $\mathbb{C}[V]$. From lemma 2.7(1) we obtain

$$\mathbb{C}[X]^{GL_n} = (\mathbb{C}[V]/I)^{GL_n} = \mathbb{C}[V]^{GL_n}/(I \cap \mathbb{C}[V]^{GL_n})$$

whence $\mathbb{C}[X]^{GL_n}$ is also an affine algebra (and generated by the images of the f_i). Define Y to be the Zariski closure of $\phi(X)$ in $\mathbb{C}^l$, then we have a diagram

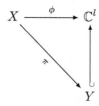

and an isomorphism $\mathbb{C}[Y] \xrightarrow{\pi} \mathbb{C}[X]^{GL_n}$. We call the morphism $X \xrightarrow{\pi} Y$ an *algebraic quotient* of X under GL_n. We will now prove some important properties of this quotient.

PROPOSITION 2.8 universal property
If $X \xrightarrow{\mu} Z$ is a morphism, which is constant along GL_n-orbits in X, then

there exists a unique factoring morphism $\overline{\mu}$

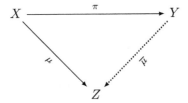

PROOF   As μ is constant along GL_n-orbits in X, we have an inclusion $\mu^*(\mathbb{C}[Z]) \subset \mathbb{C}[X]^{GL_n}$. We have the commutative diagram

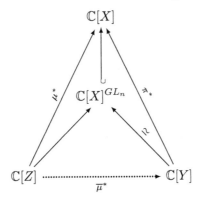

from which the existence and uniqueness of $\overline{\mu}$ follows.   ☐

As a consequence, an algebraic quotient is uniquely determined up to isomorphism (that is, we might have started from other generating invariants and still obtain the same quotient variety up to isomorphism).

PROPOSITION 2.9 onto property
The algebraic quotient $X \xrightarrow{\pi} Y$ is surjective. Moreover, if $Z \hookrightarrow X$ is a closed GL_n-stable subset, then $\pi(Z)$ is closed in Y and the morphism

$$\pi_X \mid Z : Z \longrightarrow \pi(Z)$$

is an algebraic quotient, that is, $\mathbb{C}[\pi(Z)] \simeq \mathbb{C}[Z]^{GL_n}$.

PROOF   Let $y \in Y$ with maximal ideal $M_y \lhd \mathbb{C}[Y]$. By lemma 2.7(2) we have $M_y\mathbb{C}[X] \neq \mathbb{C}[X]$ and hence there is a maximal ideal M_x of $\mathbb{C}[X]$ containing $M_y\mathbb{C}[X]$, but then $\pi(x) = y$. Let $Z = \mathbb{V}_X(I)$ for a G-stable ideal I of $\mathbb{C}[X]$, then $\overline{\pi(Z)} = \mathbb{V}_Y(I \cap \mathbb{C}[Y])$. That is, $\mathbb{C}[\overline{\pi(Z)}] = \mathbb{C}[Y]/(I \cap \mathbb{C}[Y])$. However, we have from lemma 2.7(1) that

$$\mathbb{C}[Y]/(\mathbb{C}[Y] \cap I) \simeq (\mathbb{C}[X]/I)^{GL_n} = \mathbb{C}[Z]^{GL_n}$$

and hence $\mathbb{C}[\overline{\pi(Z)}] = \mathbb{C}[Z]^{GL_n}$. Finally, surjectivity of $\pi \mid Z$ is proved as above. $\quad\square$

An immediate consequence is that the Zariski topology on Y is the quotient topology of that on X.

PROPOSITION 2.10 separation property

The quotient $X \xrightarrow{\pi} Y$ separates disjoint closed GL_n-stable subvarieties of X.

PROOF Let Z_j be closed GL_n-stable subvarieties of X with defining ideals $Z_j = \mathbb{V}_X(I_j)$. Then, $\cap_j Z_j = \mathbb{V}_X(\sum_j I_j)$. Applying lemma 2.7(3) we obtain

$$\overline{\pi(\cap_j Z_j)} = \mathbb{V}_Y((\sum_j I_j) \cap \mathbb{C}[Y]) = \mathbb{V}_Y(\sum_j (I_j \cap \mathbb{C}[Y]))$$

$$= \cap_j \mathbb{V}_Y(I_j \cap \mathbb{C}[Y]) = \cap_j \overline{\pi(Z_j)}$$

The onto property implies that $\overline{\pi(Z_j)} = \pi(Z_j)$ from which the statement follows. $\quad\square$

PROPOSITION 2.11

The algebraic quotient $X \xrightarrow{\pi} Y$ is the best continuous approximation to the orbit space. That is, points of Y parameterize the closed GL_n-orbits in X. In fact, every fiber $\pi^{-1}(y)$ contains exactly one closed orbit C and we have

$$\pi^{-1}(y) = \{x \in X \mid C \subset \overline{GL_n.x}\}$$

PROOF The fiber $F = \pi^{-1}(y)$ is a GL_n-stable closed subvariety of X. Take any orbit $GL_n.x \subset F$ then either it is closed or contains in its closure an orbit of strictly smaller dimension. Induction on the dimension then shows that $\overline{G.x}$ contains a closed orbit C. On the other hand, assume that F contains two closed orbits, then they have to be disjoint contradicting the separation property. $\quad\square$

2.6   Geometric reconstruction

In this section we will give a geometric interpretation of the reconstruction result of theorem 1.17. Let A be a Cayley-Hamilton algebra of degree n, with trace map tr_A, which is generated by at most m elements $a_1, \ldots, a_m$. We will give a functorial interpretation to the affine scheme determined by the canonical ideal $N_A \lhd \mathbb{C}[M_n^m]$ in the formulation of theorem 1.17. First, let us identify the reduced affine variety $\mathbb{V}(N_A)$. A point $m = (m_1, \ldots, m_m) \in \mathbb{V}(N_A)$ determines an algebra map $f_m : \mathbb{C}[M_n^m]/N_A \longrightarrow \mathbb{C}$ and hence an algebra map ϕ_m

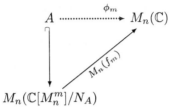

which is trace preserving. Conversely, from the universal property it follows that any trace preserving algebra morphism $A \longrightarrow M_n(\mathbb{C})$ is of this form by considering the images of the trace generators $a_1, \ldots, a_m$ of A. Alternatively, the points of $\mathbb{V}(N_A)$ classify n-dimensional *trace preserving* representations of A. That is, n-dimensional representations for which the morphism $A \longrightarrow M_n(\mathbb{C})$ describing the action is trace preserving. For this reason we will denote the variety $\mathbb{V}(N_A)$ by $trep_n$ A and call it the *trace preserving reduced representation variety* of A.

Assume that A is generated as a $\mathbb{C}$-algebra by $a_1, \ldots, a_m$ (observe that this is no restriction as trace affine algebras are affine) then clearly $I_A(n) \subset N_A$. See the following text.

LEMMA 2.8

For A a Cayley-Hamilton algebra of degree n generated by $\{a_1, \ldots, a_m\}$, the reduced trace preserving representation variety

$$trep_n \ A \lhook\joinrel\longrightarrow rep_n \ A$$

is a closed subvariety of the reduced representation variety.

It is easy to determine the additional defining equations. Write any trace monomial out in the generators

$$tr_A(a_{i_1} \ldots a_{i_k}) = \sum \alpha_{j_1 \ldots j_l} a_{j_1} \ldots a_{j_l}$$

then for a point $m = (m_1, \ldots, m_m) \in rep_n \ A$ to belong to $trep_n \ A$, it must satisfy all the relations of the form

$$tr(m_{i_1} \ldots m_{i_k}) = \sum \alpha_{j_1 \ldots j_l} m_{j_1} \ldots m_{j_l}$$

with tr the usual trace on $M_n(\mathbb{C})$. These relations define the closed subvariety $trep_n(A)$. Usually, this is a proper subvariety.

Example 2.11

Let A be a finite dimensional semisimple algebra $A = M_{d_1}(\mathbb{C}) \oplus \ldots \oplus M_{d_k}(\mathbb{C})$, then A has precisely k distinct simple modules $\{M_1, \ldots, M_k\}$ of dimensions $\{d_1, \ldots, d_k\}$. Here, M_i can be viewed as column vectors of size d_i on which the component $M_{d_i}(\mathbb{C})$ acts by left multiplication and the other factors act as zero. Because A is semisimple every n-dimensional A-representation M is isomorphic to

$$M = M_1^{\oplus e_1} \oplus \ldots \oplus M_k^{\oplus e_k}$$

for certain multiplicities e_i satisfying the numerical condition

$$n = e_1 d_1 + \ldots + e_k d_k$$

That is, $rep_n\ A$ is the disjoint union of a finite number of (closed) orbits each determined by an integral vector $(e_1, \ldots, e_k)$ satisfying the condition called the *dimension vector* of M.

$$rep_n\ A \simeq \bigsqcup_{(e_1, \ldots, e_k)} GL_n/(GL_{e_1} \times \ldots GL_{e_k})$$

Let $f_i \geq 1$ be natural numbers such that $n = f_1 d_1 + \ldots + f_k d_k$ and consider the embedding of A into $M_n(\mathbb{C})$ defined by

Via this embedding, A becomes a Cayley-Hamilton algebra of degree n when equipped with the induced trace tr from $M_n(\mathbb{C})$.

Let M be the n-dimensional A-representation with dimension vector $(e_1, \ldots, e_k)$ and choose a basis compatible with this decomposition. Let E_i be the idempotent of A corresponding to the identity matrix I_{d_i} of the i-th factor. Then, the trace of the matrix defining the action of E_i on M is clearly $e_i d_i . I_n$. On the other hand, $tr(E_i) = f_i d_i . I_n$, hence the only trace preserving n-dimensional A-representation is that of dimension vector $(f_1, \ldots, f_k)$. Therefore, $trep_n\ A$ consists of the single closed orbit determined by the integral vector $(f_1, \ldots, f_k)$.

$$trep_n\ A \simeq GL_n/(GL_{f_1} \times \ldots \times GL_{f_k})$$

Consider the scheme structure of the trace preserving representation variety $\text{trep}_n\ A$. The corresponding functor assigns to a commutative affine $\mathbb{C}$-algebra R

$$\text{trep}_n(R) = Alg_{\mathbb{C}}(\mathbb{C}[M_n^m]/N_A, R)$$

An algebra morphism $\psi : \mathbb{C}[M_n^m]/N_A \longrightarrow R$ determines uniquely an m-tuple of $n \times n$ matrices with coefficients in R by

$$r_k = \begin{bmatrix} \psi(x_{11}(k)) & \cdots & \psi(x_{1n}(k)) \\ \vdots & & \vdots \\ \psi(x_{n1}(k)) & \cdots & \psi(x_{nn}(k)) \end{bmatrix}$$

Composing with the canonical embedding

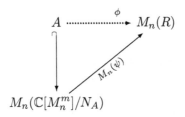

determines the trace preserving algebra morphism $\phi : A \longrightarrow M_n(R)$ where the trace map on $M_n(R)$ is the usual trace. By the universal property any trace preserving map $A \longrightarrow M_n(R)$ is also of this form.

LEMMA 2.9
Let A be a Cayley-Hamilton algebra of degree n that is generated by $\{a_1, \ldots, a_m\}$. The trace preserving representation variety $\text{trep}_n\ A$ represents the functor

$$\text{trep}_n\ A(R) = \{A \overset{\phi}{\longrightarrow} M_n(R) \mid \phi \text{ is trace preserving }\}$$

Moreover, $\text{trep}_n\ A$ is a closed subscheme of $\text{rep}_n\ A$.

Recall that there is an action of GL_n on $\mathbb{C}[M_n^m]$ and from the definition of the ideals $I_A(n)$ and N_A it is clear that they are stable under the GL_n-action. That is, there is an action by automorphisms on the quotient algebras $\mathbb{C}[M_n^m]/I_A(n)$ and $\mathbb{C}[M_n^m]/N_A$. But then, their algebras of invariants are equal to

$$\begin{cases} \mathbb{C}[\text{rep}_n\ A]^{GL_n} &= (\mathbb{C}[M_n^m]/I_A(n))^{GL_n} = \mathbb{N}_n^m/(I_A(n) \cap \mathbb{N}_n^m) \\ \mathbb{C}[\text{trep}_n\ A]^{GL_n} &= (\mathbb{C}[M_n^m]/N_A)^{GL_n} = \mathbb{N}_n^m/(N_A \cap \mathbb{N}_n^m) \end{cases}$$

That is, these rings of invariants define closed subschemes of the affine (reduced) variety associated to the necklace algebra $\mathbb{N}_n^m$. We will call these schemes the *quotient schemes* for the action of GL_n and denote them respectively by

$$\text{iss}_n \ A = \text{rep}_n \ A/GL_n \quad \text{and} \quad \text{triss}_n \ A = \text{trep}_n \ A/GL_n$$

We have seen that the geometric points of the reduced variety $\text{iss}_n \ A$ of the affine quotient scheme $\text{iss}_n \ A$ parameterize the isomorphism classes of n-dimensional semisimple A-representations. Similarly, the geometric points of the reduced variety $\text{triss}_n \ A$ of the quotient scheme $\text{triss}_n \ A$ parameterize isomorphism classes of *trace preserving* n-dimensional semisimple A-representations.

PROPOSITION 2.12

Let A be a Cayley-Hamilton algebra of degree n with trace map tr_A. Then, we have that

$$tr_A(A) = \mathbb{C}[\text{triss}_n \ A]$$

the coordinate ring of the quotient scheme $\text{triss}_n \ A$. In particular, maximal ideals of $tr_A(A)$ parameterize the isomorphism classes of trace preserving n-dimensional semisimple A-representations.

By definition, a GL_n-equivariant map between the affine GL_n-schemes

$$\text{trep}_n \ A \xrightarrow{f} M_n = \mathbb{M}_n$$

means that for any commutative affine $\mathbb{C}$-algebra R the corresponding map

$$\text{trep}_n \ A(R) \xrightarrow{f(R)} M_n(R)$$

commutes with the action of $GL_n(R)$. Alternatively, the ring of all morphisms $\text{trep}_n \ A \longrightarrow M_n$ is the matrixalgebra $M_n(\mathbb{C}[M_n^m]/N_A)$ and those that commute with the GL_n action are precisely the invariants. That is, we have the following description of A.

THEOREM 2.6

Let A be a Cayley-Hamilton algebra of degree n with trace map tr_A. Then, we can recover A as the ring of GL_n-equivariant maps

$$A = \{f : \text{trep}_n \ A \longrightarrow M_n \ \ GL_n\text{-equivariant} \ \}$$

Summarizing the results of this and the previous section we have

THEOREM 2.7

The functor

$$\text{alg@n} \xrightarrow{\text{trep}_n} \text{GL(n)-affine}$$

which assigns to a Cayley-Hamilton algebra A of degree n the GL_n-affine scheme $\texttt{trep}_n$ A *of trace preserving n-dimensional representations has a left inverse. This left inverse functor*

$$\texttt{GL(n)-affine} \xrightarrow{\Uparrow^n} \texttt{alg@n}$$

assigns to a GL_n-affine scheme $\mathbf{X}$ its witness algebra $\Uparrow^n [\mathbf{X}] = M_n(\mathbb{C}[\mathbf{X}])^{GL_n}$ which is a Cayley-Hamilton algebra of degree n.

Note however that this functor is *not* an equivalence of categories. For, there are many affine GL_n-schemes having the same witness algebra as we will see in the next section.

We will give an application of the algebraic reconstruction result, theorem 1.17, to finite dimensional algebras.

Let A be a Cayley-Hamilton algebra of degree n wit trace map tr, then we can define a *norm map* on A by

$$N(a) = \sigma_n(a) \quad \text{for all } a \in A.$$

Recall that the elementary symmetric function σ_n is a polynomial function $f(t_1, t_2, \ldots, t_n)$ in the Newton functions $t_i = \sum_{j=1}^{n} x_j^i$. Then, $\sigma(a) = f(tr(a), tr(a^2), \ldots, tr(a^n))$. Because, we have a trace preserving embedding $A \hookrightarrow M_n(\mathbb{C}[\texttt{trep}_n A])$ and the norm map N coincides with the determinant in this matrix-algebra, we have that

$$N(1) = 1 \quad \text{and} \quad \forall a, b \in A \quad N(ab) = N(a)N(b).$$

Furthermore, the norm map extends to a polynomial map on $A[t]$ and we have that $\chi_a^{(n)}(t) = N(t-a)$. In particular we can obtain the trace by polarization of the norm map. Consider a finite dimensional semisimple $\mathbb{C}$-algebra

$$A = M_{d_1}(\mathbb{C}) \oplus \ldots \oplus M_{d_k}(\mathbb{C}),$$

then all the Cayley-Hamilton structures of degree n on A with trace values in $\mathbb{C}$ are given by the following result.

LEMMA 2.10

Let A be a semisimple algebra as above and tr a trace map on A making it into a Cayley-Hamilton algebra of degree n with $tr(A) = \mathbb{C}$. Then, there exist a dimension vector $\alpha = (m_1, \ldots, m_k) \in \mathbb{N}_+^k$ such that $n = \sum_{i=1}^{k} m_i d_i$ and for any $a = (A_1, \ldots, A_k) \in A$ with $A_i \in M_{d_i}(\mathbb{C})$ we have that

$$tr(a) = m_1 Tr(A_1) + \ldots + m_k Tr(A_k)$$

where Tr are the usual trace maps on matrices.

PROOF   The norm-map N on A defined by the trace map tr induces a group morphism on the invertible elements of A

$$N : A^* = GL_{d_1}(\mathbb{C}) \times \ldots \times GL_{d_k}(\mathbb{C}) \longrightarrow \mathbb{C}^*$$

that is, a *character*. Now, any character is of the following form, let $A_i \in GL_{d_i}(\mathbb{C})$, then for $a = (A_1, \ldots, A_k)$ we must have

$$N(a) = det(A_1)^{m_1} det(A_2)^{m_2} \ldots det(A_k)^{m_k}$$

for certain integers $m_i \in \mathbb{Z}$. Since N extends to a polynomial map on the whole of A we must have that all $m_i \geq 0$. By polarization it then follows that

$$tr(a) = m_1 Tr(A_1) + \ldots + m_k Tr(A_k)$$

and it remains to show that no $m_i = 0$. Indeed, if $m_i = 0$ then tr would be the zero map on $M_{d_i}(\mathbb{C})$, but then we would have for any $a = (0, \ldots, 0, A, 0, \ldots, 0)$ with $A \in M_{d_i}(\mathbb{C})$ that

$$\chi_a^{(n)}(t) = t^n$$

whence $\chi_a^{(n)}(a) \neq 0$ whenever A is not nilpotent. This contradiction finishes the proof. $\square$

We can extend this to all finite dimensional $\mathbb{C}$-algebras. Let A be a finite dimensional algebra with radical J and assume there is a trace map tr on A making A into a Cayley-Hamilton algebra of degree n and such that $tr(A) = \mathbb{C}$. We claim that the norm map $N : A \longrightarrow \mathbb{C}$ is zero on J. Indeed, any $j \in J$ satisfies $j^l = 0$ for some l whence $N(j)^l = 0$. But then, polarization gives that $tr(J) = 0$ and we have that the semisimple algebra

$$A^{ss} = A/J = M_{d_1}(\mathbb{C}) \oplus \ldots \oplus M_{d_k}(\mathbb{C})$$

is a semisimple Cayley-hamilton algebra of degree n on which we can apply the foregoing lemma. Finally, note that $A \simeq A^{ss} \oplus J$ as $\mathbb{C}$-vector spaces. This concludes the proof as follows.

PROPOSITION 2.13
Let A be a finite dimensional $\mathbb{C}$-algebra with radical J and semisimple part

$$A^{ss} = A/J = M_{d_1}(\mathbb{C}) \oplus \ldots \oplus M_{d_k}(\mathbb{C}).$$

Let $tr : A \longrightarrow \mathbb{C} \hookrightarrow A$ be a trace map such that A is a Cayley-Hamilton algebra of degree n. Then, there exists a dimension vector $\alpha = (m_1, \ldots, m_k) \in \mathbb{N}_+^k$ such that for all $a = (A_1, \ldots, A_k, j)$ with $A_i \in M_{d_i}(\mathbb{C})$ and $j \in J$ we have

$$tr(a) = m_1 Tr(A_1) + \ldots + m_k Tr(A_k)$$

with Tr the usual traces on $M_{d_i}(\mathbb{C})$ and $\sum_i m_i d_i = n$.

Fix a trace map tr on A determined by a dimension vector $\alpha = (m_1, \ldots, m_k) \in \mathbb{N}^k$. Then, the trace preserving variety $\mathtt{trep}_n\, A$ is the scheme of A-modules of *dimension vector* α, that is, those A-modules M such that

$$M^{ss} = S_1^{\oplus m_1} \oplus \ldots \oplus S_k^{\oplus m_k}$$

where S_i is the simple A-module of dimension d_i determined by the i-th factor in A^{ss}. An immediate consequence of the reconstruction theorem 2.6 is shown below.

PROPOSITION 2.14

Let A be a finite dimensional algebra with trace map $tr : A \longrightarrow \mathbb{C}$ determined by a dimension vector $\alpha = (m_1, \ldots, m_k)$ as before with all $m_i > 0$. Then, A can be recovered from the GL_n-structure of the affine scheme $\mathtt{trep}_n\, A$ of all A-modules of dimension vector α.

Still, there can be other trace maps on A making A into a Cayley-Hamilton algebra of degree n. For example, let C be a finite dimensional commutative $\mathbb{C}$-algebra with radical N, then $A = M_n(C)$ is finite dimensional with radical $J = M_n(N)$ and the usual trace map $tr : M_n(C) \longrightarrow C$ makes A into a Cayley-Hamilton algebra of degree n such that $tr(J) = N \neq 0$. Still, if A is semisimple, the center $Z(A) = \mathbb{C} \oplus \ldots \oplus \mathbb{C}$ (as many terms as there are matrix components in A) and any subring of $Z(A)$ is of the form $\mathbb{C} \oplus \ldots \oplus \mathbb{C}$. In particular, $tr(A)$ has this form and composing the trace map with projection on the j-th component we have a trace map tr_j to which we can apply lemma 2.10.

2.7 The Gerstenhaber-Hesselink theorem

In this section we will give examples of distinct GL_n-affine schemes having the same witness algebra, proving that the left inverse of theorem 2.7 is *not* an equivalence of categories. We will study the orbits in $rep_n\, \mathbb{C}[x]$ or, equivalent, conjugacy classes of $n \times n$ matrices.

It is sometimes convenient to relax our definition of partitions to include zeroes at the tail. That is, a partition p of n is an integral n-tuple $(a_1, a_2, \ldots, a_n)$ with $a_1 \geq a_2 \geq \ldots \geq a_n \geq 0$ with $\sum_{i=1}^{n} a_i = n$. As before, we represent a partition by a Young diagram by omitting rows corresponding to zeroes.

If $q = (b_1, \ldots, b_n)$ is another partition of n we say that p *dominates* q and write

$$p > q \quad \text{if and only if} \quad \sum_{i=1}^{r} a_i \geq \sum_{i=1}^{r} b_i \quad \text{for all } 1 \leq r \leq n$$

For example, the partitions of 4 are ordered as indicated below

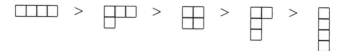

Note however that the dominance relation is *not* a total ordering whenever $n \geq 6$. For example, the following two partition of 6

are not comparable. The *dominance order* is induced by the *Young move* of throwing a row-ending box down the diagram. Indeed, let p and q be partitions of n such that $p > q$ and assume there is no partition r such that $p > r$ and $r > q$. Let i be the minimal number such that $a_i > b_i$. Then by the assumption $a_i = b_i + 1$. Let $j > i$ be minimal such that $a_j \neq b_j$, then we have $b_j = a_j + 1$ because p dominates q. But then, the remaining rows of p and q must be equal. That is, a Young move can be depicted as

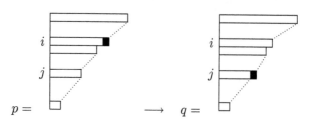

For example, the Young moves between the partitions of 4 given above are as indicated

A *Young p-tableau* is the Young diagram of p with the boxes labeled by integers from $\{1, 2, \ldots, s\}$ for some s such that each label appears at least ones. A Young p-tableau is said to be *of type q* for some partition $q = (b_1, \ldots, b_n)$ of n if the following conditions are met:

- the labels are non-decreasing along rows,

- the labels are strictly increasing along columns, and

- the label i appears exactly b_i times.

For example, if $p = (3, 2, 1, 1)$ and $q = (2, 2, 2, 1)$ then the p-tableau below

$$
\begin{array}{|c|c|c|}
\hline
1 & 1 & 3 \\
\hline
\end{array}
\begin{array}{|c|c|}
\hline
2 & 2 \\
\hline
\end{array}
\begin{array}{|c|}
\hline
3 \\
\hline
\end{array}
\begin{array}{|c|}
\hline
4 \\
\hline
\end{array}
$$

is of type q (observe that $p > q$ and even $p \to q$). In general, let $p = (a_1, \ldots, a_n)$ and $q = (b_1, \ldots, b_n)$ be partitions of n and assume that $p \to q$. Then, there is a Young p-tableau of type q. For, fill the Young diagram of q by putting 1's in the first row, 2's in the second and so on. Then, upgrade the fallen box together with its label to get a Young p-tableau of type q. In the example above

$$
\begin{array}{|c|c|c}
\hline
1 & 1 & \bullet \\
\hline
2 & 2 \\
\hline
3 & ③ \\
\hline
4 \\
\cline{1-1}
\end{array}
\implies
\begin{array}{|c|c|c|}
\hline
1 & 1 & 3 \\
\hline
2 & 2 \\
\cline{1-2}
3 \\
\cline{1-1}
4 \\
\cline{1-1}
\end{array}
$$

Conversely, assume there is a Young p-tableau of type q. The number of boxes labeled with a number $\leq i$ is equal to $b_1 + \ldots + b_i$. Further, any box with label $\leq i$ must lie in the first i rows (because the labels strictly increase along a column). There are $a_1 + \ldots + a_i$ boxes available in the first i rows whence

$$\sum_{j=1}^{i} b_i \leq \sum_{j=1}^{i} a_i \quad \text{for all} \quad 1 \leq i \leq n$$

and therefore $p > q$. After these preliminaries on partitions, let us return to nilpotent matrices.

Let A be a nilpotent matrix of type $p = (a_1, \ldots, a_n)$, that is, conjugated to a matrix with Jordan blocks (all with eigenvalue zero) of sizes a_i. We have seen before that the subspace V_l of column vectors $v \in \mathbb{C}^n$ such that $A^l.v = 0$ has dimension

$$\sum_{h=1}^{l} \#\{j \mid a_j \geq h\} = \sum_{h=1}^{l} a_h^*$$

where $p^* = (a_1^*, \ldots, a_n^*)$ is the dual partition of p. Choose a basis $\{v_1, \ldots, v_n\}$ of $\mathbb{C}^n$ such that for all l the first $a_1^* + \ldots + a_l^*$ base vectors span the subspace V_l. For example, if A is in Jordan normal form of type $p = (3, 2, 1, 1)$

$$
\begin{bmatrix}
0 & 1 & 0 & & & & \\
0 & 0 & 1 & & & & \\
0 & 0 & 0 & & & & \\
& & & 0 & 1 & & \\
& & & 0 & 0 & & \\
& & & & & 0 & \\
& & & & & & 0
\end{bmatrix}
$$

then $p^* = (4, 2, 1)$ and we can choose the standard base vectors ordered as follows

$$\{\underbrace{e_1, e_4, e_6, e_7}_{4}, \underbrace{e_2, e_5}_{2}, \underbrace{e_3}_{1}\}$$

Take a partition $q = (b_1, \ldots, b_n)$ with $p \to q$ (in particular, $p > q$), then for the dual partitions we have $q^* \to p^*$ (and thus $q^* > p^*$). But then there is a Young q^*-tableau of type p^*. In the example with $q = (2, 2, 2, 1)$ we have $q^* = (4, 3)$ and $p^* = (4, 2, 1)$ and we can take the Young q^*-tableau of type p^*

1	1	1	1
2	2	3	

Now label the boxes of this tableau by the base vectors $\{v_1, \ldots, v_n\}$ such that the boxes labeled i in the Young q^*-tableau of type p^* are filled with the base vectors from $V_i - V_{i-1}$. Call this tableau T. In the example, we can take

$$T = \begin{array}{|c|c|c|c|} \hline e_1 & e_4 & e_6 & e_7 \\ \hline e_2 & e_5 & e_3 \\ \cline{1-3} \end{array}$$

Define a linear operator F on $\mathbb{C}^n$ by the rule that $F(v_i) = v_j$ if v_j is the label of the box in T just above the box labeled v_i. In case v_i is a label of a box in the first row of T we take $F(v_i) = 0$. Obviously, F is a nilpotent $n \times n$ matrix and by construction we have that

$$rk\ F^l = n - (b_1^* + \ldots + b_l^*)$$

That is, F is nilpotent of type $q = (b_1, \ldots, b_n)$. Moreover, F satisfies $F(V_i) \subset V_{i-1}$ for all i by the way we have labeled the tableau T and defined F.

In the example above, we have $F(e_2) = e_1$, $F(e_5) = e_4$, $F(e_3) = e_6$ and all other $F(e_i) = 0$. That is, F is the matrix

$$\begin{bmatrix} 0 & 1 & & & & & \\ 0 & 0 & & & & & \\ & & 0 & & 0 & & \\ & & & 0 & 1 & & \\ & & & 0 & 0 & & \\ & & 1 & & 0 & & \\ & & & & & & 0 \end{bmatrix}$$

which is seen to be of type $(2, 2, 2, 1)$ after performing a few Jordan moves.

Returning to the general case, consider for all $\epsilon \in \mathbb{C}$ the $n \times n$ matrix:

$$F_\epsilon = (1 - \epsilon)F + \epsilon A.$$

We claim that for all but finitely many values of ϵ we have $F_\epsilon \in \mathcal{O}(A)$. Indeed, we have seen that $F(V_i) \subset V_{i-1}$ where V_i is defined as the subspace such that $A^i(V_i) = 0$. Hence, $F(V_1) = 0$ and therefore

$$F_\epsilon(V_1) = (1 - \epsilon)F + \epsilon A(V_1) = 0$$

Assume by induction that $F_\epsilon^i(V_i) = 0$ holds for all $i < l$, then we have that

$$F_\epsilon^l(V_l) = F_\epsilon^{l-1}((1 - \epsilon)F + \epsilon A)(V_l)$$
$$\subset F_\epsilon^{l-1}(V_{l-1}) = 0$$

because $A(V_l) \subset V_{l-1}$ and $F(V_l) \subset V_{l-1}$. But then we have for all l that

$$rk\ F_\epsilon^l \leq dim\ V_l = n - (a_1^* + \ldots + a_l^*) = rk\ A^l \overset{def}{=} r_l$$

Then for at least one $r_l \times r_l$ submatrix of F_ϵ^l its determinant considered it as a polynomial of degree r_l in ϵ is not identically zero (as it is nonzero for $\epsilon = 1$). But then this determinant is nonzero for all but finitely many ϵ. Hence, $rk\ F_\epsilon^l = rk\ A^l$ for all l for all but finitely many ϵ. As these numbers determine the dual partition p^* of the type of A, F_ϵ is a nilpotent $n \times n$ matrix of type p for all but finitely many values of ϵ, proving the claim. But then, $F_0 = F$, which we have proved to be a nilpotent matrix of type q, belongs to the closure of the orbit $\mathcal{O}(A)$. That is, we have proved the difficult part of the *Gerstenhaber-Hesselink theorem* .

THEOREM 2.8
Let A be a nilpotent $n \times n$ matrix of type $p = (a_1, \ldots, a_n)$ and B nilpotent of type $q = (b_1, \ldots, b_n)$. Then, B belongs to the closure of the orbit $\mathcal{O}(A)$, that is

$$B \in \overline{\mathcal{O}(A)} \quad \text{if and only if} \quad p > q$$

in the domination order on partitions of n.

To prove the theorem we only have to observe that if B is contained in the closure of A, then B^l is contained in the closure of A^l and hence $rk\ A^l \geq rk\ B^l$ (because $rk\ A^l < k$ is equivalent to vanishing of all determinants of $k \times k$ minors, which is a closed condition). But then

$$n - \sum_{i=1}^{l} a_i^* \geq n - \sum_{i=1}^{l} b_i^*$$

for all l, that is, $q^* > p^*$ and hence $p > q$. The other implication was proved above if we remember that the domination order was induced by the Young moves and clearly we have that if $B \in \overline{\mathcal{O}(C)}$ and $C \in \overline{\mathcal{O}(A)}$ then also $B \in \overline{\mathcal{O}(A)}$.

Example 2.12 Nilpotent matrices for small n
We will apply theorem 2.8 to describe the orbit-closures of nilpotent matrices
of 8×8 matrices. The following table lists all partitions (and their dual in
the other column)

The partitions of 8

a	(8)	v	(1,1,1,1,1,1,1,1)
b	(7,1)	u	(2,1,1,1,1,1,1)
c	(6,2)	t	(2,2,1,1,1,1)
d	(6,1,1)	s	(3,1,1,1,1,1)
e	(5,3)	r	(2,2,2,1,1)
f	(5,2,1)	q	(3,2,1,1,1)
g	(5,1,1,1)	p	(4,1,1,1,1)
h	(4,4)	o	(2,2,2,2)
i	(4,3,1)	n	(3,2,2,1)
j	(4,2,2)	m	(3,3,1,1)
k	(3,3,2)	k	(3,3,2)
l	(4,2,1,1)	l	(4,2,1,1)

The domination order between these partitions can be depicted as follows
where all the Young moves are from left to right

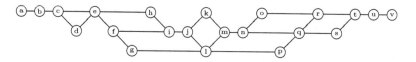

Of course, from this graph we can read off the dominance order graphs for
partitions of $n \leq 8$. The trick is to identify a partition of n with that of 8 by
throwing in a tail of ones and to look at the relative position of both partitions
in the above picture. Using these conventions we get the following graph for
partitions of 7

and for partitions of 6 the dominance order is depicted as follows

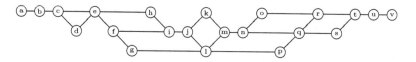

The dominance order on partitions of $n \leq 5$ is a total ordering. □

The Gerstenhaber-Hesselink theorem can be applied to describe the module
varieties of the algebras $\mathbb{C}[x]/(x^r)$.

Example 2.13 The representation variety $rep_n \frac{\mathbb{C}[x]}{(x^r)}$

Any algebra morphism from $\mathbb{C}[x]$ to M_n is determined by the image of x, whence $rep_n(\mathbb{C}[x]) = M_n$. We have seen that conjugacy classes in M_n are classified by the Jordan normalform. Let A be conjugated to a matrix in normalform

$$\begin{bmatrix} \boxed{J_1} & & & \\ & \boxed{J_2} & & \\ & & \ddots & \\ & & & \boxed{J_s} \end{bmatrix}$$

where J_i is a Jordan block of size d_i, hence $n = d_1 + d_2 + \ldots + d_s$. Then, the n-dimensional $\mathbb{C}[x]$-module M determined by A can be decomposed uniquely as

$$M = M_1 \oplus M_2 \oplus \ldots \oplus M_s$$

where M_i is a $\mathbb{C}[x]$-module of dimension d_i which is *indecomposable* , that is, cannot be decomposed as a direct sum of proper submodules.

Now, consider the quotient algebra $R = \mathbb{C}[x]/(x^r)$, then the ideal $I_R(n)$ of $\mathbb{C}[x_{11}, x_{12}, \ldots, x_{nn}]$ is generated by the n^2 entries of the matrix

$$\begin{bmatrix} x_{11} & \cdots & x_{1n} \\ \vdots & & \vdots \\ x_{n1} & \cdots & x_{nn} \end{bmatrix}^r$$

For example if $r = m = 2$, then the ideal is generated by the entries of the matrix

$$\begin{bmatrix} x_1 & x_2 \\ x_3 & x_4 \end{bmatrix}^2 = \begin{bmatrix} x_1^2 + x_2 x_3 & x_2(x_1 + x_4) \\ x_3(x_1 + x_4) & x_4^2 + x_2 x_3 \end{bmatrix}$$

That is, the ideal with generators

$$I_R = (x_1^2 + x_2 x_3, x_2(x_1 + x_4), x_3(x_1 + x_4), (x_1 - x_4)(x_1 + x_4))$$

The variety $\mathbb{V}(I_R) \hookrightarrow M_2$ consists of all matrices A such that $A^2 = 0$. Conjugating A to an upper triangular form we see that the eigenvalues of A must be zero, hence

$$rep_2 \ \mathbb{C}[x]/(x^2) = \mathcal{O}(\begin{bmatrix} 0 & 1 \\ 0 & 0 \end{bmatrix}) \cup \mathcal{O}(\begin{bmatrix} 0 & 0 \\ 0 & 0 \end{bmatrix})$$

and we have seen that this variety is a cone with top the zero matrix and defining equations

$$\mathbb{V}(x_1 + x_4, x_1^2 + x_2 x_3)$$

and we see that I_R is properly contained in this ideal. Still, we have that

$$rad(I_R) = (x_1 + x_4, x_1^2 + x_3 x_4)$$

for an easy computation shows that $\overline{x_1 + x_4}^3 = 0 \in \mathbb{C}[x_1, x_2, x_3, x_4]/I_R$. Therefore, even in the easiest of examples, the representation variety does not have to be reduced.

For the general case, observe that when J is a Jordan block of size d with eigenvalue zero an easy calculation shows that

$$
J^{d-1} = \begin{bmatrix} 0 & \cdots & 0 & d-1 \\ & \ddots & & 0 \\ & & \ddots & \vdots \\ & & & 0 \end{bmatrix} \quad \text{and} \quad J^d = \begin{bmatrix} 0 & \cdots \cdots & 0 \\ \vdots & & \vdots \\ \vdots & & \vdots \\ 0 & \cdots \cdots & 0 \end{bmatrix}
$$

Therefore, we see that the representation variety $rep_n \ \mathbb{C}[x]/(x^r)$ is the union of all conjugacy classes of matrices having 0 as only eigenvalue and all of which Jordan blocks have size $\leq r$. Expressed in module theoretic terms, any n-dimensional $R = \mathbb{C}[x]/(x^r)$-module M is isomorphic to a direct sum of indecomposables

$$
M = I_1^{\oplus e_1} \oplus I_2^{\oplus e_2} \oplus \ldots \oplus I_r^{\oplus e_r}
$$

where I_j is the unique indecomposable j-dimensional R-module (corresponding to the Jordan block of size j). Of course, the multiplicities e_i of the factors must satisfy the equation

$$
e_1 + 2e_2 + 3e_3 + \ldots + re_r = n
$$

In M we can consider the subspaces for all $1 \leq i \leq r - 1$

$$
M_i = \{ m \in M \mid x^i.m = 0 \}
$$

the dimension of which can be computed knowing the powers of Jordan blocks (observe that the dimension of M_i is equal to $n - \text{rank}(A^i)$)

$$
t_i = dim_{\mathbb{C}} \ M_i = e_1 + 2e_2 + \ldots (i-1)e_i + i(e_i + e_{i+1} + \ldots + e_r)
$$

Observe that giving n and the $r - 1$-tuple $(t_1, t_2, \ldots, t_{n-1})$ is the same as giving the multiplicities e_i because

$$
\begin{cases}
2t_1 & = t_2 + e_1 \\
2t_2 & = t_3 + t_1 + e_2 \\
2t_3 & = t_4 + t_2 + e_3 \\
\quad \vdots \\
2t_{n-2} & = t_{n-1} + t_{n-3} + e_{n-2} \\
2t_{n-1} & = n + t_{n-2} + e_{n-1} \\
n & = t_{n-1} + e_n
\end{cases}
$$

Let n-dimensional $\mathbb{C}[x]/(x^r)$-modules M and M' (or associated matrices A and A') be determined by the $r-1$-tuples $(t_1, \ldots, t_{r-1})$, respectively, $(t'_1, \ldots, t'_{r-1})$ then we have that

$$\mathcal{O}(A') \hookrightarrow \overline{\mathcal{O}(A)} \quad \text{if and only if} \quad t_1 \leq t'_1, t_2 \leq t'_2, \ldots, t_{r-1} \leq t'_{r-1}$$

Therefore, we have an inverse order isomorphism between the orbits in $rep_n(\mathbb{C}[x]/(x^r))$ and the $r-1$-tuples of natural numbers $(t_1, \ldots, t_{r-1})$ satisfying the following linear inequalities (which follow from the above system)

$$2t_1 \geq t_2, 2t_2 \geq t_3 + t_1, 2t_3 \geq t_4 + t_2, \ldots, 2t_{n-1} \geq n + t_{n-2}, n \geq t_{n-2}$$

Let us apply this general result in a few easy cases. First, consider $r = 2$, then the orbits in $rep_n \mathbb{C}[x]/(x^2)$ are parameterized by a natural number t_1 satisfying the inequalities $n \geq t_1$ and $2t_1 \geq n$, the multiplicities are given by $e_1 = 2t_1 - n$ and $e_2 = n - t_1$. Moreover, the orbit of the module $M(t'_1)$ lies in the closure of the orbit of $M(t_1)$ whenever $t_1 \leq t'_1$.

That is, if $n = 2k + \delta$ with $\delta = 0$ or 1, then $rep_n \mathbb{C}[x]/(x^2)$ is the union of $k + 1$ orbits and the orbitclosures form a linear order as follows (from big to small)

$$I_1^\delta \oplus I_2^{\oplus k} \quad\text{------}\quad I_1^{\oplus \delta+2} \oplus I_2^{\oplus k-1} \quad\text{------}\quad \cdots \quad\text{------}\quad I_1^{\oplus n}$$

If $r = 3$, orbits in $rep_n \mathbb{C}[x]/(x^3)$ are determined by couples of natural numbers (t_1, t_2) satisfying the following three linear inequalities

$$\begin{cases} 2t_1 & \geq t_2 \\ 2t_2 & \geq n + t_1 \\ n & \geq t_2 \end{cases}$$

For example, for $n = 8$ we obtain the following situation

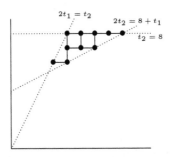

Therefore, $rep_8 \mathbb{C}[x]/(x^3)$ consists of 10 orbits with orbit closure diagram as in figure 2.2 (the nodes represent the multiplicities $[e_1 e_2 e_3]$).

Here we used the equalities $e_1 = 2t_1 - t_2$, $e_2 = 2t_2 - n - t_1$ and $e_3 = n - t_2$. For general n and r this result shows that $rep_n \mathbb{C}[x]/(x^r)$ is the closure of the orbit of the module with decomposition

$$M_{gen} = I_r^{\oplus e} \oplus I_s \quad \text{if} \quad n = er + s$$

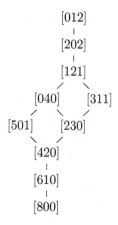

FIGURE 2.2:   Orbit closures in $reps\ \mathbb{C}[x]/(x^3)$.

⬚

We are now in a position to give the promised examples of affine GL_n-schemes having the same witness algebra.

Example 2.14

Consider the action of GL_n on M_n by conjugation and take a nilpotent matrix A. All eigenvalues of A are zero, so the conjugacy class of A is fully determined by the sizes of its Jordan blocks. These sizes determine a partition $\lambda(A) = (\lambda_1, \lambda_2, \ldots, \lambda_k)$ of n with $\lambda_1 \geq \lambda_2 \geq \ldots \geq \lambda_k$. Moreover, we have given an algorithm to determine whether an orbit $\mathcal{O}(B)$ of another nilpotent matrix B is contained in the orbit closure $\overline{\mathcal{O}(A)}$, the criterium being that

$$\mathcal{O}(B) \subset \overline{\mathcal{O}(A)} \Longleftrightarrow \lambda(B)^* \geq \lambda(A)^*$$

where λ^* denotes the dual partition. We see that the witness algebra of $\overline{\mathcal{O}(A)}$ is equal to

$$M_n(\mathbb{C}[\overline{\mathcal{O}(A)}])^{GL_n} = \mathbb{C}[X]/(X^k)$$

where k is the number of columns of the Young diagram $\lambda(A)$.

Hence, the orbit closures of nilpotent matrices such that their associated Young diagrams have equal number of columns have the same witness algebras. For example, if $n = 4$ then the closures of the orbits corresponding to

 and

have the same witness algebra, although the closure of the second is a proper closed subscheme of the closure of the first.

Recall the orbit closure diagram of conjugacy classes of nilpotent 8×8 matrices given by the Gerstenhaber-Hesselink theorem. In the picture below, the closures of orbits corresponding to connected nodes of the same color have the same witness algebra.

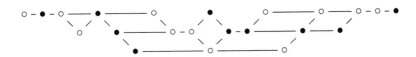

$\square$

2.8   The real moment map

In this section we will give another interpretation of the algebraic quotient variety $\texttt{triss}_n A$ with methods coming from symplectic geometry. We have an involution

$$GL_n \xrightarrow{\;i\;} GL_n \qquad \text{defined by} \qquad g \longrightarrow (g^*)^{-1}$$

where A^* is the *adjoint matrix* of g, that is, the conjugate transpose

$$M = \begin{bmatrix} m_{11} & \cdots & m_{1n} \\ \vdots & & \vdots \\ m_{n1} & \cdots & m_{nn} \end{bmatrix} \qquad M^* = \begin{bmatrix} \overline{m_{11}} & \cdots & \overline{m_{n1}} \\ \vdots & & \vdots \\ \overline{m_{1n}} & \cdots & \overline{m_{nn}} \end{bmatrix}$$

The *real points* of this involution, that is

$$(GL_n)^i = \{g \in GL_n \mid g = (g^*)^{-1}\} = U_n = \{u \in GL_n \mid uu^* = \mathbb{1}_n\}$$

is the *unitary group* . On the level of Lie algebras, the involution i gives rise to the linear map

$$M_n \xrightarrow{\;di\;} M_n \qquad \text{defined by} \qquad M \longrightarrow -M^*$$

corresponding to the fact that the Lie algebra of the unitary group, that is, the kernel of di, is the space of *skew-Hermitian matrices*

$$Lie \; U_n = \{M \in M_n \mid M = -M^*\} = iHerm_n$$

Consider the standard *Hermitian inproduct* on M_n defined by

$$(A, B) = tr(A^*B) \qquad \text{which satisfies} \qquad \begin{cases} (cA, B) & = \bar{c}(A, B) \\ (A, cB) & = c(A, B) \\ (B, A) & = \overline{(A, B)} \end{cases}$$

As a subgroup of GL_n, U_n acts on M_n by conjugation and because $(uAu^*, uBu^*) = tr(uA^*u^*uBu^*) = tr(A^*B)$, the inproduct is invariant under the U_n-action. The action of U_n on M_n induces an action of $Lie\ U_n$ on M_n given for all $h \in Lie\ U_n$ and $M \in M_n$

$$h.M = hM + Mh^* = hM - Mh$$

Using this action, we define the *real moment map* μ for the action of U_n on M_n as the map from M_n to the linear dual of the Lie algebra

$$M_n \xrightarrow{\ \mu\ } (iLie\ U_n)^* \qquad M \longrightarrow (h \mapsto i(h.M, M))$$

We will identify the inverse image of the zero map $\underline{0} : Lie\ U_n \longrightarrow 0$ under μ. Because

$$\begin{aligned} (h.M, M) &= tr((h.M - M.h)^*M) \\ &= tr(M^*h^*M - h^*M^*M) \\ &= tr(h^*(MM^* - M^*M)) \end{aligned}$$

and using the nondegeneracy of the *Killing form* on $Lie\ U_n$ we have the identification

$$\mu^{-1}(\underline{0}) = \{M \in M_n \mid MM^* = M^*M\} = Nor_n$$

the space of *normal matrices* . Alternatively, we can define the real moment map to be determined by

$$M_n \xrightarrow{\ \mu_{\mathbb{R}}\ } Lie\ U_n \qquad M \longrightarrow i(MM^* - M^*M) = i[M, M^*]$$

Recall that a matrix $M \in M_n(\mathbb{C})$ is said to be *normal* if its commutes with its adjoint. For example, diagonal matrices are normal as are unitary matrices. Further, it is clear that if M is normal and u unitary, then the conjugated matrix $uMu^{-1} = uMu^*$ is again a normal matrix, that is, we have an action of the compact Lie group U_n on the subset $Nor_n \hookrightarrow M_n(\mathbb{C})$ of normal matrices. We recall the proof of the following classical result.

THEOREM 2.9
Every U_n orbit in Nor_n contains a diagonal matrix. This gives a natural one-to-one correspondence

$$\mu^{-1}(\underline{0})/U_n = Nor_n/U_n \longleftrightarrow M_n/GL_n$$

between the U_n-orbits in Nor_n and the closed GL_n-orbits in M_n.

PROOF Equip $\mathbb{C}^n$ with the standard Hermitian form, that is

$$\langle v, w \rangle = \overline{v}^\tau . w = \overline{v_1}w_1 + \ldots + \overline{v_n}w_n$$

Take a nonzero eigenvector v of $M \in Nor_n$ and normalize it such that $\langle v, v \rangle = 1$. Extend $v = v_1$ to an orthonormal basis $\{v_1, \ldots, v_n\}$ of $\mathbb{C}^n$ and let u be the base change matrix from the standard basis. With respect to the new basis, the linear map determined by M and M^* are represented by the normal matrices

$$M_1 = uMu^* = \begin{bmatrix} a_{11} & a_{12} & \ldots & a_{1n} \\ 0 & a_{22} & \ldots & a_{2n} \\ \vdots & \vdots & & \vdots \\ 0 & a_{n2} & \ldots & a_{nn} \end{bmatrix} \qquad M_1^* = uM^*u^* = \begin{bmatrix} \overline{a_{11}} & 0 & \ldots & 0 \\ \overline{a_{12}} & \overline{a_{22}} & \ldots & \overline{a_{n2}} \\ \vdots & \vdots & & \vdots \\ \overline{a_{1n}} & \overline{a_{2n}} & \ldots & \overline{a_{nn}} \end{bmatrix}$$

Because M is normal, so is M_1. The left-hand corner of $M_1^* M_1$ is $a_{11}\overline{a_{11}}$ whereas that of $M_1 M_1^*$ is $a_{11}\overline{a_{11}} + a_{12}\overline{a_{12}} + \ldots + a_{1n}\overline{a_{1n}}$, whence

$$a_{12}\overline{a_{12}} + \ldots + a_{1n}\overline{a_{1n}} = 0$$

but as all $a_{1i}\overline{a_{1i}} = \parallel a_{1i} \parallel \geq 0$, this implies that all $a_{1i} = 0$, whence

$$M_1 = \begin{bmatrix} a_{11} & 0 & \ldots & 0 \\ 0 & a_{22} & \ldots & a_{2n} \\ \vdots & \vdots & & \vdots \\ 0 & a_{n2} & \ldots & a_{nn} \end{bmatrix}$$

and induction finishes the claim. Because permutation matrices are unitary we see that the diagonal entries are determined up to permutation, so every U_n-orbit determines a unique conjugacy class of semisimple matrices, that is, a closed GL_n-orbit in M_n. ⊔

We will generalize this classical result to m-tuples of $n \times n$ matrices, M_n^m, and then by restriction to trace preserving representation varieties. Take $A = (A_1, \ldots, A_m)$ and $B = (B_1, \ldots, B_m)$ in M_n^m and define an Hermitian inproduct on M_n^m by

$$(A, B) = tr(A_1^* B_1 + \ldots + A_m^* B_m)$$

which is again invariant under the action of U_n by simultaneous conjugation on M_n^m. The induced action of $Lie\ U_n$ on M_n^m is given by

$$h.A = (hA_1 - A_1 h, \ldots, hA_m - A_m h)$$

This allows us to define the *real moment map* μ for the action of U_n on M_n^m to be the assignment

$$M_n^m \xrightarrow{\mu} (iLie\ U_n)^* \qquad A \longrightarrow (h \mapsto i(h.A, A))$$

and again using the nondegeneracy of the Killing form on $Lie\ U_n$ we have the identification

$$\mu^{-1}(\underline{0}) = \{A \in M_n^m \mid \sum_{i=1}^{m}(A_i A_i^* - A_i^* A_i) = 0\}$$

Again, the real moment map is determined by

$$M_n^m \xrightarrow{\mu_{\mathbb{R}}} Lie\ U_n \qquad A = (A_1, \ldots, A_m) \mapsto i[A, A^*] = i\sum_{j=1}^{m}[A_j, A_j^*]$$

We will show that there is a natural one-to-one correspondence between U_n-orbits in the set $\mu^{-1}(\underline{0})$ and closed GL_n-orbits in M_n^m. We first consider the properties of the real valued function p_A defined as the norm on the orbit of any $A \in M_n^m$

$$GL_n \xrightarrow{p_A} \mathbb{R}_+ \qquad g \longrightarrow \|g.A\|^2$$

Because the Hermitian inproduct is invariant under U_n we have $p_A(ug) = p_A(g)$ for any $u \in U_n$. If $Stab(A)$ denotes the stabilizer subgroup of $A \in GL_n$, then for any $s \in Stab(A)$ we also have $p_A(gs) = p_A(g)$ hence p_A is constant along $U_n g Stab(A)$-cosets. We aim to prove that the critical points of p_A are minima and that the minimum is attained if and only if $\mathcal{O}(A)$ is a closed GL_n-orbit.

Consider the restriction of p_A to the maximal torus $T_n \hookrightarrow GL_n$ of invertible diagonal matrices. Then, $T_n \cap U_n = K = U_1 \times \ldots \times U_1$ is the subgroup

$$K = \{ \begin{bmatrix} k_1 & & 0 \\ & \ddots & \\ 0 & & k_n \end{bmatrix} \mid \forall i : |k_i| = 1 \}$$

The action by conjugation of T_n on M_n^m decomposes this space into *weight spaces*

$$M_n^m = M_n^m(0) \oplus \bigoplus_{i,j=1}^{n} M_n^m(\pi_i - \pi_j)$$

where $M_n^m(\pi_i - \pi_j) = \{A \in M_n^m \mid diag(t_1, \ldots, t_n).A = t_i t_j^{-1} A\}$. It follows from the definition of the Hermitian inproduct on M_n^m that the different weight spaces are orthogonal to each other. We decompose $A \in M_n^m$ into eigenvectors for the T_n-action as

$$A = A(0) + \sum_{i,j=1}^{n} A(i,j) \qquad \text{with} \qquad \begin{cases} A(0) & \in M_n^m(0) \\ A(i,j) & \in M_n^m(\pi_i - \pi_j) \end{cases}$$

With this convention we have for $t = diag(t_1, \ldots, t_n) \in T_n$ that

$$p_A(t) = \|A(0) + \sum_{i,j=1}^{n} t_i t_j^{-1} A(i,j)\|^2$$

$$= \|A(0)\|^2 + \sum_{i,j=1}^{n} t_i^2 t_j^{-2} \|A(i,j)\|^2$$

where the last equality follows from the orthogonality of the different weight spaces. Further, remark that the stabilizer subgroup $Stab_T(A)$ of A in T can be identified with

$$Stab_T(A) = \{t = diag(t_1, \ldots, t_n) \mid t_i = t_j \text{ if } A(i,j) \neq 0\}$$

As before, p_A induces a function on double cosets $K \backslash T_n / Stab_T(A)$, in particular p_M determines a real valued function on $K \backslash T_n \simeq \mathbb{R}^n$ (the isomorphism is given by the map $diag(t_1, \ldots, t_n) \xrightarrow{log} (log \, |t_1|, \ldots, log \, |t_n|)$). That is

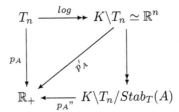

where the function p_M' is the *special function*

$$p_A'(r_1, \ldots, r_n) = e^{2log \, \|A(0)\|} + \sum_{i,j: A(i,j) \neq 0}^{n} e^{2log \, \|A(i,j)\| + 2x_i - 2x_j}$$

and where $K \backslash T_n / Stab_T(A)$ is the quotient space of $\mathbb{R}^n$ by the subspace V_A which is the image of $Stab_T(A)$ under log

$$V_A = \sum_{i: \nexists A(i,j) \neq 0} \mathbb{R}e_i + \sum_{i,j: A(i,j) \neq 0} \mathbb{R}(e_i - e_j)$$

where e_i are the standard basis vectors of $\mathbb{R}^n$. Let $\{i_1, \ldots, i_k\}$ be the minimal elements of the nonempty equivalence classes induced by the relation $i \sim j$ iff $A(i,j) \neq 0$, then

$$\begin{cases} K \backslash T_n / Stab_T(A) \simeq \sum_{j=1}^{k} \mathbb{R}e_{i_j} \\ p_A''(y_1, \ldots, y_k) = c_0 + \sum_{j=1}^{k} (\sum_{l(j)} c_{l(j)} e^{a_{l(j)} y_j}) \end{cases}$$

for certain positive real numbers $c_0, c_{l(j)}$ and real numbers $a_{l(j)}$. But then, elementary calculus shows that the $k \times k$ matrix

$$\begin{bmatrix} \frac{\partial^2 p_A''}{\partial y_1 \partial y_1}(m) & \cdots & \frac{\partial^2 p_A''}{\partial y_1 \partial y_k}(m) \\ \vdots & & \vdots \\ \frac{\partial^2 p_A''}{\partial y_k \partial y_1}(m) & \cdots & \frac{\partial^2 p_A''}{\partial y_k \partial y_k}(m) \end{bmatrix}$$

is a positive definite diagonal matrix in every point $m \in \mathbb{R}^k$. That is, p_A" is a strictly convex *Morse function* and if it has a critical point m_0 (that is, if all $\frac{\partial p_A"}{\partial y_i}(m_0) = 0$), it must be a unique minimum. Lifting this information from the double coset space $K \backslash T_n / Stab_T(A)$ to T_n we have proved

PROPOSITION 2.15

Let T_n be the maximal torus of invertible diagonal matrices in GL_n and consider the restriction of the function $GL_n \xrightarrow{\ p_A\ } \mathbb{R}_+$ to T_n for $A \in M_n^m$, then

1. *Any critical point of p_A is a point where p_A obtains its minimal value.*

2. *If p_A obtains a minimal value, then*

   - *the set V where p_A obtains this minimum consists of a single $K - Stab_T(A)$ coset in T_n and is connected.*

   - *the second order variation of p_A at a point of V in any direction not tangent to V is positive.*

The same proof applies to all maximal tori T of GL_n which are defined over $\mathbb{R}$. Recall the *Cartan decomposition* of GL_n which we proved before theorem 2.4: any $g \in GL_n$ can be written as $g = udu'$ where $u, u' \in U_n$ and d is a diagonal matrix with positive real entries. Using this fact we can now extend the above proposition to GL_n.

THEOREM 2.10

Consider the function $GL_n \xrightarrow{\ p_A\ } \mathbb{R}_+$ for $A \in M_n^m$.

1. *Any critical point of p_A is a point where p_A obtains its minimal value.*

2. *If p_A obtains its minimal value, it does so on a single $U_n - Stab(A)$-coset.*

PROOF　　(1): Because for any $h \in GL_n$ we have that $p_{h.A}(g) = p_A(gh)$ we may assume that $\mathbb{1}_n$ is the critical point of p_A. We have to prove that $p_A(g) \geq p_A(\mathbb{1}_n)$ for all $g \in GL_n$. By the Cartan decomposition $g = udu'$ whence $g = u"t$ where $u" = uu' \in U_n$ and $t = u'^{-1}du' \in T$ a maximal torus of GL_n defined over $\mathbb{R}$. Because the Hermitian inproduct is invariant under U_n we have that $p_A(g) = p_A(t)$. Because $\mathbb{1}_n$ is a critical point for the restriction of p_A to T we have by proposition 2.15 that $p_A(t) \geq p_A(\mathbb{1}_n)$, proving the claim.
　　(2): Because for any $h \in GL_n$, $p_{h.A}(g) = p_A(gh)$ and $Stab(h.A) = hStab(A)h^{-1}$ we may assume that p_A obtains its minimal value at $\mathbb{1}_n$. If V denotes the subset of GL_n where p_A obtains its minimal value we then have that $U_nStab(A) \subset V$ and we have to prove the reverse inclusion. Assume $g \in V$ and write as before $g = u"t$ with $u" \in U_n$ and $t \in T$ a maximal

torus defined over $\mathbb{R}$. Then, by unitary invariance of the inproduct, t is a point of T where the restriction of p_A to T obtains its minimal value $p_A(\mathbb{1}_n)$. By proposition 2.15 we conclude that $t \in K_T Stab_T(A)$ where $K_T = U_n \cap T$. But then,

$$V \subset U_n\left(\bigcup_T K_T Stab_T(A)\right) \subset U_n Stab(A)$$

where T runs over all maximal tori of GL_n which are defined over $\mathbb{R}$, finishing the proof. ▯

PROPOSITION 2.16
The function $p_A : GL_n \longrightarrow \mathbb{R}_+$ obtains a minimal value if and only if $\mathcal{O}(A)$ is a closed orbit in M_n^m, that is, determines a semisimple representation.

PROOF   If $\mathcal{O}(A)$ is closed then p_A clearly obtains a minimal value. Conversely, assume that $\mathcal{O}(A)$ is not closed, that is, A does not determine a semisimple n-dimensional representation M of $\mathbb{C}\langle x_1, \ldots, x_m \rangle$. By choosing a basis in M (that is, possibly going to another point in the orbit $\mathcal{O}(A)$) we have a one-parameter subgroup $\mathbb{C}^* \xrightarrow{\lambda} T_n \hookrightarrow GL_n$ corresponding to the Jordan-Hölder filtration of M with $\lim_{t\to 0} \lambda(t)A = B$ with B corresponding to the semisimplification of M. Now consider the restriction of p'_A to $U_1 \backslash \mathbb{C}^* \simeq \mathbb{R}$, then as before we can write it uniquely in the form

$$p'_A(x) = \sum_i a_i e^{l_i x} \qquad a_i > 0, \quad l_1 < l_2 < \ldots < l_z$$

for some real numbers l_i and some z. Because the above limit exists, the limit

$$\lim_{x\to-\infty} p'_A(x) \in \mathbb{R}$$

and hence none of the l_i are negative. Further, because $\mathcal{O}(A) \neq \mathcal{O}(B)$ at least one of the l_i must be positive. Therefore, p'_A is a strictly increasing function on $\mathbb{R}$ whence never obtains a minimal value, whence neither does p_A. ▯

Finally, we have to clarify the connection between the function p_A and the *real moment map*

$$\begin{cases} M_n^m \xrightarrow{\mu} (Lie\ U_n)^* & A \longrightarrow (h \mapsto (h.A, A)) \\ M_n^n \xrightarrow{\mu_\mathbb{R}} Lie\ U_n & A \longrightarrow i[A, A^*] \end{cases}$$

Assume $A \in M_n^m$ is such that p_A has a critical point, which we may assume to be $\mathbb{1}_n$ by an argument as in the proof of theorem 2.10. Then, the differential in $\mathbb{1}_n$

$$(dp_A)_{\mathbb{1}_n} : M_n = T_{\mathbb{1}_n} GL_n \longrightarrow \mathbb{R} \qquad \text{satisfies} \qquad (dp_A)_{\mathbb{1}_n}(h) = 0 \quad \forall h \in M_n$$

Let us work out this differential

$$p_A(\mathbb{1}_n) + \epsilon(dp_A)_{\mathbb{1}_n}(h) = tr((A^* + \epsilon(A^*h^* - h^*A^*)(A + \epsilon(hA - Ah))$$
$$= tr(A^*A) + \epsilon tr(A^*hA - A^*Ah + A^*h^*A - h^*A^*A)$$
$$= tr(A^*A) + \epsilon tr((AA^* - A^*A)(h - h^*))$$

But then, vanishing of the differential for all $h \in M_n$ is equivalent by the nondegeneracy of the Killing form on *Lie* U_n to

$$AA^* - A^*A = \sum_{i=1}^m A_i A_i^* - A_i^* A_i = 0$$

that is, to $A \in \mu_{\mathbb{R}}^{-1}(\underline{0})$. This concludes the proof of the main result on the real moment map for M_n^m.

THEOREM 2.11
There are natural one-to-one correspondences between

1. *isomorphism classes of semisimple n-dimensional representations of* $\mathbb{C}\langle x_1, \ldots, x_m \rangle$,

2. *closed GL_n-orbits in M_n^m,*

3. *U_n-orbits in the subset* $\mu_{\mathbb{R}}^{-1}(\underline{0}) = \{A \in M_n^m \mid \sum_{i=1}^m [A_i, A_i^*] = 0\}$.

Let $A \in \mathsf{alg@n}$ be an affine Cayley-Hamilton algebra of degree n, then we can embed the reduced variety of $\mathsf{trep}_n A$ in M_n^m and obtain the following as a consequence

THEOREM 2.12
For $A \in \mathsf{alg@n}$, there are natural one-to-one correspondences between

1. *isomorphism classes of semisimple n-dimensional trace preserving representations of A,*

2. *closed GL_n-orbits in the representation variety $\mathsf{trep}_n A$,*

3. *U_n-orbits in the intersection $\mathsf{trep}_n A \cap \mu_{\mathbb{R}}^{-1}(\underline{0})$.*

References

The generalization of the Hilbert criterium, theorem 2.2 is due to D. Birkes [9]. The connection between semisimple representations and closed orbits is due to M. Artin [2] . The geometric reconstruction result follows from theorem 1.17 and is due to C. Procesi [87] . The results on the real moment map are due to G. Kempf and L. Ness [53] . The treatment of the Hilbert criterium and invariant theory follows the textbook of H.P. Kraft [63] , that of the Gerstenhaber-Hesselink result owes to the exposition of M. Hazewinkel in [44].

Chapter 3

Etale Technology

Etale topology was introduced in algebraic geometry to bypass the coarseness of the Zariski topology for classification problems. Let us give an elementary example: the local classification of smooth varieties in the Zariski topology is a hopeless task, whereas in the étale topology there is just one local type of smooth variety in dimension d, namely, affine d-space $\mathbb{A}^d$. A major theme of this book is to generalize this result to `noncommutative geometry@n`.

Etale cohomology groups are used to classify *central simple algebras* over function fields of varieties. *Orders* in such central simple algebras (over the central structure sheaf) are an important class of Cayley-Hamilton algebras.

Over the years, one has tried to construct a suitable class of *smooth orders* that allows an étale local description. But, except in the case of curves and surfaces, no such classification is known, say, for orders of finite global dimension. In this book we introduce the class of *Cayley-smooth orders*, which does allow an étale local description in arbitrary dimensions. In this chapter we will lay the foundations for this classification by investigating étale slices of representation varieties at semisimple representations. In chapter 5 we will then show that this local structure is determined by a combinatorial gadget: a (marked) *quiver setting*.

3.1 Etale topology

A closed subvariety $X \hookrightarrow \mathbb{C}^m$ can be equipped with the *Zariski topology* or with the much finer *analytic topology*. A major disadvantage of the coarseness of the Zariski topology is the failure to have an *implicit function* theorem in algebraic geometry. Etale morphisms are introduced to bypass this problem.

We will define étale morphisms that determine the *étale topology* . This is no longer a usual topology determined by subsets, but rather a *Grothendieck topology* determined by *covers* .

DEFINITION 3.1   *A finite morphism* $A \xrightarrow{f} B$ *of commutative $\mathbb{C}$-algebras is said to be* étale *if and only if* $B = A[t_1, \ldots, t_k]/(f_1, \ldots, f_k)$ *such* *that the* Jacobian matrix

$$\begin{bmatrix} \frac{\partial f_1}{\partial t_1} & \cdots & \frac{\partial f_1}{\partial t_k} \\ \vdots & & \vdots \\ \frac{\partial f_k}{\partial t_1} & \cdots & \frac{\partial f_k}{\partial t_k} \end{bmatrix}$$

has a determinant which is a unit in B.

Recall that by spec A we denote the *prime ideal spectrum* or the *affine scheme* of a commutative $\mathbb{C}$-algebra A (even when A is not affine as a $\mathbb{C}$-algebra). That is, spec A is the set of all *prime ideals* of A equipped with the *Zariski topology* . The open subsets are of the form

$$\mathbb{X}(I) = \{P \in \text{spec } A \mid I \not\subset P\}$$

for any ideal $I \lhd A$. If A is an affine $\mathbb{C}$-algebra, the points of the corresponding affine variety correspond to *maximal ideals* of A and the induced Zariski topology coincides with the one introduced before. In this chapter, however, not all $\mathbb{C}$-algebras will be affine.

Example 3.1

Consider the morphism $\mathbb{C}[x, x^{-1}] \hookrightarrow \mathbb{C}[x, x^{-1}][\sqrt[n]{x}]$ and the induced map on the affine schemes

$$\text{spec } \mathbb{C}[x, x^{-1}][\sqrt[n]{x}] \xrightarrow{\psi} \text{spec } \mathbb{C}[x, x^{-1}] = \mathbb{C} - \{0\}$$

Clearly, every point $\lambda \in \mathbb{C} - \{0\}$ has exactly n preimages $\lambda_i = \zeta^i \sqrt[n]{\lambda}$. Moreover, in a neighborhood of λ_i, the map ψ is a diffeomorphism. Still, we do not have an inverse map in algebraic geometry as $\sqrt[n]{x}$ is not a polynomial map. However, $\mathbb{C}[x, x^{-1}][\sqrt[n]{x}]$ is an étale extension of $\mathbb{C}[x, x^{-1}]$. In this way étale morphisms can be seen as an algebraic substitute for the failure of an inverse function theorem in algebraic geometry.   ▯

PROPOSITION 3.1

Etale morphisms satisfy "sorite", that is, they satisfy the commutative diagrams of figure 3.1. In these diagrams, et denotes an étale morphism, f.f. denotes a faithfully flat morphism and the dashed arrow is the étale morphism implied by "sorite."

With these properties we can define a *Grothendieck topology* on the collection of all étale morphisms.

DEFINITION 3.2   *The* étale site *of A, which we will denote by* A_{et}*, is the category with*

- *objects: the étale extensions* $A \xrightarrow{f} B$ *of A*

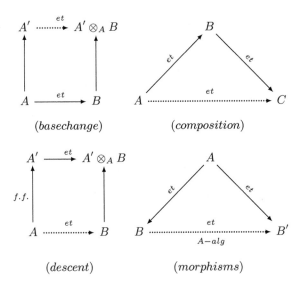

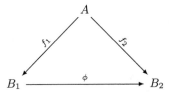

FIGURE 3.1: Sorite for étale morphisms.

- *morphisms: compatible A-algebra morphisms*

By proposition 3.1 all morphisms in $\mathsf{A_{et}}$ are étale. We can turn $\mathsf{A_{et}}$ into a Grothendieck topology by defining

- *cover: a collection $\mathcal{C} = \{B \xrightarrow{f_i} B_i\}$ in $\mathsf{A_{et}}$ such that*

$$\mathsf{spec}\ B = \cup_i\ Im\ (\mathsf{spec}\ B_i \xrightarrow{f}\ \mathsf{spec}\ B\ )$$

DEFINITION 3.3 *An étale presheaf of groups on $\mathsf{A_{et}}$ is a functor*

$$\mathbb{G} : \mathsf{A_{et}} \longrightarrow \mathsf{groups}$$

In analogy with usual (pre)sheaf notation we denote for each

- *object $B \in \mathsf{A_{et}}$ the global sections $\Gamma(B, \mathbb{G}) = \mathbb{G}(B)$*

- *morphism $B \xrightarrow{\phi} C$ in $\mathsf{A_{et}}$ the restriction map $Res^B_C = \mathbb{G}(\phi) : \mathbb{G}(B) \longrightarrow \mathbb{G}(C)$ and $g \mid C = \mathbb{G}(\phi)(g)$.*

An étale presheaf $\mathbb{G}$ *is an* étale sheaf *provided for every* $B \in \mathsf{A}_{et}$ *and every* cover $\{B \longrightarrow B_i\}$ *we have exactness of the* equalizer diagram

$$0 \longrightarrow \mathbb{G}(B) \longrightarrow \prod_i \mathbb{G}(B_i) \rightrightarrows \prod_{i,j} \mathbb{G}(B_i \otimes_B B_j)$$

Example 3.2 Constant sheaf
If G is a group, then

$$\mathbb{G} : \mathsf{A}_{et} \longrightarrow \mathbf{groups} \quad B \mapsto G^{\oplus \pi_0(B)}$$

is a sheaf where $\pi_0(B)$ is the number of connected components of $\mathbf{spec}\ B$. ☐

Example 3.3 Multiplicative group $\mathbb{G}_m$
The functor

$$\mathbb{G}_m : \mathsf{A}_{et} \longrightarrow \mathbf{groups} \quad B \mapsto B^*$$

is a sheaf on A_{et}. ☐

A sequence of sheaves of Abelian groups on A_{et} is said to be exact

$$\mathbb{G}' \xrightarrow{f} \mathbb{G} \xrightarrow{g} \mathbb{G}''$$

if for every $B \in \mathsf{A}_{et}$ and $s \in \mathbb{G}(B)$ such that $g(s) = 0 \in \mathbb{G}''(B)$ there is a cover $\{B \longrightarrow B_i\}$ in A_{et} and sections $t_i \in \mathbb{G}'(B_i)$ such that $f(t_i) = s \mid B_i$.

Example 3.4 Roots of unity μ_n
We have a sheaf morphism

$$\mathbb{G}_m \xrightarrow{(-)^n} \mathbb{G}_m$$

and we denote the kernel with μ_n. As A is a $\mathbb{C}$-algebra we can identify μ_n with the constant sheaf $\mathbb{Z}_n = \mathbb{Z}/n\mathbb{Z}$ via the isomorphism $\zeta^i \mapsto i$ after choosing a primitive n-th root of unity $\zeta \in \mathbb{C}$. ☐

LEMMA 3.1
The Kummer sequence of sheaves of Abelian groups

$$0 \longrightarrow \mu_n \longrightarrow \mathbb{G}_m \xrightarrow{(-)^n} \mathbb{G}_m \longrightarrow 0$$

is exact on A_{et} *(but not necessarily on* $\mathbf{spec}\ A$ *with the Zariski topology).*

PROOF   We only need to verify surjectivity. Let $B \in \mathsf{A}_{et}$ and $b \in \mathbb{G}_m(B) = B^*$. Consider the étale extension $B' = B[t]/(t^n - b)$ of B, then b has an n-th root over in $\mathbb{G}_m(B')$. Observe that this n-th root does not have to belong to $\mathbb{G}_m(B)$. ☐

If $\mathfrak{p}$ is a prime ideal of A we will denote with $\mathbf{k}_\mathfrak{p}$ the algebraic closure of the field of fractions of $A/\mathfrak{p}$. An *étale neighborhood* of $\mathfrak{p}$ is an étale extension $B \in A_{et}$ such that the diagram below is commutative

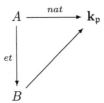

The analog of the localization $A_\mathfrak{p}$ for the étale topology is the *strict Henselization*

$$A_\mathfrak{p}^{sh} = \varinjlim B$$

where the limit is taken over all étale neighborhoods of $\mathfrak{p}$.

Recall that a local algebra L with maximal ideal m and residue map $\pi : L \longrightarrow L/m = k$ is said to be *Henselian* if the following condition holds. Let $f \in L[t]$ be a monic polynomial such that $\pi(f)$ factors as $g_0.h_0$ in $k[t]$, then f factors as $g.h$ with $\pi(g) = g_0$ and $\pi(h) = h_0$. If L is Henselian then tensoring with k induces an equivalence of categories between the étale A-algebras and the étale k-algebras.

An Henselian local algebra is said to be *strict Henselian* if and only if its residue field is algebraically closed. Thus, a strict Henselian ring has no proper finite étale extensions and can be viewed as a local algebra for the étale topology.

Example 3.5 The algebraic functions $\mathbb{C}\{x_1, \ldots, x_d\}$
Consider the local algebra of $\mathbb{C}[x_1, \ldots, x_d]$ in the maximal ideal $(x_1, \ldots, x_d)$, then the Henselization and strict Henselization are both equal to

$$\mathbb{C}\{x_1, \ldots, x_d\}$$

the ring of *algebraic functions* . That is, the subalgebra of $\mathbb{C}[[x_1, \ldots, x_d]]$ of formal power-series consisting of those series $\phi(x_1, \ldots, x_d)$ which are algebraically dependent on the coordinate functions x_i over $\mathbb{C}$. In other words, those ϕ for which there exists a nonzero polynomial $f(x_i, y) \in \mathbb{C}[x_1, \ldots, x_d, y]$ with $f(x_1, \ldots, x_d, \phi(x_1, \ldots, x_d)) = 0$.

These algebraic functions may be defined implicitly by polynomial equations. Consider a system of equations

$$f_i(x_1, \ldots, x_d; y_1, \ldots, y_m) = 0 \text{ for } f_i \in \mathbb{C}[x_i, y_j] \text{ and } 1 \leq i \leq m$$

Suppose there is a solution in $\mathbb{C}$ with

$$x_i = 0 \text{ and } y_j = y_j^o$$

such that the Jacobian matrix is nonzero

$$det\ (\frac{\partial f_i}{\partial y_j}(0,\dots,0;y_1^o,\dots,y_m^0)) \neq 0$$

Then, the system can be solved uniquely for power series $y_j(x_1,\dots,x_d)$ with $y_j(0,\dots,0) = y_j^o$ by solving inductively for the coefficients of the series. One can show that such implicitly defined series $y_j(x_1,\dots,x_d)$ are algebraic functions and that, conversely, any algebraic function can be obtained in this way.
$\square$

If $\mathbb{G}$ is a sheaf on A_{et} and $\mathfrak{p}$ is a prime ideal of A, we define the *stalk* of $\mathbb{G}$ in $\mathfrak{p}$ to be

$$\mathbb{G}_{\mathfrak{p}} = \varinjlim\ \mathbb{G}(B)$$

where the limit is taken over all étale neighborhoods of $\mathfrak{p}$. One can verify mono- epi- or isomorphisms of sheaves by checking it in all the stalks.

If A is an affine algebra defined over an algebraically closed field, then it suffices to verify it in the maximal ideals of A.

Before we define cohomology of sheaves on A_{et} let us recall the definition of *derived functors* . Let $\mathcal{A}$ be an *Abelian category* . An object I of $\mathcal{A}$ is said to be *injective* if the functor

$$\mathcal{A} \longrightarrow \texttt{abelian}\quad M \mapsto Hom_{\mathcal{A}}(M,I)$$

is exact. We say that $\mathcal{A}$ has enough injectives if, for every object M in $\mathcal{A}$, there is a monomorphism $M \hookrightarrow I$ into an injective object.

If $\mathcal{A}$ has enough injectives and $f : \mathcal{A} \longrightarrow \mathcal{B}$ is a left exact functor from $\mathcal{A}$ into a second Abelian category $\mathcal{B}$, then there is an essentially unique sequence of functors

$$R^i\ f : \mathcal{A} \longrightarrow \mathcal{B}\quad i \geq 0$$

called the *right derived functors* of f satisfying the following properties

- $R^0\ f = f$

- $R^i\ I = 0$ for I injective and $i > 0$

- For every short exact sequence in $\mathcal{A}$

$$0 \longrightarrow M' \longrightarrow M \longrightarrow M" \longrightarrow 0$$

  there are connecting morphisms $\delta^i : R^i\ f(M") \longrightarrow R^{i+1}\ f(M')$ for $i \geq 0$ such that we have a long exact sequence

$$\longrightarrow R^i\ f(M) \longrightarrow R^i\ f(M") \xrightarrow{\delta^i} R^{i+1}\ f(M') \longrightarrow R^{i+1}\ f(M) \longrightarrow$$

- For any morphism $M \longrightarrow N$ there are morphisms $R^i \, f(M) \longrightarrow R^i \, f(N)$ for $i \geq 0$

In order to compute the objects $R^i \, f(M)$ define an object N in $\mathcal{A}$ to be *f-acyclic* if $R^i \, f(M) = 0$ for all $i > 0$. If we have an *acyclic resolution* of M

$$0 \longrightarrow M \longrightarrow N_0 \longrightarrow N_1 \longrightarrow N_2 \longrightarrow \cdots$$

by *f-acyclic* objects N_i, then the objects $R^i \, f(M)$ are canonically isomorphic to the cohomology objects of the complex

$$0 \longrightarrow f(N_0) \longrightarrow f(N_1) \longrightarrow f(N_2) \longrightarrow \cdots$$

One can show that all injectives are f-acyclic and hence that derived objects of M can be computed from an *injective resolution* of M.

Now, let $\mathbf{S}^{ab}(\mathbf{A_{et}})$ be the category of all sheaves of Abelian groups on $\mathbf{A_{et}}$. This is an Abelian category having enough injectives whence we can form right derived functors of left exact functors. In particular, consider the global section functor

$$\Gamma : \mathbf{S}^{ab}(\mathbf{A_{et}}) \longrightarrow \texttt{abelian} \quad \mathbb{G} \mapsto \mathbb{G}(A)$$

which is left exact. The right derived functors of Γ will be called the *étale cohomology functors* and we denote

$$R^i \, \Gamma(\mathbb{G}) = H^i_{et}(A, \mathbb{G})$$

In particular, if we have an exact sequence of sheaves of Abelian groups $0 \longrightarrow \mathbb{G}' \longrightarrow \mathbb{G} \longrightarrow \mathbb{G}" \longrightarrow 0$, then we have a long exact cohomology sequence

$$\cdots \longrightarrow H^i_{et}(A, \mathbb{G}) \longrightarrow H^i_{et}(A, \mathbb{G}") \longrightarrow H^{i+1}_{et}(A, \mathbb{G}') \longrightarrow \cdots$$

If $\mathbb{G}$ is a sheaf of non-Abelian groups (written multiplicatively), we cannot define cohomology groups. Still, one can define a *pointed set* $H^1_{et}(A, \mathbb{G})$ as follows. Take an étale cover $\mathcal{C} = \{A \longrightarrow A_i\}$ of A and define a 1-cocycle for $\mathcal{C}$ with values in $\mathbb{G}$ to be a family

$$g_{ij} \in \mathbb{G}(A_{ij}) \text{ with } A_{ij} = A_i \otimes_A A_j$$

satisfying the cocycle condition

$$(g_{ij} \mid A_{ijk})(g_{jk} \mid A_{ijk}) = (g_{ik} \mid A_{ijk})$$

where $A_{ijk} = A_i \otimes_A A_j \otimes_A A_k$.

Two cocycles g and g' for $\mathcal{C}$ are said to be cohomologous if there is a family $h_i \in \mathbb{G}(A_i)$ such that for all $i, j \in I$ we have

$$g'_{ij} = (h_i \mid A_{ij})g_{ij}(h_j \mid A_{ij})^{-1}$$

This is an equivalence relation and the set of cohomology classes is written as $H^1_{et}(\mathcal{C}, \mathbb{G})$. It is a pointed set having as its distinguished element the cohomology class of $g_{ij} = 1 \in \mathbb{G}(A_{ij})$ for all $i, j \in I$.

We then define the non-Abelian first *cohomology pointed set* as

$$H^1_{et}(A, \mathbb{G}) = \varinjlim \; H^1_{et}(\mathcal{C}, \mathbb{G})$$

where the limit is taken over all étale coverings of A. It coincides with the previous definition in case $\mathbb{G}$ is Abelian.

A sequence $1 \longrightarrow \mathbb{G}' \longrightarrow \mathbb{G} \longrightarrow \mathbb{G}" \longrightarrow 1$ of sheaves of groups on A_{et} is said to be exact if for every $B \in A_{et}$ we have

- $\mathbb{G}'(B) = Ker \; \mathbb{G}(B) \longrightarrow \mathbb{G}"(B)$

- For every $g" \in \mathbb{G}"(B)$ there is a cover $\{B \longrightarrow B_i\}$ in A_{et} and sections $g_i \in \mathbb{G}(B_i)$ such that g_i maps to $g" \mid B_i$.

PROPOSITION 3.2
For an exact sequence of groups on A_{et}

$$1 \longrightarrow \mathbb{G}' \longrightarrow \mathbb{G} \longrightarrow \mathbb{G}" \longrightarrow 1$$

there is associated an exact sequence of pointed sets

$$1 \longrightarrow \mathbb{G}'(A) \longrightarrow \mathbb{G}(A) \longrightarrow \mathbb{G}"(A) \overset{\delta}{\longrightarrow} H^1_{et}(A, \mathbb{G}') \longrightarrow$$

$$\longrightarrow H^1_{et}(A, \mathbb{G}) \longrightarrow H^1_{et}(A, \mathbb{G}") \dashrightarrow H^2_{et}(A, \mathbb{G}')$$

where the last map exists when $\mathbb{G}'$ is contained in the center of $\mathbb{G}$ (and therefore is Abelian whence H^2 is defined).

PROOF   The connecting map δ is defined as follows. Let $g" \in \mathbb{G}"(A)$ and let $\mathcal{C} = \{A \longrightarrow A_i\}$ be an étale covering of A such that there are $g_i \in \mathbb{G}(A_i)$ that map to $g \mid A_i$ under the map $\mathbb{G}(A_i) \longrightarrow \mathbb{G}"(A_i)$. Then, $\delta(g)$ is the class determined by the one cocycle

$$g_{ij} = (g_i \mid A_{ij})^{-1}(g_j \mid A_{ij})$$

with values in $\mathbb{G}'$. The last map can be defined in a similar manner, the other maps are natural and one verifies exactness.   ◻

The main applications of this non-Abelian cohomology to noncommutative algebra is as follows. Let Λ be a not necessarily commutative A-algebra and M an A-module. Consider the sheaves of groups $\texttt{Aut}(\Lambda)$ resp. $\texttt{Aut}(M)$ on A_{et} associated to the presheaves

$$B \mapsto Aut_{B-alg}(\Lambda \otimes_A B) \text{ resp. } B \mapsto Aut_{B-mod}(M \otimes_A B)$$

for all $B \in A_{et}$. A *twisted form* of Λ (resp. M) is an A-algebra Λ' (resp. an A-module M') such that there is an étale cover $\mathcal{C} = \{A \longrightarrow A_i\}$ of A such that there are isomorphisms

$$\begin{cases} \Lambda \otimes_A A_i \xrightarrow{\phi_i} \Lambda' \otimes_A A_i \\ M \otimes_A A_i \xrightarrow{\psi_i} M' \otimes_A A_i \end{cases}$$

of A_i-algebras (resp. A_i-modules). The set of A-algebra isomorphism classes (resp. A-module isomorphism classes) of twisted forms of Λ (resp. M) is denoted by $Tw_A(\Lambda)$ (resp. $Tw_A(M)$). To a twisted form Λ' one associates a cocycle on $\mathcal{C}$

$$\alpha_{\Lambda'} = \alpha_{ij} = \phi_i^{-1} \circ \phi_j$$

with values in $\mathrm{Aut}(\Lambda)$. Moreover, one verifies that two twisted forms are isomorphic as A-algebras if their cocycles are cohomologous. That is, there are embeddings

$$\begin{cases} Tw_A(\Lambda) \hookrightarrow H^1_{et}(A, \mathrm{Aut}(\Lambda)) \\ Tw_A(M) \hookrightarrow H^1_{et}(A, \mathrm{Aut}(M)) \end{cases}$$

In favorable situations one can even show bijectivity. In particular, this is the case if the automorphisms group is a smooth affine algebraic group-scheme.

Example 3.6 Azumaya algebras
Consider $\Lambda = M_n(A)$, then the automorphism group is PGL_n and twisted forms of Λ are classified by elements of the cohomology group

$$H^1_{et}(A, \mathrm{PGL}_n)$$

These twisted forms are precisely the *Azumaya algebras* of rank n^2 with center A. When A is an affine commutative $\mathbb{C}$-algebra and Λ is an A-algebra with center A, then Λ is an *Azumaya algebra* of rank n^2 if and only if

$$\frac{\Lambda}{\Lambda \mathfrak{m} \Lambda} \simeq M_n(\mathbb{C})$$

for every maximal ideal $\mathfrak{m}$ of A. ▯

Azumaya algebras arise in representation theory as follows. Let A be a *noncommutative* affine $\mathbb{C}$-algebra and assume that the following two conditions are satisfied

- A has a simple representation of dimension n,

- $\mathrm{rep}_n A$ is an irreducible variety.

Then $\oint_n A = \mathbb{C}[\mathbf{rep}_n \, A]^{GL_n}$ is a domain (whence $\mathbf{iss}_n \, A$ is irreducible) and we have an onto trace preserving algebra map corresponding to the simple representation

$$\int_n A = M_n(\mathbb{C}[\mathbf{rep}_n A])^{GL_n} \xrightarrow{\phi} M_n(\mathbb{C})$$

Lift the standard basis e_{ij} of $M_n(\mathbb{C})$ to elements $a_{ij} \in \int_n A$ and consider the determinant d of the $n^2 \times n^2$ matrix $(tr(a_{ij}a_{kl}))_{ij,kl}$ with values in $\oint_n A$. Then $d \neq 0$ and consider the Zariski open affine subset of $\mathbf{iss}_n \, A$

$$\mathbb{X}(d) = \{ \int_n A \xrightarrow{\psi} M_n(\mathbb{C}) \mid \psi \text{ semisimple and } det(tr(\psi(a_{ij})\psi(a_{kl}))) \neq 0\}$$

If $\psi \in \mathbb{X}(d)$, then $\psi : \int_n A \longrightarrow M_n(\mathbb{C})$ is onto as the $\psi(a_{ij})$ form a basis of $M_n(\mathbb{C})$ whence ψ determines a simple n-dimensional representation.

PROPOSITION 3.3
With notations as above,

1. *The localization of $\int_n A$ at the central multiplicative set $\{1, d, d^2, \ldots\}$ is an affine Azumaya algebra with center $\mathbb{C}[\mathbb{X}(d)]$, which is the localization of $\oint_n A$ at this multiplicative set.*

2. *The restriction of the quotient map $\mathbf{rep}_n \, A \xrightarrow{\pi} \mathbf{iss}_n \, A$ to the open set $\pi^{-1}(\mathbb{X}(d))$ is a principal PGL_n-fibration and determines an element in*

$$H^1_{et}(\mathbb{C}[\mathbb{X}(d)], PGL_n)$$

giving the class of the Azumaya algebra.

PROOF   (1) : If $m = Ker \, \psi$ is the maximal ideal of $\mathbb{C}[\mathbb{X}(d)]$ corresponding to the semisimple representation $\psi : \int_n A \longrightarrow M_n(\mathbb{C})$, then we have seen that the quotient

$$\frac{\int_n A}{\int_n Am \int_n A} \simeq M_n(\mathbb{C})$$

whence $\int_n A \otimes_{\oint_n A} \mathbb{C}[\mathbb{X}(d)]$ is an Azumaya algebra. (2) will follow from the theory of Knop-Luna slices and will be proved in chapter 5.   ⬚

An Azumaya algebra over a field is a central simple algebra. Under the above conditions we have that

$$\int_n A \otimes_{\oint_n A} \mathbb{C}(\mathbf{iss}_n \, A)$$

is a central simple algebra over the function field of $\mathbf{iss}_n \, A$ and hence determines a class in its Brauer group, which is an important birational invariant.

In the following section we recall the cohomological description of Brauer groups of fields.

3.2 Central simple algebras

Let K be a field of characteristic zero, choose an algebraic closure $\mathbb{K}$ with *absolute Galois group* $G_K = Gal(\mathbb{K}/K)$.

LEMMA 3.2
The following are equivalent:

1. $K \longrightarrow A$ is étale

2. $A \otimes_K \mathbb{K} \simeq \mathbb{K} \times \ldots \times \mathbb{K}$

3. $A = \prod L_i$ where L_i/K is a finite field extension

PROOF Assume (1), then $A = K[x_1, \ldots, x_n]/(f_1, \ldots, f_n)$ where f_i have invertible Jacobian matrix. Then $A \otimes \mathbb{K}$ is a smooth commutative algebra (hence reduced) of dimension 0 so (2) holds.

Assume (2), then

$$Hom_{K-alg}(A, \mathbb{K}) \simeq Hom_{\mathbb{K}-alg}(A \otimes \mathbb{K}, \mathbb{K})$$

has $dim_{\mathbb{K}}(A \otimes \mathbb{K})$ elements. On the other hand we have by the *Chinese remainder theorem* that

$$A/Jac \, A = \prod_i L_i$$

with L_i a finite field extension of K. However,

$$dim_{\mathbb{K}}(A \otimes \mathbb{K}) = \sum_i dim_K(L_i) = dim_K(A/Jac \, A) \leq dim_K(A)$$

and as both ends are equal A is reduced and hence $A = \prod_i L_i$ whence (3).

Assume (3), then each $L_i = K[x_i]/(f_i)$ with $\partial f_i/\partial x_i$ invertible in L_i. But then $A = \prod L_i$ is étale over K whence (1). ⬜

To every finite étale extension $A = \prod L_i$ we can associate the finite set $rts(A) = Hom_{K-alg}(A, \mathbb{K})$ on which the Galois group G_K acts via a finite quotient group. If we write $A = K[t]/(f)$, then $rts(A)$ is the *set of roots* in $\mathbb{K}$ of the polynomial f with obvious action by G_K. Galois theory, in the interpretation of Grothendieck, can now be stated as follows.

PROPOSITION 3.4
The functor

$$\mathbf{K_{et}} \xrightarrow{\ rts(-)\ } \text{finite } G_K - \text{sets}$$

is an antiequivalence of categories.

We will now give a similar interpretation of the Abelian sheaves on $\mathbf{K_{et}}$. Let $\mathbb{G}$ be a presheaf on $\mathbf{K_{et}}$. Define

$$M_{\mathbb{G}} = \varinjlim \ \mathbb{G}(L)$$

where the limit is taken over all subfields $L \hookrightarrow \mathbb{K}$, which are finite over K. The Galois group G_K acts on $\mathbb{G}(L)$ on the left through its action on L whenever L/K is Galois. Hence, G_K acts an $M_{\mathbb{G}}$ and $M_{\mathbb{G}} = \cup M_{\mathbb{G}}^H$ where H runs through the *open subgroups* (that is, containing a normal subgroup having a finite quotient) of G_K. That is, $M_{\mathbb{G}}$ is a *continuous G_K-module* .

Conversely, given a continuous G_K-module M we can define a presheaf $\mathbb{G}_M$ on $\mathbf{K_{et}}$ such that

- $\mathbb{G}_M(L) = M^H$ where $H = G_L = Gal(\mathbb{K}/L)$.

- $\mathbb{G}_M(\prod L_i) = \prod \mathbb{G}_M(L_i)$.

One verifies that $\mathbb{G}_M$ is a sheaf of Abelian groups on $\mathbf{K_{et}}$.

THEOREM 3.1
There is an equivalence of categories

$$\mathbf{S}(\mathbf{K_{et}}) \rightleftarrows G_K - \text{mod}$$

induced by the correspondences $\mathbb{G} \mapsto M_{\mathbb{G}}$ *and* $M \mapsto \mathbb{G}_M$. *Here,* $G_K - \text{mod}$ *is the category of continuous G_K-modules.*

PROOF   A G_K-morphism $M \longrightarrow M'$ induces a morphism of sheaves $\mathbb{G}_M \longrightarrow \mathbb{G}_{M'}$. Conversely, if H is an open subgroup of G_K with $L = \mathbb{K}^H$, then if $\mathbb{G} \xrightarrow{\ \phi\ } \mathbb{G}'$ is a sheaf morphism, $\phi(L) : \mathbb{G}(L) \longrightarrow \mathbb{G}'(L)$ commutes with the action of G_K by functoriality of ϕ. Therefore, $\varinjlim \ \phi(L)$ is a G_K-morphism $M_{\mathbb{G}} \longrightarrow M_{\mathbb{G}'}$.

One verifies easily that $Hom_{G_K}(M, M') \longrightarrow Hom(\mathbb{G}_M, \mathbb{G}_{M'})$ is an isomorphism and that the canonical map $\mathbb{G} \longrightarrow \mathbb{G}_{M_{\mathbb{G}}}$ is an isomorphism.   ⬜

In particular, we have that $\mathbb{G}(K) = \mathbb{G}(\mathbb{K})^{G_K}$ for every sheaf $\mathbb{G}$ of Abelian groups on $\mathbf{K_{et}}$ and where $\mathbb{G}(\mathbb{K}) = M_{\mathbb{G}}$. Hence, the right derived functors of Γ and $(-)^G$ coincide for Abelian sheaves.

The category $G_K-\text{mod}$ of continuous G_K-modules is Abelian having enough injectives. Therefore, the left exact functor

$$(-)^G : G_K - \text{mod} \longrightarrow \text{abelian}$$

admits right derived functors. They are called the *Galois cohomology groups* and denoted

$$R^i \ M^G = H^i(G_K, M)$$

Therefore, we have the following.

PROPOSITION 3.5
For any sheaf of Abelian groups $\mathbb{G}$ on K_{et} we have a group isomorphism

$$H^i_{et}(K, \mathbb{G}) \simeq H^i(G_K, \mathbb{G}(\mathbb{K}))$$

Hence, étale cohomology is a natural extension of Galois cohomology to arbitrary commutative algebras. The following definition-characterization of central simple algebras is classical, see for example [84].

PROPOSITION 3.6
Let A be a finite dimensional K-algebra. The following are equivalent:

1. *A has no proper twosided ideals and the center of A is K.*

2. *$A_{\mathbb{K}} = A \otimes_K \mathbb{K} \simeq M_n(\mathbb{K})$ for some n.*

3. *$A_L = A \otimes_K L \simeq M_n(L)$ for some n and some finite Galois extension L/K.*

4. *$A \simeq M_k(D)$ for some k where D is a division algebra of dimension l^2 with center K.*

The last part of this result suggests the following definition. Call two central simple algebras A and A' equivalent if and only if $A \simeq M_k(\Delta)$ and $A' \simeq M_l(\Delta)$ with Δ a division algebra. From the second characterization it follows that the tensor product of two central simple K-algebras is again central simple. Therefore, we can equip the set of equivalence classes of central simple algebras with a product induced from the tensorproduct. This product has the class $[K]$ as unit element and $[\Delta]^{-1} = [\Delta^{opp}]$, the opposite algebra as $\Delta \otimes_K \Delta^{opp} \simeq End_K(\Delta) = M_{l^2}(K)$. This group is called the *Brauer group* and is denoted $Br(K)$. We will quickly recall its cohomological description, all of which is classical.

GL_r is an affine smooth algebraic group defined over K and is the automorphism group of a vector space of dimension r. It defines a sheaf of groups on K_{et} that we will denote by GL_r. Using the fact that the first cohomology classifies twisted forms of vector spaces of dimension r we have the following.

LEMMA 3.3

$$H^1_{et}(K, \mathrm{GL}_r) = H^1(G_K, GL_r(\mathbb{K})) = 0$$

In particular, we have "Hilbert's theorem 90"

$$H^1_{et}(K, \mathbb{G}_m) = H^1(G_K, \mathbb{K}^*) = 0$$

PROOF The cohomology group classifies K-module isomorphism classes of twisted forms of r-dimensional vector spaces over K. There is just one such class. ☐

PGL_n is an affine smooth algebraic group defined over K and it is the automorphism group of the K-algebra $M_n(K)$. It defines a sheaf of groups on K_{et} denoted by PGL_n. By proposition 3.6 we know that any central simple K-algebra Δ of dimension n^2 is a twisted form of $M_n(K)$.

LEMMA 3.4
The pointed set of K-algebra isomorphism classes of central simple algebras of dimension n^2 over K coincides with the cohomology set

$$H^1_{et}(K, \mathrm{PGL}_n) = H^1(G_K, PGL_n(\mathbb{K}))$$

THEOREM 3.2
There is a natural inclusion

$$H^1_{et}(K, \mathrm{PGL}_n) \hookrightarrow H^2_{et}(K, \mu_n) = Br_n(K)$$

where $Br_n(K)$ is the n-torsion part of the Brauer group of K. Moreover,

$$Br(K) = H^2_{et}(K, \mathbb{G}_m)$$

is a torsion group.

PROOF Consider the exact commutative diagram of sheaves of groups on K_{et} of figure 3.2. Taking cohomology of the second exact sequence we obtain

$$GL_n(K) \xrightarrow{det} K^* \longrightarrow H^1_{et}(K, \mathrm{SL}_n) \longrightarrow H^1_{et}(K, \mathrm{GL}_n)$$

where the first map is surjective and the last term is zero, whence

$$H^1_{et}(K, \mathrm{SL}_n) = 0$$

Taking cohomology of the first vertical exact sequence we get

$$H^1_{et}(K, \mathrm{SL}_n) \longrightarrow H^1_{et}(K, \mathrm{PGL}_n) \longrightarrow H^2_{et}(K, \mu_n)$$

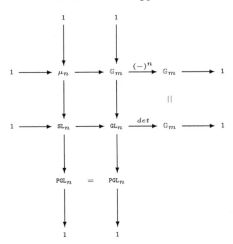

FIGURE 3.2: Brauer group diagram.

from which the first claim follows.

As for the second assertion, taking cohomology of the first exact sequence we get

$$H^1_{et}(K, \mathbb{G}_m) \longrightarrow H^2_{et}(K, \mu_n) \longrightarrow H^2_{et}(K, \mathbb{G}_m) \xrightarrow{n.} H^2_{et}(K, \mathbb{G}_m)$$

By Hilbert 90, the first term vanishes and hence $H^2_{et}(K, \mu_n)$ is equal to the n-torsion of the group

$$H^2_{et}(K, \mathbb{G}_m) = H^2(G_K, \mathbb{K}^*) = Br(K)$$

where the last equality follows from the crossed product result, see, for example, [84]. □

So far, the field K was arbitrary. If K is of transcendence degree d, this will put restrictions on the "size" of the Galois group G_K. In particular this will enable us to show in section 3.4 that $H^i(G_K, \mu_n) = 0$ for $i > d$. But first, we need to recall the definition of spectral sequences.

3.3 Spectral sequences

Let $\mathcal{A}, \mathcal{B}$ and $\mathcal{C}$ be Abelian categories such that $\mathcal{A}$ and $\mathcal{B}$ have enough injectives and consider left exact functors

$$\mathcal{A} \xrightarrow{f} \mathcal{B} \xrightarrow{g} \mathcal{C}$$

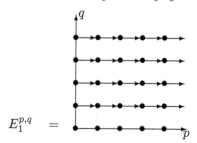

$$E_1^{p,q} \quad =$$

FIGURE 3.3:   Level 1.

Let the functors be such that f maps injectives of $\mathcal{A}$ to g-acyclic objects in $\mathcal{B}$, that is, $R^i \, g(f \, I) = 0$ for all $i > 0$. Then, there are connections between the objects

$$R^p \, g(R^q \, f(A)) \text{ and } R^n \, gf(A)$$

for all objects $A \in \mathcal{A}$. These connections can be summarized by giving a spectral sequence

THEOREM 3.3
Let $\mathcal{A}, \mathcal{B}, \mathcal{C}$ be Abelian categories with $\mathcal{A}, \mathcal{B}$ having enough injectives and left exact functors

$$\mathcal{A} \xrightarrow{\;f\;} \mathcal{B} \xrightarrow{\;g\;} \mathcal{C}$$

such that f takes injectives to g-acyclics.
    Then, for any object $A \in \mathcal{A}$ there is a spectral sequence

$$E_2^{p,q} = R^p \, g(R^q \, f(A)) \Longrightarrow R^n \, gf(A)$$

In particular, there is an exact sequence

$$0 \longrightarrow R^1 \, g(f(A)) \longrightarrow R^1 \, gf(A) \longrightarrow g(R^1 \, f(A)) \longrightarrow R^2 \, g(f(A)) \longrightarrow \ldots$$

Moreover, if f is an exact functor, then we have

$$R^p \, gf(A) \simeq R^p \, g(f(A))$$

A spectral sequence $E_2^{p,q} \Longrightarrow E^n$ (or $E_1^{p,q} \Longrightarrow E^n$) consists of the following data:

1. A family of objects $E_r^{p,q}$ in an Abelian category for $p, q, r \in \mathbb{Z}$ such that $p, q \geq 0$ and $r \geq 2$ (or $r \geq 1$).

2. A family of morphisms in the Abelian category

$$d_r^{p,q} : E_r^{p,q} \longrightarrow E_r^{p+r,q-r+1}$$

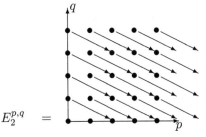

$$E_2^{p,q} \quad = $$

FIGURE 3.4: Level 2.

satisfying the complex condition

$$d_r^{p+r,q-r+1} \circ d_r^{p,q} = 0$$

and where we assume that $d_r^{p,q} = 0$ if any of the numbers $p, q, p+r$ or $q - r + 1$ is < 1. At level one we have the situation of figure 3.3. At level two we have the situation of figure 3.4

3. The objects $E_{r+1}^{p,q}$ on level $r + 1$ are derived from those on level r by taking the cohomology objects of the complexes, that is

$$E_{r+1}^p = Ker\ d_r^{p,q}\ /\ Im\ d_r^{p-r,q+r-1}$$

At each place (p, q) this process converges as there is an integer r_0 depending on (p, q) such that for all $r \geq r_0$ we have $d_r^{p,q} = 0 = d_r^{p-r,q+r-1}$. We then define

$$E_\infty^{p,q} = E_{r_0}^{p,q}(= E_{r_0+1}^{p,q} = \ldots)$$

Observe that there are injective maps $E_\infty^{0,q} \hookrightarrow E_2^{0,q}$.

4. A family of objects E^n for integers $n \geq 0$ and for each we have a filtration

$$0 \subset E_n^n \subset E_{n-1}^n \subset \ldots \subset E_1^n \subset E_0^n = E^n$$

such that the successive quotients are given by

$$E_p^n\ /\ E_{p+1}^n = E_\infty^{p,n-p}$$

That is, the terms $E_\infty^{p,q}$ are the composition terms of the *limiting terms* E^{p+q}. Pictorially,

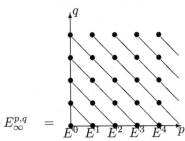

$$E_\infty^{p,q} \quad = $$

For small n one can make the relation between E^n and the terms $E_2^{p,q}$ explicit. First note that

$$E_2^{0,0} = E_\infty^{0,0} = E^0$$

Also, $E_1^1 = E_\infty^{1,0} = E_2^{1,0}$ and $E^1/E_1^1 = E_\infty^{0,1} = Ker\ d_2^{0,1}$. This gives an exact sequence

$$0 \longrightarrow E_2^{1,0} \longrightarrow E^1 \longrightarrow E_2^{0,1} \xrightarrow{d_2^{0,1}} E_2^{2,0}$$

Further, $E^2 \supset E_1^2 \supset E_2^2$ where

$$E_2^2 = E_\infty^{2,0} = E_2^{2,0}\ /\ Im\ d_2^{0,1}$$

and $E_1^2/E_2^2 = E_\infty^{1,1} = Ker\ d_2^{1,1}$ whence we can extend the above sequence to

$$\cdots \longrightarrow E_2^{0,1} \xrightarrow{d_2^{0,1}} E_2^{2,0} \longrightarrow E_1^2 \longrightarrow E_2^{1,1} \xrightarrow{d_2^{1,1}} E_2^{3,0}$$

as $E^2/E_1^2 = E_\infty^{0,2} \hookrightarrow E_2^{0,2}$ we have that $E_1^2 = Ker\ (E^2 \longrightarrow E_2^{0,2})$. If we specialize to the spectral sequence $E_2^{p,q} = R^p\ g(R^q\ f(A)) \Longrightarrow R^n\ gf(A)$ we obtain the exact sequence

$$0 \longrightarrow R^1\ g(f(A)) \longrightarrow R^1\ gf(A) \longrightarrow g(R^1\ f(A)) \longrightarrow R^2\ g(f(A)) \longrightarrow$$

$$\longrightarrow E_1^2 \longrightarrow R^1\ g(R^1\ f(A)) \longrightarrow R^3\ g(f(A))$$

where $E_1^2 = Ker\ (R^2\ gf(A) \longrightarrow g(R^2\ f(A)))$.

An important spectral sequence is the *Leray spectral sequence* . Assume we have an algebra morphism $A \xrightarrow{f} A'$ and a sheaf of groups $\mathbb{G}$ on A'_{et}. We define the *direct image* of $\mathbb{G}$ under f to be the sheaf of groups $f_*\ \mathbb{G}$ on $\mathbf{A}_{et}$ defined by

$$f_*\ \mathbb{G}(B) = \mathbb{G}(B \otimes_A A')$$

for all $B \in \mathbf{A}_{et}$ (recall that $B \otimes_A A' \in A'_{et}$ so the right-hand side is well defined).

This gives us a left exact functor

$$f_* : \mathbf{S}^{ab}(A'_{et}) \longrightarrow \mathbf{S}^{ab}(\mathbf{A}_{et})$$

and therefore we have right-derived functors of it $R^i\ f_*$. If $\mathbb{G}$ is an Abelian sheaf on A'_{et}, then $R^i\ f_*\mathbb{G}$ is a sheaf on $\mathbf{A}_{et}$. One verifies that its stalk in a prime ideal $\mathfrak{p}$ is equal to

$$(R^i\ f_*\mathbb{G})_\mathfrak{p} = H_{et}^i(A_\mathfrak{p}^{sh} \otimes_A A', \mathbb{G})$$

where the right-hand side is the direct limit of cohomology groups taken over all étale neighborhoods of $\mathfrak{p}$. We can relate cohomology of $\mathbb{G}$ and $f_*\mathbb{G}$ by the following.

THEOREM 3.4
(Leray spectral sequence) If $\mathbb{G}$ is a sheaf of Abelian groups on A'_{et} and $A \xrightarrow{\ f\ } A'$ an algebra morphism, then there is a spectral sequence

$$E_2^{p,q} = H_{et}^p(A, R^q\ f_*\mathbb{G}) \Longrightarrow H_{et}^n(A, \mathbb{G})$$

In particular, if $R^j\ f_*\mathbb{G} = 0$ for all $j > 0$, then for all $i \geq 0$ we have isomorphisms

$$H_{et}^i(A, f_*\mathbb{G}) \simeq H_{et}^i(A', \mathbb{G})$$

3.4 Tsen and Tate fields

In this section we will use spectral sequences to control the size of the Brauer group of a function field in terms of its transcendence degree.

DEFINITION 3.4 *A field K is said to be a $Tsen^d$-field if every homogeneous form of degree deg with coefficients in K and $n > deg^d$ variables has a nontrivial zero in K.*

For example, an algebraically closed field $\mathbb{K}$ is a $Tsen^0$-field as any form in n-variables defines a hypersurface in $\mathbb{P}_{\mathbb{K}}^{n-1}$. In fact, algebraic geometry tells us a stronger story

LEMMA 3.5
Let $\mathbb{K}$ be algebraically closed. If $f_1, \ldots, f_r$ are forms in n variables over $\mathbb{K}$ and $n > r$, then these forms have a common nontrivial zero in $\mathbb{K}$.

PROOF Each f_i defines a hypersurface $V(f_i) \hookrightarrow \mathbb{P}_{\mathbb{K}}^{n-1}$. The intersection of r hypersurfaces has dimension $\geq n - 1 - r$ from which the claim follows. □

We want to extend this fact to higher Tsen-fields. The proof of the following result is technical inequality manipulation, see, for example, [97].

PROPOSITION 3.7
Let K be a $Tsen^d$-field and $f_1, \ldots, f_r$ forms in n variables of degree deg. If $n > rdeg^d$, then they have a nontrivial common zero in K.

For our purposes the main interest in Tsen-fields comes from the following.

THEOREM 3.5
Let K be of transcendence degree d over an algebraically closed field $\mathbb{C}$, then K is a $Tsen^d$-field.

PROOF   First we claim that the purely transcendental field $\mathbb{C}(t_1, \ldots, t_d)$ is a $Tsen^d$-field. By induction we have to show that if L is $Tsen^k$, then $L(t)$ is $Tsen^{k+1}$.

By homogeneity we may assume that $f(x_1, \ldots, x_n)$ is a form of degree deg with coefficients in $L[t]$ and $n > deg^{k+1}$. For fixed s we introduce new variables $y_{ij}^{(s)}$ with $i \leq n$ and $0 \leq j \leq s$ such that

$$x_i = y_{i0}^{(s)} + y_{i1}^{(s)}t + \ldots + y_{is}^{(s)}t^s$$

If r is the maximal degree of the coefficients occurring in f, then we can write

$$f(x_i) = f_0(y_{ij}^{(s)}) + f_1(y_{ij}^{(s)})t + \ldots + f_{deg.s+r}(y_{ij}^{(s)})t^{deg.s+r}$$

where each f_j is a form of degree deg in $n(s+1)$-variables. By the proposition above, these forms have a common zero in L provided

$$n(s+1) > deg^k(ds + r + 1) \Longleftrightarrow (n - deg^{i+1})s > deg^i(r+1) - n$$

which can be satisfied by taking s large enough. the common nontrivial zero in L of the f_j, gives a nontrivial zero of f in $L[t]$.

By assumption, K is an algebraic extension of $\mathbb{C}(t_1, \ldots, t_d)$ which by the above argument is $Tsen^d$. As the coefficients of any form over K lie in a finite extension E of $\mathbb{C}(t_1, \ldots, t_d)$ it suffices to prove that E is $Tsen^d$.

Let $f(x_1, \ldots, x_n)$ be a form of degree deg in E with $n > deg^d$. Introduce new variables y_{ij} with

$$x_i = y_{i1}e_1 + \ldots y_{ik}e_k$$

where e_i is a basis of E over $\mathbb{C}(t_1, \ldots, t_d)$. Then,

$$f(x_i) = f_1(y_{ij})e_1 + \ldots + f_k(y_{ij})e_k$$

where the f_i are forms of degree deg in $k.n$ variables over $\mathbb{C}(t_1, \ldots, t_d)$. Because $\mathbb{C}(t_1, \ldots, t_d)$ is $Tsen^d$, these forms have a common zero as $k.n > k.deg^d$. Finding a nontrivial zero of f in E is equivalent to finding a common nontrivial zero to the $f_1, \ldots, f_k$ in $\mathbb{C}(t_1, \ldots, t_d)$, done. $\square$

A direct application of this result is *Tsen's theorem.*

THEOREM 3.6
Let K be the function field of a curve C defined over an algebraically closed field. Then, the only central simple K-algebras are $M_n(K)$. That is, $Br(K) = 1$.

PROOF Assume there exists a central division algebra Δ of dimension n^2 over K. There is a finite Galois extension L/K such that $\Delta \otimes L = M_n(L)$. If $x_1, \ldots, x_{n^2}$ is a K-basis for Δ, then the reduced norm of any $x \in \Delta$

$$N(x) = det(x \otimes 1)$$

is a form in n^2 variables of degree n. Moreover, as $x \otimes 1$ is invariant under the action of $Gal(L/K)$ the coefficients of this form actually lie in K.

By the main result, K is a $Tsen^1$-field and $N(x)$ has a nontrivial zero whenever $n^2 > n$. As the reduced norm is multiplicative, this contradicts $N(x)N(x^{-1}) = 1$. Hence, $n = 1$ and the only central division algebra is K itself. $\qquad\Box$

If K is the function field of a surface, we have the following.

PROPOSITION 3.8
Let K be the function field of a surface defined over an algebraically closed field. If Δ is a central simple K-algebra of dimension n^2, then the reduced norm map

$$N : \Delta \longrightarrow K$$

is surjective.

PROOF Let $e_1, \ldots, e_{n^2}$ be a K-basis of Δ and $k \in K$, then

$$N(\sum x_i e_i) - kx_{n^2+1}^n$$

is a form of degree n in $n^2 + 1$ variables. Since K is a $Tsen^2$ field, it has a nontrivial solution (x_i^0), but then, $\delta = (\sum x_i^0 e_i)x_{n^2+1}^{-1}$ has reduced norm equal to k. $\qquad\Box$

From the cohomological description of the Brauer group it is clear that we need to have some control on the absolute Galois group $G_K = Gal(\mathbb{K}/K)$. We will see that finite transcendence degree forces some cohomology groups to vanish.

DEFINITION 3.5 *The cohomological dimension of a group G, $cd(G) \leq d$ if and only if $H^r(G, A) = 0$ for all $r > d$ and all torsion modules $A \in G$-mod.*

DEFINITION 3.6 *A field K is said to be a $Tate^d$-field if the absolute Galois group $G_K = Gal(\mathbb{K}/K)$ satisfies $cd(G) \leq d$.*

First, we will reduce the condition $cd(G) \leq d$ to a more manageable one. To start, one can show that a profinite group G (that is, a projective limit of finite groups, see [97] for more details) has $cd(G) \leq d$ if and only if

$$H^{d+1}(G, A) = 0 \text{ for all torsion } G\text{-modules } A$$

Further, as all Galois cohomology groups of profinite groups are torsion, we can decompose the cohomology in its p-primary parts and relate their vanishing to the cohomological dimension of the p-Sylow subgroups G_p of G. This problem can then be verified by computing cohomology of finite simple G_p-modules of p-power order, but for a profinite p-group there is just one such module, namely, $\mathbb{Z}/p\mathbb{Z}$ with the trivial action.

Combining these facts we have the following manageable criterium on cohomological dimension.

PROPOSITION 3.9
$cd(G) \leq d$ if $H^{d+1}(G, \mathbb{Z}/p\mathbb{Z}) = 0$ *for the simple G-modules with trivial action* $\mathbb{Z}/p\mathbb{Z}$.

We will need the following spectral sequence in Galois cohomology

PROPOSITION 3.10
(Hochschild-Serre spectral sequence) If N is a closed normal subgroup of a profinite group G, then

$$E_2^{p,q} = H^p(G/N, H^q(N, A)) \Longrightarrow H^n(G, A)$$

holds for every continuous G-module A.

Now, we are in a position to state and prove *Tate's theorem*.

THEOREM 3.7
Let K be of transcendence degree d over an algebraically closed field, then K is a $Tate^d$-field.

PROOF   Let $\mathbb{C}$ denote the algebraically closed basefield, then K is algebraic over $\mathbb{C}(t_1, \ldots, t_d)$ and therefore

$$G_K \lhook\joinrel\longrightarrow G_{\mathbb{C}(t_1,\ldots,t_d)}$$

Thus, K is $Tate^d$ if $\mathbb{C}(t_1, \ldots, t_d)$ is $Tate^d$. By induction it suffices to prove

$$\text{If } cd(G_L) \leq k \text{ then } cd(G_{L(t)}) \leq k + 1$$

Let $\mathbb{L}$ be the algebraic closure of L and $\mathbb{M}$ the algebraic closure of $L(t)$. As $L(t)$ and $\mathbb{L}$ are linearly disjoint over L we have the following diagram of extensions

and Galois groups

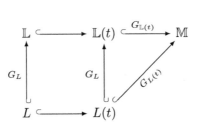

where $G_{L(t)}/G_{\mathbb{L}(t)} \simeq G_L$.

We claim that $cd(G_{\mathbb{L}(t)}) \leq 1$. Consider the exact sequence of $G_{\mathbb{L}(t)}$-modules

$$0 \longrightarrow \mu_p \longrightarrow \mathbb{M}^* \xrightarrow{(-)^p} \mathbb{M}^* \longrightarrow 0$$

where μ_p is the subgroup (of $\mathbb{C}^*$) of p-roots of unity. As $G_{\mathbb{L}(t)}$ acts trivially on μ_p it is after a choice of primitive p-th root of one isomorphic to $\mathbb{Z}/p\mathbb{Z}$. Taking cohomology with respect to the subgroup $G_{\mathbb{L}(t)}$ we obtain

$$0 = H^1(G_{\mathbb{L}(t)}, \mathbb{M}^*) \longrightarrow H^2(G_{\mathbb{L}(t)}, \mathbb{Z}/p\mathbb{Z}) \longrightarrow H^2(G_{\mathbb{L}(t)}, \mathbb{M}^*) = Br(\mathbb{L}(t))$$

But the last term vanishes by Tsen's theorem as $\mathbb{L}(t)$ is the function-field of a curve defined over the algebraically closed field $\mathbb{L}$. Therefore, $H^2(G_{\mathbb{L}(t)}, \mathbb{Z}/p\mathbb{Z}) = 0$ for all simple modules $\mathbb{Z}/p\mathbb{Z}$, whence $cd(G_{\mathbb{L}(t)}) \leq 1$.

By the inductive assumption we have $cd(G_L) \leq k$ and now we are going to use exactness of the sequence

$$0 \longrightarrow G_L \longrightarrow G_{L(t)} \longrightarrow G_{\mathbb{L}(t)} \longrightarrow 0$$

to prove that $cd(G_{L(t)}) \leq k+1$. For, let A be a torsion $G_{L(t)}$-module and consider the Hochschild-Serre spectral sequence

$$E_2^{p,q} = H^p(G_L, H^q(G_{\mathbb{L}(t)}, A)) \Longrightarrow H^n(G_{L(t)}, A)$$

By the restrictions on the cohomological dimensions of G_L and $G_{\mathbb{L}(t)}$ the level two term has following shape

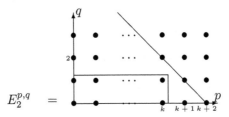

where the only nonzero groups are lying in the lower rectangular region. Therefore, all $E_\infty^{p,q} = 0$ for $p + q > k + 1$. Now, all the composition factors of $H^{k+2}(G_{L(t)}, A)$ are lying on the indicated diagonal line and hence are zero. Thus, $H^{k+2}(G_{L(t)}, A) = 0$ for all torsion $G_{L(t)}$-modules A and hence $cd(G_{L(t)}) \leq k+1$.   ⬚

THEOREM 3.8
If **A** *is a constant sheaf of an Abelian torsion group* A *on* K_{et}, *then*

$$H_{et}^i(K, \mathbf{A}) = 0$$

whenever $i > trdeg_{\mathbb{C}}(K)$.

3.5   Coniveau spectral sequence

In this section we will describe a particularly useful spectral sequence. Consider the setting $k \xleftarrow{\pi} A \xhookrightarrow{i} K$ where A is a discrete valuation ring in K with residue field $A/m = k$. As always, we will assume that A is a $\mathbb{C}$-algebra. By now we have a grip on the Galois cohomology groups

$$H_{et}^i(K, \mu_n^{\otimes l}) \text{ and } H_{et}^i(k, \mu_n^{\otimes l})$$

and we will use this information to compute the étale cohomology groups

$$H_{et}^i(A, \mu_n^{\otimes l})$$

Here, $\mu_n^{\otimes l} = \underbrace{\mu_n \otimes \ldots \otimes \mu_n}_{l}$ where the tensorproduct is as sheafs of invertible $\mathbb{Z}_n = \mathbb{Z}/n\mathbb{Z}$-modules.

We will consider the Leray spectral sequence for i and hence have to compute the derived sheaves of the direct image

LEMMA 3.6

1. $R^0 \, i_* \mu_n^{\otimes l} \simeq \mu_n^{\otimes l}$ *on* A_{et}.

2. $R^1 \, i_* \mu_n^{\otimes l} \simeq \mu_n^{\otimes l-1}$ *concentrated in* m.

3. $R^j \, i_* \mu_n^{\otimes l} \simeq 0$ *whenever* $j \geq 2$.

PROOF   The strict Henselizations of A at the two primes $\{0, m\}$ are resp.

$$A_0^{sh} \simeq \mathbb{K} \text{ and } A_m^{sh} \simeq \mathbf{k}\{t\}$$

0	0	0	...
$H^0(k, \mu_n^{\otimes l-1})$	$H^1(k, \mu_n^{\otimes l-1})$	$H^2(k, \mu_n^{\otimes l-1})$	...
$H^0(A, \mu_n^{\otimes l})$	$H^1(A, \mu_n^{\otimes l})$	$H^2(A, \mu_n^{\otimes l})$	...

FIGURE 3.5: Second term of Leray sequence.

where $\mathbb{K}$ (resp. $\mathbf{k}$) is the algebraic closure of K (resp. k). Therefore,

$$(R^j \ i_* \mu_n^{\otimes l})_0 = H_{et}^j(\mathbb{K}, \mu_n^{\otimes l})$$

which is zero for $i \geq 1$ and $\mu_n^{\otimes l}$ for $j = 0$. Further, $A_m^{sh} \otimes_A K$ is the field of fractions of $\mathbf{k}\{t\}$ and hence is of transcendence degree one over the algebraically closed field $\mathbf{k}$, whence

$$(R^j \ i_* \mu_n^{\otimes l})_m = H_{et}^j(L, \mu_n^{\otimes l})$$

which is zero for $j \geq 2$ because L is $Tate^1$.

For the field-tower $K \subset L \subset \mathbb{K}$ we have that $G_L = \hat{\mathbb{Z}} = \varprojlim \mu_m$ because the only Galois extensions of L are the Kummer extensions obtained by adjoining $\sqrt[m]{t}$. But then,

$$H_{et}^1(L, \mu_n^{\otimes l}) = H^1(\hat{Z}, \mu_n^{\otimes l}(\mathbb{K})) = Hom(\hat{Z}, \mu_n^{\otimes l}(\mathbb{K})) = \mu_n^{\otimes l-1}$$

from which the claims follow. $\qquad\Box$

THEOREM 3.9
We have a long exact sequence

$$0 \longrightarrow H^1(A, \mu_n^{\otimes l}) \longrightarrow H^1(K, \mu_n^{\otimes l}) \longrightarrow H^0(k, \mu_n^{\otimes l-1}) \longrightarrow$$
$$H^2(A, \mu_n^{\otimes l}) \longrightarrow H^2(K, \mu_n^{\otimes l}) \longrightarrow H^1(k, \mu_n^{\otimes l-1}) \longrightarrow \cdots$$

PROOF By the foregoing lemma, the second term of the Leray spectral sequence for $i_* \mu_n^{\otimes l}$ is depicted in figure 3.5 with connecting morphisms

$$H_{et}^{i-1}(k, \mu_n^{\otimes l-1}) \xrightarrow{\alpha_i} H_{et}^{i+1}(A, \mu_n^{\otimes l})$$

The spectral sequences converges to its limiting term which looks like

0	0	0	...
$Ker\ \alpha_1$	$Ker\ \alpha_2$	$Ker\ \alpha_3$	...
$H^0(A, \mu_n^{\otimes l})$	$H^1(A, \mu_n^{\otimes l})$	$Coker\ \alpha_1$	...

and the Leray sequence gives the short exact sequences

$$0 \longrightarrow H^1_{et}(A, \mu_n^{\otimes l}) \longrightarrow H^1_{et}(K, \mu_n^{\otimes l}) \longrightarrow Ker\ \alpha_1 \longrightarrow 0$$

$$0 \longrightarrow Coker\ \alpha_1 \longrightarrow H^2_{et}(K, \mu_n^{\otimes l}) \longrightarrow Ker\ \alpha_2 \longrightarrow 0$$

$$0 \longrightarrow Coker\ \alpha_{i-1} \longrightarrow H^i_{et}(K, \mu_n^{\otimes l}) \longrightarrow Ker\ \alpha_i \longrightarrow 0$$

and gluing these sequences gives us the required result.   ⬜

In particular, if A is a discrete valuation ring of K with residue field k we have for each i a connecting morphism

$$H^i_{et}(K, \mu_n^{\otimes l}) \xrightarrow{\partial_{i,A}} H^{i-1}_{et}(k, \mu_n^{\otimes l-1})$$

Like any other topology, the étale topology can be defined locally on any scheme X. That is, we call a morphism of schemes

$$Y \xrightarrow{f} X$$

an étale extension (resp. cover) if locally f has the form

$$f^a \mid U_i : A_i = \Gamma(U_i, \mathcal{O}_X) \longrightarrow B_i = \Gamma(f^{-1}(U_i), \mathcal{O}_Y)$$

with $A_i \longrightarrow B_i$ an étale extension (resp. cover) of algebras.

Again, we can construct the étale site of X locally and denote it with X_{et}. Presheaves and sheaves of groups on X_{et} are defined similarly and the right derived functors of the left exact global sections functor

$$\Gamma : \mathbf{S}^{ab}(X_{et}) \longrightarrow \texttt{abelian}$$

will be called the cohomology functors and we denote

$$R^i\ \Gamma(\mathbb{G}) = H^i_{et}(X, \mathbb{G})$$

From now on we restrict to the case when X is a smooth, irreducible projective variety of dimension d over $\mathbb{C}$. In this case, we can initiate the computation of the cohomology groups $H^i_{et}(X, \mu_n^{\otimes l})$ via Galois cohomology of function fields of subvarieties using the coniveau spectral sequence.

THEOREM 3.10
Let X be a smooth irreducible variety over $\mathbb{C}$. Let $X^{(p)}$ denote the set of irreducible subvarieties x of X of codimension p with function field $\mathbb{C}(x)$, then there exists a coniveau spectral sequence

$$E_1^{p.q} = \bigoplus_{x \in X^{(p)}} H^{q-p}_{et}(\mathbb{C}(x), \mu_n^{\otimes l-p}) \Longrightarrow H^{p+q}_{et}(X, \mu_n^{\otimes l})$$

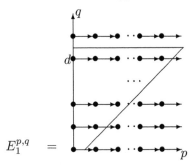

FIGURE 3.6: Coniveau spectral sequence.

In contrast to the spectral sequences used before, the existence of the coniveau spectral sequence by no means follows from general principles. In it, a lot of heavy machinery on étale cohomology of schemes is encoded. In particular,

- cohomology groups with support of a closed subscheme, see, for example [82, p. 91-94], and

- cohomological purity and duality, see [82, p. 241-252]

a detailed exposition of which would take us too far afield. For more details we refer the reader to [22].

Using the results on cohomological dimension and vanishing of Galois cohomology of $\mu_n^{\otimes k}$ when the index is larger than the transcendence degree, we see that the coniveau spectral sequence has shape as in figure 3.6 where the only nonzero terms are in the indicated region.

Let us understand the connecting morphisms at the first level, a typical instance of which is

$$\bigoplus_{x \in X^{(p)}} H^i(\mathbb{C}(x), \mu_n^{\oplus l-p}) \longrightarrow \bigoplus_{y \in X^{(p+1)}} H^{i-1}(\mathbb{C}(y), \mu_n^{\oplus l-p-1})$$

and consider one of the closed irreducible subvarieties x of X of codimension p and one of those y of codimension $p+1$. Then, either y is not contained in x in which case the component map

$$H^i(\mathbb{C}(x), \mu_n^{\oplus l-p}) \longrightarrow H^{i-1}(\mathbb{C}(y), \mu_n^{\oplus l-p-1})$$

is the zero map. Or, y is contained in x and hence defines a codimension one subvariety of x. That is, y defines a discrete valuation on $\mathbb{C}(x)$ with residue field $\mathbb{C}(y)$. In this case, the above component map is the connecting morphism defined above.

In particular, let K be the function field of X. Then we can define the unramified cohomology groups

$$F_n^{i,l}(K/\mathbb{C}) = Ker \; H^i(K, \mu_n^{\otimes l}) \xrightarrow{\oplus \partial_{i,A}} \oplus H^{i-1}(k_A, \mu_n^{\otimes l-1})$$

$\vdots$	$\vdots$	$\vdots$	$\vdots$	
0	0	0	0	$\cdots$
$H^2(\mathbb{C}(S),\mu_n)$	$\oplus_C H^1(\mathbb{C}(S),\mathbb{Z}_n)$	$\oplus_P \mu_n^{-1}$	0	$\cdots$
$H^1(\mathbb{C}(S),\mu_n)$	$\oplus_C \mathbb{Z}_n$	0	0	$\cdots$
μ_n	0	0	0	$\cdots$

FIGURE 3.7:   First term of coniveau spectral sequence for S.

where the sum is taken over all discrete valuation rings A of K (or equivalently, the irreducible codimension one subvarieties of X) with residue field k_A. By definition, this is a (stable) birational invariant of X. In particular, if X is (stably) rational over $\mathbb{C}$, then

$$F_n^{i,l}(K/\mathbb{C}) = 0 \text{ for all } i, l \geq 0$$

3.6   The Artin-Mumford exact sequence

The coniveau spectral sequence allows us to control the Brauer group of function fields of surfaces. This result, due to Michael Artin and David Mumford, was used by them to construct unirational nonrational varieties. Our main application of the description is to classify in chapter 5 the Brauer classes which do admit a Cayley-smooth noncommutative model. It will turn out that even in the case of surfaces, not every central simple algebra over the function field allows such a noncommutative model. Let S be a smooth irreducible projective surface.

DEFINITION 3.7 S is called simply connected   *if every étale   cover* $Y \longrightarrow S$ *is trivial, that is, Y is isomorphic to a finite disjoint union of copies of S.*

The first term of the coniveau spectral sequence of S has the shape of figure 3.7 where C runs over all irreducible curves on S and P over all points of S.

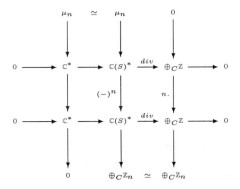

FIGURE 3.8: Divisors of rational functions on S.

LEMMA 3.7
For any smooth S we have $H^1(\mathbb{C}(S), \mu_n) \longrightarrow \oplus_C \mathbb{Z}_n$. If S is simply connected, $H^1_{et}(S, \mu_n) = 0$.

PROOF Using the Kummer sequence

$$1 \longrightarrow \mu_n \longrightarrow \mathbb{G}_m \xrightarrow{(-)} \mathbb{G}_m \longrightarrow 1$$

and Hilbert 90 we obtain that

$$H^1_{et}(\mathbb{C}(S), \mu_n) = \mathbb{C}(S)^* / \mathbb{C}(S)^{*n}$$

The first claim follows from the exact diagram describing divisors of rational functions given in figure 3.8 By the coniveau spectral sequence we have that $H^1_{et}(S, \mu_n)$ is equal to the kernel of the morphism

$$H^1_{et}(\mathbb{C}(S), \mu_n) \xrightarrow{\gamma} \oplus_C \mathbb{Z}_n$$

and in particular, $H^1(S, \mu_n) \hookrightarrow H^1(\mathbb{C}(S), \mu_n)$.

As for the second claim, an element in $H^1(S, \mu_n)$ determines a cyclic extension $L = \mathbb{C}(S) \sqrt[n]{f}$ with $f \in \mathbb{C}(S)^* / \mathbb{C}(S)^{*n}$ such that in each field component L_i of L there is an étale cover $T_i \longrightarrow S$ with $\mathbb{C}(T_i) = L_i$. By assumption no nontrivial étale covers exist whence $f = 1 \in \mathbb{C}(S)^* / \mathbb{C}(S)^{*n}$. ❑

If we invoke another major tool in étale cohomology of schemes, *Poincaré duality*, see, for example [82, VI,§11], we obtain the following information on the cohomology groups for S.

PROPOSITION 3.11
(Poincaré duality for S) If S is simply connected, then

1. $H^0_{et}(S, \mu_n) = \mu_n$

2. $H^1_{et}(S, \mu_n) = 0$

3. $H^3_{et}(S, \mu_n) = 0$

4. $H^4_{et}(S, \mu_n) = \mu_n^{-1}$

PROOF   The third claim follows from the second as both groups are dual to each other. The last claim follows from the fact that for a smooth irreducible projective variety X of dimension d

$$H^{2d}_{et}(X, \mu_n) \simeq \mu_n^{\otimes 1-d}$$

▯

We are now in a position to state and prove the important issue.

THEOREM 3.11
(Artin-Mumford exact sequence) If S is a simply connected smooth projective surface, then the sequence

$$0 \longrightarrow Br_n(S) \longrightarrow Br_n(\mathbb{C}(S)) \longrightarrow \oplus_C \mathbb{C}(C)^*/\mathbb{C}(C)^{*n} \longrightarrow$$

$$\longrightarrow \oplus_P \mu_n^{-1} \longrightarrow \mu_n^{-1} \longrightarrow 0$$

is exact.

PROOF   The top complex in the first term of the coniveau spectral sequence for S was

$$H^2(\mathbb{C}(S), \mu_n) \xrightarrow{\alpha} \oplus_C H^1(\mathbb{C}(C), \mathbb{Z}_n) \xrightarrow{\beta} \oplus_P \mu_n$$

The second term of the spectral sequence (which is also the limiting term) has the following form

$\vdots$	$\vdots$	$\vdots$	$\vdots$	
0	0	0	0	$\cdots$
$Ker\ \alpha$	$Ker\ \beta/Im\ \alpha$	$Coker\ \beta$	0	$\cdots$
$Ker\ \gamma$	$Coker\ \gamma$	0	0	$\cdots$
μ_n	0	0	0	$\cdots$

By the foregoing lemma we know that $Coker \, \gamma = 0$. By Poincare duality we know that $Ker \, \beta = Im \, \alpha$ and $Coker \, \beta = \mu_n^{-1}$. Hence, the top complex was exact in its middle term and can be extended to an exact sequence

$$0 \longrightarrow H^2(S, \mu_n) \longrightarrow H^2(\mathbb{C}(S), \mu_n) \longrightarrow \oplus_C H^1(\mathbb{C}(C), \mathbb{Z}_n) \longrightarrow$$

$$\oplus_P \mu_n^{-1} \longrightarrow \mu_n^{-1} \longrightarrow 0$$

As $\mathbb{Z}_n \simeq \mu_n$ the third term is equal to $\oplus_C \mathbb{C}(C)^*/\mathbb{C}(C)^{*n}$ by the argument given before and the second term we remember to be $Br_n(\mathbb{C}(S))$. The identification of $Br_n(S)$ with $H^2(S, \mu_n)$ will be explained below. ◻

Some immediate consequences can be drawn from this. For a smooth simply connected surface S, $Br_n(S)$ is a birational invariant (it is the birational invariant $F_n^{2,1}(\mathbb{C}(S)/\mathbb{C})$ of the foregoing section. In particular, if $S = \mathbb{P}^2$ we have that $Br_n(\mathbb{P}^2) = 0$ and as

$$0 \longrightarrow Br_n \, \mathbb{C}(x, y) \longrightarrow \oplus_C \mathbb{C}(C)^*/\mathbb{C}(C)^{*n} \longrightarrow \oplus_P \mu_n^{-1} \longrightarrow \mu_n \longrightarrow 0$$

we obtain a description of $Br_n \, \mathbb{C}(x, y)$ by a certain geocombinatorial package, which we call a $\mathbb{Z}_n$-*wrinkle* over $\mathbb{P}^2$. A $\mathbb{Z}_n$-wrinkle is determined by

- A finite collection $\mathcal{C} = \{C_1, \ldots, C_k\}$ of *irreducible curves* in $\mathbb{P}^2$, that is, $C_i = V(F_i)$ for an irreducible form in $\mathbb{C}[X, Y, Z]$ of degree d_i.

- A finite collection $\mathcal{P} = \{P_1, \ldots, P_l\}$ of *points* of $\mathbb{P}^2$ where each P_i is either an intersection point of two or more C_i or a singular point of some C_i.

- For each $P \in \mathcal{P}$ the *branch-data* $b_P = (b_1, \ldots, b_{i_P})$ with $b_i \in \mathbb{Z}_n = \mathbb{Z}/n\mathbb{Z}$ and $\{1, \ldots, i_P\}$ the different branches of $\mathcal{C}$ in P. These numbers must satisfy the admissibility condition

$$\sum_i b_i = 0 \in \mathbb{Z}_n$$

  for every $P \in \mathcal{P}$

- for each $C \in \mathcal{C}$ we fix a cyclic $\mathbb{Z}_n$-cover of smooth curves

$$D \longrightarrow\!\!\!\!\!\rightarrow \tilde{C}$$

of the desingularization $\tilde{C}$ of C, which is compatible with the branch-data. That is, if $Q \in \tilde{C}$ corresponds to a $\mathcal{C}$-branch b_i in P, then D is ramified in Q with stabilizer subgroup

$$Stab_Q = \langle b_i \rangle \subset \mathbb{Z}_n$$

For example, a portion of a $\mathbb{Z}_4$-wrinkle can have the following picture

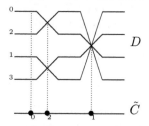

It is clear that the cover-data is the most intractable part of a $\mathbb{Z}_n$-wrinkle, so we want to have some control on the covers $D \longrightarrow \tilde{C}$. Let $\{Q_1, \ldots, Q_z\}$ be the points of $\tilde{C}$ where the cover ramifies with branch numbers $\{b_1, \ldots, b_z\}$, then D is determined by a continuous module structure (that is, a cofinite subgroup acts trivially) of

$$\pi_1(\tilde{C} - \{Q_1, \ldots, Q_z\}) \text{ on } \mathbb{Z}_n$$

where the fundamental group of the Riemann surface $\tilde{C}$ with z punctures is known (topologically) to be equal to the group

$$\langle u_1, v_1, \ldots, u_g, v_g, x_1, \ldots, x_z \rangle / ([u_1, v_1] \ldots [u_g, v_g] x_1 \ldots x_z)$$

where g is the genus of $\tilde{C}$. The action of x_i on $\mathbb{Z}_n$ is determined by multiplication with b_i. In fact, we need to use the étale fundamental group, see [82], but this group has the same finite continuous modules as the topological fundamental group.

Example 3.7 Covers of $\mathbb{P}^1$ and elliptic curves

1. If $\tilde{C} = \mathbb{P}^1$ then $g = 0$ and hence $\pi_1(\mathbb{P}^1 - \{Q_1, \ldots, Q_z\}$ is zero if $z \leq 1$ (whence no covers exist) and is $\mathbb{Z}$ if $z = 2$. Hence, there exists a unique cover $D \longrightarrow \mathbb{P}^1$ with branch-data $(1, -1)$ in say $(0, \infty)$ namely with D the normalization of $\mathbb{P}^1$ in $\mathbb{C}(\sqrt[n]{x})$.

2. If $\tilde{C} = E$ an elliptic curve, then $g = 1$. Hence, $\pi_1(C) = \mathbb{Z} \oplus \mathbb{Z}$ and there exist unramified $\mathbb{Z}_n$-covers. They are given by the isogenies

$$E' \longrightarrow E$$

where E' is another elliptic curve and $E = E'/\langle \tau \rangle$ where τ is an n-torsion point on E'.

□

Any n-fold cover $D \longrightarrow \tilde{C}$ is determined by a function $f \in \mathbb{C}(C)^*/\mathbb{C}(C)^{*n}$. This allows us to put a group-structure on the equivalence classes of $\mathbb{Z}_n$-wrinkles. In particular, we call a wrinkle *trivial* provided all coverings

$D_i \longrightarrow \tilde{C}_i$ are trivial (that is, D_i is the disjoint union of n copies of $\tilde{C}_i$). The Artin-Mumford theorem for $\mathbb{P}^2$ can now be stated as follows.

THEOREM 3.12
If Δ is a central simple $\mathbb{C}(x,y)$-algebra of dimension n^2, then Δ determines uniquely a $\mathbb{Z}_n$-wrinkle on $\mathbb{P}^2$. Conversely, any $\mathbb{Z}_n$-wrinkle on $\mathbb{P}^2$ determines a unique division $\mathbb{C}(x,y)$- algebra whose class in the Brauer group has order n.

Example 3.8
If S is not necessarily simply connected, any class in $Br(\mathbb{C}(S))_n$ still determines a $\mathbb{Z}_n$-wrinkle. ⬜

Example 3.9
If X is a smooth irreducible rational projective variety of dimension d, the obstruction to classifying $Br(\mathbb{C}(X))_n$ by $\mathbb{Z}_n$-wrinkles is given by $H^3_{et}(X, \mu_n)$. ⬜

We will give a ring theoretical interpretation of the maps in the Artin-Mumford sequence. Observe that nearly all maps are those of the top complex of the first term in the coniveau spectral sequence for S. We gave an explicit description of them using discrete valuation rings. The statements below follow from this description.

Let us consider a discrete valuation ring A with field of fractions K and residue field k. Let Δ be a central simple K-algebra of dimension n^2.

DEFINITION 3.8
An A-subalgebra Λ of Δ will be called an A-order if it is a free A-module of rank n^2 with $\Lambda.K = \Delta$. An A-order is said to be maximal *if it is not properly contained in any other order.*

In order to study maximal orders in Δ (they will turn out to be all conjugated), we consider the completion $\hat{A}$ with respect to the m-adic filtration where $m = At$ with t a uniformizing parameter of A. $\hat{K}$ will denote the field of fractions of $\hat{A}$ and $\hat{\Delta} = \Delta \otimes_K \hat{K}$.

Because $\hat{\Delta}$ is a central simple $\hat{K}$-algebra of dimension n^2 it is of the form

$$\hat{\Delta} = M_t(D)$$

where D is a division algebra with center $\hat{K}$ of dimension s^2 and hence $n = s.t$. We call t the capacity of Δ at A.

In D we can construct a unique maximal $\hat{A}$-order Γ, namely, the integral closure of $\hat{A}$ in D. We can view Γ as a discrete valuation ring extending the valuation v defined by A on K. If $v : \hat{K} \longrightarrow \mathbb{Z}$, then this extended valuation

$$w : D \longrightarrow n^{-2}\mathbb{Z} \text{ is defined as } w(a) = (\hat{K}(a) : \hat{K})^{-1} v(N_{\hat{K}(a)/\hat{K}}(a))$$

for every $a \in D$ where $\hat{K}(a)$ is the subfield generated by a and N is the norm map of fields.

The image of w is a subgroup of the form $e^{-1}\mathbb{Z} \hookrightarrow n^{-2}.\mathbb{Z}$. The number $e = e(D/\hat{K})$ is called the *ramification index* of D over $\hat{K}$. We can use it to normalize the valuation w to

$$v_D : D \longrightarrow \mathbb{Z} \text{ defined by } v_D(a) = \frac{e}{n^2}v(N_{D/\hat{K}}(a))$$

With these conventions we have that $v_D(t) = e$.

The maximal order Γ is then the subalgebra of all elements $a \in D$ with $v_D(a) \geq 0$. It has a unique maximal ideal generated by a prime element T and we have that $\overline{\Gamma} = \frac{\Gamma}{T\Gamma}$ is a division algebra finite dimensional over $\hat{A}/t\hat{A} = k$ (but not necessarily having k as its center).

The *inertial degree* of D over $\hat{K}$ is defined to be the number $f = f(D/\hat{K}) = (\overline{\Gamma} : k)$ and one shows that

$$s^2 = e.f \text{ and } e \mid s \text{ whence } s \mid f$$

After this detour, we can now take $\Lambda = M_t(\Gamma)$ as a maximal $\hat{A}$-order in $\hat{\Delta}$. One shows that all other maximal $\hat{A}$-orders are conjugated to Λ. Λ has a unique maximal ideal M with $\overline{\Lambda} = M_t(\overline{\Gamma})$.

DEFINITION 3.9   *With notations as above, we call the numbers* $e = e(D/\hat{K})$, $f = f(D/\hat{K})$ *and* t *resp. the* ramification, inertia *and* capacity *of the central simple algebra* Δ *at* A. *If* $e = 1$ *we call* Λ *an* Azumaya algebra *over* A, *or equivalently, if* $\Lambda/t\Lambda$ *is a central simple* k-algebra of dimension n^2.

Now let us consider the case of a discrete valuation ring A in K such that the residue field k is $Tsen^1$. The center of the division algebra $\overline{\Gamma}$ is a finite dimensional field extension of k and hence is also $Tsen^1$ whence has a trivial Brauer group and therefore must coincide with $\overline{\Gamma}$. Hence

$$\overline{\Gamma} = k(\overline{a})$$

a commutative field, for some $a \in \Gamma$. But then, $f \leq s$ and we have $e = f = s$ and $k(\overline{a})$ is a cyclic degree s field extension of k.

Because $s \mid n$, the cyclic extension $k(\overline{a})$ determines an element of $H^1_{et}(k, \mathbb{Z}_n)$.

DEFINITION 3.10   *Let* Z *be a normal domain with field of fractions* K *and let* Δ *be a central simple* K-algebra of dimension n^2. *A* Z-order B *is a subalgebra that is a finitely generated* Z-module. *It is called* maximal *if it is not properly contained in any other order. One can show that* B *is a maximal* Z-order *if and only if* $\Lambda = B_p$ *is a maximal order over the discrete valuation ring* $A = Z_p$ *for every height one prime ideal* p *of* Z.

Return to the situation of an irreducible smooth projective surface S. If Δ is a central simple $\mathbb{C}(S)$-algebra of dimension n^2, we define a maximal order as a sheaf $\mathcal{A}$ of $\mathcal{O}_S$-orders in Δ which for an open affine cover $U_i \hookrightarrow S$ is such that

$$A_i = \Gamma(U_i, \mathcal{A}) \text{ is a maximal } Z_i = \Gamma(U_i, \mathcal{O}_S) \text{ order in } \Delta$$

Any irreducible curve C on S defines a discrete valuation ring on $\mathbb{C}(S)$ with residue field $\mathbb{C}(C)$ which is $Tsen^1$. Hence, the above argument can be applied to obtain from $\mathcal{A}$ a cyclic extension of $\mathbb{C}(C)$, that is, an element of $\mathbb{C}(C)^*/\mathbb{C}(C)^{*n}$.

DEFINITION 3.11 *We call the union of the curves C such that $\mathcal{A}$ determines a nontrivial cyclic extension of $\mathbb{C}(C)$ the* ramification divisor *of Δ (or of $\mathcal{A}$).*

The map in the Artin-Mumford exact sequence

$$Br_n(\mathbb{C}(S)) \longrightarrow \bigoplus_C H^1_{et}(\mathbb{C}(C), \mu_n)$$

assigns to the class of Δ the cyclic extensions introduced above.

DEFINITION 3.12 *An S-Azumaya algebra (of index n) is a sheaf of maximal orders in a central simple $\mathbb{C}(S)$-algebra Δ of dimension n^2 such that it is Azumaya at each curve C, that is, such that $[\Delta]$ lies in the kernel of the above map.*

Observe that this definition of Azumaya algebra coincides with the one given in the discussion of twisted forms of matrices. One can show that if $\mathcal{A}$ and $\mathcal{A}'$ are S-Azumaya algebras of index n resp. n', then $\mathcal{A} \otimes_{\mathcal{O}_S} \mathcal{A}'$ is an Azumaya algebra of index $n.n'$. We call an Azumaya algebra trivial if it is of the form $End(\mathcal{P})$ where $\mathcal{P}$ is a vectorbundle over S. The equivalence classes of S-Azumaya algebras can be given a group-structure called the Brauer-group $Br(S)$ of the surface S.

Let us briefly sketch how Michael Artin and David Mumford used their sequence to construct unirational nonrational threefolds via the *Brauer-Severi varieties*. Let K be a field and $\Delta = (a, b)_K$ the quaternion algebra determined by $a, b \in K^*$. That is

$$\Delta = K.1 \oplus K.i \oplus K.j \oplus K.ij \quad \text{with} \quad i^2 = a \quad j^2 = b \quad \text{and} \quad ji = -ij$$

The norm map on Δ defines a conic in $\mathbb{P}^2_K$ called the Brauer-Severi variety of Δ

$$BS(\Delta) = \mathbb{V}(x^2 - ay^2 - bz^2) \hookrightarrow \mathbb{P}^2_K = \text{proj } K[x, y, z]$$

Its characteristic property is that a field extension L of K admits an L-rational point on $BS(\Delta)$ if and only if $\Delta \otimes_K L$ admits zero-divisors and hence is isomorphic to $M_2(L)$.

In general, let $\mathbb{K}$ be the algebraic closure of K, then we have seen that the Galois cohomology pointed set

$$H^1(Gal(\mathbb{K}/K), PGL_n(\mathbb{K}))$$

classifies at the same time the isomorphism classes of the following geometric and algebraic objects

- Brauer-Severi K-varieties BS, which are smooth projective K-varieties such that $BS_{\mathbb{K}} \simeq \mathbb{P}_{\mathbb{K}}^{n-1}$.

- Central simple K-algebras Δ, which are K-algebras of dimension n^2 such that $\Delta \otimes_K \mathbb{K} \simeq M_n(\mathbb{K})$.

The one-to-one correspondence between these two sets is given by associating to a central simple K-algebra Δ its Brauer-Severi variety $BS(\Delta)$, which represents the functor associating to a field extension L of K the set of left ideals of $\Delta \otimes_K L$ that have L-dimension equal to n. In particular, $BS(\Delta)$ has an L-rational point if and only if $\Delta \otimes_K L \simeq M_n(L)$ and hence the geometric object $BS(\Delta)$ encodes the algebraic splitting behavior of Δ.

Now restrict to the case when K is the function field $\mathbb{C}(X)$ of a projective variety X and let Δ be a central simple $\mathbb{C}(X)$-algebra of dimension n^2. Let $\mathcal{A}$ be a sheaf of $\mathcal{O}_X$-orders in Δ then we one can show that there is a Brauer-Severi scheme $BS(\mathcal{A})$, which is a projective space bundle over X with general fiber isomorphic to $\mathbb{P}^{n-1}(\mathbb{C})$ embedded in $\mathbb{P}^N(\mathbb{C})$ where $N = \binom{n+k-1}{k} - 1$. Over an arbitrary point of x the fiber may degenerate.

For example if $n = 2$ the $\mathbb{P}^1(\mathbb{C})$ embedded as a conic in $\mathbb{P}^2(\mathbb{C})$ can degenerate into a pair of $\mathbb{P}^1(\mathbb{C})$'s. Now, let us specialize further and consider the case when $X = \mathbb{P}^2$. Consider E_1 and E_2 two elliptic curves in $\mathbb{P}^2$ and take $\mathcal{C} = \{E_1, E_2\}$ and $\mathcal{P} = \{P_1, \ldots P_9\}$ the intersection points and all the branch data zero. Let E_i' be a twofold unramified cover of E_i, by the Artin-Mumford result there is a quaternion algebra Δ corresponding to this $\mathbb{Z}_2$-wrinkle.

Next, blow up the intersection points to get a surface S with disjoint elliptic curves C_1 and C_2. Now take a maximal $\mathcal{O}_S$ order in Δ then the relevance of the curves C_i is that they are the locus of the points $s \in S$ where $\overline{\mathcal{A}}_s \not\simeq M_2(\mathbb{C})$, the so called *ramification locus* of the order $\mathcal{A}$. The local structure of $\mathcal{A}$ in a point $s \in S$ is

- when $s \notin C_1 \cup C_2$, then $\mathcal{A}_s$ is an Azumaya $\mathcal{O}_{S,s}$-algebra in Δ

- when $s \in C_i$, then $\mathcal{A}_s = \mathcal{O}_{S,s}.1 \oplus \mathcal{O}_{S,s}.i \oplus \mathcal{O}_{S,s}.j \oplus \mathcal{O}_{S,s}.ij$ with

$$\begin{cases} i^2 &= a \\ j^2 &= bt \\ ji &= -ij \end{cases}$$

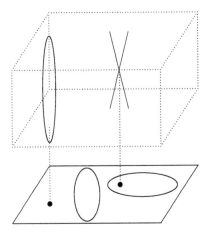

FIGURE 3.9:   The Artin-Mumford bundle.

where $t = 0$ is a local equation for C_i and a and b are units in $\mathcal{O}_{S,s}$.

In chapter 5 we will see that this is the local description of a Cayley-smooth order over a smooth surface in a quaternion algebra. Artin and Mumford then define the Brauer-Severi scheme of $\mathcal{A}$ as representing the functor that assigns to an S-scheme S' the set of left ideals of $\mathcal{A} \otimes_{\mathcal{O}_S} \mathcal{O}_{S'}$ which are locally free of rank 2. Using the local description of $\mathcal{A}$ they show that $BS(\mathcal{A})$ is a projective space bundle over S as in figure 3.9 with the properties that $BS(\mathcal{A})$ is a smooth variety and the projection morphism $BS(\mathcal{A}) \xrightarrow{\pi} S$ is flat, all of the geometric fibers being isomorphic to $\mathbb{P}^1$ (resp. to $\mathbb{P}^1 \vee \mathbb{P}^1$) whenever $s \notin C_1 \cup C_2$ (resp. $s \in C_1 \cup C_2$).

Finally, for specific starting configurations E_1 and E_2, they prove that the obtained Brauer-Severi variety $BS(\mathcal{A})$ cannot be rational because there is torsion in $H^4(BS(\mathcal{A}), \mathbb{Z}_2)$, whereas $BS(\mathcal{A})$ can be shown to be unirational.

3.7   Normal spaces

In the next section we will see that in the étale topology we can describe the local structure of representation varieties in a neighborhood of a closed orbit in terms of the normal space to this orbit. In this section we will give a representation theoretic description of this normal space.

We recall some standard facts about tangent spaces first. Let X be a not necessarily reduced affine variety with coordinate ring $\mathbb{C}[X] = \mathbb{C}[x_1, \ldots, x_n]/I$. If the origin $o = (0, \ldots, 0) \in \mathbb{V}(I)$, elements of I have no constant terms and we can write any $p \in I$ as

$$p = \sum_{i=1}^{\infty} p^{(i)} \quad \text{with } p^{(i)} \text{ homogeneous of degree } i.$$

The *order* $ord(p)$ is the least integer $r \geq 1$ such that $p^{(r)} \neq 0$. Define the following two ideals in $\mathbb{C}[x_1, \ldots, x_n]$

$$I_l = \{p^{(1)} \mid p \in I\} \quad \text{and} \quad I_m = \{p^{(r)} \mid p \in I \text{ and } ord(p) = r\}.$$

The subscripts l (respectively m) stand for *linear terms* (respectively, terms of *minimal* degree).

The *tangent space to* X *in* o , $T_o(X)$ is by definition the subscheme of $\mathbb{C}^n$ determined by I_l. Observe that

$$I_l = (a_{11}x_1 + \ldots + a_{1n}x_n, \ldots, a_{l1}x_1 + \ldots + a_{ln}x_n)$$

for some $l \times n$ matrix $A = (a_{ij})_{i,j}$ of rank l. That is, we can express all x_k as linear combinations of some $\{x_{i_1}, \ldots, x_{i_{n-l}}\}$, but then clearly

$$\mathbb{C}[T_o(X)] = \mathbb{C}[x_1, \ldots, x_n]/I_l = \mathbb{C}[x_{i_1}, \ldots, x_{i_{n-l}}]$$

In particular, $T_o(X)$ is reduced and is a linear subspace of dimension $n - l$ in $\mathbb{C}^n$ through the point o.

Next, consider an arbitrary geometric point x of X with coordinates $(a_1, \ldots, a_n)$. We can translate x to the origin o and the translate of X is then the scheme defined by the ideal

$$(f_1(x_1 + a_1, \ldots, x_n + a_n), \ldots, f_k(x_1 + a_1, \ldots, x_n + a_n))$$

Now, the linear term of the translated polynomial $f_i(x_1 + a_1, \ldots, x_n + a_n)$ is equal to

$$\frac{\partial f_i}{\partial x_1}(a_1, \ldots, a_n)x_1 + \ldots + \frac{\partial f_i}{\partial x_n}(a_1, \ldots, a_n)x_n$$

and hence the *tangent space to* X *in* x , $T_x(X)$ is the linear subspace of $\mathbb{C}^n$ defined by the set of zeroes of the linear terms

$$T_x(X) = \mathbb{V}(\sum_{j=1}^{n} \frac{\partial f_1}{\partial x_j}(x)x_j, \ldots, \sum_{j=1}^{n} \frac{\partial f_k}{\partial x_j}(x)x_j) \longrightarrow \mathbb{C}^n$$

In particular, the dimension of this linear subspace can be computed from the *Jacobian matrix* in x associated with the polynomials $(f_1, \ldots, f_k)$

$$dim \ T_x(X) = n - rk \begin{bmatrix} \frac{\partial f_1}{\partial x_1}(x) & \cdots & \frac{\partial f_1}{\partial x_n}(x) \\ \vdots & & \vdots \\ \frac{\partial f_k}{\partial x_1}(x) & \cdots & \frac{\partial f_k}{\partial x_n}(x) \end{bmatrix}.$$

Let $\mathbb{C}[\varepsilon]$ be the *algebra of dual numbers* , that is, $\mathbb{C}[\varepsilon] \simeq \mathbb{C}[y]/(y^2)$. Consider a $\mathbb{C}$-algebra morphism

$$\mathbb{C}[x_1, \ldots, x_n] \xrightarrow{\phi} \mathbb{C}[\varepsilon] \quad \text{defined by } x_i \mapsto a_i + c_i \varepsilon$$

Because $\varepsilon^2 = 0$ it is easy to verify that the image of a polynomial $f(x_1, \ldots, x_n)$ under ϕ is of the form

$$\phi(f(x_1, \ldots, x_n)) = f(a_1, \ldots, a_n) + \sum_{j=1}^{n} \frac{\partial f}{\partial x_j}(a_1, \ldots, a_n) c_j \varepsilon$$

Therefore, ϕ factors through I, that is, $\phi(f_i) = 0$ for all $1 \leq i \leq k$, if and only if $(c_1, \ldots, c_n) \in T_x(\mathbb{X})$. Hence, we can also identify the tangent space to $\mathbb{X}$ in x with the algebra morphisms $\mathbb{C}[\mathbb{X}] \xrightarrow{\phi} \mathbb{C}[\varepsilon]$ whose composition with the projection $\pi : \mathbb{C}[\varepsilon] \longrightarrow \mathbb{C}$ (sending ε to zero) is the evaluation in $x = (a_1, \ldots, a_n)$. That is, let $ev_x \in \mathbb{X}(\mathbb{C})$ be the point corresponding to evaluation in x, then

$$T_x(\mathbb{X}) = \{\phi \in \mathbb{X}(\mathbb{C}[\varepsilon]) \mid \mathbb{X}(\pi)(\phi) = ev_x\}$$

The following two examples compute the tangent spaces to the (trace preserving) representation varieties.

Example 3.10 Tangent space to $\mathbf{rep}_n$

Let A be an affine $\mathbb{C}$-algebra generated by $\{a_1, \ldots a_m\}$ and $\rho : A \longrightarrow M_n(\mathbb{C})$ an algebra morphism, that is, $\rho \in rep_n A$. We call a linear map $A \xrightarrow{D} M_n(\mathbb{C})$ a ρ-*derivation* if and only if for all $a, a' \in A$ we have that

$$D(aa') = D(a).\rho(a') + \rho(a).D(a')$$

We denote the vector space of all ρ-derivations of A by $Der_\rho(A)$. Observe that any ρ-derivation is determined by its image on the generators a_i, hence $Der_\rho(A) \subset M_n^m$. We claim that

$$T_\rho(\mathbf{rep}_n \ A) = Der_\rho(A)$$

Indeed, we know that $\mathbf{rep}_n \ A(\mathbb{C}[\varepsilon])$ is the set of algebra morphisms

$$A \xrightarrow{\phi} M_n(\mathbb{C}[\varepsilon])$$

By the functorial characterization of tangentspaces we have that $T_\rho(\mathbf{rep}_n \ A)$ is equal to

$$\{D : A \longrightarrow M_n(\mathbb{C}) \text{ linear} \mid \rho + D\varepsilon : A \longrightarrow M_n(\mathbb{C}[\varepsilon]) \text{ is an algebra map}\}.$$

Because ρ is an algebra morphism, the algebra map condition

$$\rho(aa') + D(aa')\varepsilon = (\rho(a) + D(a)\varepsilon).(\rho(a') + D(a')\varepsilon)$$

is equivalent to D being a ρ-derivation. ⬜

Example 3.11 Tangent space to trep_n

Let A be a Cayley-Hamilton algebra of degree n with trace map tr_A and trace generated by $\{a_1, \ldots, a_m\}$. Let $\rho \in \text{trep}_n A$, that is, $\rho : A \longrightarrow M_n(\mathbb{C})$ is a *trace preserving* algebra morphism. Because $\text{trep}_n A(\mathbb{C}[\varepsilon])$ is the set of all trace preserving algebra morphisms $A \longrightarrow M_n(\mathbb{C}[\varepsilon])$ (with the usual trace map tr on $M_n(\mathbb{C}[\varepsilon])$) and the previous example one verifies that

$$T_\rho(\text{trep}_n A) = Der_\rho^{tr}(A) \subset Der_\rho(A)$$

the subset of *trace preserving ρ-derivations D*, that is, those satisfying

$$D \circ tr_A = tr \circ D \qquad tr_A \qquad tr$$

$$A \xrightarrow{D} M_n(\mathbb{C})$$
$$A \xrightarrow{D} M_n(\mathbb{C})$$

Again, using this property and the fact that A is *trace* generated by $\{a_1, \ldots, a_m\}$ a trace preserving ρ-derivation is determined by its image on the a_i so is a subspace of M_n^m. ⬜

The *tangent cone to* X *in* o, $TC_o(X)$, is by definition the subscheme of $\mathbb{C}^n$ determined by I_m, that is

$$\mathbb{C}[TC_o(X)] = \mathbb{C}[x_1, \ldots, x_n]/I_m.$$

It is called a *cone* because if c is a point of the underlying variety of $TC_o(X)$, then the line $l = \overrightarrow{oc}$ is contained in this variety because I_m is a graded ideal. Further, observe that as $I_l \subset I_m$, the tangent cone is a closed subscheme of the tangent space at X in o. Again, if x is an arbitrary geometric point of X we define the *tangent cone to* X *in* x , $TC_x(X)$ as the tangent cone $TC_o(X')$ where X' is the translated scheme of X under the translation taking x to o. Both the tangent space and tangent cone contain *local information* of the scheme X in a neighborhood of x.

Let m_x be the maximal ideal of $\mathbb{C}[X]$ corresponding to x (that is, the ideal of polynomial functions vanishing in x). Then, its complement $S_x = \mathbb{C}[X] - m_x$ is a multiplicatively closed subset and the *local algebra* $\mathcal{O}_x(X)$ is the corresponding localization $\mathbb{C}[X]_{S_x}$. It has a unique maximal ideal m_x with residue field $\mathcal{O}_x(X)/m_x = \mathbb{C}$. We equip the local algebra $\mathcal{O}_x = \mathcal{O}_x(X)$ with the m_x-adic filtration that is the increasing $\mathbb{Z}$-filtration

$$\mathcal{F}_x : \quad \ldots \subset \mathfrak{m}^i \subset \mathfrak{m}^{i-1} \subset \ldots \subset \mathfrak{m} \subset \mathcal{O}_x = \mathcal{O}_x = \ldots = \mathcal{O}_x = \ldots$$

with *associated graded algebra*

$$gr(\mathcal{O}_x) = \quad \cdots \oplus \frac{\mathfrak{m}_x^i}{\mathfrak{m}_x^{i+1}} \oplus \frac{\mathfrak{m}_x^{i-1}}{\mathfrak{m}_x^i} \oplus \cdots \oplus \frac{\mathfrak{m}_x}{\mathfrak{m}_x^2} \oplus \mathbb{C} \oplus 0 \oplus \cdots \oplus 0 \oplus \cdots$$

PROPOSITION 3.12
If x is a geometric point of the affine scheme X, *then*

1. $\mathbb{C}[T_x(\mathbf{X})]$ *is isomorphic to the polynomial algebra* $\mathbb{C}[\frac{\mathfrak{m}_x}{\mathfrak{m}_x^2}]$.

2. $\mathbb{C}[TC_x(\mathbf{X})]$ *is isomorphic to the associated graded algebra* $gr(\mathcal{O}_x(\mathbf{X}))$.

PROOF   After translating we may assume that $x = o$ lies in $\mathbb{V}(I) \hookrightarrow \mathbb{C}^n$. That is

$$\mathbb{C}[\mathbf{X}] = \mathbb{C}[x_1, \ldots, x_n]/I \quad \text{and} \quad \mathfrak{m}_x = (x_1, \ldots, x_n)/I$$

(1): Under these identifications we have

$$\frac{\mathfrak{m}_x}{\mathfrak{m}_x^2} \simeq \mathfrak{m}_x/\mathfrak{m}_x^2$$
$$\simeq (x_1, \ldots, x_n)/((x_1, \ldots, x_n)^2 + I)$$
$$\simeq (x_1, \ldots, x_n)/((x_1, \ldots, x_n)^2 + I_l)$$

and as I_l is generated by linear terms it follows that the polynomial algebra on $\frac{\mathfrak{m}_x}{\mathfrak{m}_x^2}$ is isomorphic to the quotient algebra $\mathbb{C}[x_1, \ldots, x_n]/I_l$, which is by definition the coordinate ring of the tangent space.

(2): Again using the above identifications we have

$$gr(\mathcal{O}_x) \simeq \oplus_{i=0}^{\infty} \mathfrak{m}_x^i/\mathfrak{m}_x^{i+1}$$
$$\simeq \oplus_{i=0}^{\infty} \mathfrak{m}_x^i/\mathfrak{m}_x^{i+1}$$
$$\simeq \oplus_{i=0}^{\infty} (x_1, \ldots, x_n)^i/((x_1, \ldots, x_n)^{i+1} + (I \cap (x_1, \ldots, x_n)^i))$$
$$\simeq \oplus_{i=0}^{\infty} (x_1, \ldots, x_n)^i/((x_1, \ldots, x_n)^{i+1} + I_m(i))$$

where $I_m(i)$ is the homogeneous part of I_m of degree i. On the other hand, the i-th homogeneous part of $\mathbb{C}[x_1, \ldots, x_n]/I_m$ is equal to

$$\frac{(x_1, \ldots, x_n)^i}{(x_1, \ldots, x_n)^{i+1} + I_m(i)}$$

we obtain the required isomorphism. □

This gives a third interpretation of the tangent space as

$$T_x(\mathbf{X}) = Hom_{\mathbb{C}}(\frac{\mathfrak{m}_x}{\mathfrak{m}_x^2}, \mathbb{C}) = Hom_{\mathbb{C}}(\frac{\mathfrak{m}_x}{\mathfrak{m}_x^2}, \mathbb{C}).$$

Hence, we can also view the tangent space $T_x(X)$ as the space of *point derivations* $Der_x(\mathcal{O}_x)$ on $\mathcal{O}_x(X)$ (or of the point derivations $Der_x(\mathbb{C}[X])$ on $\mathbb{C}[X]$). That is, $\mathbb{C}$-linear maps $D : \mathcal{O}_x \longrightarrow \mathbb{C}$ (or $D : \mathbb{C}[X] \longrightarrow \mathbb{C}$) such that for all functions f, g we have

$$D(fg) = D(f)g(x) + f(x)D(g)$$

If we define the *local dimension of an affine scheme* X *in a geometric point* x dim_x X to be the maximal dimension of irreducible components of the reduced variety X passing through x, then

$$dim_x \; X = dim_o \; TC_x(X)$$

We say that X *is nonsingular at* x (or equivalently, that x *is a nonsingular point of* X) if the tangent cone to X in x coincides with the tangent space to X in x. An immediate consequence is discussed below.

PROPOSITION 3.13
If X *is nonsingular at* x, *then* $\mathcal{O}_x(X)$ *is a domain. That is, in a Zariski neighborhood of* x, X *is an irreducible variety.*

PROOF   If X is nonsingular at x, then

$$gr(\mathcal{O}_x) \simeq \mathbb{C}[TC_x(X)] = \mathbb{C}[T_x(X)]$$

the latter one being a polynomial algebra whence a domain. Now, let $0 \neq a, b \in \mathcal{O}_x$ then there exist k, l such that $a \in \mathfrak{m}^k - \mathfrak{m}^{k+1}$ and $b \in \mathfrak{m}^l - \mathfrak{m}^{l+1}$, that is $\bar{a}$ is a nonzero homogeneous element of $gr(\mathcal{O}_x)$ of degree $-k$ and $\bar{b}$ one of degree $-l$. But then, $\bar{a}.\bar{b} \in \mathfrak{m}^{k+l} - \mathfrak{m}^{k+l-1}$ hence certainly $a.b \neq 0$ in $\mathcal{O}_x$.

Now, consider the natural map $\phi : \mathbb{C}[X] \longrightarrow \mathcal{O}_x$. Let $\{P_1, \dots, P_l\}$ be the minimal prime ideals of $\mathbb{C}[X]$. All but one of them, say $P_1 = \phi^{-1}(0)$, extend to the whole ring $\mathcal{O}_x$. Taking the product of f functions $f_i \in P_i$ nonvanishing in x for $2 \leq i \leq l$ gives the Zariski open set $X(f)$ containing x and whose coordinate ring is a domain, whence $X(f)$ is an affine irreducible variety.   □

When restricting to nonsingular points we reduce to irreducible affine varieties. From the Jacobian condition it follows that nonsingularity is a Zariski open condition on X and by the implicit function theorem, X is a complex manifold in a neighborhood of a nonsingular point.

Let $X \xrightarrow{\phi} Y$ be a morphism of affine varieties corresponding to the algebra morphism $\mathbb{C}[Y] \xrightarrow{\phi^*} \mathbb{C}[X]$. Let x be a geometric point of X and $y = \phi(x)$. As $\phi^*(\mathfrak{m}_y) \subset \mathfrak{m}_x$, ϕ induces a linear map $\frac{\mathfrak{m}_y}{\mathfrak{m}_y^2} \longrightarrow \frac{\mathfrak{m}_x}{\mathfrak{m}_x^2}$ and taking the dual map gives the *differential of* ϕ *in* x, which is a linear map

$$d\phi_x : T_x(X) \longrightarrow T_{\phi(x)}(Y)$$

Assume X a closed subscheme of $\mathbb{C}^n$ and Y a closed subscheme of $\mathbb{C}^m$ and let ϕ be determined by the m polynomials $\{f_1, \ldots, f_m\}$ in $\mathbb{C}[x_1, \ldots, x_n]$. Then, the Jacobian matrix in x

$$J_x(\phi) = \begin{bmatrix} \frac{\partial f_1}{\partial x_1}(x) & \cdots & \frac{\partial f_m}{\partial x_1}(x) \\ \vdots & & \vdots \\ \frac{\partial f_1}{\partial x_n}(x) & \cdots & \frac{\partial f_m}{\partial x_n}(x) \end{bmatrix}$$

defines a linear map from $\mathbb{C}^n$ to $\mathbb{C}^m$ and the differential $d\phi_x$ is the induced linear map from $T_x(X) \subset \mathbb{C}^n$ to $T_{\phi(x)}(Y) \subset \mathbb{C}^m$. Let $D \in T_x(X) = Der_x(\mathbb{C}[X])$ and x_D the corresponding element of $X(\mathbb{C}[\varepsilon])$ defined by $x_D(f) = f(x) + D(f)\varepsilon$, then $x_D \circ \phi^* \in Y(\mathbb{C}[\varepsilon])$ is defined by

$$x_D \circ \phi^*(g) = g(\phi(x)) + (D \circ \phi^*)\varepsilon = g(\phi(x)) + d\phi_x(D)\varepsilon$$

giving us the ε-interpretation of the differential

$$\phi(x + v\varepsilon) = \phi(x) + d\phi_x(v)\varepsilon$$

for all $v \in T_x(X)$.

PROPOSITION 3.14
Let $X \xrightarrow{\phi} Y$ *be a dominant morphism between irreducible affine varieties. There is a Zariski open dense subset* $U \hookrightarrow X$ *such that* $d\phi_x$ *is surjective for all* $x \in U$.

PROOF We may assume that ϕ factorizes into

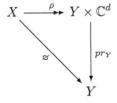

with ϕ a finite and surjective morphism. Because the tangent space of a product is the sum of the tangent spaces of the components we have that $d(pr_W)_z$ is surjective for all $z \in Y \times \mathbb{C}^d$, hence it suffices to verify the claim for a *finite* morphism ϕ. That is, we may assume that $S = \mathbb{C}[Y]$ is a finite module over $R = \mathbb{C}[X]$ and let L/K be the corresponding extension of the function fields. By the *principal element theorem* we know that $L = K[s]$ for an element $s \in L$, which is integral over R with minimal polynomial

$$F = t^n + g_{n-1}t^{n-1} + \ldots + g_1 t + g_0 \quad \text{with } g_i \in R$$

Consider the ring $S' = R[t]/(F)$, then there is an element $r \in R$ such that the localizations S'_r and S_r are isomorphic. By restricting we may assume that $X = \mathbb{V}(F) \hookrightarrow Y \times \mathbb{C}$ and that

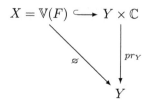

Let $x = (y, c) \in X$ then we have (again using the identification of the tangent space of a product with the sum of the tangent spaces of the components) that

$$T_x(X) = \{(v, a) \in T_y(Y) \oplus \mathbb{C} \mid c\frac{\partial F}{\partial t}(x) + vg_{n-1}c^{n-1} + \ldots + vg_1 c + vg_0 = 0\}$$

But then, $d\phi_x$ i surjective whenever $\frac{\partial F}{\partial t}(x) \neq 0$. This condition determines a *nonempty* open subset of X as otherwise $\frac{\partial F}{\partial t}$ would belong to the defining ideal of X in $\mathbb{C}[Y \times \mathbb{C}]$ (which is the principal ideal generated by F), which is impossible by a degree argument   ☐

Example 3.12 Differential of orbit map
Let X be a closed GL_n-stable subscheme of a GL_n-representation V and x a geometric point of X. Consider the orbit closure $\overline{\mathcal{O}(x)}$ of x in V. Because the orbit map

$$\mu : GL_n \longrightarrow\!\!\!\!\rightarrow GL_n.x \hookrightarrow \overline{\mathcal{O}(x)}$$

is dominant we have that $\mathbb{C}[\overline{\mathcal{O}(x)}] \hookrightarrow \mathbb{C}[GL_n]$ and therefore a domain, so $\overline{\mathcal{O}(x)}$ is an irreducible affine variety. We define the *stabilizer subgroup* $Stab(x)$ to be the fiber $\mu^{-1}(x)$, then $Stab(x)$ is a closed subgroup of GL_n. We claim that the differential of the orbit map in the identity matrix $e = \mathbb{1}_n$

$$d\mu_e : \mathfrak{gl}_n \longrightarrow T_x(\mathsf{X})$$

satisfies the following properties

$$Ker\ d\mu_e = \mathfrak{stab}(x) \quad \text{and} \quad Im\ d\mu_e = T_x(\overline{\mathcal{O}(x)})$$

By the proposition we know that there is a dense open subset U of GL_n such that $d\mu_g$ is surjective for all $g \in U$. By GL_n-equivariance of μ it follows that $d\mu_g$ is surjective for all $g \in GL_n$, in particular $d\mu_e : \mathfrak{gl}_n \longrightarrow T_x(\overline{\mathcal{O}(x)})$ is surjective. Further, all fibers of μ over $\mathcal{O}(x)$ have the same dimension. But then it follows from the *dimension formula* of proposition that

$$dim\ GL_n = dim\ Stab(x) + dim\ \overline{\mathcal{O}(x)}$$

(which, incidentally gives us an algorithm to compute the dimensions of orbit closures). Combining this with the above surjectivity, a dimension count proves that $Ker\ d\mu_e = \mathfrak{stab}(x)$, the Lie algebra of $Stab(x)$. ▯

Let A be a $\mathbb{C}$-algebra and let M and N be two A-representations of dimensions, say, m and n. An A-representation P of dimension $m + n$ is said to be an *extension of N by M* if there exists a short exact sequence of left A-modules

$$e: \qquad 0 \longrightarrow M \longrightarrow P \longrightarrow N \longrightarrow 0$$

We define an equivalence relation on extensions (P, e) of N by M : $(P, e) \cong (P', e')$ if and only if there is an isomorphism $P \xrightarrow{\phi} P'$ of left A-modules such that the diagram below is commutative

$$
\begin{array}{ccccccccc}
e: & 0 & \longrightarrow & M & \longrightarrow & P & \longrightarrow & N & \longrightarrow & 0 \\
& & & \downarrow{\scriptstyle id_M} & & \downarrow{\scriptstyle \phi} & & \downarrow{\scriptstyle id_N} & & \\
e': & 0 & \longrightarrow & M & \longrightarrow & P' & \longrightarrow & N & \longrightarrow & 0
\end{array}
$$

The set of equivalence classes of extensions of N by M will be denoted by $Ext_A^1(N, M)$.

An alternative description of $Ext_A^1(N, M)$ is as follows. Let $\rho : A \longrightarrow M_m$ and $\sigma : A \longrightarrow M_n$ be the representations defining M and N. For an extension (P, e) we can identify the $\mathbb{C}$-vector space with $M \oplus N$ and the A-module structure on P gives an algebra map $\mu : A \longrightarrow M_{m+n}$ and we can represent the action of a on P by left multiplication of the block-matrix

$$\mu(a) = \begin{bmatrix} \rho(a) & \lambda(a) \\ 0 & \sigma(a) \end{bmatrix}$$

where $\lambda(a)$ is an $m \times n$ matrix and hence defines a linear map

$$\lambda : A \longrightarrow Hom_{\mathbb{C}}(N, M)$$

The condition that μ is an algebra morphism is equivalent to the condition

$$\lambda(aa') = \rho(a)\lambda(a') + \lambda(a)\sigma(a')$$

and we denote the set of all liner maps $\lambda : A \longrightarrow Hom_{\mathbb{C}}(N, M)$ by $Z(N, M)$ and call it the space of *cycle* . The extensions of N by M corresponding to two cycles λ and λ' from $Z(N, M)$ are equivalent if and only if we have an A-module isomorphism in block form

$$\begin{bmatrix} id_M & \beta \\ 0 & id_N \end{bmatrix} \quad \text{with } \beta \in Hom_{\mathbb{C}}(N, M)$$

between them. A-linearity of this map translates into the matrix relation

$$\begin{bmatrix} id_M & \beta \\ 0 & id_N \end{bmatrix} \cdot \begin{bmatrix} \rho(a) & \lambda(a) \\ 0 & \sigma(a) \end{bmatrix} = \begin{bmatrix} \rho(a) & \lambda'(a) \\ 0 & \sigma(a) \end{bmatrix} \cdot \begin{bmatrix} id_M & \beta \\ 0 & id_N \end{bmatrix} \quad \text{for all } a \in A$$

or equivalently, that $\lambda(a) - \lambda'(a) = \rho(a)\beta - \beta\sigma(a)$ for all $a \in A$. We will now define the subspace of $Z(N, M)$ of *boundaries* $B(N, M)$

$$\{\delta \in Hom_{\mathbb{C}}(N, M) \mid \exists \beta \in Hom_{\mathbb{C}}(N, M) : \forall a \in A : \delta(a) = \rho(a)\beta - \beta\sigma(a)\}$$

We then have the description $Ext_A^1(N, M) = \frac{Z(N,M)}{B(N,M)}$.

Example 3.13 Normal space to $\mathtt{rep}_n$

Let A be an affine $\mathbb{C}$-algebra generated by $\{a_1, \ldots, a_m\}$ and $\rho : A \longrightarrow M_n(\mathbb{C})$ an algebra morphism, that is, $\rho \in \mathtt{rep}_n\, A$ determines an n-dimensional A-representation M. We claim to have the following description of the *normal space* to the orbitclosure $C_\rho = \overline{\mathcal{O}(\rho)}$ of ρ

$$N_\rho(\mathtt{rep}_n\, A) \overset{def}{=} \frac{T_\rho(\mathtt{rep}_n\, A)}{T_\rho(C_\rho)} = Ext_A^1(M, M)$$

We have already seen that the space of cycles $Z(M, M)$ is the space of ρ-derivations of A in $M_n(\mathbb{C})$, $Der_\rho(A)$, which we know to be the tangent space $T_\rho(\mathtt{rep}_n\, A)$. Moreover, we know that the differential $d\mu_e$ of the orbit map

$$GL_n \overset{\mu}{\longrightarrow} C_\rho \hookrightarrow M_n^m$$

$$d\mu_e : \quad \mathfrak{gl}_n = M_n \longrightarrow T_\rho(C_\rho)$$

is surjective. Now, $\rho = (\rho(a_1), \ldots, \rho(a_m)) \in M_n^m$ and the action of action of GL_n is given by simultaneous conjugation. But then we have for any $A \in \mathfrak{gl}_n = M_n$ that

$$(I_n + A\varepsilon).\rho(a_i).(I_n - A\varepsilon) = \rho(a_i) + (A\rho(a_i) - \rho(a_i)A)\varepsilon$$

Therefore, by definition of the differential we have that

$$d\mu_e(A)(a) = A\rho(a) - \rho(a)A \quad \text{for all } a \in A$$

that is, $d\mu_e(A) \in B(M, M)$ and as the differential map is surjective we have $T_\rho(C_\rho) = B(M, M)$ from which the claim follows. $\quad\square$

Example 3.14 Normal space to $\mathtt{trep}_n$

Let A be a Cayley-Hamilton algebra with trace map tr_A and trace generated by $\{a_1, \ldots, a_m\}$. Let $\rho \in \mathtt{trep}_n\, A$, that is, $\rho : A \longrightarrow M_n(\mathbb{C})$ is a trace preserving algebra morphism. Any cycle $\lambda : A \longrightarrow M_n(\mathbb{C})$ in $Z(M, M) = Der_\rho(A)$ determines an algebra morphism

$$\rho + \lambda\varepsilon : A \longrightarrow M_n(\mathbb{C}[\varepsilon])$$

We know that the tangent space $T_\rho(\mathtt{trep}_n\ A)$ is the subspace $Der_\rho^{tr}(A)$ of trace preserving ρ-derivations, that is, those satisfying

$$\lambda(tr_A(a)) = tr(\lambda(a)) \quad \text{for all } a \in A$$

Observe that for all boundaries $\delta \in B(M, M)$, that is, such that there is an $m \in M_n(\mathbb{C})$ with $\delta(a) = \rho(a).m - m.\rho(a)$ are trace preserving as

$$\delta(tr_A(a)) = \rho(tr_A(a)).m - m.\rho(tr_A(a)) = tr(\rho(a)).m - m.tr(\rho(a))$$
$$= \qquad 0 = tr(m.\rho(a) - \rho(a).m) = tr(\delta(a))$$

Hence, we can define the space of *trace preserving self-extensions*

$$Ext_A^{tr}(M, M) = \frac{Der_\rho^{tr}(A)}{B(M, M)}$$

and obtain as before that the *normal space* to the orbit closure $C_\rho = \overline{\mathcal{O}(\rho)}$ is equal to

$$N_\rho(\mathtt{trep}_n\ A) \overset{def}{=} \frac{T_\rho(\mathtt{trep}_n\ A)}{T_\rho(C_\rho)} = Ext_A^{tr}(M, M)$$

$$\square$$

3.8 Knop-Luna slices

Let A be an affine $\mathbb{C}$-algebra and $\xi \in \mathtt{iss}_n\ A$ a point in the quotient space corresponding to an n-dimensional semisimple representation M_ξ of A. In the next chapter we will present a method to study the étale local structure of $\mathtt{iss}_n\ A$ near ξ and the étale local GL_n-structure of the representation variety $\mathtt{rep}_n\ A$ near the closed orbit $\mathcal{O}(M_\xi) = GL_n.M_\xi$. First, we will outline the main idea in the setting of differential geometry.

Let M be a compact $\mathcal{C}^\infty$-manifold on which a compact Lie group G acts differentially. By a usual averaging process we can define a G-invariant Riemannian metric on M. For a point $m \in M$ we define

- The G-orbit $\mathcal{O}(m) = G.m$ of m in M,

- the stabilizer subgroup $H = Stab_G(m) = \{g \in G \mid g.m = m\}$ and

- the normal space N_m defined to be the orthogonal complement to the tangent space in m to the orbit in the tangent space to M. That is, we have a decomposition of H-vector spaces

$$T_m\ M = T_m\ \mathcal{O}(m) \oplus N_m$$

The normal spaces N_x when x varies over the points of the orbit $\mathcal{O}(m)$ define a vector bundle $\mathcal{N} \xrightarrow{\ p\ } \mathcal{O}(m)$ over the orbit. We can identify the bundle with the associated fiber bundle

$$\mathcal{N} \simeq G \times^H N_m$$

Any point $n \in \mathcal{N}$ in the normal bundle determines a geodesic

$$\gamma_n : \mathbb{R} \longrightarrow M \quad \text{defined by} \quad \begin{cases} \gamma_n(0) & = p(n) \\ \frac{d\gamma_n}{dt}(0) & = n \end{cases}$$

Using this geodesic we can define a G-equivariant exponential map from the normal bundle $\mathcal{N}$ to the manifold M via

$$\mathcal{N} \xrightarrow{\ exp\ } M \qquad \text{where} \qquad exp(n) = \gamma_n(1)$$

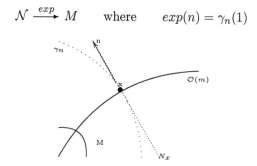

Now, take $\varepsilon > 0$ and define the $\mathcal{C}^\infty$ slice S_ε to be

$$S_\varepsilon = \{ n \in N_m \mid \ \| n \| < \varepsilon \ \}$$

then $G \times^H S_\varepsilon$ is a G-stable neighborhood of the zero section in the normal bundle $\mathcal{N} = G \times^H N_m$. But then we have a G-equivariant exponential

$$G \times^H S_\varepsilon \xrightarrow{\ exp\ } M$$

which for small enough ε gives a diffeomorphism with a G-stable tubular neighborhood U of the orbit $\mathcal{O}(m)$ in M as in figure 3.10 If we assume moreover that the action of G on M and the action of H on N_m are such that the orbit-spaces are manifolds M/G and N_m/H, then we have the situation

$$
\begin{array}{ccccc}
G \times^H S_\varepsilon & \xrightarrow[\simeq]{exp} & U & \lhook\joinrel\longrightarrow & M \\
\downarrow & & \downarrow & & \downarrow \\
S_\varepsilon/H & \xrightarrow{\simeq} & U/G & \lhook\joinrel\longrightarrow & M/G
\end{array}
$$

giving a local diffeomorphism between a neighborhood of $\bar{0}$ in N_m/H and a neighborhood of the point $\bar{m}$ in M/G corresponding to the orbit $\mathcal{O}(m)$.

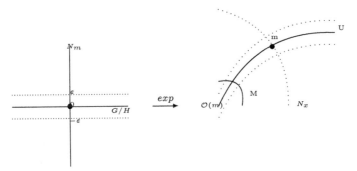

FIGURE 3.10:   Tubular neighborhood of the orbit.

Returning to the setting of the orbit $\mathcal{O}(M_\xi)$ in $\mathbf{rep}_n\ A$ we would equally like to define a GL_n-equivariant morphism from an associated fiber bundle

$$GL_n \times^{GL(\alpha)} N_\xi \xrightarrow{\ e\ } \mathbf{rep}_n\ A$$

where $GL(\xi)$ is the stabilizer subgroup of M_ξ and N_ξ is a normal space to the orbit $\mathcal{O}(M_\xi)$. Because we do not have an exponential-map in the setting of algebraic geometry, the map e will have to be an étale map. Such a map does exist and is usually called a *Luna slice* in case of a smooth point on $\mathbf{rep}_n\ A$. Later, F. Knop extended this result to allow singular points, or even points in which the scheme is not reduced.

Although the result holds for any reductive algebraic group G, we will apply it only in the case $G = GL_n$ or $GL(\alpha) = GL_{a_1} \times \ldots \times GL_{a_k}$, so restrict to the case of GL_n. We fix the setting : X and Y are (not necessarily reduced) affine GL_n-varieties, ψ is a GL_n-equivariant map

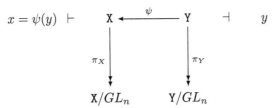

and we assume the following restrictions :

- ψ is étale in y,

- the GL_n-orbits $\mathcal{O}(y)$ in Y and $\mathcal{O}(x)$ in X are closed. For example, in representation varieties, we restrict to semisimple representations,

- the stabilizer subgroups are equal $Stab(x) = Stab(y)$. In the case of representation varieties, for a semisimple n-dimensional representation with decomposition

$$M = S_1^{\oplus e_1} \oplus \ldots \oplus S_k^{\oplus e_k}$$

into distinct simple components, this stabilizer subgroup is

$$GL(\alpha) = \begin{bmatrix} GL_{e_1}(\mathbb{C} \otimes \mathbb{1}_{d_1}) & & \\ & \ddots & \\ & & GL_{e_k}(\mathbb{C} \otimes \mathbb{1}_{d_k}) \end{bmatrix} \lhook\joinrel\longrightarrow GL_n$$

where $d_i = dim \ S_i$. In particular, the stabilizer subgroup is again reductive.

In algebraic terms: consider the coordinate rings $R = \mathbb{C}[X]$ and $S = \mathbb{C}[Y]$ and the dual morphism $R \xrightarrow{\psi^*} S$. Let $I \lhd R$ be the ideal describing $\mathcal{O}(x)$ and $J \lhd S$ the ideal describing $\mathcal{O}(y)$. With $\widehat{R}$ we will denote the I-adic completion $\varprojlim \frac{R}{I^n}$ of R and with $\widehat{S}$ the J-adic completion of S.

LEMMA 3.8

The morphism ψ^* induces for all n an isomorphism

$$\frac{R}{I^n} \xrightarrow{\ \psi^*\ } \frac{S}{J^n}$$

In particular, $\widehat{R} \simeq \widehat{S}$.

PROOF   Let $\underline{Z}$ be the closed GL_n-stable subvariety of Y where ψ is not étale. By the separation property, there is an invariant function $f \in S^{GL_n}$ vanishing on $\underline{Z}$ such that $f(y) = 1$ because the two closed GL_n-subschemes $\underline{Z}$ and $\mathcal{O}(y)$ are disjoint. Replacing S by S_f we may assume that ψ^* is an étale morphism. Because $\mathcal{O}(x)$ is smooth, $\psi^{-1} \ \mathcal{O}(x)$ is the disjoint union of its irreducible components and restricting Y if necessary we may assume that $\psi^{-1} \ \mathcal{O}(x) = \mathcal{O}(y)$. But then $J = \psi^*(I)S$ and as $\mathcal{O}(y) \xrightarrow{\simeq} \mathcal{O}(x)$ we have $\frac{R}{I} \simeq \frac{S}{J}$ so the result holds for $n = 1$.

Because étale maps are flat, we have $\psi^*(I^n)S = I^n \otimes_R S = J^n$ and an exact sequence

$$0 \longrightarrow I^{n+1} \otimes_R S \longrightarrow I^n \otimes_R S \longrightarrow \frac{I^n}{I^{n+1}} \otimes_R S \longrightarrow 0$$

But then we have

$$\frac{I^n}{I^{n+1}} = \frac{I^n}{I^{n+1}} \otimes_{R/I} \frac{S}{J} = \frac{I^n}{I^{n+1}} \otimes_R S \simeq \frac{J^n}{J^{n+1}}$$

and the result follows from induction on n and the commuting diagram

$$\begin{array}{ccccccccc}
0 & \longrightarrow & \dfrac{I^n}{I^{n+1}} & \longrightarrow & \dfrac{R}{I^{n+1}} & \longrightarrow & \dfrac{R}{I^n} & \longrightarrow & 0 \\
& & \Big\downarrow{\simeq} & & \Big\downarrow & & \Big\downarrow{\simeq} & & \\
0 & \longrightarrow & \dfrac{J^n}{J^{n+1}} & \longrightarrow & \dfrac{S}{J^{n+1}} & \longrightarrow & \dfrac{S}{J^n} & \longrightarrow & 0
\end{array}$$

▢

For an irreducible GL_n-representation s and a locally finite GL_n-module X we denote its s-isotypical component by $X_{(s)}$.

LEMMA 3.9

Let s be an irreducible GL_n-representation. There are natural numbers $m \geq 1$ (independent of s) and $n \geq 0$ such that for all $k \in \mathbb{N}$ we have

$$I^{mk+n} \cap R_{(s)} \longrightarrow (I^{GL_n})^k R_{(s)} \longrightarrow I^k \cap R_{(s)}$$

PROOF Consider $A = \oplus_{i=0}^{\infty} I^n t^n \longrightarrow R[t]$, then A^{GL_n} is affine so certainly finitely generated as R^{GL_n}-algebra, say, by

$$\{r_1 t^{m_1}, \ldots, r_z t^{m_z}\} \quad \text{with } r_i \in R \text{ and } m_i \geq 1$$

Further, $A_{(s)}$ is a finitely generated A^{GL_n}-module, say generated by

$$\{s_1 t^{n_1}, \ldots, s_y t^{n_y}\} \quad \text{with } s_i \in R_{(s)} \text{ and } n_i \geq 0$$

Take $m = max\ m_i$ and $n = max\ n_i$ and $r \in I^{mk+n} \cap R_{(s)}$, then $rt^{mk+n} \in A_{(s)}$ and

$$rt^{mk+n} = \sum_j p_j(r_1 t^{m_1}, \ldots, r_z t^{m_z}) s_j t^{n_j}$$

with p_j a homogeneous polynomial of t-degree $mk + n - n_j \geq mk$. But then each monomial in p_j occurs at least with ordinary degree $\frac{mk}{m} = k$ and therefore is contained in $(I^{GL_n})^k R_{(s)} t^{mk+n}$. ▢

Let $\widehat{R^{GL_n}}$ be the I^{GL_n}-adic completion of the invariant ring R^{GL_n} and let $\widehat{S^{GL_n}}$ be the J^{GL_n}-adic completion of S^{GL_n}.

LEMMA 3.10

The morphism ψ^* induces an isomorphism

$$R \otimes_{R^{GL_n}} \widehat{R^{GL_n}} \xrightarrow{\simeq} S \otimes_{S^{GL_n}} \widehat{S^{GL_n}}$$

PROOF Let s be an irreducible GL_n-module, then the I^{GL_n}-adic completion of $R_{(s)}$ is equal to $\widehat{R_{(s)}} = R_{(s)} \otimes_{R^{GL_n}} \widehat{R^{GL_n}}$. Moreover

$$\widehat{R_{(s)}} = \lim_{\leftarrow} (\frac{R}{I^k})_{(s)} = \lim_{\leftarrow} \frac{R_{(s)}}{(I^k \cap R_{(s)})}$$

which is the I-adic completion of $R_{(s)}$. By the foregoing lemma both topologies coincide on $R_{(s)}$ and therefore

$$\widehat{R_{(s)}} = \widehat{R}_{(s)} \quad \text{and similarly} \quad \widehat{S_{(s)}} = \widehat{S}_{(s)}$$

Because $\widehat{R} \simeq \widehat{S}$ it follows that $\widehat{R}_{(s)} \simeq \widehat{S}_{(s)}$ from which the result follows as the foregoing holds for all s. ☐

THEOREM 3.13

Consider a GL_n-equivariant map $Y \xrightarrow{\psi} X$, $y \in Y$, $x = \psi(y)$ and ψ étale in y. Assume that the orbits $\mathcal{O}(x)$ and $\mathcal{O}(y)$ are closed and that ψ is injective on $\mathcal{O}(y)$. Then, there is an affine open subset $U \hookrightarrow Y$ containing y such that

1. $U = \pi_Y^{-1}(\pi_Y(U))$ and $\pi_Y(U) = U/GL_n$

2. ψ is étale on U with affine image

3. The induced morphism $U/GL_n \xrightarrow{\overline{\psi}} X/GL_n$ is étale

4. The diagram below is commutative

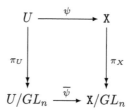

PROOF By the foregoing lemma we have $\widehat{R^{GL_n}} \simeq \widehat{S^{GL_n}}$, which means that $\overline{\psi}$ is étale in $\pi_Y(y)$. As étaleness is an open condition, there is an open affine neighborhood V of $\pi_Y(y)$ on which $\overline{\psi}$ is étale. If $\overline{R} = R \otimes_{R^{GL_n}} S^{GL_n}$ then the above lemma implies that

$$\overline{R} \otimes_{S^{GL_n}} \widehat{S^{GL_n}} \simeq S \otimes_{S^{GL_n}} \widehat{S^{GL_n}}$$

Let $S_{loc}^{GL_n}$ be the local ring of S^{GL_n} in J^{GL_n}, then as the morphism $S_{loc}^{GL_n} \longrightarrow \widehat{S^{GL_n}}$ is faithfully flat we deduce that

$$\overline{R} \otimes_{S^{GL_n}} S_{loc}^{GL_n} \simeq S \otimes_{S^{GL_n}} S_{loc}^{GL_n}$$

but then there is an $f \in S^{GL_n} - J^{GL_n}$ such that $\overline{R}_f \simeq S_f$. Now, intersect V with the open affine subset where $f \neq 0$ and let U' be the inverse image under π_Y of this set. Remains to prove that the image of ψ is affine. As $U' \xrightarrow{\psi} X$ is étale, its image is open and GL_n-stable. By the separation property we

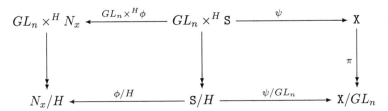

FIGURE 3.11: Etale slice diagram.

can find an invariant $h \in R^{GL_n}$ such that h is zero on the complement of the image and $h(x) = 1$. But then we take U to be the subset of U' of points u such that $h(u) \neq 0$. □

THEOREM 3.14 Slice theorem
Let X *be an affine* GL_n*-variety with quotient map* $X \xrightarrow{\pi} X/GL_n$. *Let* $x \in X$ *be such that its orbit* $\mathcal{O}(x)$ *is closed and its stabilizer subgroup* $Stab(x) = H$ *is reductive. Then, there is a locally closed affine subscheme* $S \hookrightarrow X$ *containing* x *with the following properties:*

1. S *is an affine* H*-variety*

2. *the action map* $GL_n \times S \longrightarrow X$ *induces an étale* GL_n*-equivariant morphism* $GL_n \times^H S \xrightarrow{\psi} X$ *with affine image*

3. *the induced quotient map* ψ/GL_n *is étale*

$$(GL_n \times^H S)/GL_n \simeq S/H \xrightarrow{\psi/GL_n} X/GL_n$$

and the right-hand side of figure 3.11 is commutative.

If we assume moreover that X *is smooth in* x, *then we can choose the slice* S *such that also the following properties are satisfied:*

1. S *is smooth*

2. *there is an* H*-equivariant morphism* $S \xrightarrow{\phi} T_x S = N_x$ *with* $\phi(x) = 0$ *having an affine image*

3. *the induced morphism is étale*

$$S/H \xrightarrow{\phi/H} N_x/H$$

and the left-hand side of figure 3.11 is commutative.

PROOF Choose a finite dimensional GL_n-subrepresentation V of $\mathbb{C}[X]$ that generates the coordinate ring as algebra. This gives a GL_n-equivariant embedding

$$X \xhookrightarrow{i} W = V^*$$

Choose in the vector space W an H-stable complement S_0 of $\mathfrak{gl}_n.i(x) = T_{i(x)}$ $\mathcal{O}(x)$ and denote $S_1 = i(x) + S_0$ and $S_2 = i^{-1}(S_1)$. Then, the diagram below is commutative

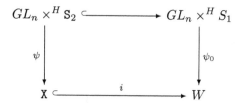

By construction we have that ψ_0 induces an isomorphism between the tangent spaces in $\overline{(1, i(x))} \in GL_n \times^H S_0$ and $i(x) \in W$, which means that ψ_0 is étale in $i(x)$, whence ψ is étale in $\overline{(1, x)} \in GL_n \times^H S_2$. By the fundamental lemma we get an affine neighborhood U, which must be of the form $U = GL_n \times^H S$ giving a slice S with the required properties.

Assume that X is smooth in x, then S_1 is transversal to X in $i(x)$ as

$$T_{i(x)} \, i(X) + S_0 = W$$

Therefore, S is smooth in x. Again using the separation property we can find an invariant $f \in \mathbb{C}[S]^H$ such that f is zero on the singularities of S (which is an H-stable closed subscheme) and $f(x) = 1$. Then replace S with its affine reduced subvariety of points s such that $f(s) \neq 0$. Let $\mathfrak{m}$ be the maximal ideal of $\mathbb{C}[S]$ in x, then we have an exact sequence of H-modules

$$0 \longrightarrow \mathfrak{m}^2 \longrightarrow \mathfrak{m} \xrightarrow{\ \alpha\ } N_x^* \longrightarrow 0$$

Choose a H-equivariant section $\phi^* : N_x^* \longrightarrow \mathfrak{m} \hookrightarrow \mathbb{C}[S]$ of α then this gives an H-equivariant morphism $S \xrightarrow{\ \phi\ } N_x$ which is étale in x. Applying again the fundamental lemma to this setting finishes the proof. $\qquad\Box$

References

More details on étale cohomology can be found in the textbook of J.S. Milne [82] . The material of Tsen and Tate fields is based on the lecture notes of S. Shatz [97]. For more details on the coniveau spectral sequence we refer to the paper [22]. The description of the Brauer group of the function field of a surface is due to M. Artin and D. Mumford [6] . The étale slices are due to D. Luna [81] and in the form presented here to F. Knop [56] . For more details we refer to the lecture notes of P. Slodowy [99].

Chapter 4

Quiver Technology

Having generalized the classical antiequivalence between commutative algebra and (affine) algebraic geometry to the pair of functors

$$
\mathtt{alg@n} \underset{\Uparrow^n}{\overset{\mathtt{trep}_n}{\rightleftarrows}} \mathtt{GL(n)\text{-}affine}
$$

where $\Uparrow^n$ is a left-inverse for $\mathtt{trep}_n$, we will define *Cayley-smooth* algebras $A \in$ $\mathtt{alg@n}$, which are analogous to smooth commutative algebras. The definition is in terms of a lifting property with respect to nilpotent ideals, following Grothendieck's characterization of regular algebras. We will prove Procesi's result that a degree n Cayley-Hamilton algebra A is Cayley-smooth if and only if $\mathtt{trep}_n A$ is a smooth (commutative) affine variety.

This result allows us, via the theory of Knop-Luna slices, to describe the étale local structure of Cayley-smooth algebras. We will prove that the local structure of A in a point $\xi \in \mathtt{triss}_n A$ is determined by a combinatorial gadget: a (marked) quiver Q (given by the simple components of the semisimple n-dimensional representation M_ξ corresponding to ξ and their (self)extensions) and a dimension vector α (given by the multiplicities of the simple factors in M_{xi}).

In the next chapter we will use this description to classify Cayley-smooth orders (as well as their central singularities) in low dimensions. In this study we will need standard results on the representation theory of quivers: the description of the simple (resp. indecomposable) dimension vectors, the canonical decomposition and the notion of semistable representations.

4.1 Smoothness

In this section we will introduce smoothness relative to a category of $\mathbb{C}$-algebras. For `comm` this notion is equivalent to the usual geometric smoothness and we will show that for `alg@n` smoothness of a Cayley-Hamilton algebra A is equivalent to $\mathtt{trep}_n A$ being a smooth affine variety. Examples of such

Cayley-smooth algebras arise as level n approximations of smooth algebras in `alg`, called *Quillen smooth algebras*.

DEFINITION 4.1   *Let* `cat` *be a category of* $\mathbb{C}$*-algebras. An object* $A \in$ $Ob(\texttt{cat})$ *is said to be* `cat`*-smooth if it satisfies the following lifting property. For* $B \in Ob(\texttt{cat})$, *a nilpotent ideal* $I \lhd B$ *such that* $B/I \in Ob(\texttt{cat})$ *and a* $\mathbb{C}$*-algebra morphism* $A \xrightarrow{\kappa} B/I$ *in* $Mor(\texttt{cat})$, *there exist a lifting*

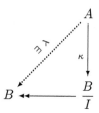

with $\lambda \in Mor(\texttt{cat})$ *making the diagram commutative. An* `alg`*-smooth algebra is called* Quillen-smooth , `comm`*-smooth algebras are called* Grothendieck-smooth *and* `alg@n`*-smooth algebras* Cayley-smooth.

To motivate these definitions, we will show that the categorical notion of `comm`-smoothness coincides with geometric smoothness. Let X be a possibly nonreduced affine variety and x a geometric point of X. As we are interested in local properties of X near x, we may assume (after translation) that $x = o$ in $\mathbb{C}^n$ and that we have a presentation

$$\mathbb{C}[\mathsf{X}] = \mathbb{C}[x_1, \ldots, x_n]/I \quad \text{with } I = (f_1, \ldots, f_m) \text{ and } m_x = (x_1, \ldots, x_n)/I$$

Denote the polynomial algebra $P = \mathbb{C}[x_1, \ldots, x_n]$ and consider the map

$$d \quad : I \longrightarrow (P dx_1 \oplus \ldots \oplus P dx_n) \otimes_P \mathbb{C}[\mathsf{X}] = \mathbb{C}[\mathsf{X}]dx_1 \oplus \ldots \oplus \mathbb{C}[\mathsf{X}]dx_n$$

where the dx_i are a formal basis of the free module of rank n and the map is defined by

$$d(f) = (\frac{\partial f}{\partial x_1}, \ldots, \frac{\partial f}{\partial x_n}) \ mod \ I$$

This gives a $\mathbb{C}[\mathsf{X}]$-linear mapping $\frac{I}{I^2} \xrightarrow{d} \mathbb{C}[\mathsf{X}]dx_1 \oplus \ldots \oplus \mathbb{C}[\mathsf{X}]dx_n$. Extending to the local algebra $\mathcal{O}_x$ at x and then quotient out the maximal ideal m_x we get a $\mathbb{C} = \mathcal{O}_x/m_x$- linear map $\frac{\mathfrak{I}}{\mathfrak{I}^2} \xrightarrow{d(x)} \mathbb{C}dx_1 \oplus \ldots \oplus \mathbb{C}dx_n$ Clearly, x is a nonsingular point of X *if and only if* the $\mathbb{C}$-linear map $d(x)$ is injective. This is equivalent to the existence of a $\mathbb{C}$-section and by the Nakayama lemma also to the existence of a $\mathcal{O}_x$-linear splitting s_x of the induced $\mathcal{O}_x$-linear map d_x

$$\frac{\mathfrak{I}}{\mathfrak{I}^2} \underset{s_x}{\overset{d_x}{\rightleftarrows}} \mathcal{O}_x dx_1 \oplus \ldots \oplus \mathcal{O}_x dx_n$$

satisfying $s_x \circ d_x = id_{\frac{\mathfrak{z}}{\mathfrak{z}^2}}$

A $\mathbb{C}$-algebra epimorphism (between commutative algebras) $R \xrightarrow{\pi} S$ with square zero kernel is called *an infinitesimal extension of S* . It is called a *trivial infinitesimal extension* if π has an algebra section $\sigma : S \hookrightarrow R$ satisfying $\pi \circ \sigma = id_S$. An infinitesimal extension $R \xrightarrow{\pi} S$ of S is said to be *versal* if for any other infinitesimal extension $R' \xrightarrow{\pi'} S$ of S there is a $\mathbb{C}$-algebra morphism

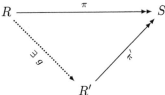

making the diagram commute. From this universal property it is clear that versal infinitesimal extensions are uniquely determined up to isomorphism. Moreover, if a versal infinitesimal extension is trivial, then so is any infinitesimal extension. By iterating, S is Grothendieck-smooth if and only if it has the lifting property with respect to nilpotent ideals I with square zero. Therefore, assume we have a *test object* (T, I) with $I^2 = 0$, then we have a commuting diagram

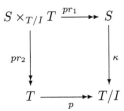

where we define the *pull-back algebra* $S \times_{T/I} T = \{(s,t) \in S \times T \mid \kappa(s) = p(t)\}$. Observe that $pr_1 : S \times_{T/I} T \twoheadrightarrow S$ is a $\mathbb{C}$-algebra epimorphism with kernel $0 \times_{T/I} I$ having square zero, that is, it is an infinitesimal extension of S. Moreover, the existence of a lifting λ of κ is equivalent to the existence of a $\mathbb{C}$-algebra section

$$\sigma : S \longrightarrow S \times_{T/I} T \quad \text{defined by } s \mapsto (s, \lambda(s))$$

Hence, S is Grothendieck-smooth if and only if a versal infinitesimal extension of S is trivial.

Returning to the situation of interest to us, we claim that the algebra epimorphism $\mathcal{O}_x(\mathbb{C}^n)/I_x^2 \xrightarrow{c_x} \mathcal{O}_x$ is a versal infinitesimal extension of $\mathcal{O}_x$. Indeed, consider any other infinitesimal extension $R \xrightarrow{\pi} \mathcal{O}_x$ then we define a $\mathbb{C}$-algebra morphism $\mathcal{O}_x(\mathbb{C}^n)/I_x^2 \longrightarrow R$ as follows: let $r_i \in R$ such that $\pi(r_i) = c_x(x_i)$ and define an algebra morphism $\mathbb{C}[x_1, \ldots, x_n] \longrightarrow R$ by sending the variable x_i to r_i. As the image of any polynomial nonvanishing in x is a unit in R, this algebra map extends to one from the local algebra $\mathcal{O}_x(\mathbb{C}^n)$ and it factors over $\mathcal{O}_x(\mathbb{C}^n)/I_x^2$ as the image of I_x lies in the kernel of π, which has square zero, proving the claim. Hence, $\mathcal{O}_x$ is Grothendieck-smooth if and only if there is a $\mathbb{C}$-algebra section

$$\mathcal{O}_x(\mathbb{C}^n)/I_x^2 \underset{r_x}{\overset{c_x}{\rightleftarrows}} \mathcal{O}_x$$

satisfying $c_x \circ r_x = id_{\mathcal{O}_x}$.

PROPOSITION 4.1
The affine scheme X *is nonsingular at the geometric point* x *if and only if the local algebra* $\mathcal{O}_x(X)$ *is Grothendieck-smooth.*

PROOF   The result will follow once we prove that there is a natural one-to-one correspondence between $\mathcal{O}_x$-module splittings s_x of d_x and $\mathbb{C}$-algebra sections r_x of c_x. This correspondence is given by assigning to an algebra section r_x the map s_x defined by

$$s_x(dx_i) = (x_i - r_x \circ c_x(x_i)) \bmod I_x^2$$

⬚

If X is an affine scheme that is smooth in *all* of its geometric points, then we have seen before that $\mathsf{X} = X$ must be reduced, that is, an affine variety. Restricting to its disjoint irreducible components we may assume that

$$\mathbb{C}[\mathsf{X}] = \cap_{x \in X} \mathcal{O}_x$$

Clearly, if $\mathbb{C}[\mathsf{X}]$ is Grothendieck-smooth, so is any of the local algebras $\mathcal{O}_x$. Conversely, if all $\mathcal{O}_x$ are Grothendieck-smooth and $\mathbb{C}[\mathsf{X}] = \mathbb{C}[x_1, \ldots, x_n]/I$ one knows that the algebra epimorphism

$$\mathbb{C}[x_1, \ldots, x_n]/I^2 \overset{c}{\twoheadrightarrow} \mathbb{C}[\mathsf{X}]$$

has local sections in every x, but then there is an algebra section. Because c is clearly a versal infinitesimal deformation of $\mathbb{C}[\mathsf{X}]$, it follows that $\mathbb{C}[\mathsf{X}]$ is Grothendieck-smooth.

PROPOSITION 4.2
Let X *be an affine scheme. Then,* $\mathbb{C}[\mathsf{X}]$ *is Grothendieck-smooth if and only if* X *is nonsingular in all of its geometric points. In this case,* X *is a reduced affine variety.*

However, Grothendieck-smooth algebras do not have to be **cat**-smooth for more general categories of $\mathbb{C}$-algebras.

Example 4.1

Consider the polynomial algebra $\mathbb{C}[x_1, \ldots, x_d]$ and the 4-dimensional noncommutative local algebra

$$T = \frac{\mathbb{C}\langle x, y \rangle}{(x^2, y^2, xy + yx)} = \mathbb{C} \oplus \mathbb{C}x \oplus \mathbb{C}y \oplus \mathbb{C}xy$$

Consider the one-dimensional nilpotent ideal $I = \mathbb{C}(xy - yx)$ of T, then the 3-dimensional quotient $\frac{T}{I}$ is commutative and we have a morphism $\mathbb{C}[x_1, \ldots, x_d] \xrightarrow{\phi} \frac{T}{I}$ by $x_1 \mapsto x, x_2 \mapsto y$ and $x_i \mapsto 0$ for $i \geq 2$. This morphism admits no lift to T as for any potential lift the commutator

$$[\tilde{\phi}(x), \tilde{\phi}(y)] \neq 0 \quad \text{in } T$$

Therefore, $\mathbb{C}[x_1, \ldots, x_d]$ can only be Quillen smooth if $d = 1$. □

Because $\texttt{comm} = \texttt{alg@1}$, it is natural to generalize the foregoing to Cayley-smooth algebras. Let B be a Cayley-Hamilton algebra of degree n with trace map tr_B and trace generated by m elements, say, $\{b_1, \ldots, b_m\}$. Then, we can write

$$B = \mathbb{T}_n^m / T_B \quad \text{with } T_B \text{ closed under traces}$$

Now, consider the extended ideal

$$E_B = M_n(\mathbb{C}[M_n^m]).T_B.M_n(\mathbb{C}[M_n^m]) = M_n(N_B)$$

and we have seen that $\mathbb{C}[\texttt{trep}_n \ B] = \mathbb{C}[M_n^m]/N_B$. We need the following technical result.

LEMMA 4.1

With notations as above, we have for all k that

$$E_B^{kn^2} \cap \mathbb{T}_n^m \subset T_B^k$$

PROOF Let $\mathbb{T}_n^m$ be the trace algebra on the generic $n \times n$ matrices $\{X_1, \ldots, X_m\}$ and $\mathbb{T}_n^{l+m}$ the trace algebra on the generic matrices $\{Y_1, \ldots, Y_l, X_1, \ldots, X_m\}$. Let $\{U_1, \ldots, U_l\}$ be elements of $\mathbb{T}_n^m$ and consider the trace preserving map $\mathbb{T}_n^{l+m} \xrightarrow{u} \mathbb{T}_n^m$ induced by the map defined by sending Y_i to U_i. Then, by the universal property we have a commutative diagram of Reynold operators

$$
\begin{array}{ccc}
M_n(\mathbb{C}[M_n^{l+m}]) & \xrightarrow{\tilde{u}} & M_n(\mathbb{C}[M_n^m]) \\
\Big\downarrow{R} & & \Big\downarrow{R} \\
\mathbb{T}_n^{l+m} & \xrightarrow{\quad u \quad} & \mathbb{T}_n^m
\end{array}
$$

Now, let $A_1, \ldots, A_{l+1}$ be elements from $M_n(\mathbb{C}[M_n^m])$, then we can calculate $R(A_1 U_1 A_2 U_2 A_3 \ldots A_l U_l A_{l+1})$ by first computing

$$r = R(A_1 Y_1 A_2 Y_2 A_3 \ldots A_l Y_l A_{l+1})$$

and then substituting the Y_i with U_i. The Reynolds operator preserves the degree in each of the generic matrices, therefore r will be linear in each of the Y_i and is a sum of trace algebra elements. By our knowledge of the generators of necklaces and the trace algebra we can write each term of the sum as an expression

$$tr(M_1) tr(M_2) \ldots tr(M_z) M_{z+1}$$

where each of the M_i is a monomial of degree $\leq n^2$ in the generic matrices $\{Y_1, \ldots, Y_l, X_1, \ldots, X_m\}$. Now, look at how the generic matrices Y_i are distributed among the monomials M_j. Each M_j contains *at most* n^2 of the Y_i's, hence the monomial M_{z+1} contains *at least* $l - vn^2$ of the Y_i where $v \leq z$ is the number of M_i with $i \leq z$ containing at least one Y_j.

Now, assume all the U_i are taken from the ideal $T_B \lhd \mathbb{T}_n^m$ that is closed under taking traces, then it follows that

$$R(A_1 U_1 A_2 U_2 A_3 \ldots A_l U_l A_{l+1}) \in T_B^{v+(l-vn^2)} \subset T_B^k$$

if we take $l = kn^2$ as $v + (k - v)n^2 \geq k$. But this finishes the proof of the required inclusion. $\square$

Let B be a Cayley-Hamilton algebra of degree n with trace map tr_B and I a two-sided ideal of B that is closed under taking traces. We will denote by $E(I)$ the extended ideal with respect to the universal embedding, that is

$$E(I) = M_n(\mathbb{C}[\text{trep}_n \, B]) I M_n(\mathbb{C}[\text{trep}_n \, B])$$

Then, for all powers k we have the inclusion $E(I)^{kn^2} \cap B \subset I^k$.

THEOREM 4.1

Let A be a Cayley-Hamilton algebra of degree n with trace map tr_A. Then, A is Cayley-smooth if and only if the trace preserving representation variety $\text{trep}_n \, A$ is nonsingular in all points (in particular, $\text{trep}_n \, A$ is reduced).

PROOF   Let A be Cayley-smooth, then we have to show that $\mathbb{C}[\text{trep}_n \, A]$ is Grothendieck-smooth. Take a commutative test-object (T, I) with I nilpotent and an algebra map $\kappa : \mathbb{C}[\text{trep}_n \, A] \longrightarrow T/I$. Composing with the universal embedding i_A we obtain a trace preserving morphism μ_0

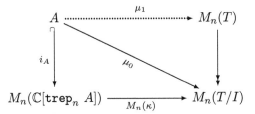

Because $M_n(T)$ with the usual trace is a Cayley-Hamilton algebra of degree n and $M_n(I)$ a trace stable ideal and A is Cayley-smooth there is a trace preserving algebra map μ_1. But then, by the universal property of the embedding i_A there exists a $\mathbb{C}$-algebra morphism

$$\lambda : \mathbb{C}[\mathbf{trep}_n \ A] \longrightarrow T$$

such that $M_n(\lambda)$ completes the diagram. The morphism λ is the required lift.

Conversely, assume that $\mathbb{C}[\mathbf{trep}_n \ A]$ is Grothendieck-smooth. Assume we have a Cayley-Hamilton algebra of degree n with trace map tr_T and a trace-stable nilpotent ideal I of T and a trace preserving $\mathbb{C}$-algebra map $\kappa : A \longrightarrow T/I$. If we combine this test-data with the universal embeddings we obtain a diagram

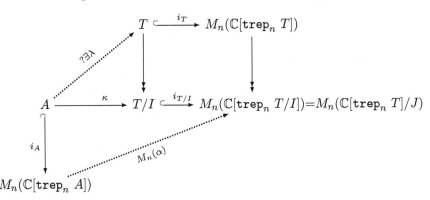

Here, $J = M_n(\mathbb{C}[\mathbf{trep}_n \ T]) I M_n(\mathbb{C}[\mathbf{trep}_n \ T])$ and we know already that $J \cap T = I$. By the universal property of the embedding i_A we obtain a $\mathbb{C}$-algebra map

$$\mathbb{C}[\mathbf{trep}_n \ A] \xrightarrow{\ \alpha\ } \mathbb{C}[\mathbf{trep}_n \ T]/J$$

which we would like to lift to $\mathbb{C}[\mathbf{trep}_n \ T]$. This does *not* follow from Grothendieck-smoothness of $\mathbb{C}[\mathbf{trep}_n \ A]$ as J is usually not nilpotent. However, as I is a nilpotent ideal of T there is some h such that $I^h = 0$. As I is closed under taking traces we know by the remark preceding the theorem that

$$E(I)^{hn^2} \cap T \subset I^h = 0$$

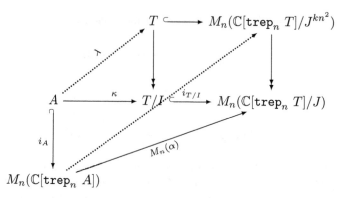

FIGURE 4.1:   The lifted morphism.

Now, by definition $E(I) = M_n(\mathbb{C}[\text{trep}_n\ T])IM_n(\mathbb{C}[\text{trep}_n\ T])$, which is equal to $M_n(J)$. That is, the inclusion can be rephrased as $M_n(J)^{hn^2} \cap T = 0$, whence there is a trace preserving embedding $T \hookrightarrow M_n(\mathbb{C}[\text{trep}_n\ T]/J^{hn^2})$. Now, we are in the situation of figure 4.1 This time we *can* lift α to a $\mathbb{C}$-algebra morphism

$$\mathbb{C}[\text{trep}_n\ A] \longrightarrow \mathbb{C}[\text{trep}_n\ T]/J^{hn^2}$$

This in turn gives us a trace preserving morphism

$$A \xrightarrow{\ \lambda\ } M_n(\mathbb{C}[\text{trep}_n\ T]/J^{hn^2})$$

the image of which is contained in the algebra of GL_n-invariants. Because $T \hookrightarrow M_n(\mathbb{C}[\text{trep}_n\ T]/J^{hn^2})$ and by surjectivity of invariants under surjective maps, the GL_n-equivariants are equal to T, giving the required lift λ. □

For an affine $\mathbb{C}$-algebra A recall the construction of its *level n approximation*

$$\int_n A = \frac{\int A}{(tr(1) - n, \chi_a^{(n)}(a)\ \forall a \in A)} = M_n(\mathbb{C}[\text{rep}_n\ A])^{GL_n}$$

In general, it may happen that $\int_n A = 0$ for example if A has no n-dimensional representations. The characteristic feature of $\int_n A$ is that any $\mathbb{C}$-algebra map $A \longrightarrow B$ with B a Cayley-Hamilton algebra of degree n factors through $\int_n A$

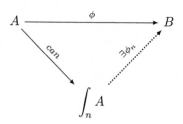

with ϕ_n a trace preserving algebra morphism. From this universal property we deduce the following.

PROPOSITION 4.3
If A is Quillen-smooth, then for every integer n, the Cayley-Hamilton algebra of degree n, $\int_n A$, is Cayley-smooth. Moreover

$$\mathtt{rep}_n\ A \simeq \mathtt{trep}_n\ \int_n A$$

is a smooth affine GL_n-variety.

This result allows us to study a Quillen-smooth algebra locally in the étale topology. We know that the algebra $\int_n A$ is given by the GL_n-equivariant maps from $\mathtt{rep}_n\ A = \mathtt{trep}_n\ \int_n A$ to $M_n(\mathbb{C})$. As this representation variety is smooth we can apply the full strength of the slice theorem to determine the local structure of the GL_n-variety $\mathtt{trep}_n\ \int_n A$ and hence of $\int_n A$. In the next section we will prove that this local structure is fully determined by a quiver setting.

Therefore, let us recall the definition of *quivers* and their path algebras and show that these algebras are all Quillen-smooth.

DEFINITION 4.2   *A quiver Q is a directed graph determined by*

- *a finite set $Q_v = \{v_1, \ldots, v_k\}$ of vertices, and*

- *a finite set $Q_a = \{a_1, \ldots, a_l\}$ of arrows where we allow multiple arrows between vertices and loops in vertices.*

Every arrow $\overset{a}{\circ\!\!\leftarrow\!\!\circ}$ *has a starting vertex $s(a) = i$ and a terminating vertex $t(a) = j$. Multiplication in the path algebra $\mathbb{C}Q$ is induced by (left) concatenation of paths. More precisely, $1 = v_1 + \ldots + v_k$ is a decomposition of 1 into mutually orthogonal idempotents and further we define*

- $v_j.a$ *is always zero unless* $\overset{a}{\circ\!\!\leftarrow\!\!\circ}$ *in which case it is the path a,*

- $a.v_i$ *is always zero unless* $\overset{a}{\circ\!\!\leftarrow\!\!\circ}$ *in which case it is the path a,*

- $a_i.a_j$ *is always zero unless* $\overset{a_i}{\circ\!\!\leftarrow\!\!\circ}\overset{a_j}{\leftarrow\!\!\circ}$ *in which case it is the path $a_i a_j$.*

Consider the commutative $\mathbb{C}$-algebra

$$C_k = \mathbb{C}[e_1, \ldots, e_k]/(e_i^2 - e_i, e_i e_j, \sum_{i=1}^{k} e_i - 1)$$

C_k is the universal $\mathbb{C}$-algebra in which 1 is decomposed into k orthogonal idempotents, that is, if R is any $\mathbb{C}$-algebra such that $1 = r_1 + \ldots + r_k$ with $r_i \in R$ idempotents satisfying $r_i r_j = 0$, then there is an embedding $C_k \hookrightarrow R$ sending e_i to r_i.

PROPOSITION 4.4

C_k is Quillen smooth. That is, if I be a nilpotent ideal of a $\mathbb{C}$-algebra T and if $1 = \bar{e}_1 + \ldots + \bar{e}_k$ is a decomposition of 1 into orthogonal idempotents $\bar{e}_i \in T/I$. Then, we can lift this decomposition to $1 = e_1 + \ldots + e_k$ for orthogonal idempotents $e_i \in T$ such that $\pi(e_i) = \bar{e}_i$ where $T \xrightarrow{\pi} T/I$ is the canonical projection.

PROOF   Assume that $I^l = 0$, clearly any element $1 - i$ with $i \in I$ is invertible in T as

$$(1 - i)(1 + i + i^2 + \ldots + i^{l-1}) = 1 - i^l = 1$$

If $\bar{e}$ is an idempotent of T/I and $x \in T$ such that $\pi(x) = \bar{e}$. Then, $x - x^2 \in I$ whence

$$0 = (x - x^2)^l = x^l - lx^{l+1} + \binom{l}{2} x^{l+2} - \ldots + (-1)^l x^{2l}$$

and therefore $x^l = ax^{l+1}$ where $a = l - \binom{l}{2} x + \ldots + (-1)^{l-1} x^{l-1}$ and so $ax = xa$. If we take $e = (ax)^l$, then e is an idempotent in T as

$$e^2 = (ax)^{2l} = a^l(a^l x^{2l}) = a^l x^l = e$$

the next to last equality follows from $x^l = ax^{l+1} = a^2 x^{l+2} = \ldots = a^l x^{2l}$. Moreover,

$$\pi(e) = \pi(a)^l \pi(x)^l = \pi(a)^l \pi(x)^{2l} = \pi(a^l x^{2l}) = \pi(x)^l = \bar{e}$$

If $\bar{f}$ is another idempotent in T/I such that $\bar{e}\bar{f} = 0 = \bar{f}\bar{e}$ then as above we can lift $\bar{f}$ to an idempotent f' of T. As $f'e \in I$ we can form the element

$$f = (1 - e)(1 - f'e)^{-1} f'(1 - f'e)$$

Because $f'(1 - f'e) = f'(1 - e)$ one verifies that f is idempotent, $\pi(f) = \bar{f}$ and $e.f = 0 = f.e$. Assume by induction that we have already lifted the pairwise orthogonal idempotents $\bar{e}_1, \ldots, \bar{e}_{k-1}$ to pairwise orthogonal idempotents $e_1, \ldots, e_{k-1}$ of R, then $e = e_1 + \ldots + e_{k-1}$ is an idempotent of T such that $\bar{e}\bar{e}_k = 0 = \bar{e}_k \bar{e}$. Hence, we can lift $\bar{e}_k$ to an idempotent $e_k \in T$ such that $ee_k = 0 = e_k e$. But then also

$$e_i e_k = (e_i e) e_k = 0 = e_k (e e_i) = e_k e_i$$

Finally, as $e_1 + \ldots + e_k - 1 = i \in I$ we have that

$$e_1 + \ldots + e_k - 1 = (e_1 + \ldots + e_k - 1)^l = i^l = 0$$

finishing the proof. □

PROPOSITION 4.5
For any quiver Q, the path algebra $\mathbb{C}Q$ is Quillen smooth.

PROOF Take an algebra T with a nilpotent two-sided ideal $I \triangleleft T$ and consider

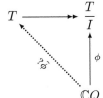

The decomposition $1 = \phi(v_1) + \ldots + \phi(v_k)$ into mutually orthogonal idempotents in $\frac{T}{I}$ can be lifted up the nilpotent ideal I to a decomposition $1 = \tilde{\phi}(v_1) + \ldots + \tilde{\phi}(v_k)$ into mutually orthogonal idempotents in T. But then, taking for every arrow a

①⟵$\overset{a}{\qquad}$①   an arbitrary element $\tilde{\phi}(a) \in \tilde{\phi}(v_j)(\phi(a) + I)\tilde{\phi}(v_i)$

gives a required lifted algebra morphism $\mathbb{C}Q \overset{\tilde{\phi}}{\longrightarrow} T$. □

Recall that a *representation V* of the quiver Q is given by

- a finite dimensional $\mathbb{C}$-vector space V_i for each vertex $v_i \in Q_v$, and

- a linear map $V_j \overset{V_a}{\longleftarrow} V_i$ for every arrow ①⟵$\overset{a}{\qquad}$① in Q_a.

If $dim\ V_i = d_i$ we call the integral vector $\alpha = (d_1, \ldots, d_k) \in \mathbb{N}^k$ the *dimension vector* of V and denote it with $dim\ V$. A *morphism $V \overset{\phi}{\longrightarrow} W$* between two representations V and W of Q is determined by a set of linear maps

$$V_i \overset{\phi_i}{\longrightarrow} W_i \quad \text{for all vertices } v_i \in Q_v$$

satisfying the following compatibility conditions for every arrow ①⟵$\overset{a}{\qquad}$① in Q_a

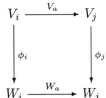

Clearly, *composition* of morphisms $V \xrightarrow{\phi} W \xrightarrow{\psi} X$ is given by the rule that $(\psi \circ \phi)_i = \psi_i \circ \phi_i$ and one readily verifies that this is again a morphism of representations of Q. In this way we form a *category* **rep** Q of all finite dimensional representations of the quiver Q.

PROPOSITION 4.6
The category **rep** Q *is equivalent to the category of finite dimensional* $\mathbb{C}Q$-*representations* $\mathbb{C}Q - \text{mod}$.

PROOF   Let M be an n-dimensional $\mathbb{C}Q$-representation. Then, we construct a representation V of Q by taking

- $V_i = v_i M$, and for any arrow $\xymatrix{⃝_j & \ar[l]_a ⃝_i}$ in Q_a define
- $V_a : V_i \longrightarrow V_j$ by $V_a(x) = v_j a x$.

Observe that the dimension vector $dim(V) = (d_1, \ldots, d_k)$ satisfies $\sum d_i = n$. If $\phi : M \longrightarrow N$ is $\mathbb{C}Q$-linear, then we have a linear map $V_i = v_i M \xrightarrow{\phi_i} W_i = v_i N$, which clearly satisfies the compatibility condition.

Conversely, let V be a representation of Q with dimension vector $dim(V) = (d_1, \ldots, d_k)$. Then, consider the $n = \sum d_i$-dimensional space $M = \oplus_i V_i$, which we turn into a $\mathbb{C}Q$-representation as follows. Consider the canonical injection and projection maps $V_j \xhookrightarrow{i_j} M \xrightarrow{\pi_j} V_j$. Then, define the action of $\mathbb{C}Q$ by fixing the action of the algebra generators v_j and a_l to be

$$\begin{cases} v_j m & = i_j(\pi_j(m)) \\ a_l m & = i_j(V_a(\pi_i(m))) \end{cases}$$

for all arrows $\xymatrix{⃝_j & \ar[l]_{a_l} ⃝_i}$. A computation verifies that these two operations are inverse to each other and induce an equivalence of categories.   ▯

4.2   Local structure

In this section we give some applications of the slice theorem to the local structure of quotient varieties of representation spaces. We will first handle the case of an affine $\mathbb{C}$-algebra A leading to a local description of $\int_n A$. Next, we will refine this slightly to prove similar results for an arbitrary affine $\mathbb{C}$-algebra B in **alg@n**.

When A is an affine $\mathbb{C}$-algebra generated by m elements $\{a_1, \ldots, a_m\}$, its level n approximation $\int_n A$ is trace generated by m determining a trace preserving epimorphism $\mathbb{T}_n^m \longrightarrow \int_n A$. Thus we have a GL_n-equivariant closed embedding of affine schemes

$$\mathbf{rep}_n\, A = \mathbf{trep}_n \int_n A \xrightarrow{\ \psi\ } \mathbf{trep}_n\, \mathbb{T}_n^m = M_n^m$$

Take a point ξ of the quotient scheme $\mathbf{iss}_n\, A = \mathbf{trep}_n \int_n A/GL_n$. We know that ξ determines the isomorphism class of a semisimple n-dimensional representation of A, say

$$M_\xi = S_1^{\oplus e_1} \oplus \ldots \oplus S_k^{\oplus e_k}$$

where the S_i are distinct simple A-representations, say, of dimension d_i and occurring in M_ξ with multiplicity e_i. These numbers determine the *representation type* $\tau(\xi)$ of ξ (or of the semisimple representation M_ξ), that is

$$\tau(\xi) = (e_1, d_1; e_2, d_2; \ldots; e_k, d_k)$$

Choosing a basis of M_ξ adapted to this decomposition gives us a point $x = (X_1, \ldots, X_m)$ in the orbit $\mathcal{O}(M_\xi)$ such that each $n \times n$ matrix X_i is of the form

$$X_i = \begin{bmatrix} m_1^{(i)} \otimes \mathbb{1}_{e_1} & 0 & \cdots & 0 \\ 0 & m_2^{(i)} \otimes \mathbb{1}_{e_2} & \cdots & 0 \\ \vdots & \vdots & \ddots & \vdots \\ 0 & 0 & \cdots & m_k^{(i)} \otimes \mathbb{1}_{e_k} \end{bmatrix}$$

where each $m_j^{(i)} \in M_{d_j}(\mathbb{C})$. Using this description we can compute the stabilizer subgroup $Stab(x)$ of GL_n consisting of those invertible matrices $g \in GL_n$ commuting with every X_i. That is, $Stab(x)$ is the multiplicative group of units of the centralizer of the algebra generated by the X_i. It is easy to verify that this group is isomorphic to

$$Stab(x) \simeq GL_{e_1} \times GL_{e_2} \times \ldots \times GL_{e_k} = GL(\alpha_\xi)$$

for the dimension vector $\alpha_\xi = (e_1, \ldots, e_k)$ determined by the multiplicities and with embedding $Stab(x) \hookrightarrow GL_n$ given by

$$\begin{bmatrix} GL_{e_1}(\mathbb{C} \otimes \mathbb{1}_{d_1}) & 0 & \cdots & 0 \\ 0 & GL_{e_2}(\mathbb{C} \otimes \mathbb{1}_{d_2}) & \cdots & 0 \\ \vdots & \vdots & \ddots & \vdots \\ 0 & 0 & \cdots & GL_{e_k}(\mathbb{C} \otimes \mathbb{1}_{d_k}) \end{bmatrix}$$

A different choice of point in the orbit $\mathcal{O}(M_\xi)$ gives a subgroup of GL_n conjugated to $Stab(x)$.

We know that the normal space N_x^{sm} can be identified with the self-extensions $Ext_A^1(M, M)$ and we will give a quiver-description of this space. The idea is to describe first the $GL(\alpha)$-module structure of N_x^{big}, the normal

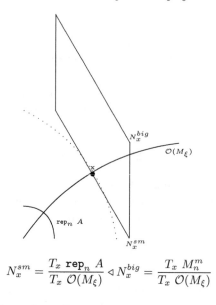

$$N_x^{sm} = \frac{T_x \ \mathbf{rep}_n \ A}{T_x \ \mathcal{O}(M_\xi)} \triangleleft N_x^{big} = \frac{T_x \ M_n^m}{T_x \ \mathcal{O}(M_\xi)}$$

FIGURE 4.2: Big and small normal spaces to the orbit.

space to the orbit $\mathcal{O}(M_\xi)$ in M_n^m (see figure 4.2) and then to identify the direct summand N_x^{sm}. The description of N_x^{big} follow from a bookkeeping operation involving $GL(\alpha)$-representations. For $x = (X_1, \ldots, X_m)$, the tangent space $T_x \ \mathcal{O}(M_{xi})$ in M_n^m to the orbit is equal to the image of the linear map

$$\begin{array}{ccc} \mathfrak{gl}_n = M_n & \longrightarrow & M_n \oplus \ldots \oplus M_n = T_x \ M_n^m \\ A & \mapsto & ([A, X_1], \ldots, [A, X_m]) \end{array}$$

Observe that the kernel of this map is the centralizer of the subalgebra generated by the X_i, so we have an exact sequence of $Stab(x) = GL(\alpha)$-modules

$$0 \longrightarrow \mathfrak{gl}(\alpha) = Lie \ GL(\alpha) \longrightarrow \mathfrak{gl}_n = M_n \longrightarrow T_x \ \mathcal{O}(x) \longrightarrow 0$$

Because $GL(\alpha)$ is a reductive group every $GL(\alpha)$-module is completely reducible and so the sequence splits. But then, the normal space in $M_n^m = T_x \ M_n^m$ to the orbit is isomorphic as $GL(\alpha)$-module to

$$N_x^{big} = \underbrace{M_n \oplus \ldots \oplus M_n}_{m-1} \oplus \mathfrak{gl}(\alpha)$$

with the action of $GL(\alpha)$ (embedded as above in GL_n) is given by simultaneous conjugation. If we consider the $GL(\alpha)$-action on M_n depicted in figure 4.2 we see that it decomposes into a direct sum of subrepresentations

- for each $1 \leq i \leq k$ we have d_i^2 copies of the $GL(\alpha)$-module M_{e_i} on which GL_{e_i} acts by conjugation and the other factors of $GL(\alpha)$ act trivially,

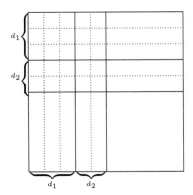

FIGURE 4.3: The $GL(\alpha)$-action on M_n

- for all $1 \le i, j \le k$ we have $d_i d_j$ copies of the $GL(\alpha)$-module $M_{e_i \times e_j}$ on which $GL_{e_i} \times GL_{e_j}$ acts via $g.m = g_i m g_j^{-1}$ and the other factors of $GL(\alpha)$ act trivially.

These $GL(\alpha)$ components are precisely the modules appearing in representation spaces of quivers.

THEOREM 4.2
Let ξ be of representation type $\tau = (e_1, d_1; \ldots; e_k, d_k)$ and let $\alpha = (e_1, \ldots, e_k)$. Then, the $GL(\alpha)$-module structure of the normal space N_x^{big} in M_n^m to the orbit of the semisimple n-dimensional representation $\mathcal{O}(M_\xi)$ is isomorphic to

$$\mathrm{rep}_\alpha \; Q_\xi^{big}$$

where the quiver Q_ξ^{big} has k vertices (the number of distinct simple summands of M_ξ) and the subquiver on any two vertices v_i, v_j for $1 \le i \ne j \le k$ has the following shape

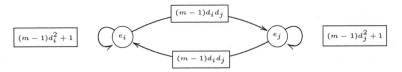

That is, in each vertex v_i there are $(m-1)d_i^2 + 1$-loops and there are $(m-1)d_i d_j$ arrows from vertex v_i to vertex v_j for all $1 \le i \ne j \le k$.

Example 4.2
If $m = 2$ and $n = 3$ and the representation type is $\tau = (1, 1; 1, 1; 1, 1)$ (that is, M_ξ is the direct sum of three distinct one-dimensional simple representations),

then the quiver Q_ξ is

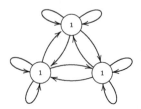

We have GL_n-equivariant embeddings $\mathcal{O}(M_\xi) \longhookrightarrow \mathbf{trep}_n \int_n A \longhookrightarrow M_n^m$ and corresponding embeddings of the tangent spaces in x

$$T_x \, \mathcal{O}(M_\xi) \longhookrightarrow T_x \, \mathbf{trep}_n \int_n A \longhookrightarrow T_x \, M_n^m$$

Because $GL(\alpha)$ is reductive we then obtain that the normal spaces to the orbit is a direct summand of $GL(\alpha)$-modules.

$$N_x^{sm} = \frac{T_x \, \mathbf{trep}_n \int_n A}{T_x \, \mathcal{O}(M_\xi)} \lhd N_x^{big} = \frac{T_x \, M_n^m}{T_x \, \mathcal{O}(M_\xi)}$$

As we know the isotypical decomposition of N_x^{big} as the $GL(\alpha)$-module $\mathbf{rep}_\alpha \, Q_\xi$ this allows us to control N_x^{sm}. We only have to observe that arrows in Q_ξ correspond to simple $GL(\alpha)$-modules, whereas a loop at vertex v_i decomposes as $GL(\alpha)$-module into the simples

$$M_{e_i} = M_{e_i}^0 \oplus \mathbb{C}_{triv}$$

where $\mathbb{C}_{triv}$ is the one-dimensional simple with trivial $GL(\alpha)$-action and $M_{e_i}^0$ is the space of trace zero matrices in M_{e_i}. Any $GL(\alpha)$-submodule of N_x^{big} can be represented by a *marked quiver* using the dictionary

- a loop at vertex v_i corresponds to the $GL(\alpha)$-module M_{e_i} on which GL_{e_i} acts by conjugation and the other factors act trivially,

- a marked loop at vertex v_i corresponds to the simple $GL(\alpha)$-module $M_{e_i}^0$ on which GL_{e_i} acts by conjugation and the other factors act trivially,

- an arrow from vertex v_i to vertex v_j corresponds to the simple $GL(\alpha)$-module $M_{e_i \times e_j}$ on which $GL_{e_i} \times GL_{e_j}$ acts via $g.m = g_i m g_j^{-1}$ and the other factors act trivially,

Combining this with the calculation that the normal space is the space of self-extensions $Ext_A^1(M_\xi, M_\xi)$ or the trace preserving self-extensions $Ext_B^{tr}(M_\xi, M_{xi})$ (in case $B \in Ob(\mathbf{alg@n})$) we have the following.

THEOREM 4.3
Consider the marked quiver on k vertices such that the full marked subquiver on any two vertices $v_i \neq v_j$ has the form

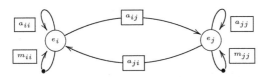

where these numbers satisfy $a_{ij} \leq (m-1)d_id_j$ and $a_{ii} + m_{ii} \leq (m-1)d_i^2 + 1$. Then,

1. *Let A be an affine $\mathbb{C}$-algebra generated by m elements, let M_ξ be an n-dimensional semisimple A-module of representation-type $\tau = (e_1, d_1; \ldots; e_k, d_k)$ and let $\alpha = (e_1, \ldots, e_k)$. Then, the normal space N_x^{sm} in a point $x \in \mathcal{O}(M_\xi)$ to the orbit with respect to the representation space $\mathbf{rep}_n A$ is isomorphic to the $GL(\alpha)$-module of quiver-representations $\mathbf{rep}_\alpha Q_\xi$ of above type with*

   - *$a_{ii} = dim_\mathbb{C} \, Ext_A^1(S_i, S_i)$ and $m_{ii} = 0$ for all $1 \leq i \leq k$.*
   - *$a_{ij} = dim_\mathbb{C} \, Ext_A^1(S_i, S_j)$ for all $1 \leq i \neq j \leq n$.*

2. *Let B be a Cayley-Hamilton algebra of degree n, trace generated by m elements, let M_ξ be a trace preserving n-dimensional semisimple B-module of representation type $\tau = (e_1, d_1; \ldots; e_k, d_k)$ and let $\alpha = (e_1, \ldots, e_k)$. Then, the normal space N_x^{tr} in a point $x \in \mathcal{O}(M_\xi)$ to the orbit with respect to the trace preserving representation space $\mathbf{trep}_n B$ is isomorphic to the $GL(\alpha)$-module of marked quiver-representations $\mathbf{rep}_\alpha Q_\xi^\bullet$ of above type with*

   - *$a_{ij} = dim_\mathbb{C} \, Ext_B^1(S_i, S_j)$ for all $1 \leq i \neq j \leq k$*

   and the (marked) vertex loops further determine the structure of $Ext_B^{tr}(M_\xi, M_\xi)$.

By a *marked quiver-representation* we mean a representation of the underlying quiver (that is, forgetting the marks) subject to the condition that the matrices corresponding to marked loops have trace zero.

Consider the slice diagram of figure 4.4 for the representation space $\mathbf{rep}_n A$. The left-hand side exists when x is a smooth point of $\mathbf{rep}_n A$, the right-hand side exists always. The horizontal maps are étale and the upper ones GL_n-equivariant.

DEFINITION 4.3 *A point $\xi \in \mathtt{iss}_n A$ is said to belong to the n-smooth locus of A iff the representation space $\mathbf{rep}_n A$ is smooth in $x \in \mathcal{O}(M_\xi)$. The n-smooth locus of A will be denoted by $Sm_n(A)$.*

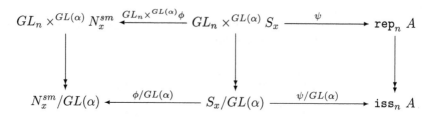

FIGURE 4.4:   Slice diagram for representation space.

To determine the étale local structure of Cayley-Hamilton algebras in their n-smooth locus, we need to investigate the special case of *quiver orders*. We will do this in the next section and, at its end, draw some consequences about the étale local structure. We end this section by explaining the remarkable success of these local quiver settings and suggest that one can extend this using the theory of A_∞-algebras.

The category **alg** has a topological origin. Consider the *tiny interval operad* D_1, that is, let $D_1(n)$ be the collection of all configurations

consisting of the unit interval with n closed intervals removed, each gap given a label i_j where $(i_1, i_2, \ldots, i_n)$ is a permutation of $(1, 2, \ldots, n)$. Clearly, $D_1(n)$ is a real $2n$-dimensional $\mathcal{C}^\infty$-manifold having $n!$ connected components, each of which is a contractible space. The operad structure comes from the collection of composition maps

$$D_1(n) \times (D_1(m_1) \times \ldots D_1(m_n)) \xrightarrow{m_{(n, m_1, \ldots, m_n)}} D_1(m_1 + \ldots + m_n)$$

defined by resizing the configuration in the $D_1(m_i)$-component such that it fits precisely in the i-th gap of the configuration of the $D_1(n)$-component, see figure 4.5. We obtain a unit interval having $m_1 + \ldots + m_n$ gaps which are labeled in the natural way, that is the first m_1 labels are for the gaps in the $D_1(m_1)$-configuration fitted in gap 1, the next m_2 labels are for the gaps in the $D_1(m_2)$-configuration fitted in gap 2 and so on. The tiny interval operad D_1 consists of

- a collection of topological spaces $D_1(n)$ for $n \geq 0$,

- a continuous action of S_n on $D_1(n)$ by relabeling, for every n,

- an identity element $id \in D_1(1)$,

- the continuous composition maps $m_{(n, m_1, \ldots, m_n)}$, which satisfy associativity and equivariance with respect to the symmetric group actions.

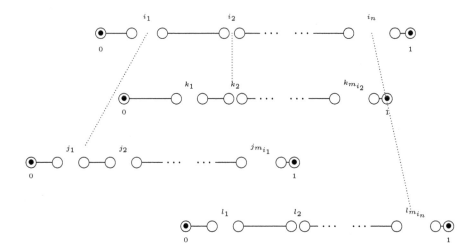

FIGURE 4.5: The tiny interval operad.

By taking the *homology groups* of these manifolds $D_1(n)$ we obtain a *linear operad* assoc. Because $D_1(n)$ has $n!$ contractible components we can identify $\mathsf{assoc}(n)$ with the subspace of the free algebra $\mathbb{C}\langle x_1, \ldots, x_n \rangle$ spanned by the *multilinear monomials*. $\mathsf{assoc}(n)$ has dimension $n!$ with basis $x_{\sigma(1)} \cdots x_{\sigma(n)}$ for $\sigma \in S_n$. Each $\mathsf{assoc}(n)$ has a natural action of S_n and as S_n-representation it is isomorphic to the regular representation. The composition maps $m_{(n, m_1, \ldots, m_n)}$ induce on the homology level linear composition maps

$$\mathsf{assoc}(n) \otimes \mathsf{assoc}(m_1) \otimes \ldots \otimes \mathsf{assoc}(m_n) \xrightarrow{\gamma_{(n, m_1, \ldots, m_n)}} \mathsf{assoc}(m_1 + \ldots + m_n)$$

obtained by substituting the multilinear monomials $\phi_i \in \mathsf{assoc}(m_i)$ in the place of the variable x_i into the multilinear monomial $\psi \in \mathsf{assoc}(n)$.

In general, a $\mathbb{C}$-linear operad P consists of a family of vector spaces $\mathsf{P}(n)$ each equipped with an S_n-action, $\mathsf{P}(1)$ contains an identity element and there are composition linear morphisms

$$\mathsf{P}(n) \otimes \mathsf{P}(m_1) \otimes \ldots \otimes \mathsf{P}(m_n) \xrightarrow{c_{(n, m_1, \ldots, m_n)}} \mathsf{P}(m_1 + \ldots + m_n)$$

satisfying the same compatibility relations as the maps $\gamma_{(n, m_1, \ldots, m_n)}$ above. An example is the *endomorphism operad* end_V for a vector space V defined by taking

$$\mathsf{end}_V(n) = Hom_{\mathbb{C}}(V^{\otimes n}, V)$$

with compositions and S_n-action defined in the obvious way and unit element $\mathbb{1}_V \in \mathsf{end}_V(1) = End(V)$. A *morphism* of linear operads $\mathsf{P} \xrightarrow{f} \mathsf{P}'$ is a collection of linear maps, which are equivariant with respect to the S_n-action, commute with the composition maps and take the identity element of P to the identity element of P'.

DEFINITION 4.4   *Let* P *be a* $\mathbb{C}$-*linear operad. A* P-*algebra is a vector space* A *equipped with a morphism of operads* P $\xrightarrow{f}$ end$_A$.

For example, assoc-algebras are just associative $\mathbb{C}$-algebras, explaining the topological origin of alg. Instead of considering the *homology operad* assoc of the tiny intervals D_1 we can consider its *chain operad* chain. For a topological space X, let chains(X) be the complex concentrated in nonpositive degrees, whose $-k$-component consists of the finite formal additive combinations $\sum c_i.f_i$ where $c_i \in \mathbb{C}$ and $f_i : [0,1]^k \longrightarrow X$ is a continuous map (a *singular cube in* X) modulo the following relations

- For any $\sigma \in S_k$ acting on $[0,1]^k$ by permutation, we have $f \circ \sigma = sg(\sigma)f$.

- For $pr_{k-1}^k : [0,1]^k \xrightarrow{k-1}$ the projection on the first $k-1$ coordinates and any continuous map $[0,1]^{k-1} \xrightarrow{f'} X$ we have $f' \circ pr_{k-1}^k = 0$.

Then, chain is the collection of complexes chains$(D_1(n))$ and is an operad in the category of complexes of vector spaces with cohomology the homology operad assoc. Again, we can consider chain-algebras, this time as complexes of vector spaces. These are the A_∞-algebras.

DEFINITION 4.5   *An* A_∞-*algebra is a* $\mathbb{Z}$-*graded complex vector space*

$$B = \oplus_{p \in \mathbb{Z}} B_p$$

endowed with homogeneous $\mathbb{C}$-*linear maps*

$$m_n : B^{\otimes n} \longrightarrow B$$

of degree $2 - n$ *for all* $n \geq 1$, *satisfying the following relations*

- *We have* $m_1 \circ m_1 = 0$, *that is* (B, m_1) *is a differential complex*

$$\cdots \xrightarrow{m_1} B_{i-1} \xrightarrow{m_1} B_i \xrightarrow{m_1} B_{i+1} \xrightarrow{m_1} \cdots$$

- *We have the equality of maps* $B \otimes B \longrightarrow B$

$$m_1 \circ m_2 = m_2 \circ (m_1 \otimes \mathbb{1} + \mathbb{1} \otimes m_1)$$

where $\mathbb{1}$ *is the identity map on the vector space* B. *That is,* m_1 *is a derivation with respect to the multiplication* $B \otimes B \xrightarrow{m_2} B$.

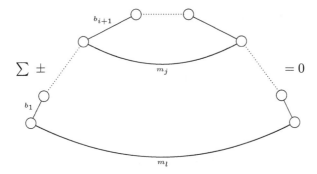

FIGURE 4.6: A_∞-identities.

- *We have the equality of maps $B \otimes B \otimes B \longrightarrow B$*

$$m_2 \circ (\mathbb{1} \otimes m_2 - m_2 \otimes \mathbb{1})$$
$$= m_1 \circ m_3 + m_3 \circ (m_1 \otimes \mathbb{1} \otimes \mathbb{1} + \mathbb{1} \otimes m_1 \otimes \mathbb{1} + \mathbb{1} \otimes \mathbb{1} \otimes m_1)$$

  where the right second expression is the associator for the multiplication m_2 and the first is a boundary of m_3, implying that m_2 is associative up to homology.

- *More generally, for $n \geq 1$ we have the relations*

$$\sum (-1)^{i+j+k} m_l \circ (\mathbb{1}^{\otimes i} \otimes m_j \otimes \mathbb{1}^{\otimes k}) = 0$$

  where the sum runs over all decompositions $n = i + j + k$ and where $l = i + 1 + k$. These identities are pictorially represented in figure 4.6.

Observe that an A_∞-algebra B is in general not associative for the multiplication m_2, but its homology

$$H^* B = H^*(B, m_2)$$

is an associative graded algebra for the multiplication induced by m_2. Further, if $m_n = 0$ for all $n \geq 3$, then B is an associative *differentially graded algebra* and conversely every differentially graded algebra yields an A_∞-algebra with $m_n = 0$ for all $n \geq 3$.

Let A be an associative $\mathbb{C}$-algebra and M a left A-module. Choose an *injective resolution* of M

$$0 \longrightarrow M \longrightarrow I^0 \longrightarrow I^1 \longrightarrow \cdots$$

with the I^k injective left A-modules and denote by $I^\bullet$ the complex

$$I^\bullet : 0 \longrightarrow I^0 \xrightarrow{d} I^1 \xrightarrow{d} \cdots$$

Let $B = HOM_A^\bullet(I^\bullet, I^\bullet)$ be the morphism complex. That is, its n-th component are the graded A-linear maps $I^\bullet \longrightarrow I^\bullet$ of degree n. This space can be equipped with a differential

$$d(f) = d \circ f - (-1)^n f \circ d \quad \text{for } f \text{ in the } n\text{-th part}$$

Then, B is a differentially graded algebra where the multiplication is the natural composition of graded maps. The homology algebra

$$H^* B = Ext_A^*(M, M)$$

is the *extension algebra* of M. Generalizing the description of $Ext_A^1(M, M)$ given in section 4.3, an element of $Ext_A^k(M, M)$ is an equivalence class of exact sequences of A-modules

$$0 \longrightarrow M \longrightarrow P_1 \longrightarrow P_2 \longrightarrow \ldots \longrightarrow P_k \longrightarrow M \longrightarrow 0$$

and the algebra structure on the extension algebra is induced by concatenation of such sequences. This extension algebra has a *canonical* structure of A_∞-algebra with $m_1 = 0$ and m_2 the usual multiplication.

Now, let $M_1, \ldots, M_k$ be A-modules (for example, finite dimensional representations) and with $filt(M_1, \ldots, M_k)$ we denote the full subcategory of all A-modules whose objects admit finite filtrations with subquotients among the M_i. We have the following result, for a proof and more details we refer to the excellent notes by B. Keller [51, §6].

THEOREM 4.4

Let $M = M_1 \oplus \ldots \oplus M_k$. The canonical A_∞-structure on the extension algebra $Ext_A^*(M, M)$ contains enough information to reconstruct the category $filt(M_1, \ldots, M_k)$.

If we specialize to the case when M is a semisimple n-dimensional representation of A of representation type $\tau = (e_1, d_1; \ldots; e_k, d_k)$, say, with decomposition

$$M_\xi = S_1^{\oplus e_1} \oplus \ldots \oplus S_k^{\oplus e_k}$$

Then, the first two terms of the extension algebra $Ext_A^*(M_\xi, M_\xi)$ are

- $Ext_A^0(M_\xi, M_\xi) = End_A(M_\xi) = M_{e_1}(\mathbb{C}) \oplus \ldots \oplus M_{e_k}(\mathbb{C})$ because by *Schur's lemma* $Hom_A(S_i, S_j) = \delta_{ij}\mathbb{C}$. Hence, the 0-th part of $Ext_A^*(M_\xi, M_\xi)$ determine the dimension vector $\alpha = (e_1, \ldots, e_k)$.

- $Ext_A^1(M_\xi, M_\xi) = \oplus_{i,j=1}^k M_{e_j \times e_i}(Ext_A^1(S_i, S_j))$ and we have seen that $dim_\mathbb{C} Ext_A^1(S_i, S_j)$ is the number of arrows from vertex v_i to v_j in the local quiver Q_ξ.

A brief summary follows.

PROPOSITION 4.7
Let $\xi \in Sm_n(A)$, then the first two terms of the extension algebra $Ext_A^*(M_\xi, M_\xi)$ contain enough information to determine the étale local structure of $\mathbf{rep}_n A$ and $\mathbf{iss}_n A$ near M_ξ.

If one wants to extend this result to noncommutative singular points $\xi \notin Sm_n(A)$, one will have to consider the canonical A_∞-structure on the full extension algebra $Ext_A^*(M_\xi, M_\xi)$.

4.3 Quiver orders

In this section and the next we will construct a large class of central simple algebras controlled by combinatorial data, using the setting of proposition 3.3.

The description of the quiver Q can be encoded in an integral $k \times k$ matrix

$$
\chi_Q = \begin{bmatrix} \chi_{11} & \cdots & \chi_{1k} \\ \vdots & & \vdots \\ \chi_{k1} & \cdots & \chi_{kk} \end{bmatrix} \quad \text{with} \quad \chi_{ij} = \delta_{ij} - \# \{ \, \textcircled{\scriptsize j} \longleftarrow \textcircled{\scriptsize i} \, \}
$$

Example 4.3
Consider the quiver Q

Then, with the indicated ordering of the vertices we have that the integral matrix is

$$
\chi_Q = \begin{bmatrix} 1 & 0 & 0 \\ -2 & 1 & -1 \\ 0 & 0 & 0 \end{bmatrix}
$$

and the path algebra of Q is isomorphic to the block-matrix algebra

$$
\mathbb{C}Q' \simeq \begin{bmatrix} \mathbb{C} & \mathbb{C} \oplus \mathbb{C} & 0 \\ 0 & \mathbb{C} & 0 \\ 0 & \mathbb{C}[x] & \mathbb{C}[x] \end{bmatrix}
$$

where x is the loop in vertex v_3.

The subspace $\mathbb{C}Qv_i$ has as its basis the paths starting in vertex v_i and because $\mathbb{C}Q = \oplus_i \mathbb{C}Qv_i$, $\mathbb{C}Qv_i$ is a projective left ideal of $\mathbb{C}Q$. Similarly, $v_i\mathbb{C}Q$ has as its basis the paths ending at v_i and is a projective right ideal of $\mathbb{C}Q$. The subspace $v_i\mathbb{C}Qv_j$ has as its basis the paths starting at v_j and ending at v_i and $\mathbb{C}Qv_i\mathbb{C}Q$ is the two-sided ideal of $\mathbb{C}Q$ having as its basis all paths passing through v_i. If $0 \neq f \in \mathbb{C}Qv_i$ and $0 \neq g \in v_i\mathbb{C}Q$, then $f.g \neq 0$ for let p be a longest path occurring in f and q a longest path in g, then the coefficient of $p.q$ in $f.g$ cannot be zero. As a consequence we have the following.

LEMMA 4.2
The projective left ideals $\mathbb{C}Qv_i$ are indecomposable and pairwise non-isomorphic.

PROOF   If $\mathbb{C}Qv_i$ is not indecomposable, then there exists a projection idempotent $f \in Hom_{\mathbb{C}Q}(\mathbb{C}Qv_i, \mathbb{C}Qv_i) \simeq v_i\mathbb{C}Qv_i$. But then, $f^2 = f = f.v_i$ whence $f.(f - v_i) = 0$, contradicting the remark above. Further, for any left $\mathbb{C}Q$-module M we have that $Hom_{\mathbb{C}Q}(\mathbb{C}Qv_i, M) \simeq v_iM$. So, if $\mathbb{C}Qv_i \simeq \mathbb{C}Qv_j$ then the isomorphism gives elements $f \in v_i\mathbb{C}Qv_j$ and $g \in v_j\mathbb{C}Qv_i$ such that $f.g = v_i$ and $g.f = v_j$. But then, $v_i \in \mathbb{C}Qv_j\mathbb{C}Q$, a contradiction unless $i = j$ as this space has as its basis all paths passing through v_j.   ▯

Example 4.4
Let Q be a quiver, then the following properties hold:

1. $\mathbb{C}Q$ is finite dimensional if and only if Q has no oriented cycles.

2. $\mathbb{C}Q$ is prime (that is, $I.J \neq 0$ for all two-sided ideals $I, J \neq 0$) if and only if Q is *strongly connected*, that is, for all vertices v_i and v_j there is a path from v_i to v_j.

3. $\mathbb{C}Q$ is Noetherian (that is, satisfies the ascending chain condition on left (or right) ideals) if and only if for every vertex v_i belonging to an oriented cycle there is only one arrow starting at v_i and only one arrow terminating at v_i.

4. The radical of $\mathbb{C}Q$ has as its basis all paths from v_i to v_j for which there is no path from v_j to v_i.

5. The center of $\mathbb{C}Q$ is of the form $\mathbb{C} \times \ldots \times \mathbb{C} \times \mathbb{C}[x] \times \ldots \times \mathbb{C}[x]$ with one factor for each connected component C of Q (that is, connected component for the underlying graph forgetting the orientation) and this factor is isomorphic to $\mathbb{C}[x]$ if and only if C is one oriented cycle.

▯

The *Euler form* of the quiver Q is the bilinear form on $\mathbb{Z}^k$

$$\chi_Q(.,.) : \mathbb{Z}^k \times \mathbb{Z}^k \longrightarrow \mathbb{Z} \quad \text{defined by} \quad \chi_Q(\alpha, \beta) = \alpha.\chi_Q.\beta^\tau$$

for all row vectors $\alpha, \beta \in \mathbb{Z}^k$.

THEOREM 4.5
Let V and W be two representations of Q, then

$$dim_\mathbb{C} \ Hom_{\mathbb{C}Q}(V, W) - dim_\mathbb{C} \ Ext^1_{\mathbb{C}Q}(V, W) = \chi_Q(dim(V), dim(W))$$

PROOF We claim that there exists an exact sequence of $\mathbb{C}$-vector spaces

$$0 \longrightarrow Hom_{\mathbb{C}Q}(V, W) \overset{\gamma}{\longrightarrow} \oplus_{v_i \in Q_v} Hom_\mathbb{C}(V_i, W_i) \overset{d^V_W}{\longrightarrow}$$

$$\overset{d^V_W}{\longrightarrow} \oplus_{a \in Q_a} Hom_\mathbb{C}(V_{s(a)}, W_{t(a)}) \overset{\epsilon}{\longrightarrow} Ext^1_{\mathbb{C}Q}(V, W) \longrightarrow 0$$

Here, $\gamma(\phi) = (\phi_1, \ldots, \phi_k)$ and d^V_W maps a family of linear maps $(f_1, \ldots, f_k)$ to the linear maps $\mu_a = f_{t(a)} V_a - W_a f_{s(a)}$ for any arrow a in Q, that is, to the obstruction of the following diagram to be commutative

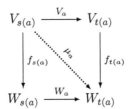

By the definition of morphisms between representations of Q it is clear that the kernel of d^V_W coincides with $Hom_{\mathbb{C}Q}(V, W)$.

Further, the map ϵ is defined by sending a family of maps $(g_1, \ldots, g_s) = (g_a)_{a \in Q_a}$ to the equivalence class of the exact sequence

$$0 \longrightarrow W \overset{i}{\longrightarrow} E \overset{p}{\longrightarrow} V \longrightarrow 0$$

where for all $v_i \in Q_v$ we have $E_i = W_i \oplus V_i$ and the inclusion i and projection map p are the obvious ones and for each generator $a \in Q_a$ the action of a on E is defined by the matrix

$$E_a = \begin{bmatrix} W_a & g_a \\ 0 & V_a \end{bmatrix} : E_{s(a)} = W_{s(a)} \oplus V_{s(a)} \longrightarrow W_{t(a)} \oplus V_{t(a)} = E_{t(a)}$$

Clearly, this makes E into a $\mathbb{C}Q$-module and one verifies that the above short exact sequence is one of $\mathbb{C}Q$-modules. Remains to prove that the cokernel of d^V_W can be identified with $Ext^1_{\mathbb{C}Q}(V, W)$.

A set of algebra generators of $\mathbb{C}Q$ is given by $\{v_1, \ldots, v_k, a_1, \ldots, a_l\}$. A cycle is given by a linear map $\lambda : \mathbb{C}Q \longrightarrow Hom_{\mathbb{C}}(V, W)$ such that for all $f, f' \in \mathbb{C}Q$ we have the condition

$$\lambda(ff') = \rho(f)\lambda(f') + \lambda(f)\sigma(f')$$

where ρ determines the action on W and σ that on V. First, consider v_i then the condition says $\lambda(v_i^2) = \lambda(v_i) = p_i^W \lambda(v_i) + \lambda(v_i)p_i^V$ whence $\lambda(v_i) :$ $V_i \longrightarrow W_i$ but then applying again the condition we see that $\lambda(v_i) = 2\lambda(v_i)$ so $\lambda(v_i) = 0$. Similarly, using the condition on $a = v_{t(a)}a = av_{s(a)}$ we deduce that $\lambda(a) : V_{s(a)} \longrightarrow W_{t(a)}$. That is, we can identify $\oplus_{a \in Q_a} Hom_{\mathbb{C}}(V_{s(a)}, W_{t(a)})$ with $Z(V, W)$ under the map ϵ. Moreover, the image of δ gives rise to a family of morphisms $\lambda(a) = f_{t(a)}V_a - W_a f_{s(a)}$ for a linear map $f = (f_i) : V \longrightarrow W$ so this image coincides precisely to the subspace of boundaries $B(V, W)$ proving that indeed the cokernel of d_W^V is $Ext_{\mathbb{C}Q}^1(V, W)$ finishing the proof of exactness of the long sequence of vector spaces. But then, if $dim(V) = (r_1, \ldots, r_k)$ and $dim(W) = (s_1, \ldots, s_k)$, we have that $dim\ Hom(V, W) - dim\ Ext^1(V, W)$ is equal to

$$\sum_{v_i \in Q_v} dim\ Hom_{\mathbb{C}}(V_i, W_i) - \sum_{a \in Q_a} dim\ Hom_{\mathbb{C}}(V_{s(a)}, W_{t(a)})$$

$$= \sum_{v_i \in Q_v} r_i s_i - \sum_{a \in Q_a} r_{s(a)} s_{t(a)}$$

$$= (r_1, \ldots, r_k) M_Q (s_1, \ldots, s_k)^\tau = \chi_Q(dim(V), dim(W))$$

finishing the proof. ▯

Fix a dimension vector $\alpha = (d_1, \ldots, d_k) \in \mathbb{N}^k$ and consider the set $\mathbf{rep}_\alpha\ Q$ of all representations V of Q such that $dim(V) = \alpha$. Because V is completely determined by the linear maps

$$V_a : V_{s(a)} = \mathbb{C}^{d_{s(a)}} \longrightarrow \mathbb{C}^{d_{t(a)}} = V_{t(a)}$$

we see that $\mathbf{rep}_\alpha\ Q$ is the affine space

$$\mathbf{rep}_\alpha\ Q = \bigoplus_{\textcircled{j} \xleftarrow{a} \textcircled{i}} M_{d_j \times d_i}(\mathbb{C}) \simeq \mathbb{C}^r$$

where $r = \sum_{a \in Q_a} d_{s(a)} d_{t(a)}$. On this affine space we have an action of the algebraic group $GL(\alpha) = GL_{d_1} \times \ldots \times GL_{d_k}$ by conjugation. That is, if $g = (g_1, \ldots, g_k) \in GL(\alpha)$ and if $V = (V_a)_{a \in Q_a}$ then $g.V$ is determined by the matrices

$$(g.V)_a = g_{t(a)} V_a g_{s(a)}^{-1}$$

If V and W in $\mathbf{rep}_\alpha\ Q$ are isomorphic as representations of Q, such an isomorphism is determined by invertible matrices $g_i : V_i \longrightarrow W_i \in GL_{d_i}$ such that for every arrow $\textcircled{j}\xleftarrow{\quad a\quad}\textcircled{i}$ we have a commutative diagram

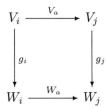

or equivalently, $g_j V_a = W_a g_i$. That is, two representations are isomorphic if and only if they belong to the same orbit under $GL(\alpha)$. In particular, we see that

$$Stab_{GL(\alpha)}\ V \simeq Aut_{\mathbb{C}Q}\ V$$

and the latter is an open subvariety of the affine space $End_{\mathbb{C}Q}(V) = Hom_{\mathbb{C}Q}(V,V)$ whence they have the same dimension. The dimension of the orbit $\mathcal{O}(V)$ of V in $\mathbf{rep}_\alpha\ Q$ is equal to

$$dim\ \mathcal{O}(V) = dim\ GL(\alpha) - dim\ Stab_{GL(\alpha)}\ V$$

But then we have a geometric reformulation of the above theorem.

LEMMA 4.3
Let $V \in \mathbf{rep}_\alpha\ Q$, then

$$dim\ \mathbf{rep}_\alpha\ Q - dim\ \mathcal{O}(V) = dim\ End_{\mathbb{C}Q}(V) - \chi_Q(\alpha,\alpha) = dim\ Ext^1_{\mathbb{C}Q}(V,V)$$

PROOF We have seen that $dim\ \mathbf{rep}_\alpha\ Q - dim\ \mathcal{O}(V)$ is equal to

$$\sum_a d_{s(a)}d_{t(a)} - (\sum_i d_i^2 - dim\ End_{\mathbb{C}Q}(V)) = dim\ End_{\mathbb{C}Q}(V) - \chi_Q(\alpha,\alpha)$$

and the foregoing theorem asserts that the latter term is equal to $dim\ Ext^1_{\mathbb{C}Q}(V,V)$. $\qquad\square$

In particular it follows that the orbit $\mathcal{O}(V)$ is *open* in $\mathbf{rep}_\alpha\ Q$ if and only if V has no self-extensions. Moreover, as $\mathbf{rep}_\alpha\ Q$ is irreducible there can be at most one isomorphism class of a representation without self-extensions.

For every dimension vector $\alpha = (d_1, \ldots, d_k)$ we will construct a *quiver order* $\mathbb{T}_\alpha Q$ which is a Cayley-Hamilton algebra of degree n where $n = d_1 + \ldots + d_k$. First, we describe the n-dimensional representations of the Quillen-smooth algebra C_k.

PROPOSITION 4.8

Let $C_k = \mathbb{C}[e_1, \ldots, e_k]/(e_i^2 - e_i, e_i e_j, \sum_{i=1}^k e_i - 1)$, then $\mathbf{rep}_n\ C_k$ is reduced and is the disjoint union of the homogeneous varieties

$$\mathbf{rep}_n\ C_k = \bigcup_\alpha GL_n/(GL_{d_1} \times \ldots \times GL_{d_k})$$

where the union is taken over all $\alpha = (d_1, \ldots, d_k)$ such that $n = \sum_i d_i$.

PROOF   As C_k is Quillen smooth we will see in section 4.1 that all its representation spaces $\mathbf{rep}_n\ C_k$ are smooth varieties, hence in particular reduced. Therefore, it suffices to describe the points. For any n-dimensional representation

$$C_k \xrightarrow{\phi} M_n(\mathbb{C})$$

the image is a commutative semisimple algebra with orthogonal idempotents $f_i = \phi(e_i)$ of rank d_i. Because $\sum_i e_i = 1_n$ we must have that $\sum_i d_i = n$. Alternatively, the corresponding n-dimensional representation $M = \oplus_i M_i$ where $M_i = e_i \mathbb{C}^n$ has dimension d_i. The stabilizer subgroup of M is equal to $GL(\alpha) = GL_{d_1} \times \ldots \times GL_{d_k}$, proving the claim.   ▯

The algebra embedding $C_k \xrightarrow{\phi} \mathbb{C}Q$ obtained by $\phi(e_i) = v_i$ determines a morphism

$$\mathbf{rep}_n\ \mathbb{C}Q \xrightarrow{\pi} \mathbf{rep}_n\ C_k = \cup_\alpha \mathcal{O}(\alpha) = \cup_\alpha GL_n/GL(\alpha)$$

where the disjoint union is taken over all the dimension vectors $\alpha = (d_1, \ldots, d_k)$ such that $n = \sum d_i$. Consider the point $p_\alpha \in \mathcal{O}(\alpha)$ determined by sending the idempotents e_i to the canonical diagonal idempotents

$$\sum_{j=\sum_{l=1}^{i-1} d_l + 1}^{\sum_{l=1}^i d_i} e_{jj} \in M_n(\mathbb{C})$$

We denote by $C_k(\alpha)$ this semisimple commutative subalgebra of $M_n(\mathbb{C})$. As $\mathbf{rep}_\alpha\ Q$ can be identified with the variety of n-dimensional representations of $\mathbb{C}Q$ in block form determined by these idempotents we see that $\mathbf{rep}_\alpha\ Q = \pi^{-1}(p)$.

We define the *quiver trace algebra* $\mathbb{T}Q$ to be the path algebra of Q over the polynomial algebra R in the variables t_p where p is a word in the arrows $a_j \in Q_a$ and is determined only up to cyclic permutation. As a consequence we only retain the variables t_p where p is an oriented cycle in Q (as all the others have a cyclic permutation that is the zero element in $\mathbb{C}Q$). We define a formal trace map tr on $\mathbb{T}Q$ by $tr(p) = t_p$ if p is an oriented cycle in Q and $tr(p) = 0$ otherwise.

For a fixed dimension vector $\alpha = (d_1, \ldots, d_k)$ with $\sum_i d_i = n$ we define $\mathbb{T}_\alpha Q$ to be the quotient

$$\mathbb{T}_\alpha Q = \frac{\mathbb{T}Q}{(\chi_a^{(n)}(a), tr(v_i) - d_i)}$$

by dividing out the substitution invariant two-sided ideal generated by all the evaluations of the formal Cayley-Hamilton algebras of degree n, $\chi_a^{(n)}(a)$ for $a \in \mathbb{T}Q$ together with the additional relations that $tr(v_i) = d_i$. $\mathbb{T}_\alpha Q$ is a Cayley-Hamilton algebra of degree n with a decomposition $1 = e_1 + \ldots + e_k$ into orthogonal idempotents such that $tr(e_i) = d_i$.

More generally, let A be a Cayley-Hamilton algebra of degree n with decomposition $1 = a_1 + \ldots + a_n$ into orthogonal idempotents such that $tr(a_i) = d_i \in \mathbb{N}_+$ and $\sum d_i = n$. Then, we have a trace preserving embedding $C_k(\alpha) \overset{i}{\hookrightarrow} A$ making A into a $C_k(\alpha) = \times_{i=1}^k \mathbb{C}$-algebra. We have a trace preserving embedding $C_k(\alpha) \overset{i'}{\hookrightarrow} M_n(\mathbb{C})$ by sending the idempotent e_i to the diagonal idempotent $E_i \in M_n(\mathbb{C})$ with ones on the diagonal from position $\sum_{j=1}^{i-1} d_j - 1$ to $\sum_{j=1}^{i} d_i$. This calls for the introduction of a *restricted representation space* of all trace preserving algebra morphisms χ such that the diagram below is commutative

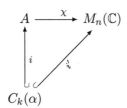

that is, such that $\chi(a_i) = E_i$. This again determines an affine scheme $\mathbf{rep}_\alpha^{res} A$, which is in fact a closed subscheme of $\mathbf{trep}_n A$. The functorial description of the restricted module scheme is as follows. Let C be any commutative $\mathbb{C}$-algebra, then $M_n(C)$ is a $C_k(\alpha)$-algebra and the idempotents E_i allow for a block decomposition

$$M_n(C) = \oplus_{i,j} E_i M_n(C) E_j = \begin{bmatrix} E_1 M_n(C) E_1 & \ldots & E_1 M_n(C) E_k \\ \vdots & & \vdots \\ E_k M_n(C) E_1 & \ldots & E_k M_n(C) E_k \end{bmatrix}$$

The scheme $\mathbf{rep}_\alpha^{res} A$ assigns to the algebra C the set of all trace preserving algebra maps

$$A \overset{\phi}{\longrightarrow} M_n(B) \quad \text{such that } \phi(a_i) = E_i$$

Equivalently, the idempotents a_i decompose A into block form $A = \oplus_{i,j} a_i A a_j$ and then $\mathbf{rep}_\alpha^{res} A(C)$ are the trace preserving algebra morphisms $A \longrightarrow M_n(B)$ compatible with the block decompositions.

Still another description of the restricted representation scheme is therefore that $\mathbf{rep}_\alpha^{res}\ A$ is the scheme theoretic fiber $\pi^{-1}(p_\alpha)$ of the point p_α under the GL_n-equivariant morphism

$$\mathbf{trep}_n\ A \xrightarrow{\ \pi\ } \mathbf{trep}_n\ C_k(\alpha)$$

Hence, the stabilizer subgroup of p acts on $\mathbf{rep}_\alpha^{res}\ A$. This stabilizer is the subgroup $GL(\alpha) = GL_{m_1} \times \ldots \times GL_{m_k}$ embedded in GL_n along the diagonal

$$GL(\alpha) = \begin{bmatrix} GL_{m_1} & & \\ & \ddots & \\ & & GL_{m_k} \end{bmatrix} \hookrightarrow GL_n$$

Clearly, $GL(\alpha)$ acts via this embedding by conjugation on $M_n(\mathbb{C})$.

THEOREM 4.6
Let A be a Cayley-Hamilton algebra of degree n such that $1 = a_1 + \ldots + a_k$ is a decomposition into orthogonal idempotents with $tr(a_i) = m_i \in \mathbb{N}_+$. Then, A is isomorphic to the ring of $GL(\alpha)$-equivariant maps

$$\mathbf{rep}_\alpha^{res}\ A \longrightarrow M_n.$$

PROOF   We know that A is the ring of GL_n-equivariant maps $\mathbf{trep}_n\ A \longrightarrow M_n$. Further, we have a GL_n-equivariant map

$$\mathbf{trep}_n\ A \xrightarrow{\ \pi\ } \mathbf{rep}_n tr\ C_k(\alpha) = GL_n.p \simeq GL_n/GL(\alpha)$$

Thus, the GL_n-equivariant maps from $\mathbf{trep}_n\ A$ to M_n coincide with the $Stab(p) = GL(\alpha)$-equivariant maps from the fiber $\pi^{-1}(p) = \mathbf{rep}_\alpha^{res}\ A$ to M_n.   ⧠

That is, we have a block matrix decomposition for A. Indeed, we have

$$A \simeq (\mathbb{C}[\mathbf{rep}_\alpha^{res}\ A] \otimes M_n(\mathbb{C}))^{GL(\alpha)}$$

and this isomorphism is clearly compatible with the block decomposition and thus we have for all i, j that

$$a_i A a_j \simeq (\mathbb{C}[\mathbf{rep}_\alpha^{res}\ A] \otimes M_{m_i \times m_j}(\mathbb{C}))^{GL(\alpha)}$$

where $M_{m_i \times m_j}(\mathbb{C})$ is the space of rectangular $m_i \times m_j$ matrices M with coefficients in $\mathbb{C}$ on which $GL(\alpha)$ acts via

$$g.M = g_i M g_j^{-1} \quad \text{where } g = (g_1, \ldots, g_k) \in GL(\alpha).$$

If we specialize this result to the case of quiver orders we have

$$\mathbf{rep}_\alpha^{res} \, \mathbb{T}_\alpha Q \simeq \mathbf{rep}_\alpha \, Q$$

as $GL(\alpha)$-varieties and we deduce

THEOREM 4.7
With notations as before,

1. $\mathbb{T}_\alpha Q$ *is the algebra of* $GL(\alpha)$*-equivariant maps from* $\mathbf{rep}_\alpha \, Q$ *to* M_n*, that is,*

$$\mathbb{T}_\alpha Q = M_n(\mathbb{C}[\mathbf{rep}_\alpha \, Q])^{GL(\alpha)}$$

2. *The* quiver necklace algebra

$$\mathbb{N}_\alpha Q = \mathbb{C}[\mathbf{rep}_\alpha \, Q]^{GL(\alpha)}$$

is generated by traces along oriented cycles in the quiver Q *of length bounded by* $n^2 + 1$.

A concrete realization of these algebras is as follows. To an arrow $\overset{j}{\bigcirc} \xleftarrow{\quad a \quad} \overset{i}{\bigcirc}$ corresponds a $d_j \times d_i$ matrix of variables from $\mathbb{C}[\mathbf{rep}_\alpha \, Q]$

$$\boxed{M_a} = \begin{bmatrix} x_{11}(a) & \cdots\cdots & x_{1d_i}(a) \\ \vdots & & \vdots \\ x_{d_j 1}(a) & \cdots\cdots & x_{d_j d_i}(a) \end{bmatrix}$$

where $x_{ij}(a)$ are the coordinate functions of the entries of V_a of a representation $V \in \mathbf{rep}_\alpha \, Q$. Let $p = a_1 a_2 \ldots a_r$ be an oriented cycle in Q, then we can compute the following matrix

$$M_p = M_{a_r} \ldots M_{a_2} M_{a_1}$$

over $\mathbb{C}[\mathbf{rep}_\alpha \, Q]$. As we have that $s(a_r) = t(a_1) = v_i$, this is a square $d_i \times d_i$ matrix with coefficients in $\mathbb{C}[\mathbf{rep}_\alpha \, Q]$ and we can take its ordinary trace

$$Tr(M_p) \in \mathbb{C}[\mathbf{rep}_\alpha \, Q].$$

Then, $\mathbb{N}_\alpha \, Q$ is the $\mathbb{C}$-subalgebra of $\mathbb{C}[\mathbf{rep}_\alpha \, Q]$ generated by these elements. Consider the block structure of $M_n(\mathbb{C}[\mathbf{rep}_\alpha \, Q])$ with respect to the idempotents e_i

$$\begin{bmatrix} M_{d_1}(S) & \cdots & \cdots M_{d_1 \times d_k}(S) \\ \vdots & & \vdots \\ \vdots & M_{d_j \times d_i}(S) & \vdots \\ M_{d_k \times d_i}(S) & \cdots & \cdots M_{d_k}(S) \end{bmatrix}$$

where $S = \mathbb{C}[\text{rep}_\alpha \, Q]$. Then, we can also view the matrix M_a for an arrow $\textcircled{j} \xleftarrow{\;\;a\;\;} \textcircled{i}$ as a block matrix in $M_n(\mathbb{C}[\text{rep}_\alpha \, Q])$

$$
\begin{bmatrix}
0 & \cdots & \cdots & 0 \\
\vdots & & & \vdots \\
\vdots & \boxed{M_a} & & \vdots \\
0 & \cdots & \cdots & 0
\end{bmatrix}
$$

Then, $\mathbb{T}_\alpha \, Q$ is the $C_k(\alpha)$-subalgebra of $M_n(\mathbb{C}[\text{rep}_\alpha \, Q])$ generated by $N_\alpha \, Q$ and these block matrices for all arrows $a \in Q_a$. $\mathbb{T}_\alpha \, Q$ itself has a block decomposition

$$
\mathbb{T}_\alpha \, Q =
\begin{bmatrix}
P_{11} & \cdots & \cdots & P_{1k} \\
\vdots & & & \vdots \\
\vdots & P_{ij} & & \vdots \\
P_{k1} & \cdots & \cdots & P_{kk}
\end{bmatrix}
$$

where P_{ij} is the $N_\alpha \, Q$-module spanned by all matrices M_p where p is a path from v_i to v_j of length bounded by n^2.

Example 4.5

Consider the path algebra $\mathbb{M}$ of the quiver which we will encounter in chapter 8 in connection with the Hilbert scheme of points in the plane and with the Calogero-Moser system

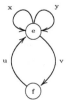

and take as dimension vector $\alpha = (n, 1)$. The total dimension is in this case $\overline{n} = n + 1$ and we fix the embedding $C_2 = \mathbb{C} \times \mathbb{C} \hookrightarrow \mathbb{M}$ given by the decomposition $1 = e + f$. Then, the above realization of $\mathbb{T}_\alpha \, \mathbb{M}$ consists in taking the following $\overline{n} \times \overline{n}$ matrices

$$
e_n =
\begin{bmatrix}
1 & & & 0 \\
& \ddots & & \vdots \\
& & 1 & 0 \\
0 & \cdots & 0 & 0
\end{bmatrix}
\quad
f_n =
\begin{bmatrix}
0 & \cdots & 0 & 0 \\
\vdots & & \vdots & \vdots \\
0 & \cdots & 0 & 0 \\
0 & \cdots & 0 & 1
\end{bmatrix}
\quad
x_n =
\begin{bmatrix}
x_{11} & \cdots & x_{1n} & 0 \\
\vdots & & \vdots & \vdots \\
x_{n1} & \cdots & x_{nn} & 0 \\
0 & \cdots & 0 & 0
\end{bmatrix}
$$

$$
y_n =
\begin{bmatrix}
y_{11} & \cdots & y_{1n} & 0 \\
\vdots & & \vdots & \vdots \\
y_{n1} & \cdots & y_{nn} & 0 \\
0 & \cdots & 0 & 0
\end{bmatrix}
\quad
u_n =
\begin{bmatrix}
0 & \cdots & 0 & u_1 \\
\vdots & & \vdots & \vdots \\
0 & \cdots & 0 & u_n \\
0 & \cdots & 0 & 0
\end{bmatrix}
\quad
v_n =
\begin{bmatrix}
0 & \cdots & 0 & 0 \\
\vdots & & \vdots & \vdots \\
0 & \cdots & 0 & 0 \\
v_1 & \cdots & v_n & 0
\end{bmatrix}
$$

In order to determine the ring of $GL(\alpha)$-polynomial invariants of $\mathbf{rep}_\alpha\ \mathbb{M}$ we have to consider the traces along oriented cycles in the quiver. Any nontrivial such cycle must pass through the vertex e and then we can decompose the cycle into factors x, y and uv (observe that if we wanted to describe circuits based at the vertex f they are of the form $c = vc'u$ with c' a circuit based at e and we can use the cyclic property of traces to bring it into the claimed form). That is, all relevant oriented cycles in the quiver can be represented by a necklace word w

where each bead is one of the elements

$$\boxed{\bullet} = x \qquad \boxed{\circ} = y \quad \text{and} \quad \boxed{\blacktriangledown} = uv$$

In calculating the trace, we first have to replace each occurrence of x, y, u or v by the relevant $\overline{n} \times \overline{n}$-matrix above. This results in replacing each of the beads in the necklace by one of the following $n \times n$ matrices

$$\boxed{\bullet} = \begin{bmatrix} x_{11} & \cdots & x_{1n} \\ \vdots & & \vdots \\ x_{n1} & \cdots & x_{nn} \end{bmatrix} \quad \boxed{\circ} = \begin{bmatrix} y_{11} & \cdots & y_{1n} \\ \vdots & & \vdots \\ y_{n1} & \cdots & y_{nn} \end{bmatrix} \quad \boxed{\blacktriangledown} = \begin{bmatrix} u_1 v_1 & \cdots & u_1 v_n \\ \vdots & & \vdots \\ u_n v_1 & \cdots & u_n v_n \end{bmatrix}$$

and taking the trace of the $n \times n$ matrix obtained after multiplying these bead-matrices cyclically in the indicated orientation. This concludes the description of the invariant ring $\mathbb{N}_\alpha\ \mathbb{Q}$. The algebra $\mathbb{T}_\alpha\ \mathbb{M}$ of $GL(\alpha)$-equivariant maps from $\mathbf{rep}_\alpha\ \mathbb{M}$ to $M_{\overline{n}}$ is then the subalgebra of $M_{\overline{n}}(\mathbb{C}[\mathbf{rep}_\alpha\ \mathbb{M}])$ generated as $C_2(\alpha)$-algebra (using the idempotent $\overline{n} \times \overline{n}$ matrices corresponding to e and f) by $\mathbb{N}_\alpha\ \mathbb{M}$ and the $\overline{n} \times \overline{n}$-matrices corresponding to x, y, u and v. $\qquad\square$

After these preliminaries, let us return to the local quiver setting (Q_ξ, α) associated to a point $\xi \in Sm_n(A)$ as described in the previous section. Above, we have seen that quiver necklace algebra $\mathbb{N}_\alpha\ Q_\xi$ is the coordinate ring of $N_x/GL(\alpha)$. $\mathbb{N}_\alpha\ Q_\xi$ is a graded algebra and is generated by all traces along oriented cycles in the quiver Q_ξ. Let $\mathfrak{m}_0$ be the graded maximal ideal of $\mathbb{N}_\alpha\ Q_\xi$, that is corresponding to the closed orbit of the trivial representation. With $\widehat{\mathbb{T}_\xi}$ (respectively $\widehat{\mathbb{N}_\alpha}$) we will denote the $\mathfrak{m}_0$-adic filtration of the quiver-order $\mathbb{T}_\alpha\ Q_\xi$ (respectively of the quiver necklace algebra $\mathbb{N}_\alpha\ Q_\xi$). Recall that the quiver-order $\mathbb{T}_\alpha\ Q_\xi$ has a block-decomposition determined by oriented paths in the quiver Q_ξ. A consequence of the slice theorem and the description of Cayley-Hamilton algebras and their algebra of traces by geometric data we deduce.

THEOREM 4.8

Let $\xi \in Sm_n(A)$. Let $N = tr \int_n A$, let $\mathfrak{m}$ be the maximal ideal of N corresponding to ξ and denote $T = \int_n A$, then we have the isomorphism and Morita equivalence

$$\widehat{N_{\mathfrak{m}}} \simeq \widehat{\mathbb{N}_\alpha} \quad \text{and} \quad \widehat{T_{\mathfrak{m}}} \underset{Morita}{\sim} \widehat{\mathbb{T}_\alpha}$$

We have an explicit description of the algebras on the right in terms of the quiver setting (Q_ξ, α) and the Morita equivalence is determined by the embedding $GL(\alpha) \hookrightarrow GL_n$.

Let $Q^\bullet$ be a marked quiver with underlying quiver Q and let $\alpha = (d_1, \ldots, d_k)$ be a dimension vector. We define the *marked quiver-necklace algebra* $\mathbb{N}_\alpha \, Q^\bullet$ to be the ring of $GL(\alpha)$-polynomial invariants on the representation space $\mathbf{rep}_\alpha \, Q^\bullet$, that is, $\mathbb{N}_\alpha \, Q^\bullet$ is the coordinate ring of the quotient variety $\mathbf{rep}_\alpha \, Q^\bullet / GL(\alpha)$. The *marked quiver-order* $\mathbb{T}_\alpha \, Q^\bullet$ is defined to be the algebra of $GL(\alpha)$-equivariant polynomial maps from $\mathbf{rep}_\alpha \, Q^\bullet$ to $M_d(\mathbb{C})$ where $d = \sum_i d_i$. Because we can separate traces, it follows that

$$\mathbb{N}_\alpha Q^\bullet = \frac{\mathbb{N}_\alpha \, Q}{(tr(m_1), \ldots, tr(m_l))} \quad \text{and} \quad \mathbb{T}_\alpha Q^\bullet = \frac{\mathbb{T}_\alpha \, Q}{(tr(m_1), \ldots, tr(m_l))}$$

where $\{m_1, \ldots, m_l\}$ is the set of all marked loops in $Q^\bullet$.

Let B be a Cayley-Hamilton algebra of degree n and let M_ξ be a trace preserving semisimple B-representation of type $\tau = (e_1, d_1; \ldots; e_k, d_k)$ corresponding to the point ξ in the quotient variety $\mathbf{iss}_n^{tr} B$.

DEFINITION 4.6   *A point $\xi \in \mathbf{iss}_n^{tr} B$ is said to belong to the smooth locus of B iff the trace preserving representation space $\mathbf{trep}_n B$ is smooth in $x \in \mathcal{O}(M_\xi)$. The smooth locus of the Cayley-Hamilton algebra B of degree n will be denotes by $Sm_{tr}(B)$.*

THEOREM 4.9

Let $\xi \in Sm_{tr}(B)$ and $N = tr \, B$. Let $\mathfrak{m}$ be the maximal ideal of N corresponding to ξ, then we have the isomorphism and Morita equivalence

$$\widehat{N_{\mathfrak{m}}} \simeq \widehat{\mathbb{N}_\alpha^\bullet} \quad \text{and} \quad \widehat{B_{\mathfrak{m}}} \underset{Morita}{\sim} \widehat{\mathbb{T}_\alpha^\bullet}$$

where we have an explicit description of the algebras on the right in terms of the quiver setting (Q_ξ, α) and where the Morita equivalence is determined by the embedding $GL(\alpha) \hookrightarrow GL(n)$.

Even if the left hand sides of the slice diagrams are not defined when ξ is not contained in the smooth locus, the dimension of the normal spaces (that is, the (trace preserving) self-extensions of M_ξ) allow us to have a numerical measure of the 'badness' of this noncommutative singularity.

DEFINITION 4.7 *Let A be an affine $\mathbb{C}$-algebra and $\xi \in$ iss$_n$ A of type $\tau = (e_1, d_1; \ldots; e_k, d_k)$. The measure of singularity in ξ is given by the non-negative number*

$$ms(\xi) = n^2 + dim_{\mathbb{C}} \, Ext_A^1(M_\xi, M_\xi) - e_1^2 - \ldots - e_k^2 - dim_{M_\xi} \, \text{rep}_n \, A$$

Let B be a Cayley-Hamilton algebra of degree n and $\xi \in$ iss$_n^{tr}$ B of type $\tau = (e_1, d_1; \ldots; e_k, d_k)$. The measure of singularity in ξ is given by the nonnegative number

$$ms(\xi) = n^2 + dim_{\mathbb{C}} \, Ext_B^{tr}(M_\xi, M_\xi) - e_1^2 - \ldots - e_k^2 - dim_{M_\xi} \, \text{trep}_n \, A$$

Clearly, $\xi \in Sm_n(A)$ (respectively, $\xi \in Sm_{tr}(B)$) if and only if $ms(\xi) = 0$.

As an application to the slice theorem, let us prove the connection between Azumaya algebras and principal fibrations. The *Azumaya locus* of an algebra A will be the open subset U_{Az} of iss$_n$ A consisting of the points ξ of type $(1, n)$. Let $\text{rep}_n \, A \xrightarrow{\pi} \text{iss}_n \, A$ be the quotient map.

PROPOSITION 4.9
The quotient $\pi^{-1}(U_{Az}) \longrightarrow U_{Az}$ is a principal PGL_n-fibration in the étale topology, that is determines an element in $H_{et}^1(U_{Az}, PGL_n)$.

PROOF Let $\xi \in U_{Az}$ and $x = M_\xi$ a corresponding simple representation. Let S_x be the slice in x for the PGL_n-action on $\text{rep}_n \, A$. By taking traces of products of a lifted basis from $M_n(\mathbb{C})$ we find a PGL_n-affine open neighborhood U_ξ of ξ contained in U_{Az} and hence by the slice result a commuting diagram

$$
\begin{array}{ccc}
PGL_n \times S_x & \xrightarrow{\psi} & \pi^{-1}(U_\xi) \\
\downarrow & & \downarrow{\scriptstyle \pi} \\
S_x & \xrightarrow{\psi/PGL_n} & U_\xi
\end{array}
$$

where ψ and ψ/PGL_n are étale maps. That is, ψ/PGL_n is an étale neighborhood of ξ over which π is trivialized. As this holds for all points $\xi \in U_{Az}$ the result follows. $\square$

4.4 Simple roots

In this section we will use proposition 3.3 to construct quiver orders $\mathbb{T}_\alpha Q$ that determine central simple algebras over the functionfield of the quotient variety $\text{iss}_\alpha Q = \text{rep}_\alpha Q/GL(\alpha)$. With $\text{PGL}(\alpha)$ we denote the group scheme corresponding to the algebraic group

$$PGL(\alpha) = GL(\alpha)/\mathbb{C}^*(\mathbb{1}_{d_1}, \ldots, \mathbb{1}_{d_k})$$

If C is a commutative $\mathbb{C}$-algebra, then using the embedding $PGL(\alpha) \hookrightarrow PGL_n$, the pointed cohomology set

$$H^1_{et}(C, \text{PGL}(\alpha)) \hookrightarrow H^1_{et}(C, \text{PGL}_n)$$

classifies Azumaya algebras A over C with a distinguished embedding $C_k \hookrightarrow A$ that are split by an étale cover such that on this cover the embedding of C_k in matrices is conjugate to the standard embedding $C_k(\alpha)$. Modifying the argument of proposition 3.3 we have the following.

PROPOSITION 4.10

If α is the dimension vector of a simple representation of Q, then

$$\mathbb{T}_\alpha Q \otimes_{\mathbb{N}_\alpha Q} \mathbb{C}(\text{iss}_\alpha Q)$$

is a central simple algebra over the function field of the quotient variety $\text{iss}_\alpha Q$.

Remains to classify the *simple roots* α, that is, the dimension vectors of simple representations of the quiver Q. Consider the vertex set $Q_v = \{v_1, \ldots, v_k\}$. To a subset $S \hookrightarrow Q_v$ we associate the full subquiver Q_S of Q, that is, Q_S has as set of vertices the subset S and as set of arrows all arrows $\textcircled{j} \xleftarrow{\ a\ } \textcircled{i}$ in Q_a such that v_i and v_j belong to S. A full subquiver Q_S is said to be *strongly connected* if and only if for all $v_i, v_j \in V$ there is an oriented cycle in Q_S passing through v_i and v_j. We can partition

$$Q_v = S_1 \sqcup \ldots \sqcup S_s$$

such that the Q_{S_i} are maximal strongly connected components of Q. Clearly, the direction of arrows in Q between vertices in S_i and S_j is the same by the maximality assumption and can be used to define an orientation between S_i and S_j. The *strongly connected component quiver* $SC(Q)$ is then the quiver on s vertices $\{w_1, \ldots, w_s\}$ with w_i corresponding to S_i and there is one arrow from w_i to w_j if and only if there is an arrow in Q from a vertex in S_i to a vertex in S_j. Observe that when the underlying graph of Q is connected, then so is the underlying graph of $SC(Q)$ and $SC(Q)$ is a quiver without oriented cycles.

Vertices with specific in- and out-going arrows are given names as in figure 4.7 If $\alpha = (d_1, \ldots, d_k)$ is a dimension vector, we define the *support* of α to be $supp(\alpha) = \{v_i \in Q_v \mid d_i \neq 0\}$.

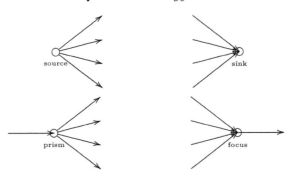

FIGURE 4.7: Vertex terminology.

LEMMA 4.4

If α is the dimension vector of a simple representation of Q, then $Q_{supp(\alpha)}$ is a strongly connected subquiver.

PROOF If not, we consider the strongly connected component quiver $SC(Q_{supp(\alpha)})$ and by assumption there must be a sink in it corresponding to a proper subset $S \overset{\neq}{\hookrightarrow} Q_v$. If $V \in \mathbf{rep}_\alpha Q$ we can then construct a representation W by

- $W_i = V_i$ for $v_i \in S$ and $W_i = 0$ if $v_i \notin S$

- $W_a = V_a$ for an arrow a in Q_S and $W_a = 0$ otherwise

One verifies that W is a proper subrepresentation of V, so V cannot be simple, a contradiction. □

The second necessary condition involves the Euler form of Q. With ϵ_i be denote the dimension vector of the simple representation having a one-dimensional space at vertex v_i and zero elsewhere and all arrows zero matrices.

LEMMA 4.5

If α is the dimension vector of a simple representation of Q, then

$$\begin{cases} \chi_Q(\alpha, \epsilon_i) & \leq 0 \\ \chi_Q(\epsilon_i, \alpha) & \leq 0 \end{cases}$$

for all $v_i \in supp(\alpha)$.

PROOF Let V be a simple representation of Q with dimension vector $\alpha = (d_1, \ldots, d_k)$. One verifies that

$$\chi_Q(\epsilon_i, \alpha) = d_i - \sum_{\textcircled{j} \longleftarrow \textcircled{i}} d_j$$

Assume that $\chi_Q(\epsilon_i, \alpha) > 0$, then the natural linear map

$$\bigoplus_{\substack{\text{\scriptsize\textcircled{\scriptsize j}}\xleftarrow{a}\text{\scriptsize\textcircled{\scriptsize i}}}} V_a \; : \; V_i \longrightarrow \bigoplus_{\substack{\text{\scriptsize\textcircled{\scriptsize j}}\xleftarrow{a}\text{\scriptsize\textcircled{\scriptsize i}}}} V_j$$

has a nontrivial kernel, say K. But then we consider the representation W of Q determined by

- $W_i = K$ and $W_j = 0$ for all $j \neq i$,

- $W_a = 0$ for all $a \in Q_a$.

It is clear that W is a proper subrepresentation of V, a contradiction.

Similarly, assume that $\chi_Q(\alpha, \epsilon_i) = d_i - \sum_{\text{\scriptsize\textcircled{\scriptsize i}}\xleftarrow{}\text{\scriptsize\textcircled{\scriptsize j}}} d_j > 0$, then the linear map

$$\bigoplus_{\substack{\text{\scriptsize\textcircled{\scriptsize i}}\xleftarrow{a}\text{\scriptsize\textcircled{\scriptsize j}}}} V_a \; : \; \bigoplus_{\substack{\text{\scriptsize\textcircled{\scriptsize i}}\xleftarrow{a}\text{\scriptsize\textcircled{\scriptsize j}}}} V_j \longrightarrow V_i$$

has an image I that is a proper subspace of V_i. The representation W of Q determined by

- $W_i = I$ and $W_j = V_j$ for $j \neq i$,

- $W_a = V_a$ for all $a \in Q_a$.

is a proper subrepresentation of V, a contradiction finishing the proof.   □

Example 4.6

The necessary conditions of the foregoing two lemmas are not sufficient. Consider the extended Dynkin quiver of type $\tilde{A}_k$ with cyclic orientation.

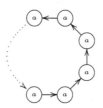

and dimension vector $\alpha = (a, \ldots, a)$. For a simple representation all arrow matrices must be invertible but then, under the action of $GL(\alpha)$, they can be diagonalized. Hence, the only simple representations (which are not the trivial simples concentrated in a vertex) have dimension vector $(1, \ldots, 1)$.   □

Nevertheless, we will show that these are the only exceptions. A vertex v_i is said to be *large* with respect to a dimension vector $\alpha = (d_1, \ldots, d_k)$ whenever d_i is maximal among the d_j. The vertex v_i is said to be *good* if v_i is large and has no direct successor, which is a large prism nor a direct predecessor that is a large focus.

LEMMA 4.6

Let Q be a strongly connected quiver, not of type $\tilde{A}_k$, then one of the following hold

1. *Q has a good vertex, or,*

2. *Q has a large prism having no direct large prism successors, or*

3. *Q has a large focus having no direct large focus predecessors.*

PROOF If neither of the cases hold, we would have an oriented cycle in Q consisting of prisms (or consisting of focusses). Assume $(v_{i_1}, \ldots, v_{i_l})$ is a cycle of prisms, then the unique incoming arrow of v_{i_j} belongs to the cycle. As $Q \neq \tilde{A}_k$ there is at least one extra vertex v_a not belonging to the cycle. But then, there can be no oriented path from v_a to any of the v_{i_j}, contradicting the assumption that Q is strongly connected. □

If we are in one of the two last cases, let a be the maximum among the components of the dimension vector α and assume that α satisfies $\chi_Q(\alpha, \epsilon_i) \leq 0$ and $\chi_Q(\epsilon_i, \alpha) \leq 0$ for all $1 \leq i \leq k$, then we have the following subquiver in Q

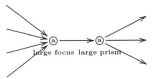

We can reduce to a quiver situation with strictly less vertices.

LEMMA 4.7

Assume Q is strongly connected and we have a vertex v_i, which is a prism with unique predecessor the vertex v_j, which is a focus. Consider the dimension vector $\alpha = (d_1, \ldots, d_k)$ with $d_i = d_j = a \neq 0$. Then, α is the dimension of a simple representation of Q if and only if

$$\alpha' = (d_1, \ldots, d_{i-1}, d_{i+1}, \ldots, d_k) \in \mathbb{N}^{k-1}$$

is the dimension vector of a simple representation of the quiver Q' on $k-1$ vertices, obtained from Q by identifying the vertices v_i and v_j, that is, the *above subquiver in Q is simplified to the one below in Q'*

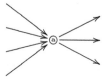

PROOF   If b is the unique arrow from v_j to v_i and if $V \in \mathbf{rep}_\alpha Q$ is a simple representation then V_b is an isomorphism, so we can identify V_i with V_j and obtain a simple representation of Q'. Conversely, if $V' \in \mathbf{rep}_{\alpha'} Q'$ is a simple representation, define $V \in \mathbf{rep}_\alpha Q$ by $V_i = V_j'$ and $V_z = V_z'$ for $z \neq i$, $V_{b'} = V_{b'}'$ for all arrows $b' \neq b$ and $V_b = \mathbb{1}_a$. Clearly, V is a simple representation of Q.   ☐

THEOREM 4.10
$\alpha = (d_1, \ldots, d_k)$ *is the dimension vector of a simple representation of Q if and only if one of the following two cases holds*

1. $supp(\alpha) = \tilde{A}_k$, *the extended Dynkin quiver on k vertices with cyclic orientation and $d_i = 1$ for all $1 \leq i \leq k$*

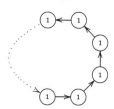

2. $supp(\alpha) \neq \tilde{A}_k$. *Then, $supp(\alpha)$ is strongly connected and for all $1 \leq i \leq k$ we have*

$$\begin{cases} \chi_Q(\alpha, \epsilon_i) & \leq 0 \\ \chi_Q(\epsilon_i, \alpha) & \leq 0 \end{cases}$$

PROOF   We will use induction, both on the number of vertices k in $supp(\alpha)$ and on the total dimension $n = \sum_i d_i$ of the representation. If $supp(\alpha)$ does not possess a good vertex, then the above lemma finishes the proof by induction on k. Observe that the Euler-form conditions are preserved in passing from Q to Q' as $d_i = d_j$.

Hence, assume v_i is a good vertex in $supp(\alpha)$. If $d_i = 1$ then all $d_j = 1$ for $v_j \in supp(\alpha)$ and we can construct a simple representation by taking $V_b = 1$ for all arrows b in $supp(\alpha)$. Simplicity follows from the fact that $supp(\alpha)$ is strongly connected.

If $d_i > 1$, consider the dimension vector $\alpha' = (d_1, \ldots, d_{i-1}, d_i - 1, d_{i+1}, \ldots, d_k)$. Clearly, $supp(\alpha') = supp(\alpha)$ is strongly connected and we claim that the Euler-form conditions still hold for α'. the only vertices v_l where things might go wrong are direct predecessors or direct successors of v_i. Assume for one of them $\chi_Q(\epsilon_l, \alpha) > 0$ holds, then

$$ d_l = d_l' > \sum_{\substack{m \xleftarrow{a} i}} d_m' \geq d_i' = d_i - 1 $$

But then, $d_l = d_i$ whence v_l is a large vertex of α and has to be also a focus with end vertex v_i (if not, $d_l > d_i$), contradicting goodness of v_i.

Hence, by induction on n we may assume that there is a simple representation $W \in \mathbf{rep}_{\alpha'} Q$. Consider the space $\mathbf{rep}_W$ of representations $V \in \mathbf{rep}_\alpha Q$ such that $V \mid \alpha' = W$. That is, for every arrow

$$ \underset{j}{\bigcirc} \xleftarrow{\quad a \quad} \underset{i}{\bigcirc} \qquad V_a = \boxed{\begin{array}{c} W_a \\ \hline v_1 \ \cdots \ v_{d_j} \end{array}} $$

$$ \underset{i}{\bigcirc} \xleftarrow{\quad a \quad} \underset{j}{\bigcirc} \qquad V_a = \boxed{\begin{array}{c|c} W_a & \begin{array}{c} v_1 \\ \vdots \\ v_{d_j} \end{array} \end{array}} $$

Hence, $\mathbf{rep}_W$ is an affine space consisting of all representations degenerating to $W \oplus S_i$ where S_i is the simple one-dimensional representation concentrated in v_i. As $\chi_Q(\alpha', \epsilon_i) < 0$ and $\chi_Q(\epsilon_i, \alpha') < 0$ we have that $Ext^1(W, S_i) \neq 0 \neq Ext^1(S_i, W)$ so there is an open subset of representations which are not isomorphic to $W \oplus S_i$.

As there are simple representations of Q having a one-dimensional component at each vertex in $supp(\alpha)$ and as the subset of simple representations in $\mathbf{rep}_{\alpha'} Q$ is open, we can choose W such that $\mathbf{rep}_W$ contains representations V such that a trace of an oriented cycle differs from that of $W \oplus S_i$. Hence, by the description of the invariant ring $\mathbb{C}[\mathbf{rep}_\alpha Q]^{GL(\alpha)}$ as being generated by traces along oriented cycles and by the identification of points in the quotient variety as isomorphism classes of semisimple representations, it follows that the Jordan-Hölder factors of V are different from W and S_i. In view of the definition of $\mathbf{rep}_W$, this can only happen if V is a simple representation, finishing the proof of the theorem. $\quad\square$

4.5 Indecomposable roots

Throughout, Q will be a quiver on k vertices $\{v_1, \ldots, v_k\}$ with Euler form χ_Q. For a dimension vector $\alpha = (d_1, \ldots, d_k)$, any $V \in \mathbf{rep}_\alpha Q$ decomposes uniquely into

$$V = W_1^{\oplus f_1} \oplus \ldots \oplus W_z^{\oplus f_z}$$

where the W_i are *indecomposable* representations . This follows from the fact that $End(V)$ is finite dimensional. Recall also that a representation W of Q is indecomposable if and only if $End(W)$ is a *local algebra* , that is, the nilpotent endomorphisms in $End_{CQ}(W)$ form an ideal of codimension one. Equivalently, the maximal torus of the stabilizer subgroup $Stab_{GL(\alpha)}(W) = Aut_{CQ}(W)$ is one-dimensional, which means that every semisimple element of $Aut_{CQ}(W)$ lies in $\mathbb{C}^*(\mathbb{1}_{d_1}, \ldots, \mathbb{1}_{d_k})$. More generally, decomposing a representation V into indecomposables corresponds to choosing a maximal torus in the stabilizer subgroup $Aut_{CQ}(V)$. Let T be such a maximal torus, we define a decomposition of the vertexspaces

$$V_i = \oplus_\chi V_i(\chi) \qquad \text{where} \qquad V_i(\chi) = \{v \in V_i \mid t.v = \chi(t)v \ \forall t \in T\}$$

where χ runs over all characters of T. One verifies that each $V(\chi) = \oplus_i V_i(\chi)$ is a subrepresentation of V giving a decomposition $V = \oplus_\chi V(\chi)$. Because T acts by scalar multiplication on each component $V(\chi)$ we have that $\mathbb{C}^*$ is the maximal torus of $Aut_{CQ}(V(\chi))$, whence $V(\chi)$ is indecomposable. Conversely, if $V = W_1 \oplus \ldots \oplus W_r$ is a decomposition with the W_i indecomposable, then the product of all the one-dimensional maximal tori in $Aut_{CQ}(W_i)$ is a maximal torus of $Aut_{CQ}(V)$.

In this section we will give a classification of the *indecomposable roots* , that is, the dimension vectors of indecomposable representations. As the name suggests, these dimension vectors will form a *root system* .

The *Tits form* of a quiver Q is the symmetrization of its Euler form, that is,

$$T_Q(\alpha, \beta) = \chi_Q(\alpha, \beta) + \chi_Q(\beta, \alpha)$$

This symmetric bilinear form is described by the *Cartan matrix*

$$C_Q = \begin{bmatrix} c_{11} & \cdots & c_{1k} \\ \vdots & & \vdots \\ c_{k1} & \cdots & c_{kk} \end{bmatrix} \qquad \text{with} c_{ij} = 2\delta_{ij} - \#\{ \ ⓙ\!\!-\!\!-\!\!-\!\!-\!\!ⓘ \ \}$$

where we count all arrows connecting v_i with v_j *forgetting the orientation*. The corresponding *quadratic form* $q_Q(\alpha) = \frac{1}{2}\chi_Q(\alpha, \alpha)$ on $\mathbb{Q}^k$ is defined to be

$$q_Q(x_1, \ldots, x_k) = \sum_{i=1}^k x_i^2 - \sum_{a \in Q_a} x_{t(a)} x_{h(a)}$$

Hence, $q_Q(\alpha) = dim \ GL(\alpha) - dim \ \mathbf{rep}_\alpha \ Q$. With Γ_Q we denote the underlying *graph* of Q, that is, forgetting the orientation of the arrows. The following classification result is classical, see for example [15]. A quadratic form q on $\mathbb{Z}^k$ is said to be *positive definite* if $0 \neq \alpha \in \mathbb{Z}^k$ implies $q(\alpha) > 0$. It is

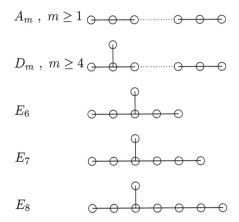

A_m , $m \geq 1$

D_m , $m \geq 4$

E_6

E_7

E_8

FIGURE 4.8: The Dynkin diagrams.

called positive *semi-definite* if $q(\alpha) \geq 0$ for all $\alpha \in \mathbb{Z}^k$. The *radical* of q is $rad(q) = \{\alpha \in \mathbb{Z}^k \mid T(\alpha, -) = 0\}$. Recall that when Q is a connected and $\alpha \geq \underline{0}$ is a nonzero radical vector, then α is *sincere* (that is, all components of α are nonzero) and q_Q is positive semidefinite. There exist a minimal $\delta_Q \geq \underline{0}$ with the property that $q_Q(\alpha) = 0$ if and only if $\alpha \in \mathbb{Q}\delta_Q$ if and only if $\alpha \in rad(q_Q)$. If the quadratic form q is neither positive definite nor semidefinite, it is called *indefinite*.

THEOREM 4.11
Let Q be a connected quiver with Tits form q_Q, Cartan matrix C_Q and underlying graph Γ_Q. Then,

1. *q_Q is positive definite if and only if Γ_Q is a Dynkin diagram , that is one of the graphs of figure 4.8. The number of vertices is m.*

2. *q_Q is semidefinite if and only if Γ_Q is an extended Dynkin diagram, that is one of the graphs of figure 4.9 and δ_Q is the indicated dimension vector. The number of vertices is $m + 1$.*

Let $V \in \mathbf{rep}_\alpha Q$ be decomposed into indecomposables

$$V = W_1^{\oplus f_1} \oplus \ldots \oplus W_z^{\oplus f_z}$$

If $dim(W_i) = \gamma_i$ we say that V is of type $(f_1, \gamma_1; \ldots; f_z, \gamma_z)$.

PROPOSITION 4.11
For any dimension vector α, there exists a unique type $\tau_{can} = (e_1, \beta_1; \ldots; e_l, \beta_l)$ with $\alpha = \sum_i e_i \beta_i$ such that the set $\mathbf{rep}_\alpha(\tau_{can}) =$

$$\{V \in \mathbf{rep}_\alpha Q \mid V \simeq W_1^{\oplus e_1} \oplus \ldots \oplus W_l^{\oplus e_l}, \ dim(W_i) = \beta_i, \ W_i \text{ is indecomposable }\}$$

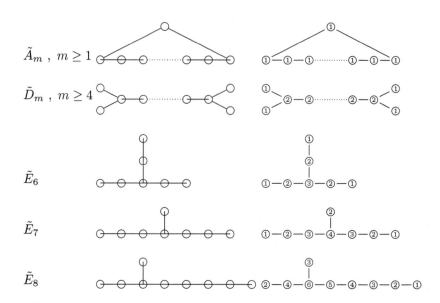

FIGURE 4.9: The extended Dynkin diagrams.

contains a dense open set of $\mathbf{rep}_\alpha\ Q$.

PROOF   Recall from example 2.4 that for any dimension vector β the sub-set $\mathbf{rep}_\beta^{ind}\ Q$ of *indecomposable* representations of dimension β is constructible. Consider for a type $\tau = (f_1, \gamma_1, ; \ldots ; f_z, \gamma_z)$ the subset $\mathbf{rep}_\alpha(\tau) =$

$$\{V \in \mathbf{rep}_\alpha\ Q \mid V \simeq W_1^{\oplus f_1} \oplus \ldots \oplus W_z^{\oplus f_z},\ dim(W_i) = \gamma_i, W_i\ \text{indecomposable}\ \}$$

then $\mathbf{rep}_\alpha(\tau)$ is a constructible subset of $\mathbf{rep}_\alpha\ Q$ as it is the image of the constructible set

$$GL(\alpha) \times \mathbf{rep}_{\gamma_1}^{ind}\ Q \times \ldots \times \mathbf{rep}_{\gamma_z}^{ind}\ Q$$

under the map sending $(g, W_1, \ldots, W_z)$ to $g.(W_1^{\oplus f_1} \oplus \ldots \oplus W_z^{\oplus f_z})$. Because of the uniqueness of the decomposition into indecomposables we have a finite disjoint decomposition

$$\mathbf{rep}_\alpha\ Q = \bigsqcup_\tau \mathbf{rep}_\alpha(\tau)$$

and by irreducibility of $\mathbf{rep}_\alpha\ Q$ precisely one of the $\mathbf{rep}_\alpha(\tau)$ contains a dense open set of $\mathbf{rep}_\alpha\ Q$.   ∎

We call τ_{can} the *canonical decomposition* of α. In the next section we will give an algorithm to compute the canonical decomposition. Consider the action morphisms $GL(\alpha) \times \textbf{rep}_\alpha\ Q \xrightarrow{\phi} \textbf{rep}_\alpha\ Q$. By Chevalley's theorem 2.1 we know that the function

$$V \mapsto dim\ Stab_{GL(\alpha)}(V)$$

is upper semi-continuous. Because $dim\ GL(\alpha) = dim\ Stab_{GL(\alpha)}(V) + dim\ \mathcal{O}(V)$ we conclude that for all m, the subset

$$\textbf{rep}_\alpha(m) = \{V \in \textbf{rep}_\alpha\ Q \mid dim\ \mathcal{O}(V) \geq m\}$$

is Zariski open. In particular, $\textbf{rep}_\alpha(max)$ the union of all orbits of maximal dimension is open and dense in $\textbf{rep}_\alpha\ Q$. A representation $V \in \textbf{rep}_\alpha\ Q$ lying in the intersection

$$\textbf{rep}_\alpha(\tau_{can}) \cap \textbf{rep}_\alpha(max)$$

is called a *generic representation* of dimension α.

Assume that Q is a connected quiver of *finite representation type*, that is, there are only a finite number of isomorphism classes of indecomposable representations. Let α be an arbitrary dimension vector. Since any representation of Q can be decomposed into a direct sum of indecomposables, $\textbf{rep}_\alpha\ Q$ contains only finitely many orbits. Hence, one orbit $\mathcal{O}(V)$ must be dense and have th same dimension as $\textbf{rep}_\alpha\ Q$, but then

$$dim\ \textbf{rep}_\alpha\ Q = dim\ \mathcal{O}(V) \leq dim\ GL(\alpha) - 1$$

as any representation has $\mathbb{C}^*(\mathbb{1}_{a_1}, \dots, \mathbb{1}_{a_k})$ in its stabilizer subgroup. That is, for every $\alpha \in \mathbb{N}^k$ we have $q_Q(\alpha) \geq 1$. Because all off-diagonal entries of the Cartan matrix C_Q are non-positive, it follows that q_Q is positive definite on $\mathbb{Z}^k$ whence Γ_Q must be a Dynkin diagram. It is well known that to a Dynkin diagram one associates a simple Lie algebra and a corresponding *root system*. We will generalize the notion of a root system to an arbitrary quiver Q.

Let $\epsilon_i = (\delta_{1i}, \dots, \delta_{ki})$ be the standard basis of $\mathbb{Q}^k$. The *fundamental set of roots* is defined to be the following set of dimension vectors

$$F_Q = \{\alpha \in \mathbb{N}^k - \underline{0} \mid T_Q(\alpha, \epsilon_i) \leq 0 \text{ and } supp(\alpha) \text{ is connected }\}$$

Recall that it follows from the description of dimension vectors of simple representations given in section 4.4 that any simple root lies in the fundamental set.

LEMMA 4.8
Let $\alpha = \beta_1 + \dots + \beta_s \in F_Q$ with $\beta_i \in \mathbb{N}^k - \underline{0}$ for $1 \leq i \leq s \geq 2$. If $q_Q(\alpha) \geq q_Q(\beta_1) + \dots + q_Q(\beta_s)$, then $supp(\alpha)$ is a tame quiver (that is, its underlying graph is an extended Dynkin diagram) and $\alpha \in \mathbb{N}\delta_{supp(\alpha)}$.

PROOF Let $s = 2$, $\beta_1 = (c_1, \dots, c_k)$ and $\beta_2 = (d_1, \dots, d_k)$ and we may assume that $supp(\alpha) = Q$. By assumption $T_Q(\beta_1, \beta_2) = q_Q(\alpha) - q_Q(\beta_1) -$

$q_Q(\beta_2) \geq 0$. Using that C_Q is symmetric and $\alpha = \beta_1 + \beta_2$ we have

$$0 \leq T_Q(\beta_1, \beta_2) = \sum_{i,j} c_{ij} c_i d_i$$

$$= \sum_j \frac{c_j d_j}{a_j} \sum_i c_{ij} a_i + \frac{1}{2} \sum_{i \neq j} c_{ij} \left(\frac{c_i}{a_i} - \frac{c_j}{a_j}\right)^2 a_i a_j$$

and because $T_Q(\alpha, \epsilon_i) \leq 0$ and $c_{ij} \leq 0$ for all $i \neq j$, we deduce that

$$\frac{c_i}{a_i} = \frac{c_j}{a_j} \qquad \text{for all } i \neq j \text{ such that } c_{ij} \neq 0$$

Because Q is connected, α and β_1 are proportional. But then, $T_Q(\alpha, \epsilon_i) = 0$ and hence $C_Q \alpha = \underline{0}$. By the classification result, q_Q is semidefinite whence Γ_Q is an extended Dynkin diagram and $\alpha \in \mathbb{N}\delta_Q$. Finally, if $s > 2$, then

$$T_Q(\alpha, \alpha) = \sum_i T_Q(\alpha, \beta_i) \geq \sum_i T_Q(\beta_i, \beta_i)$$

whence $T_Q(\alpha - \beta_i, \beta_i) \geq 0$ for some i and then we can apply the foregoing argument to β_i and $\alpha - \beta_i$.   ⬚

DEFINITION 4.8   *If G is an algebraic group acting on a variety Y and if $X \hookrightarrow Y$ is a G-stable subset, then we can decompose $X = \bigcup_d X_{(d)}$ where $X_{(d)}$ is the union of all orbits $\mathcal{O}(x)$ of dimension d. The number of parameters of X is*

$$\mu(X) = \max_d \ (\dim X_{(d)} - d)$$

where $\dim X_{(d)}$ denotes the dimension of the Zariski closure of $X_{(d)}$.
    In the special case of $GL(\alpha)$ acting on $\mathbf{rep}_\alpha Q$, we denote $\mu(\mathbf{rep}_\alpha(max)) = p_Q(\alpha)$ and call it the number of parameters of α. For example, if α is a Schur root, then $p(\alpha) = \dim \mathbf{rep}_\alpha Q - (\dim GL(\alpha) - 1) = 1 - q_Q(\alpha)$.

Recall that a matrix $m \in M_n(\mathbb{C})$ is *unipotent* if some power $m^k = \mathbb{1}_n$. It follows from the Jordan normal form that $GL(\alpha)$ and $PGL(\alpha) = GL(\alpha)/\mathbb{C}^*$ contain only finitely many conjugacy classes of unipotent matrices.

THEOREM 4.12
If α lies in the fundamental set and $\mathrm{supp}(\alpha)$ is not tame, then

$$p_Q(\alpha) = \mu(\mathbf{rep}_\alpha(max)) = \mu(\mathbf{rep}_\alpha^{ind} Q) = 1 - q_Q(\alpha) > \mu(\mathbf{rep}_\alpha^{ind}(d))$$

for all $d > 1$ where $\mathbf{rep}_\alpha^{ind}(d)$ is the union of all indecomposable orbits of dimension d.

PROOF A representation $V \in \mathbf{rep}_\alpha Q$ is indecomposable if and only if its stabilizer subgroup $Stab_{GL(\alpha)}(V)$ is a *unipotent group* , that is all its elements are unipotent elements. By proposition 4.14 we know that $\mathbf{rep}_\alpha(max) \hookrightarrow \mathbf{rep}_\alpha^{ind} Q$ and that $p_Q(\alpha) = \mu(\mathbf{rep}_\alpha(max)) = 1 - q_Q(\alpha)$. Denote $\mathbf{rep}_\alpha(sub) = \mathbf{rep}_\alpha Q - \mathbf{rep}_\alpha(max)$. We claim that for any unipotent element $u \neq \mathbb{1}$ we have that

$$dim \ \mathbf{rep}_\alpha(sub)(u) - dim \ cen_{GL(\alpha)}(u) + 1 < 1 - q_Q(\alpha)$$

where $\mathbf{rep}_\alpha(sub)(g)$ denotes the representations in $\mathbf{rep}_\alpha(sub)$ having g in their stabilizer subgroup. In fact, for any $g \in GL(\alpha) - \mathbb{C}^*$ we have

$$dim \ cen_{GL(\alpha)}(g) - dim \ \mathbf{rep}_\alpha(g) > q_Q(\alpha)$$

Indeed, we may reduce to g being a semisimple element, see [61, lemma 3.4]. then, if $\alpha = \alpha_1 + \ldots + \alpha_s$ is the decomposition of α obtained from the eigenspace decompositions of g (we have $s \geq 2$ as $g \notin \mathbb{C}^*$), then

$$cen_{GL(\alpha)}(g) = \prod_i GL(\alpha_i) \quad \text{and} \quad \mathbf{rep}_\alpha(g) = \prod_i \mathbf{rep}_{\alpha_i}(g)$$

whence $dim \ cen_{GL(\alpha)}(g) - dim \ \mathbf{rep}_\alpha(g) = \sum_i q_Q(\alpha_i) > q_Q(\alpha)$, proving the claim. Further, we claim that

$$\mu(\mathbf{rep}_\alpha(sub)) \leq \max_u \ (dim \ \mathbf{rep}_\alpha(sub)(u) - dim \ cen_{GL(\alpha)}(u) + 1)$$

Let $Z = \mathbf{rep}_\alpha(sub)$ and consider the closed subvariety of $PGL(\alpha) \times Z$

$$L = \{(g, z) \mid g.z = z\}$$

For $z \in Z$ we have $pr_1^{-1}(z) = Stab_{PGL(\alpha)}(z) \times \{z\}$ and if z is indecomposable with orbit dimension d then $dim \ Stab_{PGL(\alpha)}(z) = dim \ PGL(\alpha) - d$, whence

$$dim \ pr_1^{-1}(\mathbf{rep}_\alpha^{ind})_{(d)} = dim \ (\mathbf{rep}_\alpha^{ind})_{(d)} + dim \ PGL(\alpha) - d$$

But then,

$$p_Q(\alpha) = \max_d (dim \ (\mathbf{rep}_\alpha^{ind})_{(d)} - d)$$
$$= -dim \ PGL(\alpha) + \max_d \ dim \ pr_1^{-1}((\mathbf{rep}_\alpha^{ind})_{(d)})$$
$$= -dim \ PGL(\alpha) + dim \ pr_1^{-1}(\mathbf{rep}_\alpha^{ind} Q)$$

By the characterization of indecomposables, we have $pr_1^{-1}(\mathbf{rep}_\alpha^{ind} Q) \subset pr_2^{-1}(U)$ where U consists of the (finitely many) conjugacy classes C_u of conjugacy classes of unipotent $u \in PGL(\alpha)$. But then,

$$p_Q(\alpha) \leq -dim \ PGL(\alpha) + \max_u \ dim \ pr_2^{-1}(C_u)$$
$$= -dim \ PGL(\alpha) + \max_u dim \ \mathbf{rep}_\alpha(sub)(u) + dim \ PGL(\alpha) - dim \ cen_{PGL(\alpha)}(u)$$

proving the claim. Finally, as $dim\ \mathbf{rep}_\alpha(sub) - dim\ PGL(\alpha) < dim\ \mathbf{rep}_\alpha\ Q - dim\ GL(\alpha) + 1 < 1 - q_Q(\alpha)$, we are done.   □

We will now extend this result to arbitrary *roots* using *reflection functors* .
Let v_i be a *source vertex* of Q and let $\alpha = (a_1, \ldots, a_k)$ be a dimension vector
such that $\sum_{t(a)=v_i} a_{h(a)} \geq a_i$, then we can consider the subset

$$\mathbf{rep}_\alpha^{mono}(i) = \{V \in \mathbf{rep}_\alpha\ Q \mid\ \oplus V_a : V_i \longrightarrow \oplus_{t(a)=v_i} V_{s(a)} \text{ is injective }\}$$

Clearly, all indecomposable representations are contained in $\mathbf{rep}_\alpha^{mono}(i)$. Construct the *reflected quiver* $R_i Q$ obtained from Q by reversing the direction of
all arrows with tail v_i. The *reflected dimension vector* $R_i\alpha = (r1, \ldots, r_k)$ is
defined to be

$$r_j = \begin{cases} a_j & \text{if } j \neq i \\ \sum_{t(a)=i} a_{s(a)} - a_i & \text{if } j = i \end{cases}$$

then clearly we have in the reflected quiver $R_i Q$ that $\sum_{h(a)=i} r_{t(a)} \geq r_i$ and
we define the subset

$$\mathbf{rep}_{R_i\alpha}^{epi}(i) = \{V \in \mathbf{rep}_{R_i\alpha}\ R_i Q \mid\ \oplus V_a : \oplus_{s(a)=i} V_{t(a)} \longrightarrow V_i \text{ is surjective }\}$$

Before stating the main result on reflection functors, we need to recall the
definition of the Grassmann manifolds.

Let $k \leq l$ be integers, then the points of the Grassmannian $Grass_k(l)$ are in
one-to-one correspondence with k-dimensional subspaces of $\mathbb{C}^l$. For example,
if $k = 1$ then $Grass_1(l) = \mathbb{P}^{l-1}$. We know that projective space can be covered
by affine spaces defining a manifold structure on it. Also Grassmannians admit
a cover by affine spaces.

Let W be a k-dimensional subspace of $\mathbb{C}^l$ then fixing a basis $\{w_1, \ldots, w_k\}$
of W determines an $k \times l$ matrix M having as i-th row the coordinates of w_i
with respect to the standard basis of $\mathbb{C}^l$. Linear independence of the vectors
w_i means that there is a barcode design I on M

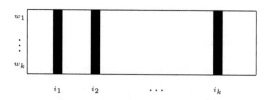

where $I = 1 \leq i_1 < i_2 < \ldots < i_k \leq l$ such that the corresponding $k \times k$ minor
M_I of M is invertible. Observe that M can have several such designs.

Conversely, given a $k \times l$ matrix M of rank k determines a k-dimensional
subspace of l spanned by the transposed rows. Two $k \times l\ M$ and M' matrices of rank k determine the same subspace provided there is a basechange
matrix $g \in GL_k$ such that $gM = M'$. That is, we can identify $Grass_k(l)$
with the orbit space of the linear action of GL_k by left multiplication on the

open set $M_{k \times l}^{max}(\mathbb{C})$ of $M_{k \times l}(\mathbb{C})$ of matrices of maximal rank. Let I be a barcode design and consider the subset of $Grass_k(l)(I)$ of subspaces having a matrix representation M having I as barcode design. Multiplying on the left with M_I^{-1} the GL_k-orbit $\mathcal{O}_M$ has a unique representant N with $N_I = \mathbb{1}_k$. Conversely, any matrix N with $N_I = \mathbb{1}_k$ determines a point in $Grass_k(l)(I)$. Thus, $Grass_k(l)(I)$ depends on $k(l - k)$ free parameters (the entries of the negative of the barcode)

and we have an identification $Grass_k(l)(I) \xrightarrow{\pi_I} \mathbb{C}^{k(l-k)}$. For a different barcode design I' the image $\pi_I(Grass_k(l)(I) \cap Grass_k(l)(I'))$ is an open subset of $\mathbb{C}^{k(l-k)}$ (one extra nonsingular minor condition) and $\pi_{I'} \circ \pi_I^{-1}$ is a diffeomorphism on this set. That is, the maps π_I provide us with an atlas and determine a manifold structure on $Grass_k(l)$.

THEOREM 4.13

For the quotient Zariski topology, we have an homeomorphism

$$\mathbf{rep}_\alpha^{mono}(i)/GL(\alpha) \xrightarrow{\simeq} \mathbf{rep}_{R_i\alpha}^{epi}(i)/GL(R_i\alpha)$$

such that corresponding representations have isomorphic endomorphism rings.

In particular, the number of parameters as well as the number of irreducible components of maximal dimension coincide for $(\mathbf{rep}_\alpha^{ind} Q)_{(d)}$ and $\mathbf{rep}_{R_i\alpha}^{ind} R_iQ)_{(d)}$ for all dimensions d.

PROOF Let $m = \sum_{t(a)=i} a_i$, $\overline{rep} = \oplus_{t(a)\neq i} M_{a_{s(a)} \times a_{t(a)}}(\mathbb{C})$ and $\overline{GL} = \prod_{j \neq i} GL_{a_j}$. We have the following isomorphisms

$$\mathbf{rep}_\alpha^{mono}(i)/GL_{a_i} \xrightarrow{\simeq} \overline{rep} \times Gass_{a_i}(m)$$

defined by sending a representation V to its restriction to $\overline{rep}$ and $im \oplus_{t(a)=i} V_a$. In a similar way, sending a representation V to its restriction and $ker \oplus_{s(a)=i} V_a$ we have

$$\mathbf{rep}_{R_i\alpha}^{epi}(i)/GL_{r_i} \xrightarrow{\simeq} \overline{rep} \times Grass_{a_i}(m)$$

But then, the first claim follows from the diagram of figure 4.10. If $V \in \mathbf{rep}_\alpha Q$

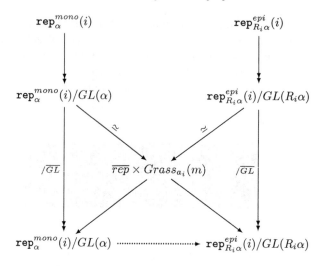

FIGURE 4.10: Reflection functor diagram.

and $V' \in \mathbf{rep}_{R_i\alpha} R_iQ$ with images respectively v and v' in $\overline{rep} \times Grass_{a_i}(m)$, we have isomorphisms

$$\begin{cases} Stab_{\overline{GL} \times GL_{a_i}}(V) & \xrightarrow{\simeq} Stab_{\overline{GL}}(v) \\ Stab_{\overline{GL} \times GL_{r_i}}(V') & \xrightarrow{\simeq} Stab_{\overline{GL}}(v') \end{cases}$$

from which the claim about endomorphisms follows. □

A similar results holds for *sink vertices*, hence we can apply these *Bernstein-Gelfand- Ponomarev reflection functors* iteratively using a sequence of *admissible vertices* (that is, either a source or a sink).

To a vertex v_i in which Q has no loop, we define a *reflection* $\mathbb{Z}^k \xrightarrow{r_i} \mathbb{Z}^k$ by

$$r_i(\alpha) = \alpha - T_Q(\alpha, \epsilon_i)$$

The *Weyl group of the quiver Q $Weyl_Q$* is the subgroup of $GL_k(\mathbb{Z})$ generated by all reflections r_i.

A *root* of the quiver Q is a dimension vector $\alpha \in \mathbb{N}^k$ such that $\mathbf{rep}_\alpha Q$ contains indecomposable representations. All roots have connected support. A root is said to be

$$\begin{cases} real & \text{if } \mu(\mathbf{rep}_\alpha^{ind} Q) = 0 \\ imaginary & \text{if } \mu(\mathbf{rep}_\alpha^{ind} Q) \geq 1 \end{cases}$$

For a fixed quiver Q we will denote the set of all roots, real roots and imaginary roots respectively by Δ, Δ_{re} and Δ_{im}. With Π we denote the set

$\{\epsilon_i \mid v_i$ has no loops $\}$. the main result on indecomposable representations is due to V. Kac .

THEOREM 4.14

With notations as before, we have

1. $\Delta_{re} = Weyl_Q.\Pi \cap \mathbb{N}^k$ *and if* $\alpha \in \Delta_{re}$, *then* $\mathbf{rep}_\alpha^{ind} Q$ *is one orbit*

2. $\Delta_{im} = Weyl.F_Q \cap \mathbb{N}^k$ *and if* $\alpha \in \Delta_{im}$ *then*

$$p_Q(\alpha) = \mu(\mathbf{rep}_\alpha^{ind} Q) = 1 - q_Q(\alpha)$$

For a sketch of the proof we refer to [35, §7], full details can be found in the lecture notes [61].

4.6   Canonical decomposition

In this section we will determine the canonical decomposition. We need a technical result.

LEMMA 4.9

Let $W \longrightarrow\!\!\!\!\rightarrow W'$ *be an epimorphism of* $\mathbb{C}Q$*-representations. Then, for any* $\mathbb{C}Q$*-representation* V *we have that the canonical map*

$$Ext_{\mathbb{C}Q}^1(V, W) \longrightarrow\!\!\!\!\rightarrow Ext_{\mathbb{C}Q}^1(V, W')$$

is surjective. If $W \hookrightarrow W'$ *is a monomorphism of* $\mathbb{C}Q$*-representations, then the canonical map*

$$Ext_{\mathbb{C}Q}^1(W', V) \longrightarrow\!\!\!\!\rightarrow Ext_{\mathbb{C}Q}^1(W, V)$$

is surjective.

PROOF   From the proof of theorem 4.5 we have the exact diagram

$$
\begin{array}{ccccccc}
\bigoplus_{v_i \in Q_v} Hom_\mathbb{C}(V_i, W_i) & \xrightarrow{d_W^V} & \bigoplus_{a \in Q_a} Hom_\mathbb{C}(V_{s(a)}, W_{t(a)}) & \longrightarrow & Ext_{\mathbb{C}Q}^1(V, W) & \longrightarrow & 0 \\
\downarrow & & \downarrow & & \vdots & & \\
\bigoplus_{v_i \in Q_v} Hom_\mathbb{C}(V_i, W_i') & \xrightarrow{d_{W'}^V} & \bigoplus_{a \in Q_a} Hom_\mathbb{C}(V_{s(a)}, W_{t(a)}') & \longrightarrow & Ext_{\mathbb{C}Q}^1(V, W') & \longrightarrow & 0
\end{array}
$$

and applying the *snake lemma* gives the result. The second part is proved similarly. $\square$

LEMMA 4.10
If $V = V' \oplus V'' \in \mathbf{rep}_\alpha(max)$, *then* $Ext^1_{\mathbb{C}Q}(V', V'') = 0$.

PROOF   Assume $Ext^1(V', V'') \neq 0$, that is, there is a nonsplit exact sequence

$$0 \longrightarrow V'' \longrightarrow W \longrightarrow V' \longrightarrow 0$$

then it follows from section 2.3 that $\mathcal{O}(V) \subset \overline{\mathcal{O}(W)} - \mathcal{O}(W)$, whence $dim\ \mathcal{O}(W) > dim\ \mathcal{O}(V)$ contradicting the assumption that $V \in \mathbf{rep}_\alpha(max)$.
$\square$

LEMMA 4.11
If W, W' *are indecomposable representation with* $Ext^1_{\mathbb{C}Q}(W, W') = 0$, *then any nonzero map* $W' \overset{\phi}{\longrightarrow} W$ *is an epimorphism or a monomorphism. In particular, if* W *is indecomposable with* $Ext^1_{\mathbb{C}Q}(W, W) = 0$, *then* $End_{\mathbb{C}Q}(W) \simeq \mathbb{C}$.

PROOF   Assume ϕ is neither mono- nor epimorphism then decompose ϕ into

$$W' \overset{\epsilon}{\twoheadrightarrow} U \overset{\mu}{\hookrightarrow} W$$

As ϵ is epi, we get a surjection from lemma 4.9

$$Ext^1_{\mathbb{C}Q}(W/U, W') \longrightarrow\!\!\!\!\!\rightarrow Ext^1_{\mathbb{C}Q}(W/U, U)$$

giving a representation V fitting into the exact diagram of extensions

$$
\begin{array}{ccccccccc}
0 & \longrightarrow & W' & \overset{\mu'}{\longrightarrow} & V & \longrightarrow & W'/U & \longrightarrow & 0 \\
 & & \downarrow{\scriptstyle \epsilon} & & \downarrow{\scriptstyle \epsilon'} & & \downarrow{\scriptstyle id} & & \\
0 & \longrightarrow & U & \overset{\mu}{\longrightarrow} & W & \longrightarrow & W'/U & \longrightarrow & 0
\end{array}
$$

from which we construct an exact sequence of representations

$$0 \longrightarrow W' \overset{\begin{bmatrix} \epsilon \\ -\mu' \end{bmatrix}}{\longrightarrow} U \oplus V \overset{[\mu\ \epsilon']}{\longrightarrow} W \longrightarrow 0$$

This sequence cannot split as otherwise we would have $W \oplus W' \simeq U \oplus V$ contradicting uniqueness of decompositions, whence $Ext^1_{\mathbb{C}Q}(W, W') \neq 0$, a contradiction.

For the second part, as W is finite dimensional it follows that $End_{\mathbb{C}Q}(W)$ is a (finite dimensional) division algebra whence it must be $\mathbb{C}$. □

DEFINITION 4.9 *A representation $V \in \mathbf{rep}_\alpha Q$ is said to be a* Schur representation *if $End_{\mathbb{C}Q}(V) = \mathbb{C}$. The dimension vector α of a Schur representation is said to be a* Schur root.

THEOREM 4.15
α *is a Schur root if and only if there is a Zariski open subset of $\mathbf{rep}_\alpha Q$ consisting of indecomposable representations.*

PROOF If $V \in \mathbf{rep}_\alpha Q$ is a Schur representation, $V \in \mathbf{rep}_\alpha(max)$ and therefore all representations in the dense open subset $\mathbf{rep}_\alpha(max)$ have endomorphism ring $\mathbb{C}$ and are therefore indecomposable. Conversely, let $Ind \hookrightarrow \mathbf{rep}_\alpha Q$ be an open subset of indecomposable representations and assume that for $V \in Ind$ we have $Stab_{GL(\alpha)}(V) \neq \mathbb{C}^*$ and consider $\phi_0 \in Stab_{GL(\alpha)}(V) - \mathbb{C}^*$. For any $g \in GL(\alpha)$ we define the *set of fixed elements*

$$\mathbf{rep}_\alpha(g) = \{W \in \mathbf{rep}_\alpha Q \mid g.W = W\}$$

Define the subset of $GL(\alpha)$

$$S = \{g \in GL(\alpha) \mid dim\ \mathbf{rep}_\alpha(g) = dim\ \mathbf{rep}_\alpha(\phi_0)\}$$

which has no intersection with $\mathbb{C}^*(\mathbb{1}_{d_1}, \ldots, \mathbb{1}_{d_k})$ as $\phi_0 \notin \mathbb{C}^*$. Consider the subbundle of the trivial vectorbundle over S

$$\mathcal{B} = \{(s, W) \in S \times \mathbf{rep}_\alpha Q \mid s.W = W\} \hookrightarrow S \times \mathbf{rep}_\alpha Q \xrightarrow{p} S$$

As all fibers have equal dimension, the restriction of p to $\mathcal{B}$ is a *flat morphism* whence *open* . In particular, the image of the open subset $\mathcal{B} \cap S \times Ind$

$$S' = \{g \in S \mid \exists W \in Ind : g.W = W\}$$

is an open subset of S. Now, S contains a dense set of semisimple elements, see for example [61, (2.5)], whence so does $S' = \cup_{W \in Ind} End_{\mathbb{C}Q}(W) \cap S$. But then one of the $W \in Ind$ must have a torus of rank greater than one in its stabilizer subgroup contradicting indecomposability. □

Schur roots give rise to principal $PGL(\alpha) = GL(\alpha)/\mathbb{C}^*$-fibrations, and hence to quiver orders and division algebras.

PROPOSITION 4.12
If $\alpha = (a_1, \ldots, a_k)$ is a Schur root, then there is a $GL(\alpha)$-stable affine open subvariety U_α of $\mathbf{rep}_\alpha Q$ such that generic orbits are closed in U.

PROOF   Let $T_k = \mathbb{C}^* \times \ldots \times \mathbb{C}^*$ the k-dimensional torus in $GL(\alpha)$. Consider the semisimple subgroup $SL(\alpha) = SL_{a_1} \times \ldots \times SL_{a_k}$ and consider the corresponding quotient map

$$\mathbf{rep}_\alpha \, Q \xrightarrow{\pi_s} \mathbf{rep}_\alpha \, Q/SL(\alpha)$$

As $GL(\alpha) = T_k SL(\alpha)$, T_k acts on $\mathbf{rep}_\alpha \, Q/SL(\alpha)$ and the generic stabilizer subgroup is trivial by the Schurian condition. Hence, there is a T_k-invariant open subset U_1 of $\mathbf{rep}_\alpha \, Q/SL(\alpha)$ such that T_k-orbits are closed. But then, according to [52, §2, Thm.5] there is a T_k-invariant *affine* open U_2 in U_1. Because the quotient map ψ_s is an affine map, $U = \psi_s^{-1}(U_2)$ is an affine $GL(\alpha)$-stable open subvariety of $\mathbf{rep}_\alpha \, Q$. Let x be a generic point in U, then its orbit

$$\mathcal{O}(x) = GL(\alpha).x = T_k SL(\alpha).x = T_k(\psi_s^{-1}(\psi_s(x))) = \psi_s^{-1}(T_k.\psi_s(x))$$

is the inverse image under the quotient map of a closed set, hence is itself closed. $\square$

If we define $\mathbb{T}_\alpha^s \, Q$ to be the ring of $GL(\alpha)$-equivariant maps from U_α to $M_n(\mathbb{C})$, then this *Schurian quiver order* has *simple* α-dimensional representations. Then, extending the argument of proposition 4.9 we have that the quotient map $\mathbf{rep}_\alpha \, Q \longrightarrow \mathbf{iss}_\alpha \, Q$ is a principal $PGL(\alpha)$-fibration in the étale topology over the Azumaya locus of the Schurian quiver order $\mathbb{T}_\alpha^s \, Q$. Recall that $H_{et}^1(X, PGL(\alpha))$ classifies twisted forms of $M_n(\mathbb{C})$ (where $n = \sum_a a_i$) as C_k-algebra. That is, Azumaya algebras over X with a distinguished embedding of C_k that are split by an étale cover on which this embedding is conjugate to the standard α-embedding of C_k in $M_n(\mathbb{C})$. The class in the Brauer group of the functionfield of $\mathbf{iss}_\alpha \, \mathbb{T}_\alpha^s \, Q$ determined by the quiver order $\mathbb{T}_\alpha^s \, Q$ is rather special.

PROPOSITION 4.13
If $\alpha = (a_1, \ldots, a_k)$ is a Schur root of Q such that $gcd(a_1, \ldots, a_k) = 1$, then $\mathbb{T}_\alpha^s \, Q$ determines the trivial class in the Brauer group.

PROOF   Let A be an Azumaya localization of $\mathbb{T}_\alpha^s \, Q$. By assumption, the natural map between the K-groups $K_0(C_k) \longrightarrow K_0(M_n(\mathbb{C}))$ is surjective, whence the same is true for A proving that the class of A is split by a Zariski cover, that is $\mathbf{rep}_\alpha \, Q \simeq X \times PGL(\alpha)$ where $X = \mathbf{iss}_\alpha \, A$. $\square$

PROPOSITION 4.14
If α lies in the fundamental region F_Q and $supp(\alpha)$ is not a tame quiver. then, α is a Schur root.

PROOF   Let $\alpha = \beta_1 + \ldots + \beta_s$ be the canonical decomposition of α (some β_i may occur with higher multiplicity) and assume that $s \geq 2$. By definition, the image of

$$GL(\alpha) \times (\mathbf{rep}_{\beta_1} Q \times \ldots \times \mathbf{rep}_{\beta_s} Q) \xrightarrow{\phi} \mathbf{rep}_\alpha Q$$

is dense and ϕ is constant on orbits of the *free action* of $GL(\alpha)$ on the left hand side given by $h.(g, V) = (gh^{-1}, h.V)$. But then

$$dim\ GL(\alpha) + \sum_i dim\ \mathbf{rep}_{\beta_i} Q - \sum_i dim\ GL(\beta_i) \geq dim\ \mathbf{rep}_\alpha Q$$

whence $q_Q(\alpha) \geq \sum_i q_Q(\beta_i)$ and lemma 4.8 finishes the proof.   □

Next, we want to describe morphisms between quiver-representations. Let $\alpha = (a_1, \ldots, a_k)$ and $\beta = (b_1, \ldots, b_k)$ and $V \in \mathbf{rep}_\alpha Q$, $W \in \mathbf{rep}_\beta Q$. Consider the closed subvariety

$$Hom_Q(\alpha, \beta) \hookrightarrow M_{a_1 \times b_1} \oplus \ldots \oplus M_{a_k \times b_k} \oplus \mathbf{rep}_\alpha Q \oplus \mathbf{rep}_\beta Q$$

consisting of the triples (ϕ, V, W) where $\phi = (\phi_1, \ldots, \phi_k)$ is a morphism of quiver-representations $V \longrightarrow W$. Projecting to the two last components we have an onto morphism between affine varieties

$$Hom_Q(\alpha, \beta) \xrightarrow{h} \mathbf{rep}_\alpha Q \oplus \mathbf{rep}_\beta Q$$

In theorem 2.1 we have proved that the dimension of fibers is an upper-semicontinuous function. That is, for every natural number d, the set

$$\{\Phi \in Hom_Q(\alpha, \beta) \mid dim_\Phi\ h^{-1}(h(\Phi)) \leq d\}$$

is a Zariski open subset of $Hom_Q(\alpha, \beta)$. As the target space $\mathbf{rep}_\alpha Q \oplus \mathbf{rep}_\beta Q$ is irreducible, it contains a nonempty open subset hom_{min} where the dimension of the fibers attains a minimal value. This minimal fiber dimension will be denoted by $hom(\alpha, \beta)$.

Similarly, we could have defined an affine variety $Ext_Q(\alpha, \beta)$ where the fiber over a point $(V, W) \in \mathbf{rep}_\alpha Q \oplus \mathbf{rep}_\beta Q$ is given by the extensions $Ext^1_{\mathbb{C}Q}(V, W)$. If χ_Q is the Euler-form of Q we recall that for all $V \in \mathbf{rep}_\alpha Q$ and $W \in \mathbf{rep}_\beta Q$ we have

$$dim_{\mathbb{C}}\ Hom_{\mathbb{C}Q}(V, W) - dim_{\mathbb{C}}\ Ext^1_Q(V, W) = \chi_Q(\alpha, \beta)$$

Hence, there is also an open set ext_{min} of $\mathbf{rep}_\alpha Q \oplus \mathbf{rep}_\beta Q$ where the dimension of $Ext^1(V, W)$ attains a minimum. This minimal value we denote by $ext(\alpha, \beta)$. As $hom_{min} \cap ext_{min}$ is a nonempty open subset we have the numerical equality

$$hom(\alpha, \beta) - ext(\alpha, \beta) = \chi_Q(\alpha, \beta).$$

In particular, if $hom(\alpha, \alpha + \beta) > 0$, there will be an open subset where the morphism $V \xrightarrow{\phi} W$ is a monomorphism. Hence, there will be an open subset of $\mathbf{rep}_{\alpha+\beta} Q$ consisting of representations containing a subrepresentation of dimension vector α. We say that α is a general subrepresentation of $\alpha + \beta$ and denote this with $\alpha \longrightarrow \alpha + \beta$. We want to characterize this property. To do this, we introduce the quiver-Grassmannians

$$Grass_\alpha(\alpha + \beta) = \prod_{i=1}^{k} Grass_{a_i}(a_i + b_i)$$

which is a projective manifold.

Consider the following diagram of morphisms of reduced varieties

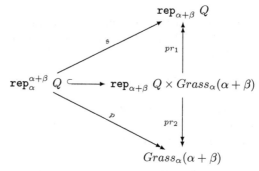

with the following properties

- $\mathbf{rep}_{\alpha+\beta} Q \times Grass_\alpha(\alpha + \beta)$ is the trivial vectorbundle with fiber $\mathbf{rep}_{\alpha+\beta} Q$ over the projective smooth variety $Grass_\alpha(\alpha+\beta)$ with structural morphism pr_2.

- $\mathbf{rep}_\alpha^{\alpha+\beta} Q$ is the subvariety of $\mathbf{rep}_{\alpha+\beta} Q \times Grass_\alpha(\alpha + \beta)$ consisting of couples (W, V) where V is a subrepresentation of W (observe that this is for fixed W a linear condition). Because $GL(\alpha + \beta)$ acts transitively on the Grassmannian $Grass_\alpha(\alpha + \beta)$ (by multiplication on the right) we see that $\mathbf{rep}_\alpha^{\alpha+\beta} Q$ is a sub-vectorbundle over $Grass_\alpha(\alpha + \beta)$ with structural morphism p. In particular, $\mathbf{rep}_\alpha^{\alpha+\beta} Q$ is a reduced variety.

- The morphism s is a projective morphism, that is, can be factored via the natural projection

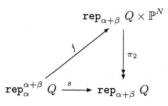

where f is the composition of the inclusion $\mathbf{rep}_\alpha^{\alpha+\beta} Q \longrightarrow \mathbf{rep}_{\alpha+\beta} Q \times Grass_\alpha(\alpha+\beta)$ with the natural inclusion of Grassmannians in projective spaces recalled in the previous section $Grass_\alpha(\alpha + \beta) \hookrightarrow \prod_{i=1}^{k} \mathbb{P}^{n_i}$ with the Segre embedding $\prod_{i=1}^{k} \mathbb{P}^{n_i} \hookrightarrow \mathbb{P}^N$. In particular, s is proper by [43, Thm. II.4.9], that is, maps closed subsets to closed subsets.

We are interested in the scheme-theoretic fibers of s. If $W \in \mathbf{rep}_{\alpha+\beta} Q$ lies in the image of s, we denote the fiber $s^{-1}(W)$ by $\mathbf{Grass}_\alpha(W)$. Its geometric points are couples (W, V) where V is an α-dimensional subrepresentation of W. Whereas $\mathbf{Grass}_\alpha(W)$ is a projective scheme, it is in general neither smooth, nor irreducible nor even reduced. Therefore, in order to compute the tangent space in a point (W, V) of $\mathbf{Grass}_\alpha(W)$ we have to clarify the functor it represents on the category *commalg* of commutative $\mathbb{C}$-algebras.

Let C be a commutative $\mathbb{C}$-algebra, a representation $\mathcal{R}$ of the quiver Q over C consists of a collection $\mathcal{R}_i = P_i$ of projective C-modules of finite rank and a collection of C-module morphisms for every arrow a in Q

$$\overset{a}{\textcircled{\scriptsize j} \longleftarrow \textcircled{\scriptsize i}} \qquad \mathcal{R}_j = P_j \overset{\mathcal{R}_a}{\longleftarrow} P_i = \mathcal{R}_i$$

The dimension vector of the representation $\mathcal{R}$ is given by the k-tuple $(rk_C\ \mathcal{R}_1, \ldots, rk_C\ \mathcal{R}_k)$. A subrepresentation $\mathcal{S}$ of $\mathcal{R}$ is determined by a collection of projective subsummands (and not merely submodules) $\mathcal{S}_i \lhd \mathcal{R}_i$. In particular, for $W \in \mathbf{rep}_{\alpha+\beta} Q$ we define the representation $\mathcal{W}_C$ of Q over the commutative ring C by

$$\begin{cases} (\mathcal{W}_C)_i &= C \otimes_{\mathbb{C}} W_i \\ (\mathcal{W}_C)_a &= id_C \otimes_{\mathbb{C}} W_a \end{cases}$$

With these definitions, we can now define the functor represented by $\mathbf{Grass}_\alpha(W)$ as the functor assigning to a commutative $\mathbb{C}$-algebra C the set of all subrepresentations of dimension vector α of the representation $\mathcal{W}_C$.

LEMMA 4.12
Let $x = (W, V)$ be a geometric point of $\mathbf{Grass}_\alpha(W)$, then

$$T_x\ \mathbf{Grass}_\alpha(W) = Hom_{\mathbb{C}Q}(V, \frac{W}{V})$$

PROOF The tangent space in $x = (W, V)$ are the $\mathbb{C}[\epsilon]$-points of $\mathbf{Grass}_\alpha(W)$ lying over (W, V). To start, let $V \overset{\psi}{\longrightarrow} \frac{W}{V}$ be a homomorphism of representations of Q and consider a $\mathbb{C}$-linear lift of this map $\tilde{\psi} : V \longrightarrow W$. Consider the $\mathbb{C}$-linear subspace of $\mathcal{W}_{\mathbb{C}[\epsilon]} = \mathbb{C}[\epsilon] \otimes W$ spanned by the sets

$$\{v + \epsilon \otimes \tilde{\psi}(v)\ |\ v \in V\} \quad \text{and} \quad \epsilon \otimes V$$

This determines a $\mathbb{C}[\epsilon]$-subrepresentation of dimension vector α of $\mathcal{W}_{\mathbb{C}[\epsilon]}$ lying over (W, V) and is independent of the chosen linear lift $\tilde{\psi}$.

Conversely, if $\mathcal{S}$ is a $\mathbb{C}[\epsilon]$-subrepresentation of $\mathcal{W}_{\mathbb{C}[\epsilon]}$ lying over (W, V), then $\frac{\mathcal{S}}{\epsilon \mathcal{S}} = V \hookrightarrow W$. But then, a $\mathbb{C}$-linear complement of $\epsilon \mathcal{S}$ is spanned by elements of the form $v + \epsilon \psi(v)$ where $\psi(v) \in W$ and $\epsilon \otimes \psi$ is determined modulo an element of $\epsilon \otimes V$. But then, we have a $\mathbb{C}$-linear map $\tilde{\psi} : V \longrightarrow \frac{W}{V}$ and as $\mathcal{S}$ is a $\mathbb{C}[\epsilon]$-subrepresentation, $\tilde{\psi}$ must be a homomorphism of representations of Q.   □

THEOREM 4.16
The following are equivalent

1. $\alpha \hookrightarrow \alpha + \beta$

2. *Every representation* $W \in \mathbf{rep}_{\alpha + \beta} Q$ *has a subrepresentation* V *of dimension* α

3. $ext(\alpha, \beta) = 0$

PROOF   Assume 1., then the image of the proper map

$$s : \mathbf{rep}_\alpha^{\alpha + \beta} Q \longrightarrow \mathbf{rep}_{\alpha + \beta} Q$$

contains a Zariski open subset. As properness implies that the image of s must also be a closed subset of $\mathbf{rep}_{\alpha + \beta} Q$ it follows that $Im \ s = \mathbf{rep}_{\alpha + \beta} Q$, that is 2. holds. Conversely, 2. clearly implies 1. so they are equivalent.

We compute the dimension of the vectorbundle $\mathbf{rep}_\alpha^{\alpha + \beta} Q$ over $Grass_\alpha(\alpha + \beta)$. Using that the dimension of a Grassmannians $Grass_k(l)$ is $k(l - k)$ we know that the base has dimension $\sum_{i=1}^k a_i b_i$. Now, fix a point $V \hookrightarrow W$ in $Grass_\alpha(\alpha + \beta)$, then the fiber over it determines all possible ways in which this inclusion is a subrepresentation of quivers. That is, for every arrow in Q of the form $\textcircled{\scriptsize j} \xleftarrow{\ a\ } \textcircled{\scriptsize i}$ we need to have a commuting diagram

$$\begin{array}{ccc} V_i & \longrightarrow & V_j \\ \cap\downarrow & & \cap\downarrow \\ W_i & \longrightarrow & W_j \end{array}$$

Here, the vertical maps are fixed. If we turn $V \in \mathbf{rep}_\alpha Q$, this gives us the $a_i a_j$ entries of the upper horizontal map as degrees of freedom, leaving only freedom for the lower horizontal map determined by a linear map $\frac{W_i}{V_i} \longrightarrow W_j$, that is, having $b_i(a_j + b_j)$ degrees of freedom. Hence, the dimension of the vector space-fibers is

$$\sum_{\textcircled{\scriptsize j} \xleftarrow{\ a\ } \textcircled{\scriptsize i}} (a_i a_j + b_i(a_j + b_j))$$

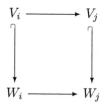

giving the total dimension of the reduced variety $\mathbf{rep}_\alpha^{\alpha+\beta} Q$. But then

$$dim \ \mathbf{rep}_\alpha^{\alpha+\beta} Q - dim \ \mathbf{rep}_{\alpha+\beta} Q = \sum_{i=1}^{k} a_i b_i + \sum_{①\overset{a}{\leftarrow}①} (a_i a_j + b_i(a_j + b_j))$$

$$- \sum_{①\overset{a}{\leftarrow}①} (a_i + b_i)(a_j + b_j)$$

$$= \sum_{i=1}^{k} a_i b_i - \sum_{①\overset{a}{\leftarrow}①} a_i b_j = \chi_Q(\alpha, \beta)$$

Assume that 2. holds, then the proper map $\mathbf{rep}_\alpha^{\alpha+\beta} \overset{s}{\longrightarrow} \mathbf{rep}_{\alpha+\beta} Q$ is onto and as both varieties are reduced, the general fiber is a reduced variety of dimension $\chi_Q(\alpha, \beta)$, whence the general fiber contains points such that their tangentspaces have dimension $\chi_Q(\alpha, \beta)$. By the foregoing lemma we can compute the dimension of this tangentspace as $dim \ Hom_{\mathbb{C}Q}(V, \frac{W}{V})$. But then, as

$$\chi_Q(\alpha, \beta) = dim_{\mathbb{C}} \ Hom_{\mathbb{C}Q}(V, \frac{W}{V}) - dim_{\mathbb{C}} \ Ext_{\mathbb{C}Q}^1(V, \frac{W}{V})$$

it follows that $Ext^1(V, \frac{W}{V}) = 0$ for some representation V of dimension vector α and $\frac{W}{V}$ of dimension vector β. But then, $ext(\alpha, \beta) = 0$, that is, 3. holds.

Conversely, assume that $ext(\alpha, \beta) = 0$. Then, for a general point $W \in \mathbf{rep}_{\alpha+\beta} Q$ in the image of s and for a general point in its fiber $(W, V) \in \mathbf{rep}_\alpha^{\alpha+\beta} Q$ we have $dim_{\mathbb{C}} \ Ext_{\mathbb{C}Q}^1(V, \frac{W}{V}) = 0$ whence $dim_{\mathbb{C}} \ Hom_{\mathbb{C}Q}(V, \frac{W}{V}) = \chi_Q(\alpha, \beta)$. But then, the general fiber of s has dimension $\chi_Q(\alpha, \beta)$ and as this is the difference in dimension between the two irreducible varieties, the map is generically onto. Finally, properness of s then implies that it is onto, giving 2. and finishing the proof. ∎

PROPOSITION 4.15
Let α be a Schur root such that $\chi_Q(\alpha, \alpha) < 0$, then for any integer n we have that $n\alpha$ is a Schur root.

PROOF There are infinitely many nonisomorphic Schur representations of dimension vector α. Pick n of them $\{W_1, \ldots, W_n\}$ and from $\chi_Q(\alpha, \alpha) < 0$ we deduce

$$Hom_{\mathbb{C}Q}(W_i, W_j) = \delta_{ij}\mathbb{C} \qquad \text{and} \qquad Ext_{\mathbb{C}Q}^1(W_i, W_j) \neq 0$$

By lemma 4.9 we can construct a representation V_n having a filtration

$$0 = V_0 \subset V_1 \subset \ldots \subset V_n \quad \text{with} \quad \frac{V_j}{V_{j-1}} \simeq W_j$$

and such that the short exact sequences $0 \longrightarrow V_{j-1} \longrightarrow V_j \longrightarrow W_j \longrightarrow 0$ do not split. By induction on n we may assume that $End_{\mathbb{C}Q}(V_{n-1}) = \mathbb{C}$ and we have that $Hom_{\mathbb{C}Q}(V_{n-1}, W_n) = 0$. But then, the restriction of any endomorphism ϕ of V_n to V_{n-1} must be an endomorphism of V_{n-1} and therefore a scalar $\lambda \mathbb{1}$. Hence, $\phi - \lambda \mathbb{1} \in End_{\mathbb{C}Q}(V_n)$ is trivial on V_{n-1}. As $Hom_{\mathbb{C}Q}(W_n, V_{n-1}) = 0$, $End_{\mathbb{C}Q}(W_n) = \mathbb{C}$ and nonsplitness of the sequence $0 \longrightarrow V_{n-1} \longrightarrow V_n \longrightarrow W_n \longrightarrow 0$ we must have $\phi - \lambda \mathbb{1} = 0$ whence $End_{\mathbb{C}Q}(V_n) = \mathbb{C}$, that is, $n\alpha$ is a Schur root. ▯

We say that a dimension vector α is *left orthogonal* to β if $hom(\alpha, \beta) = 0$ and $ext(\alpha, \beta) = 0$.

DEFINITION 4.10 *An ordered sequence $C = (\beta_1, \ldots, \beta_s)$ of dimension vectors is said to be a* compartment *for Q if and only if*

1. *for all i, β_i is a Schur root*

2. *for al $i < j$, β_i is left orthogonal to β_j*

3. *for all $i < j$ we have $\chi_Q(\beta_j, \beta_i) \geq 0$*

THEOREM 4.17
Suppose that $C = (\beta_1, \ldots, \beta_s)$ is a compartment for Q and that there are nonnegative integers $e_1, \ldots, e_s$ such that $\alpha = e_1\beta_1 + \ldots + e_s\beta_s$. Assume that $e_i = 1$ whenever $\chi_Q(\beta_i, \beta_i) < 0$. Then,

$$\tau_{can} = (e_1, \beta_1; \ldots; e_s, \beta_s)$$

is the canonical decomposition of the dimension vector α.

PROOF Let V be a generic representation of dimension vector α with decomposition into indecomposables

$$V = W_1^{\oplus e_1} \oplus \ldots \oplus W_s^{\oplus e_s} \quad \text{with} \quad dim(W_i) = \beta_i$$

we will show that (after possibly renumbering the factors $(\beta_1, \ldots, \beta_s)$ is a compartment for Q. To start, it follows from lemma 4.10 that for all $i \neq j$ we have $Ext^1_{\mathbb{C}Q}(W_i, W_j) = 0$. From lemma 4.11 we deduce a partial ordering $i \rightarrow j$ on the indices whenever $Hom_{\mathbb{C}Q}(W_i, W_j) \neq 0$. Indeed, any nonzero morphism $W_i \longrightarrow W_j$ is either a mono- or an epimorphism, assume $W_i \twoheadrightarrow W_j$ then there can be no monomorphism $W_j \hookrightarrow W_k$ as the composition $W_i \longrightarrow W_k$ would be neither mono nor epi. That is, all nonzero morphisms from W_j to factors must be (proper) epi and we cannot obtain cycles in this way by counting dimensions. If $W_i \hookrightarrow W_j$, a similar argument proves the claim. From now on we assume that the chosen index-ordering of the factors is (reverse) compatible with the partial ordering $i \rightarrow j$, that is $Hom(W_i, W_j) = 0$ whenever $i < j$, that is, β_i is left orthogonal to β_j whenever $i < j$. As $Ext^1_{CQ}(W_j, W_i) = 0$, it follows that $\chi_Q(\beta_j, \beta_i) \geq 0$. As generic representations are open it follows that all $\mathbf{rep}_{\beta_i} Q$ have an open subset of indecomposables, proving that the β_i are Schur roots. Finally, it follows from proposition 4.15 that a Schur root β_i with $\chi_Q(\beta_i, \beta_i)$ can occur only with multiplicity one in any canonical decomposition.

Conversely, assume that $(\beta_1, \ldots, \beta_s)$ is a compartment for Q, $\alpha = \sum_i e_i \beta_i$ satisfying the requirements on multiplicities. Choose Schur representations $W_i \in \mathbf{rep}_{\beta_i} Q$, then we have to prove that

$$V = W_1^{\oplus e_1} \oplus \ldots \oplus W_s^{\oplus e_s}$$

is a generic representation of dimension vector α. In view of the properties of the compartment we already know that $Ext^1_{CQ}(W_i, W_j) = 0$ for all $i < j$ and we need to show that $Ext^1_{CQ}(W_j, W_i) = 0$. Indeed, if this condition is satisfied we have

$$dim \ \mathbf{rep}_\alpha Q - dim \ \mathcal{O}(V) = dim_{\mathbb{C}} Ext^1(V, V)$$
$$= \sum_i e_i^2 dim_{\mathbb{C}} Ext^1(W_i, W_i) = \sum_i e_i^2 (1 - q_Q(\beta_i)$$

We know that the Schur representations of dimension vector β_i depend on $1 - q_Q(\beta_i)$ parameters by Kac s theorem 4.14 and $e_i = 1$ unless $q_Q(\beta_i) = 1$. Therefore, the union of all orbits of representations with the same Schur-decomposition type as V contain a dense open set of $\mathbf{rep}_\alpha Q$ and so this must be the canonical decomposition.

If this extension space is nonzero, $Hom_{CQ}(W_j, W_i) \neq 0$ as $\chi_Q(\beta_j, \beta_i) \geq 0$. But then by lemma 4.11 any nonzero homomorphism from W_j to W_i must be either a mono or an epi. Assume it is a mono, so $\beta_j < \beta_i$, so in particular a general representation of dimension β_i contains a subrepresentation of dimension β_j and hence by theorem 4.16 we have $ext(\beta_j, \beta_i - \beta_j) = 0$. Suppose that β_j is a real Schur root, then $Ext^1_{CQ}(W_j, W_j) = 0$ and therefore also $ext(\beta_j, \beta_i) = 0$ as $Ext^1_{CQ}(W_j, W_j \oplus (W_j/W_i)) = 0$. If β is not a real root, then for a general representation $S \in \mathbf{rep}_{\beta_j} Q$ take a representation $R \in \mathbf{rep}_{\beta_i} Q$ in the open set where $Ext^1_{CQ}(S, R) = 0$, then there is a monomorphism $S \hookrightarrow R$. Because $Ext^1_{CQ}(S, S) \neq 0$ we deduce from lemma 4.9 that $Ext^1_{CQ}(R, S) \neq 0$ contradicting the fact that $ext(\beta_i, \beta_j) = 0$. If the nonzero morphism $W_j \longrightarrow W_i$ is epi one has a similar argument. $\quad\square$

This result can be used to obtain a fairly efficient algorithm to compute the canonical decomposition in case the quiver Q has *no oriented cycles*. Fortunately, one can reduce the general problem to that of quiver without oriented cycles using the *bipartite double* Q^b of Q. We double the vertex-set of Q in a left and right set of vertices, that is

$$Q_v^b = \{v_1^l, \ldots, v_k^l, v_1^r, \ldots, v_k^r\}$$

To every arrow $a \in Q_a$ from v_i to v_j we assign an arrow $\tilde{a} \in Q_b^b$ from v_i^l to v_j^r. In addition, we have for each $1 \leq i \leq k$ one extra arrow $\tilde{i}$ in Q_a^b from v_i^l to v_i^r. If $\alpha = (a_1, \ldots, a_k)$ is a dimension vector for Q, the associated dimension vector $\tilde{\alpha}$ for Q^b has components

$$\tilde{\alpha} = (a_1, \ldots, a_k, a_1, \ldots, a_k)$$

Example 4.7

Consider the quiver Q and dimension vector $\alpha = (a, b)$ on the left-hand side, then

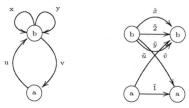

the bipartite quiver situation Q^b and $\tilde{\alpha}$ is depicted on the right-hand side. □

If the canonical decomposition of α for Q is $\tau_{can} = (e_1, \beta_1; \ldots; e_s, \beta_s)$, then the canonical decomposition of $\tilde{\alpha}$ for Q^b is $(e_1, \tilde{\beta}_1; \ldots; e_s, \tilde{\beta}_s)$ as for a general representation of Q^b of dimension vector $\tilde{\alpha}$ the morphisms corresponding to $\tilde{i}$ for $1 \leq i \leq k$ are all invertible matrices and can be used to identify the left and right vertex sets, that is, there is an equivalence of categories between representations of Q^b where all the maps $\tilde{i}$ are invertible and representations of the quiver Q. That is, the algorithm below can be applied to $(Q^b, \tilde{\alpha})$ to obtain the canonical decomposition of α for an arbitrary quiver Q.

Let Q be a quiver *without oriented cycles* then we can order the vertices $\{v_1, \ldots, v_k\}$ such that there are no oriented paths from v_i to v_j whenever $i < j$ (start with a sink of Q, drop it and continue recursively). For example, for the bipartite quiver Q^b we first take all the right vertices and then the left ones.

input: quiver Q, ordered set of vertices as above, dimension vector $\alpha = (a_1, \ldots, a_k)$ and type $\tau = (a_1, \vec{v_1}; \ldots; a_k, \vec{v_k})$ where $\vec{v_i} = (\delta_{ij})_j = dim\ v_i$ is the canonical basis. By the assumption on the ordering of vertices we have that τ is a *good type* for α. We say that a type $(f_1, \gamma_1; \ldots; f_s, \gamma_s)$ is a good type for α if $\alpha = \sum_i f_i \gamma_i$ and the following properties are satisfied

1. $f_i \geq 0$ for all i

2. γ_i is a Schur root

3. for each $i < j$, γ_i is left orthogonal to γ_j

4. $f_i = 1$ whenever $\chi_Q(\gamma_i, \gamma_i) < 0$

A type is said to be *excellent* provided that, in addition to the above, we also have that for all $i < j$, $\chi_Q(\alpha_j, \alpha_i) \geq 0$. In view of theorem 4.17 the purpose of the algorithm is to transform the good type τ into the excellent type τ_{can}. We will describe the main loop of the algorithm on a good type $(f_1, \gamma_1; \ldots; f_s, \gamma_s)$.

step 1: Omit all couples (f_i, γ_i) with $f_i = 0$ and verify whether the remaining type is excellent. If it is, *stop* and output this type. If not, proceed.

step 2: Reorder the type as follows, choose i and j such that $j - i$ is minimal and $\chi_Q(\beta_j, \beta_i) < 0$. Partition the intermediate entries $\{i+1, \ldots, j-1\}$ into the sets

- $\{k_1, \ldots, k_a\}$ such that $\chi_Q(\gamma_j, \gamma_{k_m}) = 0$
- $\{l_1, \ldots, l_b\}$ such that $\chi_Q(\gamma_j, \gamma_{l_m}) > 0$

Reorder the couples in the type in the sequence

$$(1, \ldots, i-1, k_1, \ldots, k_a, i, j, l_1, \ldots, l_b, j+1, \ldots, s)$$

define $\mu = \gamma_i$, $\nu = \gamma_j$, $p = f_i$, $q = f_j$, $\zeta = p\mu + q\nu$ and $t = -\chi_Q(\nu, \mu)$, then proceed.

step 3: Change the part $(p, \mu; q, \nu)$ of the type according to the following scheme

- If μ and ν are real Schur roots, consider the subcases
  1. $\chi_Q(\zeta, \zeta) > 0$, replace (p, μ, q, ν) by $(p', \mu'; q', \nu')$ where ν' and ν' are nonnegative combinations of ν and μ such that μ' is left orthogonal to ν', $\chi_Q(\nu', \mu') = t \geq 0$ and $\zeta = p'\mu' + q'\nu'$ for nonnegative integers p', q'.
  2. $\chi_Q(\zeta, \zeta) = 0$, replace $(p, \mu; q, \nu)$ by (k, ζ') with $\zeta = k\zeta'$, k positive integer, and ζ' an indivisible root.
  3. $\chi_Q(\zeta, \zeta) < 0$, replace $(p, \mu; q, \nu)$ by $(1, \zeta)$.

- If μ is a real root and ν is imaginary, consider the subcases
  1. If $p + q\chi_Q(\nu, \mu) \geq 0$, replace $(p, \mu; q, \nu)$ by $(q, \nu - \chi_Q(\nu, \mu)\mu; p + q\chi_Q(\nu, \mu), \mu)$.
  2. If $p + q\chi_Q(\nu, \mu) < 0$, replace $(p, \mu; q, \nu)$ by $(1, \zeta)$.

- If μ is an imaginary root and ν is real, consider the subcases
  1. If $q + p\chi_Q(\nu, \mu) \geq 0$, replace $(p, \mu; q, \nu)$ by $(q + p\chi_Q(\nu, \mu), \nu; p, \mu - \chi_Q(\nu, \mu)\nu)$.
  2. If $q + p\chi_Q(\nu, \mu) < 0$, replace $(p, \mu; q, \nu)$ by $(1, \zeta)$.

- If μ and ν are imaginary roots, replace $(p, \mu; q, \nu)$ by $(1, \zeta)$.

then go to step 1.

One can show that in every loop of the algorithm the number $\sum_i f_i$ decreases, so the algorithm must stop, giving the canonical decomposition of α. A consequence of this algorithm is that $r(\alpha) + 2i(\alpha) \leq k$ where $r(\alpha)$ is the number of real Schur roots occurring in the canonical decomposition of α, $i(\alpha)$ the number of imaginary Schur roots and k the number of vertices of Q. For more details we refer to [30].

4.7   General subrepresentations

Often, we will need to determine the dimension vectors of *general subrepresentations* . It follows from theorem 4.16 that this problem is equivalent to the calculation of $ext(\alpha, \beta)$. An inductive algorithm to do this was discovered by A. Schofield [93].

Recall that $\alpha \hookrightarrow \beta$ iff a general representation $W \in \mathbf{rep}_\beta \, Q$ contains a subrepresentation $S \hookrightarrow W$ of dimension vector α. Similarly, we denote $\beta \twoheadrightarrow \gamma$ if and only if a general representation $W \in \mathbf{rep}_\beta \, Q$ has a quotient-representation $W \twoheadrightarrow T$ of dimension vector γ. As before, Q will be a quiver on k-vertices $\{v_1, \ldots, v_k\}$ and we denote dimension vectors $\alpha = (a_1, \ldots, a_k)$, $\beta = (b_1, \ldots, b_k)$ and $\gamma = (c_1, \ldots, c_k)$. We will first determine the *rank of a general homomorphism* $V \longrightarrow W$ between representations $V \in \mathbf{rep}_\alpha \, Q$ and $W \in \mathbf{rep}_\beta \, Q$. We denote

$$Hom(\alpha, \beta) = \oplus_{i=1}^k M_{b_i \times a_i} \quad \text{and} \quad Hom(V, \beta) = Hom(\alpha, \beta) = Hom(\alpha, W)$$

for any representations V and W as above. With these conventions we have the following.

LEMMA 4.13

There is an open subset $Hom_m(\alpha, \beta) \hookrightarrow \mathbf{rep}_\alpha \, Q \times \mathbf{rep}_\beta \, Q$ and a dimension vector $\gamma \stackrel{def}{=} rk \, hom(\alpha, \beta)$ such that for all $(V, W) \in Hom_{min}(\alpha, \beta)$

- *$dim_{\mathbb{C}} \, Hom_{\mathbb{C}Q}(V, W)$ is minimal, and*

- *$\{\phi \in Hom_{\mathbb{C}Q}(V, W) \mid rk \, \phi = \gamma\}$ is a nonempty Zariski open subset of $Hom_{\mathbb{C}Q}(V, W)$.*

PROOF Consider the subvariety $Hom_Q(\alpha, \beta)$ of the trivial vectorbundle

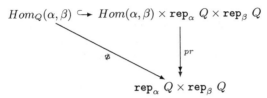

of triples (ϕ, V, W) such that $V \xrightarrow{\phi} W$ is a morphism of representations of Q. The fiber $\Phi^{-1}(V, W) = Hom_{\mathbb{C}Q}(V, W)$. As the fiber dimension is upper semicontinuous, there is an open subset $Hom_{min}(\alpha, \beta)$ of $\mathbf{rep}_\alpha Q \times \mathbf{rep}_\beta Q$ consisting of points (V, W) where $dim_{\mathbb{C}} Hom_{\mathbb{C}Q}(V, W)$ is minimal. For given dimension vector $\delta = (d_1, \ldots, d_k)$ we consider the subset

$$Hom_Q(\alpha, \beta, \delta) = \{(\phi, V, W) \in Hom_Q(\alpha, \beta) \mid rk\ \phi = \delta\} \hookrightarrow Hom_Q(\alpha, \beta)$$

This is a constructible subset of $Hom_Q(\alpha, \beta)$ and hence there is a dimension vector γ such that $Hom_Q(\alpha, \beta, \gamma) \cap \Phi^{-1}(Hom_{min}(\alpha, \beta))$ is constructible and dense in $\Phi^{-1}(Hom_{min}(\alpha, \beta))$. But then

$$\Phi(Hom_Q(\alpha, \beta, \gamma) \cap \Phi^{-1}(Hom_{min}(\alpha, \beta)))$$

is constructible and dense in $Hom_{min}(V, W)$. Therefore it contains an open subset $Hom_m(V, W)$ satisfying the requirements of the lemma. ☐

LEMMA 4.14
Assume we have short exact sequences of representations of Q

$$\begin{cases} 0 \longrightarrow S \longrightarrow V \longrightarrow X \longrightarrow 0 \\ 0 \longrightarrow Y \longrightarrow W \longrightarrow T \longrightarrow 0 \end{cases}$$

then there is a natural onto map

$$Ext^1_{\mathbb{C}Q}(V, W) \longrightarrow\!\!\!\!\rightarrow Ext^1_{\mathbb{C}Q}(S, T)$$

PROOF By lemma 4.9 we have surjective maps

$$Ext^1_{\mathbb{C}Q}(V, W) \longrightarrow\!\!\!\!\rightarrow Ext^1_{\mathbb{C}Q}(V, T) \longrightarrow\!\!\!\!\rightarrow Ext^1_{\mathbb{C}Q}(S, T)$$

from which the assertion follows. ☐

THEOREM 4.18
Let $\gamma = rk\ hom(\alpha, \beta)$ (with notations as in lemma 4.13), then

1. $\alpha - \gamma \hookrightarrow \alpha \longrightarrow\!\!\!\!\rightarrow \gamma \hookrightarrow \beta \longrightarrow\!\!\!\!\rightarrow \beta - \gamma$

2. $ext(\alpha,\beta) = -\chi_Q(\alpha - \gamma, \beta - \gamma) = ext(\alpha - \gamma, \beta - \gamma)$

PROOF   The first statement is obvious from the definitions. The strategy of the proof of the second statement is to compute the dimension of the subvariety of $Hom(\alpha,\beta) \times \mathbf{rep}_\alpha \times \mathbf{rep}_\beta \times \mathbf{rep}_\gamma$ defined by

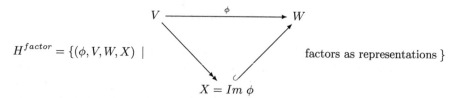

$$H^{factor} = \{(\phi,V,W,X) \mid \quad\quad\quad\quad\quad\quad\quad\quad \text{factors as representations} \}$$

in two different ways. Consider the intersection of the open set $Hom_m(\alpha,\beta)$ determined by lemma 4.13 with the open set of couples (V,W) such that $dim\ Ext(V,W) = ext(\alpha,\beta)$ and let (V,W) lie in this intersection. In the previous section we have proved that

$$dim\ \mathbf{Grass}_\gamma(W) = \chi_Q(\gamma, \beta - \gamma)$$

Let H be the subbundle of the trivial vectorbundle over $\mathbf{Grass}_\gamma(W)$

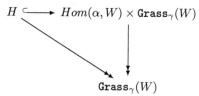

consisting of triples (ϕ, W, U) with $\phi : \oplus_i \mathbb{C}^{\oplus a_i} \longrightarrow W$ a linear map such that $Im(\phi)$ is contained in the subrepresentation $U \hookrightarrow W$ of dimension γ. That is, the fiber over (W,U) is $Hom(\alpha, U)$ and therefore has dimension $\sum_{i=1}^k a_i c_i$. With H^{full} we consider the open subvariety of H of triples (ϕ, W, U) such that $Im\ \phi = U$. We have

$$dim\ H^{full} = \sum_{i=1}^k a_i c_i + \chi_Q(\gamma, \beta - \gamma)$$

But then, H^{factor} is the subbundle of the trivial vectorbundle over H^{full}

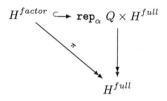

consisting of quadruples (V, ϕ, W, X) such that $V \xrightarrow{\phi} W$ is a morphism of representations, with image the subrepresentation X of dimension γ. The fiber of π over a triple (ϕ, W, X) is determined by the property that for each arrow $(j) \xleftarrow{\ a\ } (i)$ the following diagram must be commutative, where we decompose the vertex spaces $V_i = X_i \oplus K_i$ for $K = Ker\ \phi$

$$
\begin{array}{ccc}
X_i \oplus K_i & \xrightarrow{\ \begin{bmatrix} A & B \\ C & D \end{bmatrix}\ } & X_j \oplus K_j \\
{\scriptstyle \begin{bmatrix} 1\!\!1_{c_i} & 0 \end{bmatrix}} \downarrow & & \downarrow {\scriptstyle \begin{bmatrix} 1\!\!1_{c_j} & 0 \end{bmatrix}} \\
X_i & \xrightarrow{\quad A \quad} & X_j
\end{array}
$$

where A is fixed, giving the condition $B = 0$ and hence the fiber has dimension equal to

$$
\sum_{(j)\xleftarrow{a}(i)} (a_i - c_i)(a_j - c_j) + \sum_{(j)\xleftarrow{a}(i)} c_i(a_j - c_j) = \sum_{(j)\xleftarrow{a}(i)} a_i(a_j - c_j)
$$

This gives our first formula for the dimension of H^{factor}

$$
H^{factor} = \sum_{i=1}^{k} a_i c_i + \chi_Q(\gamma, \beta - \gamma) + \sum_{(j)\xleftarrow{a}(i)} a_i(a_j - c_j)
$$

On the other hand, we can consider the natural map $H^{factor} \xrightarrow{\ \Phi\ } \mathbf{rep}_\alpha\ Q$ defined by sending a quadruple (V, ϕ, W, X) to V. the fiber in V is given by all quadruples (V, ϕ, W, X) such that $V \xrightarrow{\ \phi\ } W$ is a morphism of representations with $Im\ \phi = X$ a representation of dimension vector γ, or equivalently

$$
\Phi^{-1}(V) = \{ V \xrightarrow{\ \phi\ } W \mid rk\ \phi = \gamma \}
$$

Now, recall our restriction on the couple (V, W) giving at the beginning of the proof. There is an open subset max of $\mathbf{rep}_\alpha\ Q$ of such V and by construction $max \hookrightarrow Im\ \Phi$, $\Phi^{-1}(max)$ is open and dense in H^{factor} and the fiber $\Phi^{-1}(V)$ is open and dense in $Hom_{\mathbb{C}Q}(V, W)$. This provides us with the second formula for the dimension of H^{factor}

$$
dim\ H^{factor} = dim\ \mathbf{rep}_\alpha\ Q + hom(\alpha, W) = \sum_{(j)\xleftarrow{a}(i)} a_i a_j + hom(\alpha, \beta)
$$

Equating both formulas we obtain the equality

$$
\chi_Q(\gamma, \beta - \gamma) + \sum_{i=1}^{k} a_i c_i - \sum_{(j)\xleftarrow{a}(i)} a_i c_j = hom(\alpha, \beta)
$$

which is equivalent to

$$\chi_Q(\gamma, \beta - \gamma) + \chi_Q(\alpha, \gamma) - \chi_Q(\alpha, \beta) = ext(\alpha, \beta)$$

Now, for our (V, W) we have that $Ext(V, W) = ext(\alpha, \beta)$ and we have exact sequences of representations

$$0 \longrightarrow S \longrightarrow V \longrightarrow X \longrightarrow 0 \qquad 0 \longrightarrow X \longrightarrow W \longrightarrow T \longrightarrow 0$$

and using lemma 4.14 this gives a surjection $Ext(V, W) \longrightarrow\!\!\!\!\rightarrow Ext(S, T)$. On the other hand we always have from the homological interpretation of the Euler form the first inequality

$$dim_\mathbb{C} \ Ext(S, T) \geq -\chi_Q(\alpha - \gamma, \beta - \gamma) = \chi_Q(\gamma, \beta - \gamma) - \chi_Q(\alpha, \beta) + \chi_Q(\alpha, \gamma)$$
$$= ext(\alpha, \beta)$$

As the last term is $dim_\mathbb{C} \ Ext(V, W)$, this implies that the above surjection must be an isomorphism and that

$$dim_\mathbb{C} \ Ext(S, T) = -\chi_Q(\alpha - \gamma, \beta - \gamma) \quad \text{whence} \quad dim_\mathbb{C} \ Hom(S, T) = 0$$

But this implies that $hom(\alpha - \gamma, \beta - \gamma) = 0$ and therefore $ext(\alpha - \gamma, \beta - \gamma) = -\chi_Q(\alpha - \gamma, \beta - \gamma)$. Finally,

$$ext(\alpha - \gamma, \beta - \gamma) = dim \ Ext(S, T) = dim \ Ext(V, W) = ext(\alpha, \beta)$$

finishing the proof.   □

THEOREM 4.19

For all dimension vectors α and β we have

$$ext(\alpha, \beta) = \max_{\substack{\alpha' \hookrightarrow \alpha \\ \beta \twoheadrightarrow \beta'}} -\chi_Q(\alpha', \beta')$$

$$= \max_{\beta \twoheadrightarrow \beta''} -\chi_Q(\alpha, \beta'')$$

$$= \max_{\alpha'' \hookrightarrow \alpha} -\chi_Q(\alpha'', \beta)$$

PROOF   Let V and W be representation of dimension vector α and β such that $dim \ Ext(V, W) = ext(\alpha, \beta)$. Let $S \hookrightarrow V$ be a subrepresentation of dimension α' and $W \twoheadrightarrow T$ a quotient representation of dimension vector β'. Then, we have

$$ext(\alpha, \beta) = dim_\mathbb{C} \ Ext(V, W) \geq dim_\mathbb{C} \ Ext(S, T) \geq -\chi_Q(\alpha', \beta')$$

where the first inequality is lemma 4.14 and the second follows from the interpretation of the Euler form. Therefore, $ext(\alpha, \beta)$ is greater or equal than all the terms in the statement of the theorem. The foregoing theorem asserts the first equality, as for $rk\ hom(\alpha,\beta) = \gamma$ we do have that $ext(\alpha,\beta) = -\chi_Q(\alpha-\gamma,\beta-\gamma)$.

In the proof of the above theorem, we have found for sufficiently general V and W an exact sequence of representations

$$0 \longrightarrow S \longrightarrow V \longrightarrow W \longrightarrow T \longrightarrow 0$$

where S is of dimension $\alpha - \gamma$ and T of dimension $\beta - \gamma$. Moreover, we have a commuting diagram of surjections

$$
\begin{array}{ccc}
Ext(V,W) & \twoheadrightarrow & Ext(V,T) \\
\downarrow & \searrow & \downarrow \\
Ext(S,W) & \twoheadrightarrow & Ext(S,T)
\end{array}
$$

and the dashed map is an isomorphism, hence so are all the epimorphisms. Therefore, we have

$$\begin{cases} ext(\alpha,\beta-\gamma) & \leq dim\ Ext(V,T) = dim\ Ext(V,W) = ext(\alpha,\beta) \\ ext(\alpha-\gamma,\beta) & \leq dim\ Ext(S,W) = dim\ Ext(V,W) = ext(\alpha,\beta) \end{cases}$$

Further, let T' be a sufficiently general representation of dimension $\beta - \gamma$, then it follows from $Ext(V,T') \longrightarrow Ext(S,T)$ that

$$ext(\alpha-\gamma,\beta-\gamma) \leq dim\ Ext(S,T') \leq dim\ Ext(V,T') = ext(\alpha,\beta-\gamma)$$

but the left term is equal to $ext(\alpha,\beta)$ by the above theorem. But then, we have $ext(\alpha,\beta) = ext(\alpha,\beta-\gamma)$. Now, we may assume by induction that the theorem holds for $\beta - \gamma$. That is, there exists $\beta - \gamma \longrightarrow \beta"$ such that $ext(\alpha,\beta-\gamma) = -\chi_Q(\alpha,\beta")$. Whence, $\beta \longrightarrow \beta"$ and $ext(\alpha,\beta) = -\chi_Q(\alpha,\beta")$ and the middle equality of the theorem holds. By a dual argument so does the last. $\qquad\square$

By induction we therefore have that $\beta \hookrightarrow \alpha$ if and only if

$$0 = ext(\beta,\alpha-\beta) = \underset{\beta' \hookrightarrow \beta}{max}\ -\chi_Q(\beta',\alpha-\beta)$$

4.8 Semistable representations

Let Q be a quiver on k vertices $\{v_1,\ldots,v_k\}$ and fix a dimension vector α. So far, we have considered the algebraic quotient map

$$rep_\alpha\ Q \longrightarrow iss_\alpha\ Q$$

classifying closed $GL(\alpha)$-orbits in $\mathbf{rep}_\alpha\ Q$, that is, isomorphism classes of semisimple representations of dimension α. We have seen that the invariant polynomial maps are generated by traces along oriented cycles in the quiver. Hence, if Q has no oriented cycles, the quotient variety $\mathbf{iss}_\alpha\ Q$ is reduced to one point corresponding to the semisimple

$$S_1^{\oplus a_1} \oplus \ldots \oplus S_k^{\oplus a_k}$$

where S_i is the trivial one-dimensional simple concentrated in vertex v_i. Still, in these cases one can often classify nice families of representations.

Example 4.8

Consider the quiver setting

Then, $\mathbf{rep}_\alpha\ Q = \mathbb{C}^3$ and the action of $GL(\alpha) = \mathbb{C}^* \times \mathbb{C}^*$ is given by $(\lambda, \mu).(x, y, z) = (\frac{\lambda}{\mu}x, \frac{\lambda}{\mu}y, \frac{\lambda}{\mu}z)$. The only closed $GL(\alpha)$-orbit in $\mathbb{C}^3$ is $(0, 0, 0)$ as the one-parameter subgroup $\lambda(t) = (t, 1)$ has the property

$$\lim_{t \to 0} \lambda(t).(x, y, z) = (0, 0, 0)$$

so $(0, 0, 0) \in \overline{\mathcal{O}(x, y, z)}$ for any representation (x, y, z). Still, if we throw away the zero-representation, then we have a nice quotient map

$$\mathbb{C}^3 - \{(0, 0, 0)\} \xrightarrow{\ \pi\ } \mathbb{P}^2 \qquad (x, y, z) \mapsto [x : y : z]$$

and as $\mathcal{O}(x, y, z) = \mathbb{C}^*(x, y, z)$ we see that every $GL(\alpha)$-orbit is closed in this complement $\mathbb{C}^3 - \{(0, 0, 0)\}$. We will generalize such settings to arbitrary quivers.   □

A *character* of $GL(\alpha)$ is an algebraic group morphism $\chi : GL(\alpha) \longrightarrow \mathbb{C}^*$. They are fully determined by an integral k-tuple $\theta = (t_1, \ldots, t_k) \in \mathbb{Z}^k$ where

$$GL(\alpha) \xrightarrow{\ \chi_\theta\ } \mathbb{C}^* \qquad (g_1, \ldots, g_k) \mapsto det(g_1)^{t_1}.\ \ldots\ .det(g_k)^{t_k}$$

For a fixed θ we can extend the $GL(\alpha)$-action to the space $\mathbf{rep}_\alpha \oplus \mathbb{C}$ by

$$GL(\alpha) \times \mathbf{rep}_\alpha\ Q \oplus \mathbb{C} \longrightarrow \mathbf{rep}_\alpha\ Q \oplus \mathbb{C} \qquad g.(V, c) = (g.V, \chi_\theta^{-1}(g)c)$$

The coordinate ring $\mathbb{C}[\mathbf{rep}_\alpha\ Q \oplus \mathbb{C}] = \mathbb{C}[\mathbf{rep}_\alpha][t]$ can be given a $\mathbb{Z}$-gradation by defining $deg(t) = 1$ and $deg(f) = 0$ for all $f \in \mathbb{C}[\mathbf{rep}_\alpha\ Q]$. The induced action of $GL(\alpha)$ on $\mathbb{C}[\mathbf{rep}_\alpha\ Q \oplus \mathbb{C}]$ preserves this gradation. Therefore, the ring of invariant polynomial maps

$$\mathbb{C}[\mathbf{rep}_\alpha\ Q \oplus \mathbb{C}]^{GL(\alpha)} = \mathbb{C}[\mathbf{rep}_\alpha\ Q][t]^{GL(\alpha)}$$

is also graded with homogeneous part of degree zero the ring of invariants $\mathbb{C}[\mathrm{rep}_\alpha]^{GL(\alpha)}$. An invariant of degree n, say, ft^n with $f \in \mathbb{C}[\mathrm{rep}_\alpha Q]$ has the characteristic property that

$$f(g.V) = \chi_\theta^n(g)f(V)$$

that is, f is a *semi-invariant* of weight χ_θ^n. That is, the graded decomposition of the invariant ring is

$$\mathbb{C}[\mathrm{rep}_\alpha Q \oplus \mathbb{C}]^{GL(\alpha)} = R_0 \oplus R_1 \oplus \dots \quad \text{with} \quad R_i = \mathbb{C}[\mathrm{rep}_\alpha Q]^{GL(\alpha),\chi^n\theta}$$

DEFINITION 4.11   *With notations as above, the* moduli space of semi-stable quiver representations of dimension α *is the projective variety*

$$M_\alpha^{ss}(Q,\theta) = \mathrm{proj}\ \mathbb{C}[\mathrm{rep}_\alpha Q \oplus \mathbb{C}]^{GL(\alpha)} = \mathrm{proj}\ \oplus_{n=0}^\infty \mathbb{C}[\mathrm{rep}_\alpha Q]^{GL(\alpha),\chi^n\theta}$$

Recall that for a positively graded affine commutative $\mathbb{C}$-algebra $R = \oplus_{i=0}^\infty R_i$, the geometric points of the projective scheme $\mathrm{proj}\ R$ correspond to graded-maximal ideals $\mathfrak{m}$ not containing the positive part $R_+ = \oplus_{i=1}^\infty R_i$. Intersecting $\mathfrak{m}$ with the part of degree zero R_0 determines a point of $\mathrm{spec}\ R_0$, the affine variety with coordinate ring R_0 and defines a structural morphism

$$\mathrm{proj}\ R \longrightarrow \mathrm{spec}\ R_0$$

The *Zariski closed subsets* of $\mathrm{proj}\ R$ are of the form

$$\mathbb{V}(I) = \{\mathfrak{m} \in \mathrm{proj}\ R \mid I \subset \mathfrak{m}\}$$

for a homogeneous ideal $I \lhd R$. Further, recall that $\mathrm{proj}\ R$ can be covered by affine varieties of the form $\mathbb{X}(f)$ with f a homogeneous element in R_+. The coordinate ring of this affine variety is the part of degree zero of the *graded localization* R_f^g. We refer to [43, II.2] for more details.

Example 4.9
Consider again the quiver-situation

and character $\theta = (-1,1)$, then the three coordinate functions x, y and z of $\mathbb{C}[\mathrm{rep}_\alpha Q]$ are semi-invariants of weight χ_θ. It is clear that the invariant ring is equal to

$$\mathbb{C}[\mathrm{rep}_\alpha Q \oplus \mathbb{C}]^{GL(\alpha)} = \mathbb{C}[xt, yt, zt]$$

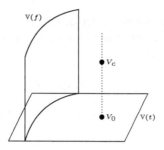

FIGURE 4.11:   The projection map.

where the three generators all have degree one. That is,

$$M_\alpha^{ss}(Q, \theta) = \text{proj } \mathbb{C}[xt, yt, zt] = \mathbb{P}^2$$

as desired.   ⬜

We will now investigate which orbits in $\text{rep}_\alpha Q$ are parameterized by the moduli space $M_\alpha^{ss}(Q, \theta)$.

DEFINITION 4.12   *We say that a representation $V \in \text{rep}_\alpha Q$ is χ_θ-semistable if and only if there is a semi-invariant $f \in \mathbb{C}[\text{rep}_\alpha Q]^{GL(\alpha), \chi^n \theta}$ for some $n \geq 1$ such that $f(V) \neq 0$.*

The subset of $\text{rep}_\alpha Q$ consisting of all χ_θ-semistable representations will be denoted by $\text{rep}_\alpha^{ss}(Q, \theta)$.

Observe that $\text{rep}_\alpha^{ss}(Q, \theta)$ is Zariski open (but it may be empty for certain (α, θ)). We can lift a representation $V \in \text{rep}_\alpha Q$ to points $V_c = (V, c) \in \text{rep}_\alpha Q \oplus \mathbb{C}$ and use $GL(\alpha)$-invariant theory on this larger $GL(\alpha)$-module see figure 4.8 Let $c \neq 0$ and assume that the orbit closure $\overline{\mathcal{O}(V_c)}$ does not intersect $\mathbb{V}(t) = \text{rep}_\alpha Q \times \{0\}$. As both are $GL(\alpha)$-stable closed subsets of $\text{rep}_\alpha Q \oplus \mathbb{C}$ we know from the separation property of invariant theory, proposition 2.10, that this is equivalent to the existence of a $GL(\alpha)$-invariant function $g \in \mathbb{C}[\text{rep}_\alpha Q \oplus \mathbb{C}]^{GL(\alpha)}$ such that $g(\overline{\mathcal{O}(V_c)}) \neq 0$ but $g(\mathbb{V}(t)) = 0$.

We have seen that the invariant ring is graded, hence we may assume g to be homogeneous, that is, of the form $g = ft^n$ for some n. But then, f is a semi-invariant on $\text{rep}_\alpha Q$ of weight χ_θ^n and we see that V must be χ_θ-semistable. Moreover, we must have that $\theta(\alpha) = \sum_{i=1}^k t_i a_i = 0$, for the one-dimensional central torus of $GL(\alpha)$

$$\mu(t) = (t \mathbb{1}_{a_1}, \dots, t \mathbb{1}_{a_k}) \longrightarrow GL(\alpha)$$

acts trivially on $\text{rep}_\alpha Q$ but acts on $\mathbb{C}$ via multiplication with $\prod_{i=1}^k t^{-a_i t_i}$ hence if $\theta(\alpha) \neq 0$ then $\overline{\mathcal{O}(V_c)} \cap \mathbb{V}(t) \neq \emptyset$.

More generally, we have from the strong form of the Hilbert criterion proved in theorem 2.2 that $\overline{\mathcal{O}(V_c)} \cap \mathbb{V}(t) = \emptyset$ if and only if for every one-parameter subgroup $\lambda(t)$ of $GL(\alpha)$ we must have that $\lim_{t \to 0} \lambda(t).V_c \notin \mathbb{V}(t)$. We can also formulate this in terms of the $GL(\alpha)$-action on $\mathbf{rep}_\alpha Q$. The composition of a one-parameter subgroup $\lambda(t)$ of $GL(\alpha)$ with the character

$$\mathbb{C}^* \xrightarrow{\ \lambda(t)\ } GL(\alpha) \xrightarrow{\ \chi_\theta\ } \mathbb{C}^*$$

is an algebraic group morphism and is therefore of the form $t \longmapsto t^m$ for some $m \in \mathbb{Z}$ and we denote this integer by $\theta(\lambda) = m$. Assume that $\lambda(t)$ is a one-parameter subgroup such that $\lim_{t \to 0} \lambda(t).V = V'$ exists in $\mathbf{rep}_\alpha Q$, then as

$$\lambda(t).(V, c) = (\lambda(t).V, t^{-m}c)$$

we must have that $\theta(\lambda) \geq 0$ for the orbitclosure $\overline{\mathcal{O}(V_c)}$ not to intersect $\mathbb{V}(t)$.

That is, we have the following characterization of χ_θ-semistable representations.

PROPOSITION 4.16
The following are equivalent

1. $V \in \mathbf{rep}_\alpha Q$ *is* χ_θ-*semistable.*

2. *For* $c \neq 0$, *we have* $\overline{\mathcal{O}(V_c)} \cap \mathbb{V}(t) = \emptyset$.

3. *For every one-parameter subgroup* $\lambda(t)$ *of* $GL(\alpha)$ *we have* $\lim_{t \to 0} \lambda(t).V_c \notin$ $\mathbb{V}(t) = \mathbf{rep}_\alpha Q \times \{0\}$.

4. *For every one-parameter subgroup* $\lambda(t)$ *of* $GL(\alpha)$ *such that* $\lim_{t \to 0} \lambda(t).V$ *exists in* $\mathbf{rep}_\alpha Q$ *we have* $\theta(\lambda) \geq 0$.

Moreover, these cases can only occur if $\theta(\alpha) = 0$.

Assume that $g = ft^n$ is a homogeneous invariant function for the $GL(\alpha)$-action on $\mathbf{rep}_\alpha Q \oplus \mathbb{C}$ and consider the affine open $GL(\alpha)$-stable subset $\mathbb{X}(g)$. The construction of the algebraic quotient and the fact that the invariant rings here are graded asserts that the closed $GL(\alpha)$-orbits in $\mathbb{X}(g)$ are classified by the points of the graded localization at g which is of the form

$$(\mathbb{C}[\mathbf{rep}_\alpha Q \oplus \mathbb{C}]^{GL(\alpha)})_g = R_f[h, h^{-1}]$$

for some homogeneous invariant h and where R_f is the coordinate ring of the affine open subset $\mathbb{X}(f)$ in $M_\alpha^{ss}(Q, \theta)$ determined by the semi-invariant f of weight χ_θ^n. Because the moduli space is covered by such open subsets we have

PROPOSITION 4.17
The moduli space of θ-semistable representations of $\mathbf{rep}_\alpha Q$

$$M_\alpha^{ss}(Q, \theta)$$

classifies closed $GL(\alpha)$-orbits in the open subset $\mathbf{rep}_\alpha^{ss}(Q, \theta)$ *of all χ_θ-semistable representations of Q of dimension vector α.*

Example 4.10
In the foregoing example $\mathbf{rep}_\alpha^{ss}(Q, \theta) = \mathbb{C}^3 - \{(0,0,0)\}$ as for all these points one of the semi-invariant coordinate functions is nonzero. For $\theta = (-1, 1)$ the lifted $GL(\alpha) = \mathbb{C}^* \times \mathbb{C}^*$-action to $\mathbf{rep}_\alpha Q \oplus \mathbb{C} = \mathbb{C}^4$ is given by

$$(\lambda, \mu).(x, y, z, t) = (\frac{\mu}{\lambda}x, \frac{\mu}{\lambda}y, \frac{\mu}{\lambda}z, \frac{\lambda}{\mu}t)$$

We have seen that the ring of invariants is $\mathbb{C}[xt, yt, zt]$. Consider the affine open set $\mathbb{X}(xt)$ of $\mathbb{C}^4$, then the closed orbits in $\mathbb{X}(xt)$ are classified by

$$\mathbb{C}[xt, yt, zt]_{xt}^g = \mathbb{C}[\frac{y}{x}, \frac{z}{x}][xt, \frac{1}{xt}]$$

and the part of degree zero $\mathbb{C}[\frac{y}{x}, \frac{z}{x}]$ is the coordinate ring of the open set $\mathbb{X}(x)$ in $\mathbb{P}^2$.   ∏

We have seen that closed GL_n-orbits in $\mathbf{rep}_n A$ correspond to semisimple n-dimensional representations. We will now give a representation theoretic interpretation of closed $GL(\alpha)$-orbits in $\mathbf{rep}_\alpha^{ss}(Q, \theta)$.

Again, the starting point is that one-parameter subgroups $\lambda(t)$ of $GL(\alpha)$ correspond to filtrations of representations. Let us go through the motions one more time. For $\lambda : \mathbb{C}^* \longrightarrow GL(\alpha)$ a one-parameter subgroup and $V \in \mathbf{rep}_\alpha Q$ we can decompose for every vertex v_i the vertex-space in weight spaces

$$V_i = \oplus_{n \in \mathbb{Z}} V_i^{(n)}$$

where $\lambda(t)$ acts on the weight space $V_i^{(n)}$ as multiplication by t^n. This decomposition allows us to define a filtration

$$V_i^{(\geq n)} = \oplus_{m \geq n} V_i^{(m)}$$

For every arrow $\circled{i} \xleftarrow{\quad a \quad} \circled{j}$, $\lambda(t)$ acts on the components of the arrow maps

$$V_i^{(n)} \xrightarrow{V_a^{m,n}} V_j^{(m)}$$

by multiplication with t^{m-n}. That is, a limit $\lim_{t \to 0} V_a$ exists if and only if $V_a^{m,n} = 0$ for all $m < n$, that is, if V_a induces linear maps

$$V_i^{(\geq n)} \xrightarrow{V_a} V_j^{(\geq n)}$$

Hence, a limiting representation exists if and only if the vertex-filtration spaces $V_i^{(\geq n)}$ determine a subrepresentation $V_n \hookrightarrow V$ for all n. That is, a one-parameter subgroup λ such that $\lim\limits_{t \to} \lambda(t).V$ exists determines a decreasing filtration of V by subrepresentations

$$\ldots \longleftarrow V_n \longleftarrow V_{n+1} \longleftarrow \ldots$$

Further, the limiting representation is then the associated graded representation

$$\lim_{t \to 0} \lambda(t).V = \oplus_{n \in \mathbb{Z}} \frac{V_n}{V_{n+1}}$$

where of course only finitely many of these quotients can be nonzero. For the given character $\theta = (t_1, \ldots, t_k)$ and a representation $W \in \mathbf{rep}_\beta Q$ we denote

$$\theta(W) = t_1 b_1 + \ldots + t_k b_k \quad \text{where} \quad \beta = (b_1, \ldots, b_k)$$

Assume that $\theta(V) = 0$, then with the above notations, we have an interpretation of $\theta(\lambda)$ as

$$\theta(\lambda) = \sum_{i=1}^k t_i \sum_{n \in Z} n \dim_{\mathbb{C}} V_i^{(n)} = \sum_{n \in \mathbb{Z}} n\theta(\frac{V_n}{V_{n+1}}) = \sum_{n \in \mathbb{Z}} \theta(V_n)$$

DEFINITION 4.13   *A representation $V \in \mathbf{rep}_\alpha Q$ is said to be*

- *θ-semistable if $\theta(V) = 0$ and for all subrepresentations $W \hookrightarrow V$ we have $\theta(W) \geq 0$.*

- *θ-stable if V is θ-semistable and if the only subrepresentations $W \hookrightarrow V$ such that $\theta(W) = 0$ are V and 0.*

PROPOSITION 4.18
For $V \in \mathbf{rep}_\alpha Q$ the following are equivalent

1. *V is χ_θ-semistable.*

2. *V is θ-semistable.*

PROOF $(1) \Rightarrow (2)$: Let W be a subrepresentation of V and let λ be the one-parameter subgroup associated to the filtration $V \longleftarrow W \longleftarrow 0$, then $\lim\limits_{t \to 0} \lambda(t).V$ exists whence by proposition 4.16.4 we have $\theta(\lambda) \geq 0$, but we have

$$\theta(\lambda) = \theta(V) + \theta(W) = \theta(W)$$

$(2) \Rightarrow (1)$: Let λ be a one-parameter subgroup of $GL(\alpha)$ such that $\lim\limits_{t \to 0} \lambda(t).V$ exists and consider the induced filtration by subrepresentations V_n defined above. By assumption all $\theta(V_n) \geq 0$, whence

$$\theta(\lambda) = \sum_{n \in Z} \theta(V_n) \geq 0$$

and again proposition 4.16.4 finishes the proof. $\square$

LEMMA 4.15
Let $V \in \mathbf{rep}_\alpha Q$ and $W \in \mathbf{rep}_\beta Q$ be both θ-semistable and

$$V \xrightarrow{\ f\ } W$$

a morphism of representations. Then, $Ker\ f$, $Im\ f$ and $Coker\ f$ are θ-semistable representations.

PROOF   Consider the two short exact sequences of representations of Q

$$\begin{cases} 0 \longrightarrow Ker\ f \longrightarrow V \longrightarrow Im\ f \longrightarrow 0 \\ 0 \longrightarrow Im\ f \longrightarrow W \longrightarrow Coker\ f \longrightarrow 0 \end{cases}$$

As $\theta(-)$ is additive, we have $0 = \theta(V) = \theta(Ker\ f) + \theta(Im\ f)$ and as both are subrepresentations of θ-semistable representations V resp. W, the right-hand terms are ≥ 0 whence are zero. But then, from the second sequence also $\theta(Coker\ f) = 0$. Being submodules of θ-semistable representations, $Ker\ f$ and $Im\ f$ also satisfy $\theta(S) \geq 0$ for all their subrepresentations U. Finally, a subrepresentation $T' \hookrightarrow Coker\ f$ can be lifted to a subrepresentation $T' \hookrightarrow W$ and $\theta(T) \geq 0$ follows from the short exact sequence $0 \longrightarrow Im\ f \longrightarrow T' \longrightarrow T \longrightarrow 0$. $\square$

That is, the full subcategory $rep^{ss}(Q, \theta)$ of $rep\ Q$ consisting of all θ-semistable representations is an Abelian subcategory and clearly the simple objects in $rep^{ss}(Q, \theta)$ are precisely the θ-stable representations. As this Abelian subcategory has the necessary finiteness conditions, one can prove a version of the Jordan-Hölder theorem. That is, every θ-semistable representation V has a finite filtration

$$V = V_0 \hookleftarrow V_1 \hookleftarrow \ldots \hookleftarrow V_z = 0$$

of subrepresentation such that every factor $\frac{V_i}{V_{i+1}}$ is θ-stable. Moreover, the unordered set of these θ-stable factors are uniquely determined by V.

THEOREM 4.20
For a θ-semistable representation $V \in \mathbf{rep}_\alpha Q$ the following are equivalent

   1. The orbit $\mathcal{O}(V)$ is closed in $\mathbf{rep}_\alpha^{ss}(Q, \alpha)$.

2. $V \simeq W_1^{\oplus e_1} \oplus \ldots \oplus W_l^{\oplus e_l}$ with every W_i a θ-stable representation.

That is, the geometric points of the moduli space $M_\alpha^{ss}(Q, \theta)$ are in natural one-to-one correspondence with isomorphism classes of α-dimensional representations which are direct sums of θ-stable subrepresentations. The quotient map

$$\mathbf{rep}_\alpha^{ss}(Q, \theta) \longrightarrow M_\alpha^{ss}(Q, \theta)$$

maps a θ-semistable representation V to the direct sum of its Jordan-Hölder factors in the Abelian category $rep^{ss}(Q, \theta)$.

PROOF Assume that $\mathcal{O}(V)$ is closed in $\mathbf{rep}_\alpha^{ss}(Q, \theta)$ and consider the θ-semistable representation $W = gr_{ss}\, V$, the direct sum of the Jordan-Hölder factors in $rep^{ss}(Q, \theta)$. As W is the associated graded representation of a filtration on V, there is a one-parameter subgroup λ of $GL(\alpha)$ such that $\lim_{t \to 0} \lambda(t).V \simeq W$, that is $\mathcal{O}(W) \subset \overline{\mathcal{O}(V)} = \mathcal{O}(V)$, whence $W \simeq V$ and 2. holds.

Conversely, assume that V is as in 2. and let $\mathcal{O}(W)$ be a closed orbit contained in $\overline{\mathcal{O}(V)}$ (one of minimal dimension). By the Hilbert criterium there is a one-parameter subgroup λ in $GL(\alpha)$ such that $\lim_{t \to 0} \lambda(t).V \simeq W$. Hence, there is a finite filtration of V with associated graded θ-semistable representation W. As none of the θ-stable components of V admits a proper quotient which is θ-semistable (being a direct summand of W), this shows that $V \simeq W$ and so $\mathcal{O}(V) = \mathcal{O}(W)$ is closed. The other statements are clear from this. $\qquad\Box$

Remains to determine the situations (α, θ) such that the corresponding moduli space $M_\alpha^{ss}(Q, \theta)$ is non-empty, or equivalently, such that the Zariski open subset $\mathbf{rep}_\alpha^{ss}(Q, \theta) \hookrightarrow \mathbf{rep}_\alpha\, Q$ is non-empty.

THEOREM 4.21
Let α be a dimension vector such that $\theta(\alpha) = 0$. Then

1. *$\mathbf{rep}_\alpha^{ss}(Q, \alpha)$ is a non-empty Zariski open subset of $\mathbf{rep}_\alpha\, Q$ if and only if for every $\beta \hookrightarrow \alpha$ we have that $\theta(\beta) \geq 0$.*

2. *The θ-stable representations $\mathbf{rep}_\alpha^s(Q, \alpha)$ are a non-empty Zariski open subset of $\mathbf{rep}_\alpha\, Q$ if and only if for every $0 \neq \beta \hookrightarrow \alpha$ we have that $\theta(\beta) > 0$*

The algorithm at the end of the last section gives an inductive procedure to calculate these conditions.

The graded algebra $\mathbb{C}[\mathbf{rep}_\alpha \oplus \mathbb{C}]^{GL(\alpha)}$ of all semi-invariants on $\mathbf{rep}_\alpha\, Q$ of weight χ_θ^n for some $n \geq 0$ has as degree zero part the ring of polynomial invariants $\mathbb{C}[\mathbf{rep}_\alpha\ Q]^{GL(\alpha)}$. This embedding determines a *proper morphism*

$$M_\alpha^{ss}(Q,\theta) \xrightarrow{\ \pi\ } \mathrm{iss}_\alpha\ Q$$

which is onto whenever $\mathbf{rep}_\alpha^{ss}(Q,\alpha)$ is non-empty. In particular, if Q is a quiver without oriented cycles, then the moduli space of θ-semistable representations of dimension vector α, $M_\alpha^{ss}(Q,\theta)$, is a projective variety.

References

The étale slices are due to D. Luna [81] and in the form presented here to F. Knop [56] . For more details we refer to the lecture notes of P. Slodowy [99]. The geometric interpretation of Cayley-smoothness is due to C. Procesi [87]. The local structure description with marked quivers is due to L. Le Bruyn [74]. The description of the division algebras of quiver orders is due to L. Le Bruyn and A. Schofield [77]. The material on A_∞-algebras owes to the notes by B. Keller [51] and the work of M. Kontsevich [60] and [59]. The determination of indecomposable dimension vectors is due to V. Kac [48], [49]. The present description owes to the lecture notes of H. Kraft and Ch. Riedtmann [61], W. Crawley-Boevey [24] and the book by P. Gabriel and A.V. Roiter [35]. The main results on the canonical decomposition are due to A. Schofield [93], lemma 4.10 and lemma 4.11 are due to D. Happel and C.M. Ringel [42]. The algorithm is due to H. Derksen and J. Weyman [30] leaning on work of A. Schofield [94].

Chapter 5

Semisimple Representations

For a Cayley-Hamilton algebra $A \in \texttt{alg@n}$ we have seen that the quotient scheme

$$\texttt{triss}_n\, A = \texttt{trep}_n\, A/GL_n$$

classifies isomorphism classes of (trace preserving) semisimple n-dimensional representations. A point $\xi \in \texttt{triss}_n\, A$ is said to lie in the Cayley-smooth locus of A if $\texttt{trep}_n\, A$ is a smooth variety in the semisimple module M_ξ determined by ξ. In this case, the étale local structure of A and its central subalgebra $tr(A)$ are determined by a marked quiver setting.

We will extend some results on quotient varieties of representations of quivers to the setting of marked quivers. We will give a computational method to verify whether ξ belongs to the Cayley-smooth locus of A and develop reduction steps for the corresponding marked quiver setting that preserve geometric information, such as the type of singularity.

In low dimensions we can give a complete classification of all marked quiver settings that can arise for a Cayley-smooth order, allowing us to determine the classes in the Brauer group of the function field of a projective smooth surface, which allow a noncommutative smooth model.

In the arbitrary (central) dimension we are able to determine the smooth locus of the center as well as to classify the occurring singularities up to smooth equivalence.

5.1 Representation types

In this section we will determine the étale local structure of quotient varieties of marked quivers, characterize their dimension vectors of simples and introduce the *representation type* of a representation.

We fix a quiver Q and dimension vector α. Closed $GL(\alpha)$-orbits is $\texttt{rep}_\alpha\, Q$ correspond to isomorphism classes of semisimple representations of Q of dimension vector α. We have a quotient map

$$\texttt{rep}_\alpha\, Q \xrightarrow{\pi} \texttt{rep}_\alpha\, Q/GL(\alpha) = \texttt{iss}_\alpha\, Q$$

and we know that the coordinate ring $\mathbb{C}[\text{iss}_\alpha Q]$ is generated by traces along oriented cycles in the quiver Q. Consider a point $\xi \in \text{iss}_\alpha Q$ and assume that the corresponding semisimple representation V_ξ has a decomposition

$$V_\xi = V_1^{\oplus e_1} \oplus \ldots \oplus V_z^{\oplus e_z}$$

into distinct simple representations V_i of dimension vector, say, α_i and occurring in V_ξ with multiplicity e_i. We then say that ξ is a point of *representation-type*

$$\tau = t(\xi) = (e_1, \alpha_1; \ldots, e_z, \alpha_z) \quad \text{with} \quad \alpha = \sum_{i=1}^{z} e_i \alpha_i$$

We want to apply the slice theorem to obtain the étale $GL(\alpha)$-local structure of the representation space $\text{rep}_\alpha Q$ in a neighborhood of V_ξ and the étale local structure of the quotient variety $\text{iss}_\alpha Q$ in a neighborhood of ξ. We have to calculate the normal space N_ξ to the orbit $\mathcal{O}(V_\xi)$ as a representation over the stabilizer subgroup $GL(\alpha)_\xi = Stab_{GL(\alpha)}(V_\xi)$.

Denote $a_i = \sum_{j=1}^{k} a_{ij}$ where $\alpha_i = (a_{i1}, \ldots, a_{ik})$, that is, $a_i = dim \ V_i$. We will choose a basis of the underlying vectorspace

$$\oplus_{v_i \in Q_v} \mathbb{C}^{\oplus e_i a_i} \quad \text{of} \quad V_\xi = V_1^{\oplus e_1} \oplus \ldots \oplus V_z^{\oplus e_z}$$

as follows: the first $e_1 a_1$ vectors give a basis of the vertex spaces of all simple components of type V_1, the next $e_2 a_2$ vectors give a basis of the vertex spaces of all simple components of type V_2, and so on. If $n = \sum_{i=1}^{k} e_i d_i$ is the total dimension of V_ξ, then with respect to this basis, the subalgebra of $M_n(\mathbb{C})$ generated by the representation V_ξ has the following block-decomposition

$$\begin{bmatrix} M_{a_1}(\mathbb{C}) \otimes \mathbb{1}_{e_1} & 0 & \cdots & 0 \\ 0 & M_{a_2}(\mathbb{C}) \otimes \mathbb{1}_{e_2} & & 0 \\ \vdots & & \ddots & \vdots \\ 0 & 0 & \cdots & M_{a_z}(\mathbb{C}) \otimes \mathbb{1}_{e_z} \end{bmatrix}$$

But then, the stabilizer subgroup

$$Stab_{GL(\alpha)}(V_\xi) \simeq GL_{e_1} \times \ldots \times GL_{e_z}$$

embedded in $GL(\alpha)$ with respect to this particular basis as

$$\begin{bmatrix} GL_{e_1}(\mathbb{C} \otimes \mathbb{1}_{a_1}) & 0 & \cdots & 0 \\ 0 & GL_{e_2}(\mathbb{C} \otimes \mathbb{1}_{a_2}) & & 0 \\ \vdots & & \ddots & \vdots \\ 0 & 0 & \cdots & GL_{e_z}(\mathbb{C} \otimes \mathbb{1}_{a_z}) \end{bmatrix}$$

The tangent space to the $GL(\alpha)$-orbit in V_ξ is equal to the image of the natural linear map

$$Lie \ GL(\alpha) \longrightarrow \text{rep}_\alpha Q$$

sending a matrix $m \in Lie\, GL(\alpha) \simeq M_{e_1} \oplus \ldots \oplus M_{e_k}$ to the representation determined by the commutator $[m, V_\xi] = mV_\xi - V_\xi m$. By this we mean that the matrix $[m, V_\xi]_a$ corresponding to an arrow a is obtained as the commutator in $M_n(\mathbb{C})$ using the canonical embedding with respect to the above choice of basis. The kernel of this linear map is the centralizer subalgebra. That is, we have an exact sequence of $GL(\alpha)_\xi$-modules

$$0 \longrightarrow C_{M_n(\mathbb{C})}(V_\xi) \longrightarrow Lie\, GL(\alpha) \longrightarrow T_{V_\xi}\, \mathcal{O}(V_\xi) \longrightarrow 0$$

where

$$C_{M_n(\mathbb{C})}(V_\xi) = \begin{bmatrix} M_{e_1}(\mathbb{C} \otimes \mathbb{1}_{a_1}) & 0 & \cdots & 0 \\ 0 & M_{e_2}(\mathbb{C} \otimes \mathbb{1}_{a_2}) & & 0 \\ \vdots & & \ddots & \vdots \\ 0 & 0 & \cdots & M_{e_z}(\mathbb{C} \otimes \mathbb{1}_{a_z}) \end{bmatrix}$$

and the action of $GL(\alpha)_{V_\xi}$ is given by conjugation on $M_n(\mathbb{C})$ via the above embedding. We will now engage in a bookkeeping operation counting the factors of the relevant $GL(\alpha)_\xi$-spaces. We identify the factors by the action of the GL_{e_i}-components of $GL(\alpha)_\xi$

1. The centralizer $C_{M_n(\mathbb{C})}(V_\xi)$ decomposes as a $GL(\alpha)_\xi$-module into

   - one factor M_{e_i} on which GL_{e_1} acts via conjugation and the other factors act trivially

   $$\vdots$$

   - one factor M_{e_z} on which GL_{e_z} acts via conjugation and the other factors act trivially

2. Recall the notation $\alpha_i = (a_{i1}, \ldots, a_{ik})$, then the Lie algebra $Lie\, GL(\alpha)$ decomposes as a $GL(\alpha)_\xi$-module into

   - $\sum_{j=1}^{k} a_{1j}^2$ factors M_{e_1} on which GL_{e_1} acts via conjugation and the other factors act trivially

   $$\vdots$$

   - $\sum_{j=1}^{k} a_{zj}^2$ factors M_{e_z} on which GL_{e_z} acts via conjugation and the other factors act trivially

   - $\sum_{j=1}^{k} a_{1j} a_{2j}$ factors $M_{e_1 \times e_2}$ on which $GL_{e_1} \times GL_{e_2}$ acts via $\gamma_1 . m . \gamma_2^{-1}$ and the other factors act trivially

   $$\vdots$$

   - $\sum_{j=1}^{k} a_{zj} a_{z-1\, j}$ factors $M_{e_z \times e_{z-1}}$ on which $GL_{e_z} \times GL_{e_{z-1}}$ acts via $\gamma_z . m . \gamma_{z-1}^{-1}$ and the other factors act trivially

3. The representation space $\mathbf{rep}_\alpha\, Q$ decomposes as a $GL(\alpha)_\xi$-modulo into the following factors, for every arrow $\textcircled{j}\xleftarrow{\quad a \quad}\textcircled{i}$ in Q (or every loop in v_i by setting $i=j$ in the expressions below) we have

- $a_{1i}a_{1j}$ factors M_{e_1} on which GL_{e_1} acts via conjugation and the other factors act trivially

- $a_{1i}a_{2j}$ factors $M_{e_1\times e_2}$ on which $GL_{e_1}\times GL_{e_2}$ acts via $\gamma_1.m.\gamma_2^{-1}$ and the other factors act trivially

$$\vdots$$

- $a_{zi}a_{z-1\,j}$ factors $M_{e_z\times e_{z-1}}$ on which $GL_{e_z}\times GL_{e_{z-1}}$ act via $\gamma_z.m.\gamma_{z-1}^{-1}$ and the other factors act trivially

- $a_{zi}a_{zj}$ factors M_{e_z} on which GL_{e_z} acts via conjugation and the other factors act trivially

Removing the factors of 1. from those of 2. we obtain a description of the tangent space to the orbit $T_{V_\xi}\,\mathcal{O}(V_\xi)$. But then, removing these factors from those of 3, we obtain the description of the normal space N_{V_ξ} as a $GL(\alpha)_\xi$-module as there is an exact sequence of $GL(\alpha)_\xi$-modules

$$0 \longrightarrow T_{V_\xi}\,\mathcal{O}(V_\xi) \longrightarrow \mathbf{rep}_\alpha\, Q \longrightarrow N_{V_\xi} \longrightarrow 0$$

This proves that the normal space to the orbit in V_ξ depends only on the representation type $\tau = t(\xi)$ of the point ξ and can be identified with the representation space of a *local quiver* Q_τ.

THEOREM 5.1
Let $\xi \in \mathbf{iss}_\alpha\, Q$ be a point of representation type

$$\tau = t(\xi) = (e_1,\alpha_1;\ldots,e_z,\alpha_z)$$

Then, the normal space N_{V_ξ} to the orbit, as a module over the stabilizer subgroup, is identical to the representation space of a local quiver situation

$$N_{V_\xi} \simeq \mathbf{rep}_{\alpha_\tau}\, Q_\tau$$

where Q_τ is the quiver on z vertices (the number of distinct simple components of V_ξ) say $\{w_1,\ldots,w_z\}$ such that in Q_τ

$$\#\; \textcircled{j}\xleftarrow{\quad a \quad}\textcircled{i} \;=\; -\chi_Q(\alpha_i,\alpha_j) \quad \text{for } i \neq j, \text{ and}$$

$$\#\; \underset{\textcircled{i}}{\circlearrowleft} \;=\; 1 - \chi_Q(\alpha_i,\alpha_i)$$

and such that the dimension vector $\alpha_\tau = (e_1, \ldots, e_z)$ (the multiplicities of the simple components in V_ξ).

We can repeat this argument in the case of a marked quiver $Q^\bullet$. The only difference is the description of the factors of $\text{rep}_\alpha Q^\bullet$ where we need to replace the factors M_{e_j} in the description of a loop in v_i by $M_{e_i}^0$ (trace zero matrices) in case the loop gets a mark in $Q^\bullet$. We define the *Euler form* of the marked quiver $Q^\bullet$

$$
\chi_{Q^\bullet}^1 = \begin{bmatrix} 1 - a_{11} & \chi_{12} & \cdots & \chi_{1k} \\ \chi_{21} & 1 - a_{22} & \cdots & \chi_{2k} \\ \vdots & \vdots & \ddots & \vdots \\ \chi_{k1} & \chi_{k2} & \cdots & 1 - a_{kk} \end{bmatrix} \qquad \chi_{Q^\bullet}^2 = \begin{bmatrix} -m_{11} & & & \\ & -m_{22} & & \\ & & \ddots & \\ & & & -m_{kk} \end{bmatrix}
$$

such that $\chi_Q = \chi_{Q^\bullet}^1 + \chi_{Q^\bullet}^2$ where Q is the underlying quiver of $Q^\bullet$.

THEOREM 5.2
Let $\xi \in \text{iss}_\alpha Q^\bullet$ be a point of representation type

$$\tau = t(\xi) = (e_1, \alpha_1; \ldots, e_z, \alpha_z)$$

Then, the normal space N_{V_ξ} to the orbit, as a module over the stabilizer subgroup, is identical to the representation space of a local marked quiver situation

$$N_{V_\xi} \simeq \text{rep}_{\alpha_\tau} Q_\tau^\bullet$$

where $Q_\tau^\bullet$ is the quiver on z vertices (the number of distinct simple components of V_ξ) say $\{w_1, \ldots, w_z\}$ such that in $Q_\tau^\bullet$

$$\# \; \underset{\textcircled{\scriptsize i}}{\overset{a}{\longleftarrow}} \; \textcircled{\scriptsize j} \quad = \quad -\chi_Q(\alpha_i, \alpha_j) \quad \text{for } i \neq j, \text{ and}$$

$$\# \; \textcircled{\scriptsize i}\,\circlearrowright \quad = \quad 1 - \chi_{Q^\bullet}^1(\alpha_i, \alpha_i)$$

$$\# \; \textcircled{\scriptsize i}\,\overset{\bullet}{\circlearrowright} \quad = \quad -\chi_{Q^\bullet}^2(\alpha_i, \alpha_i)$$

and such that the dimension vector $\alpha_\tau = (e_1, \ldots, e_z)$ (the multiplicities of the simple components in V_ξ).

PROPOSITION 5.1
If $\alpha = (d_1, \ldots, d_k)$ is the dimension vector of a simple representation of $Q^\bullet$, then the dimension of the quotient variety $\text{iss}_\alpha Q^\bullet$ is equal to

$$1 - \chi_{Q^\bullet}^1(\alpha, \alpha)$$

PROOF   There is a Zariski open subset of $\text{iss}_\alpha \, Q^\bullet$ consisting of points ξ such that the corresponding semisimple module V_ξ is simple, that is, ξ has representation type $\tau = (1, \alpha)$. But then the local quiver setting (Q_τ, α_τ) is

where $a = 1 - \chi^1_{Q^\bullet}(\alpha, \alpha)$ and $b = -\chi^2_{Q^\bullet}(\alpha, \alpha)$. The corresponding representation space has the coordinate ring

$$\mathbb{C}[\text{rep}_{\alpha_\tau} \, Q^\bullet_\tau] = \mathbb{C}[x_1, \ldots, x_a]$$

on which $GL(\alpha_\tau) = \mathbb{C}^*$ acts trivially. That is, the quotient variety is

$$\text{rep}_{\alpha_\tau} \, Q^\bullet_\tau / GL(\alpha_\tau) = \text{rep}_{\alpha_\tau} \, Q^\bullet_\tau \simeq \mathbb{C}^a$$

By the slice theorem, $\text{iss}_\alpha \, Q^\bullet$ has the same local structure near ξ as this quotient space near the origin and the result follows. $\square$

We can extend the classifications of simple roots of a quiver to the setting of marked quivers. Let Q be the underlying quiver of a marked quiver $Q^\bullet$. If $\alpha = (a_1, \ldots, a_k)$ is a simple root of Q and if l is a marked loop in a vertex v_i with $a_i > 1$, then we can replace the matrix V_l of a simple representation $V \in \text{rep}_\alpha \, Q$ by $V'_l = V_l - \frac{1}{d_i} \mathbb{1}_{d_i}$ and retain the property that V' is a simple representation. Things are different, however, for a marked loop in a vertex v_i with $a_i = 1$ as this 1×1-matrix factor is removed from the representation space. That is, we have the following characterization result.

THEOREM 5.3
$\alpha = (a_1, \ldots, a_k)$ *is the dimension vector of a simple representation of a marked quiver $Q^\bullet$ if and only if $\alpha = (a_1, \ldots, a_k)$ is the dimension vector of a simple representation of the quiver Q' obtained from the underlying quiver Q of $Q^\bullet$ after removing the loops in Q, which are marked in $Q^\bullet$ in all vertices v_i where $a_i = 1$.*

We draw some consequences from the description of the local quiver. We state all results in the setting of marked quivers. Often, the quotient varieties $\text{iss}_\alpha \, Q^\bullet = \text{rep}_\alpha \, Q^\bullet / GL(\alpha)$ classifying isomorphism classes of semisimple α-dimensional representations have singularities. Still, we can decompose these quotient varieties into smooth pieces according to representation types.

PROPOSITION 5.2
Let $\text{iss}_\alpha \, Q^\bullet(\tau)$ be the set of points $\xi \in \text{iss}_\alpha \, Q^\bullet$ of representation type

$$\tau = (e_1, \alpha_1; \ldots; e_z, \alpha_z)$$

Then, $\text{iss}_\alpha \, Q^\bullet(\tau)$ *is a locally closed smooth subvariety of* $\text{iss}_\alpha \, Q^\bullet$ *and*

$$\text{iss}_\alpha \, Q^\bullet = \bigsqcup_\tau \text{iss}_\alpha \, Q^\bullet(\tau)$$

is a finite smooth stratification *of the quotient variety.*

PROOF Let $Q_\tau^\bullet$ be the local marked quiver in ξ. Consider a nearby point ξ'. If some trace of an oriented cycles of length > 1 in $Q_\tau^\bullet$ is nonzero in ξ', then ξ' cannot be of representation type τ as it contains a simple factor composed of vertices of that cycle. That is, locally in ξ the subvariety $\text{iss}_\alpha \, Q^\bullet(\tau)$ is determined by the traces of unmarked loops in vertices of the local quiver $Q_\tau^\bullet$ and hence is locally in the étale topology an affine space whence smooth. All other statements are direct. □

Given a *stratification of a topological space* , one wants to determine which strata make up the boundary of a given stratum. For the above stratification of $\text{iss}_\alpha \, Q^\bullet$ we have a combinatorial solution to this problem. Two representation types

$$\tau = (e_1, \alpha_1; \dots; e_z, \alpha_z) \quad \text{and} \quad \tau' = (e_1', \alpha_1'; \dots; e_{z'}', \alpha_{z'}')$$

are said to be direct *successors* $\tau < \tau'$ if and only if one of the following two cases occurs

- (splitting of one simple): $z' = z + 1$ and for all but one $1 \le i \le z$ we have that $(e_i, \alpha_i) = (e_j', \alpha_j')$ for a uniquely determined j and for the remaining i_0 we have that the remaining couples of τ' are

$$(e_i, \alpha_u'; e_i, \alpha_v') \quad \text{with} \quad \alpha_i = \alpha_u' + \alpha_v'$$

- (combining two simple types): $z' = z - 1$ and for all but one $1 \le i \le z'$ we have that $(e_i', \alpha_i') = (e_j, \alpha_j)$ for a uniquely determined j and for the remaining i we have that the remaining couples of τ are

$$(e_u, \alpha_i'; e_v, \alpha_i') \quad \text{with} \quad e_u + e_v = e_i'$$

This direct successor relation $<$ induces an ordering that we will denote with $<<$. Observe that $\tau << \tau'$ if and only if the stabilizer subgroup $GL(\alpha)_\tau$ is conjugated to a subgroup of $GL(\alpha)_{\tau'}$. The following result either follows from general theory, see, for example, [96, lemma 5.5], or from the description of the local marked quivers.

PROPOSITION 5.3
The stratum $\text{iss}_\alpha \, Q^\bullet(\tau')$ *lies in the closure of the stratum* $\text{iss}_\alpha \, Q^\bullet$ *if and only if* $\tau << \tau'$.

Proposition 5.1 gives us the dimensions of the different strata $\text{iss}_\alpha\ Q^\bullet(\tau)$.

PROPOSITION 5.4

Let $\tau = (e_1, \alpha_1; \ldots; e_z, \alpha_z)$ *a representation type of* α. *Then*

$$dim\ \text{iss}_\alpha\ Q^\bullet(\tau) = \sum_{j=1}^{z}(1 - \chi_{Q^\bullet}^1(\alpha_j, \alpha_j))$$

Because $\text{rep}_\alpha\ Q^\bullet$ and hence $\text{iss}_\alpha\ Q^\bullet$ is an irreducible variety, there is a unique representation type τ_{gen}^{ss} such that $\text{iss}_\alpha\ Q^\bullet(\tau_{gen}^{ss})$ is Zariski open in the quotient variety $\text{iss}_\alpha\ Q^\bullet$. We call τ_{gen}^{ss} the *generic semisimple representation type* for α. The generic semisimple representation type can be determined by the following algorithm.

input : A quiver Q, a dimension vector $\alpha = (a_1, \ldots, a_k)$ and a semisimple representation type

$$\tau = (e_1, \alpha_1; \ldots; e_l, \alpha_l)$$

with $\alpha = \sum +i = 1^l e_i \alpha_i$ and all α_i simple roots for Q. For example, one can always start with the type $(a_1, \vec{v_1}; \ldots; a_k, \vec{v_k})$.

step 1: Compute the local quiver Q_τ on l vertices and the dimension vector α_τ. If the only oriented cycles in Q_τ are vertex-loops, stop and output this type. If not, proceed.

step 2: Take a proper oriented cycle $C = (j_1, \ldots, j_r)$ with $r \geq 2$ in Q_τ where j_s is the vertex in Q_τ determined by the dimension vector α_{j_s}. Set $\beta = \alpha_{j_1} + \ldots + \alpha_{j_r}$, $e'_i = e_i - \delta_{iC}$ where $\delta_{iC} = 1$ if $i \in C$ and is 0 otherwise. replace τ by the new semisimple representation type

$$\tau' = (e'_1, \alpha_1; \ldots; e'_l, \alpha_l; 1, \beta)$$

delete the terms (e'_i, α_i) with $e'_i = 0$ and set τ to be the resulting type. go to step 1.

The same algorithm extends to marked quivers with the modified construction of the local marked quiver $Q_\tau^\bullet$ in that case. We can give an A_∞-interpretation of the characterization of the *canonical decomposition* and the generic semisimple representation type . Let

$$\tau = (e_1, \alpha_1; \ldots; e_z, \alpha_z) \qquad \alpha = \sum_{i=1}^{z} e_i \alpha_i$$

be a decomposition of α with all the α_i *roots*. We define $\alpha_\tau = (e_1, \ldots, e_z)$ and construct two quivers Q_τ^0 and Q_τ^1 on z vertices determined by the rules

$$\text{in } Q_\tau^0 : \qquad \# \ ⓙ \xleftarrow{\ \ a\ \ } ⓘ \quad = \quad dim_{\mathbb{C}}\ Hom_{\mathbb{C}Q}(V_i, V_j)$$

in Q_τ^1 : $\quad \# \, \textcircled{j} \xleftarrow{\quad a \quad} \textcircled{i} \;\; = \;\; dim_\mathbb{C} \; Ext^1_{\mathbb{C}Q}(V_i, V_j)$

where V_i is a general representation of Q of dimension vector α_i.

THEOREM 5.4
With notations as above, we have:

1. *The canonical decomposition τ_{can} is the unique type $\tau = (e_1, \alpha_1; \ldots; e_z, \alpha_z)$ such that all α_i are Schur roots, Q_τ^0 has no (non-loop) oriented cycles and Q_τ^1 has no arrows and loops only in vertices where $e_i = 1$.*

2. *The generic semisimple representation type τ_{gen}^{ss} is the unique type $\tau = (e_1, \alpha_1; \ldots; e_z, \alpha_z)$ such that all α_i are simple roots, Q_τ^0 has only loops and Q_τ^1 has no (nonloop) oriented cycles.*

5.2 Cayley-smooth locus

Let A be a Cayley-Hamilton algebra of degree n equipped with a trace map $A \xrightarrow{tr} A$ and consider the quotient map

$$\mathtt{trep}_n \, A \xrightarrow{\;\pi\;} \mathtt{triss}_n \, A$$

Let ξ be a geometric point of he quotient scheme $\mathtt{triss}_n \, A$ with corresponding n-dimensional trace preserving semisimple representation V_ξ with decomposition

$$V_\xi = S_1^{\oplus e_1} \oplus \ldots \oplus S_k^{\oplus e_k}$$

where the S_i are distinct simple representations of A of dimension d_i such that $n = \sum_{i=1}^k d_i e_i$.

DEFINITION 5.1 *The Cayley-smooth locus of A is the subset of $\mathtt{triss}_n \, A$*

$$Sm_{tr} \, A = \{\xi \in \mathtt{triss}_n \, A \mid \mathtt{trep}_n \, A \text{ is smooth along } \pi^{-1}(\xi) \}$$

As the singular locus of $\mathtt{trep}_n \, A$ is a GL_n-stable closed subscheme of $\mathtt{trep}_n \, A$ this is equivalent to

$$Sm_{tr} \, A = \{\xi \in \mathtt{triss}_n \, A \mid \mathtt{trep}_n \, A \text{ is smooth in } V_\xi \}$$

We will give some numerical conditions on ξ to be in the smooth locus $Sm_{tr} \, A$. To start, $\mathtt{trep}_n \, A$ is smooth in V_ξ if and only if the dimension of the tangent space in V_ξ is equal to the local dimension of $\mathbf{trep}_n\, A$ in V_ξ. From example 3.11 we know that the tangent space is the set of trace preserving derivations $A \xrightarrow{\;D\;} M_n(\mathbb{C})$ satisfying

$$D(aa') = D(a)\rho(a') + \rho(a)D(a')$$

where $A \xrightarrow{\;\rho\;} M_n(\mathbb{C})$ is the $\mathbb{C}$-algebra morphism determined by the action of A on V_ξ. The $\mathbb{C}$-vectorspace of such derivations is denoted by $Der_\rho^t\, A$. Therefore

$$\xi \in Sm_{tr}\, A \iff dim_\mathbb{C}\, Der_\rho^t\, A = dim_{V_\xi}\, \mathbf{trep}_n\, A$$

Next, if $\xi \in Sm_{tr}\, A$, then we know from the slice theorem that the local GL_n-structure of $\mathbf{trep}_n\, A$ near V_ξ is determined by a local marked quiver setting $(Q_\xi^\bullet, \alpha_\xi)$ as defined in theorem 4.3. We have local étale isomorphisms between the varieties

$$GL_n \times^{GL(\alpha_\xi)} \mathbf{rep}_{\alpha_\xi}\, Q_\xi^\bullet \xleftrightarrow{\;et\;} \mathbf{trep}_n\, A \quad \text{and} \quad \mathbf{rep}_{\alpha_\xi}\, Q_\xi^\bullet / GL(\alpha_\xi) \xleftrightarrow{\;et\;} \mathbf{triss}_n\, A$$

which gives us the following numerical restrictions on $\xi \in Sm_{tr}\, A$:

PROPOSITION 5.5
$\xi \in Sm_{tr}\, A$ *if and only if the following two equalities hold*

$$\begin{cases} dim_{V_\xi}\, \mathbf{trep}_n\, A &= n^2 - (e_1^2 + \ldots + e_k^2) + dim_\mathbb{C}\, Ext_A^{tr}(V_\xi, V_\xi) \\ dim_\xi\, \mathbf{triss}_n\, A &= dim_{\overline{0}}\, \mathbf{rep}_{\alpha_\xi}\, Q_\xi^\bullet / GL(\alpha_\xi) = dim_{\underline{0}}\, \mathbf{iss}_{\alpha_\xi}\, Q_\xi^\bullet \end{cases}$$

Moreover, if $\xi \in Sm_{tr}\, A$, *then* $\mathbf{trep}_n\, A$ *is a normal variety (that is, the coordinate ring is integrally closed) in a neighborhood of* ξ.

PROOF   The last statement follows from the fact that $\mathbb{C}[\mathbf{rep}_{\alpha_\xi}\, Q_\xi^\bullet]^{GL(\alpha_\xi)}$ is integrally closed and this property is preserved under the étale map.   ▯

In general, the difference between these numbers gives a measure for the noncommutative singularity of A in ξ.

Example 5.1   Quantum plane of order 2
Consider the affine $\mathbb{C}$-algebra $A = \frac{\mathbb{C}\langle x,y\rangle}{(xy+yx)}$ then $u = x^2$ and $v = y^2$ are central elements of A and A is a free module of rank 4 over $\mathbb{C}[u,v]$. In fact, A is a $\mathbb{C}[u,v]$-order in the quaternion division algebra

$$\Delta = \begin{pmatrix} u & v \\ & \mathbb{C}(u,v) \end{pmatrix}$$

and the reduced trace map on Δ makes A into a Cayley-Hamilton algebra of degree 2. More precisely, tr is the linear map on A such that

$$\begin{cases} tr(x^i y^j) = 0 & \text{if either } i \text{ or } j \text{ are odd, and} \\ tr(x^i y^j) = 2x^i y^j & \text{if } i \text{ and } j \text{ are even.} \end{cases}$$

In particular, a trace preserving 2-dimensional representation is determined by a couple of 2×2 matrices

$$\rho = \left(\begin{bmatrix} x_1 & x_2 \\ x_3 & -x_1 \end{bmatrix}, \begin{bmatrix} x_4 & x_5 \\ x_6 & -x_4 \end{bmatrix} \right) \quad \text{with} \quad tr\left(\begin{bmatrix} x_1 & x_2 \\ x_3 & -x_1 \end{bmatrix} \cdot \begin{bmatrix} x_4 & x_5 \\ x_6 & -x_4 \end{bmatrix} \right) = 0$$

That is, $\mathbf{trep}_2 A$ is the hypersurface in $\mathbb{C}^6$ determined by the equation

$$\mathbf{trep}_2 A = \mathbb{V}(2x_1x_4 + x_2x_6 + x_3x_5) \hookrightarrow \mathbb{C}^6$$

and is therefore irreducible of dimension 5 with an isolated singularity at $p = (0, \ldots, 0)$. The image of the trace map is equal to the center of A, which is $\mathbb{C}[u, v]$, and the quotient map

$$\mathbf{trep}_2 A \xrightarrow{\pi} \mathbf{triss}_2 A = \mathbb{C}^2 \qquad \pi(x_1, \ldots, x_6) = (x_1^2 + x_2x_3, x_4^2 + x_5x_6)$$

There are three different representation types to consider. Let $\xi = (a, b) \in \mathbb{C}^2 = \mathbf{triss}_2 A$ with $ab \neq 0$, then $\pi^{-1}(\xi)$ is a closed GL_2-orbit and a corresponding simple A-module is given by the matrixcouple

$$\left(\begin{bmatrix} i\sqrt{a} & 0 \\ 0 & -i\sqrt{a} \end{bmatrix}, \begin{bmatrix} 0 & \sqrt{b} \\ -\sqrt{b} & 0 \end{bmatrix} \right)$$

That is, ξ is of type $(1, 2)$ and the stabilizer subgroup are the scalar matrixes $\mathbb{C}^* \mathbb{1}_2 \hookrightarrow GL_2$. So, the action on both the tangent space to $\mathbf{trep}_2 A$ and the tangent space to the orbit are trivial. As they have, respectively, dimension 5 and 3, the normal space corresponds to the quiver setting

$$N_\xi = \quad \text{①}$$

which is compatible with the numerical restrictions. Next, consider a point $\xi = (0, b)$ (or similarly, $(a, 0)$), then ξ is of type $(1, 1; 1, 1)$ and the corresponding semisimple representation is given by the matrices

$$\left(\begin{bmatrix} 0 & 0 \\ 0 & 0 \end{bmatrix}, \begin{bmatrix} i\sqrt{b} & 0 \\ 0 & -i\sqrt{b} \end{bmatrix} \right)$$

The stabilizer subgroup is in this case the maximal torus of diagonal matrices $\mathbb{C}^* \times \mathbb{C}^* \hookrightarrow GL_2$. The tangent space in this point to $\mathbf{trep}_2 A$ are the 6-tuples $(a_1, \ldots, a_6)$ such that

$$tr\left(\begin{bmatrix} 0 & 0 \\ 0 & 0 \end{bmatrix} + \epsilon \begin{bmatrix} a_1 & a_2 \\ a_3 & -a_1 \end{bmatrix} \right) \cdot \left(\begin{bmatrix} i\sqrt{b} & 0 \\ 0 & -i\sqrt{b} \end{bmatrix} + \epsilon \begin{bmatrix} b_4 & b_5 \\ b_6 & -b_4 \end{bmatrix} \right) = 0 \quad \text{where } \epsilon^2 = 0$$

This leads to the condition $a_1 = 0$, so the tangent space are the matrix couples

$$\left(\begin{bmatrix} 0 & a_2 \\ a_3 & 0 \end{bmatrix}, \begin{bmatrix} a_4 & a_5 \\ a_6 & -a_4 \end{bmatrix} \right) \quad \text{on which the stabilizer } \begin{bmatrix} \lambda & 0 \\ 0 & \mu \end{bmatrix}$$

acts via conjugation. That is, the tangent space corresponds to the quiver setting

Moreover, the tangent space to the orbit is the image of the linear map

$$(1\!\!1_2 + \epsilon \begin{bmatrix} m_1 & m_2 \\ m_3 & m_4 \end{bmatrix}). \left(\begin{bmatrix} 0 & 0 \\ 0 & 0 \end{bmatrix}, \begin{bmatrix} \sqrt{b} & 0 \\ 0 & -\sqrt{b} \end{bmatrix} \right). (1\!\!1_2 - \begin{bmatrix} m_1 & m_2 \\ m_3 & m_4 \end{bmatrix})$$

which is equal to

$$\left(\begin{bmatrix} 0 & 0 \\ 0 & 0 \end{bmatrix}, \begin{bmatrix} \sqrt{b} & 0 \\ 0 & -\sqrt{b} \end{bmatrix} + \epsilon \begin{bmatrix} 0 & -2m_2\sqrt{b} \\ 2m_3\sqrt{b} & 0 \end{bmatrix} \right)$$

on which the stabilizer acts again via conjugation giving the quiver setting

Therefore, the normal space to the orbit corresponds to the quiver setting

which is again compatible with the numerical restrictions. Finally, consider $\xi = (0,0)$ which is of type $(2,1)$ and whose semisimple representation corresponds to the zero matrix-couple. The action fixes this point, so the stabilizer is GL_2 and the tangent space to the orbit is the trivial space. Hence, the tangent space to $\mathbf{trep}_2$ A coincides with the normal space to the orbit and both spaces are acted on by GL_2 via simultaneous conjugation leading to the quiver setting

$$N_\xi =$$

This time, the data is not compatible with the numerical restriction as

$$5 = \dim \mathbf{trep}_2 \, A \neq n^2 - e^2 + \dim \mathbf{rep}_\alpha \, Q^\bullet = 4 - 4 + 6$$

consistent with the fact that the zero matrix-couple is a (in fact, the only) singularity on $\mathbf{trep}_2 \, A$.

We will put additional conditions on the Cayley-Hamilton algebra A. Let X be a normal affine variety with coordinate ring $\mathbb{C}[X]$ and function field $\mathbb{C}(X)$. Let Δ be a central simple $\mathbb{C}(X)$-algebra of dimension n^2 which is a Cayley-Hamilton algebra of degree n using the reduced trace map tr. Let A be a $\mathbb{C}[X]$-order in Δ, that is, the center of A is $\mathbb{C}[X]$ and $A \otimes_{\mathbb{C}[X]} \mathbb{C}(X) \simeq \Delta$. Because $\mathbb{C}[X]$ is integrally closed, the restriction of the reduced trace tr to A has its image in $\mathbb{C}[X]$, that is, A is a Cayley-Hamilton algebra of degree n and

$$tr(A) = \mathbb{C}[X]$$

Consider the quotient morphism for the representation variety

$$\mathtt{trep}_n\, A \xrightarrow{\;\pi\;} \mathtt{triss}_n\, A$$

then the above argument shows that $X \simeq \mathtt{triss}_n\, A$ and in particular the quotient scheme is reduced.

PROPOSITION 5.6
Let A be a Cayley-Hamilton order of degree n over $\mathbb{C}[X]$. Then, its smooth locus $Sm_{tr}\, A$ is a nonempty Zariski open subset of X. In particular, the set X_{az} of Azumaya points, that is, of points $x \in X = \mathtt{triss}_n\, A$ of representation type $(1, n)$ is a nonempty Zariski open subset of X and its intersection with the Zariski open subset X_{reg} of smooth points of X satisfies

$$X_{az} \cap X_{reg} \lhook\joinrel\longrightarrow Sm_{tr}\, A$$

PROOF   Because $A\mathbb{C}(X) = \Delta$, there is an $f \in \mathbb{C}[X]$ such that $A_f = A \otimes_{\mathbb{C}[X]} \mathbb{C}[X]_f$ is a free $\mathbb{C}[X]_f$-module of rank n^2, say, with basis $\{a_1, \ldots, a_{n^2}\}$. Consider the $n^2 \times n^2$ matrix with entries in $\mathbb{C}[X]_f$

$$R = \begin{bmatrix} tr(a_1 a_1) & \ldots & tr(a_1 a_{n^2}) \\ \vdots & & \vdots \\ tr(a_{n^2} a_1) & \ldots & tr(a_{n^2} a_{n^2}) \end{bmatrix}$$

The determinant $d = det\, R$ is nonzero in $\mathbb{C}[X]_f$. For, let $\mathbb{K}$ be the algebraic closure of $\mathbb{C}(X)$ then $A_f \otimes_{\mathbb{C}[X]_f} \mathbb{K} \simeq M_n(\mathbb{K})$ and for any $\mathbb{K}$-basis of $M_n(\mathbb{K})$ the corresponding matrix is invertible (for example, verify this on the matrixes e_{ij}). As $\{a_1, \ldots, a_{n^2}\}$ is such a basis, $d \neq 0$. Next, consider the Zariski open subset $U = \mathbb{X}(f) \cap \mathbb{X}(d) \lhook\joinrel\longrightarrow X$. For any $x \in X$ with maximal ideal $\mathfrak{m}_x \lhd \mathbb{C}[X]$ we claim that

$$\frac{A}{A\mathfrak{m}_x A} \simeq M_n(\mathbb{C})$$

Indeed, the images of the a_i give a $\mathbb{C}$-basis in the quotient such that the $n^2 \times n^2$-matrix of their product-traces is invertible. This property is equivalent to the quotient being $M_n(\mathbb{C})$. The corresponding semisimple representation of A is simple, proving that X_{az} is a nonempty Zariski open subset of X. But then, over U the restriction of the quotient map

$$\mathbf{trep}_n \; A \; | \; \pi^{-1}(U) \longrightarrow U$$

is a principal PGL_n-fibration. In fact, this restricted quotient map determines an element in $H^1_{et}(U, PGL_n)$ determining the class of the central simple $\mathbb{C}(X)$-algebra Δ in $H^1_{et}(\mathbb{C}(X), PGL_n)$. Restrict this quotient map further to $U \cap X_{reg}$, then the PGL_n-fibration

$$\mathbf{trep}_n \; A \; | \; \pi^{-1}(U \cap X_{reg}) \longrightarrow U \cap X_{reg}$$

has a smooth base and therefore also the total space is smooth. But then, $U \cap X_{reg}$ is a nonempty Zariski open subset of $Sm_{tr} \; A$.   ☐

Observe that the normality assumption on X is no restriction as the quotient scheme is locally normal in a point of $Sm_{tr} \; A$. Our next result limits the local dimension vectors α_ξ.

PROPOSITION 5.7
Let A be a Cayley-Hamilton order and $\xi \in Sm_{tr} \; A$ such that the normal space to the orbit of the corresponding semisimple n-dimensional representation is

$$N_\xi = \mathbf{rep}_{\alpha_\xi} \; Q_\xi^{\bullet}$$

Then, α_ξ is the dimension vector of a simple representation of $Q_\xi^{\bullet}$.

PROOF   Let V_ξ be the semisimple representation of A determined by ξ. Let S_ξ be the slice variety in V_ξ then by the slice theorem we have the following diagram of étale GL_n-equivariant maps

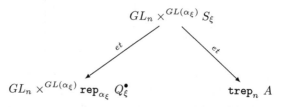

linking a neighborhood of V_ξ with one of $\overline{(\mathbb{1}_n, 0)}$. Because A is an order, every Zariski neighborhood of V_ξ in $\mathbf{trep}_n \; A$ contains simple n-dimensional representations, that is, closed GL_n-orbits with stabilizer subgroup isomorphic to $\mathbb{C}^*$. Transporting this property via the GL_n-equivariant étale maps, every Zariski neighborhood of $\overline{(\mathbb{1}_n, 0)}$ contains closed GL_n-orbits with stabilizer $\mathbb{C}^*$. By the correspondence of orbits is associated fiber bundles, every Zariski neighborhood of the trivial representation $0 \in \mathbf{rep}_{\alpha_\xi} \; Q_\xi^{\bullet}$ contains closed $GL(\alpha_\xi)$-orbits with stabilizer subgroup $\mathbb{C}^*$. We have seen that closed

$GL(\alpha_\xi)$-orbits correspond to semisimple representations of $Q_\xi^\bullet$. However, if the stabilizer subgroup of a semisimple representation is $\mathbb{C}^*$ this representation must be simple. $\quad\square$

THEOREM 5.5

Let A be a Cayley-Hamilton order of degree n with center $\mathbb{C}[X]$, X a normal variety of dimension d. For $\xi \in X = \mathtt{triss}_n\, A$ with corresponding semisimple representation

$$V_\xi = S_1^{\oplus e_1} \oplus \ldots \oplus S_k^{\oplus e_k}$$

and normal space to the orbit $\mathcal{O}(V_\xi)$ isomorphic to $\mathtt{rep}_{\alpha_\xi}\, Q_\xi^\bullet$ as $GL(\alpha_\xi)$-modules where $\alpha_\xi = (e_1, \ldots, e_k)$. Then, $\xi \in Sm_{tr}\, A$ if and only if the following two conditions are met

$$\begin{cases} \alpha_\xi & \text{is the dimension vector of a simple representation of } Q^\bullet, \text{ and} \\ d & = 1 - \chi_Q(\alpha_\xi, \alpha_\xi) - \sum_{i=1}^k m_{ii} \end{cases}$$

where Q is the underlying quiver of $Q_\xi^\bullet$ and m_{ii} is the number of marked loops in $Q_\xi^\bullet$ in vertex v_i.

PROOF By the slice theorem we have étale maps

$$\mathtt{rep}_{\alpha_\xi}\, Q_\xi^\bullet / GL(\alpha_\xi) \xleftarrow{\ et\ } S_\xi / GL(\alpha_\xi) \xrightarrow{\ et\ } \mathtt{triss}_n\, A = X$$

connecting a neighborhood of $\xi \in X$ with one of the trivial semisimple representation $\overline{0}$. By definition of the Euler-form of Q we have that

$$\chi_Q(\alpha_\xi, \alpha_\xi) = -\sum_{i \neq j} e_i e_j \chi_{ij} + \sum_i e_i^2 (1 - a_{ii} - m_{ii})$$

On the other hand we have

$$dim\ \mathtt{rep}_\alpha\, Q_{\alpha_\xi}^\bullet = \sum_{i \neq j} e_i e_j \chi_{ij} + \sum_i e_i^2 (a_{ii} + m_{ii}) - \sum_i m_{ii}$$

$$dim\ GL(\alpha_\xi) = \sum_i e_i^2$$

As any Zariski open neighborhood of ξ contains an open set where the quotient map is a $PGL(\alpha_\xi) = \frac{GL(\alpha_\xi)}{\mathbb{C}^*}$-fibration we see that the quotient variety $\mathtt{rep}_{\alpha_\xi}\, Q_\xi^\bullet$ has dimension equal to

$$dim\ \mathtt{rep}_{\alpha_\xi}\, Q_\xi^\bullet - dim\ GL(\alpha_\xi) + 1$$

and plugging in the above information we see that this is equal to $1 - \chi_Q(\alpha_\xi, \alpha_\xi) - \sum_i m_{ii}$. $\quad\square$

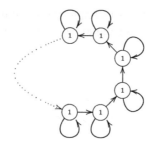

FIGURE 5.1:   Ext-quiver of quantum plane.

Example 5.2 Quantum plane
We will generalize the discussion of example 5.1 to the algebra

$$A = \frac{\mathbb{C}\langle x, y \rangle}{(yx - qxy)}$$

where q is a primitive n-th root of unity. Let $u = x^n$ and $v = y^n$ then it is easy to see that A is a free module of rank n^2 over its center $\mathbb{C}[u, v]$ and is a Cayley-Hamilton algebra of degree n with the trace determined on the basis

$$tr(x^i y^j) = \begin{cases} 0 & \text{when either } i \text{ or } j \text{ is not a multiple of } n \\ nx^i y^j & \text{when } i \text{ and } j \text{ are multiples of } n \end{cases}$$

Let $\xi \in \text{iss}_n A = \mathbb{C}^2$ be a point (a^n, b) with $a.b \neq 0$, then ξ is of representation type $(1, n)$ as the corresponding (semi)simple representation V_ξ is determined by (if m is odd, for even n we replace a by ia and b by $-b$)

$$\rho(x) = \begin{bmatrix} a & & & \\ & qa & & \\ & & \ddots & \\ & & & q^{n-1}a \end{bmatrix} \quad \text{and} \quad \rho(y) = \begin{bmatrix} 0 & 1 & 0 & \dots & 0 \\ 0 & 0 & 1 & & 0 \\ \vdots & \vdots & & \ddots & \\ 0 & 0 & 0 & \dots & 1 \\ b & 0 & 0 & \dots & 0 \end{bmatrix}$$

One computes that $Ext_A^1(V_\xi, V_\xi) = \mathbb{C}^2$ where the algebra map $A \xrightarrow{\phi} M_n(\mathbb{C}[\varepsilon])$ corresponding to (α, β) is given by

$$\begin{cases} \phi(x) = \rho(x) + \varepsilon \, \alpha \mathbb{1}_n \\ \phi(y) = \rho(y) + \varepsilon \, \beta \mathbb{1}_n \end{cases}$$

and all these algebra maps are trace preserving. That is, $Ext_A^1(V_\xi, V_\xi) = Ext_A^{tr}(V_\xi, V_\xi)$ and as the stabilizer subgroup is $\mathbb{C}^*$ the marked quiver-setting $(Q_\xi^\bullet, \alpha_\xi)$ is and $d = 1 - \chi_Q(\alpha, \alpha) - \sum_i m_{ii}$ as $2 = 1 - (-1) + 0$, compatible with the fact that over these points the quotient map is a principal PGL_n-fibration.

Next, let $\xi = (a^n, 0)$ with $a \neq 0$ (or, by a similar argument $(0, b^n)$ with $b \neq 0$). Then, the representation type of ξ is $(1, 1; \ldots; 1, 1)$ because

$$V_\xi = S_1 \oplus \ldots \oplus S_n$$

where the simple one-dimensional representation S_i is given by

$$\begin{cases} \rho(x) &= q^i a \\ \rho(y) &= 0 \end{cases}$$

One verifies that

$$Ext^1_A(S_i, S_i) = \mathbb{C} \quad \text{and} \quad Ext^1_A(S_i, S_j) = \delta_{i+1,j}\, \mathbb{C}$$

and as the stabilizer subgroup is $\mathbb{C}^* \times \ldots \times \mathbb{C}^*$, the Ext-quiver setting is depicted in figure 5.1. The algebra map $A \xrightarrow{\phi} M_n(\mathbb{C}[\varepsilon])$ corresponding to the extension $(\alpha_1, \beta_1, \ldots, \alpha_n, \beta_n) \in Ext^1_A(V_\xi, V_\xi)$ is given by

$$\begin{cases} \phi(x) &= \begin{bmatrix} a + \varepsilon\, \alpha_1 & & & \\ & qa + \varepsilon\, \alpha_2 & & \\ & & \ddots & \\ & & & q^{n-1}a + \varepsilon\, \alpha_n \end{bmatrix} \\[30pt] \phi(y) &= \varepsilon \begin{bmatrix} 0 & \beta_1 & 0 & \ldots & 0 \\ 0 & 0 & \beta_2 & & 0 \\ \vdots & \vdots & & \ddots & \vdots \\ 0 & 0 & 0 & & \beta_{n-1} \\ \beta_n & 0 & 0 & \ldots & 0 \end{bmatrix} \end{cases}$$

The conditions $tr(x^j) = 0$ for $1 \leq i < n$ impose $n - 1$ linear conditions among the α_j, whence the space of trace preserving extensions $Ext^{tr}_A(V_\xi, V_\xi)$ corresponds to the quiver setting

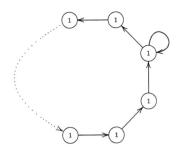

The Euler-form of this quiver $Q^\bullet$ is given by the $n \times n$ matrix

$$\begin{bmatrix} 0 & -1 & 0 & \dots & 0 \\ & 1 & -1 & & 0 \\ & & \ddots & \ddots & \\ & & & 1 & -1 \\ -1 & & & & 1 \end{bmatrix}$$

giving the numerical restriction as $\alpha_\xi = (1, \dots, 1)$

$$1 - \chi_Q(\alpha, \alpha) - \sum_i m_{ii} = 1 - (-1) - 0 = 2 = dim\ \mathbf{triss}_n\ A$$

so $\xi \in Sm_{tr}\ A$. Finally, the only remaining point is $\xi = (0,0)$. This has representation type $(n, 1)$ as the corresponding semisimple representation V_ξ is the trivial one. The stabilizer subgroup is GL_n and the (trace preserving) extensions are given by

$$Ext_A^1(V_\xi, V_\xi) = M_n \oplus M_n \quad \text{and} \quad Ext_A^{tr}(V_\xi, V_\xi) = M_n^0 \oplus M_n^0$$

determined by the algebra maps $A \xrightarrow{\ \phi\ } M_n(\mathbb{C}[\varepsilon])$ given by

$$\begin{cases} \phi(x) & = \varepsilon\ m_1 \\ \phi(y) & = \varepsilon\ m_2 \end{cases}$$

That is, the relevant quiver setting $(Q_\xi^\bullet, \alpha_\xi)$ is in this point

This time, $\xi \notin Sm_{tr}\ A$ as the numerical condition fails

$$1 - \chi_Q(\alpha, \alpha) - \sum_i m_{ii} = 1 - (-n^2) - 0 \neq 2 = dim\ \mathbf{triss}_n\ A$$

unless $n = 1$. That is, $Sm_{tr}\ A = \mathbb{C}^2 - \{(0,0)\}$.   ☐

5.3 Reduction steps

If we want to study the local structure of Cayley-Hamilton orders A of degree n over a central normal variety X of dimension d, we have to compile a list of admissible marked quiver settings, that is, couples $(Q^\bullet, \alpha)$ satisfying the two properties

$$\begin{cases} \alpha & \text{is the dimension vector of a simple representation of } Q^\bullet, \text{ and} \\ d & = 1 - \chi_Q(\alpha, \alpha) - \sum_i m_i \end{cases}$$

In this section, we will give two methods to start this classification project.

The first idea is to shrink a marked quiver-setting to its simplest form and classify these simplest forms for given d. By *shrinking* we mean the following process. Assume $\alpha = (e_1, \ldots, e_k)$ is the dimension vector of a simple representation of $Q^\bullet$ and let v_i and v_j be two vertices connected with an arrow such that $e_i = e_j = e$. That is, locally we have the following situation

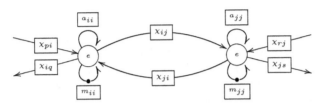

We will use one of the arrows connecting v_i with v_j to identify the two vertices. That is, we form the shrinked marked quiver-setting $(Q_s^\bullet, \alpha_s)$ where $Q_s^\bullet$ is the marked quiver on $k-1$ vertices $\{v_1, \ldots, \hat{v}_i, \ldots, v_k\}$ and α_s is the dimension vector with e_i removed. $Q_s^\bullet$ has the following form in a neighborhood of the contracted vertex

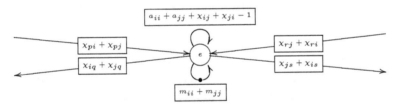

In $Q_s^\bullet$ we have for all $k, l \neq i, j$ that $\chi_{kl}^s = \chi_{kl}$, $a_{kk}^s = a_{kk}$, $m_{kk}^s = m_{kk}$ and the number of arrows and (marked) loops connected to v_j are determined as follows

- $\chi_{jk}^s = \chi_{ik} + \chi_{jk}$

- $\chi_{kj}^s = \chi_{ki} + \chi_{kj}$

- $a_{jj}^s = a_{ii} + a_{jj} + \chi_{ij} + \chi_{ji} - 1$

- $m_{jj}^s = m_{ii} + m_{jj}$

LEMMA 5.1
α is the dimension vector of a simple representation of $Q^\bullet$ if and only if α_s is the dimension vector of a simple representation of $Q_s^\bullet$. Moreover

$$\dim \operatorname{rep}_\alpha Q^\bullet / GL(\alpha) = \dim \operatorname{rep}_{\alpha_s} Q_s^\bullet / GL(\alpha_s)$$

PROOF Fix an arrow $\textcircled{j} \xleftarrow{\; a \;} \textcircled{i}$. As $e_i = e_j = e$ there is a Zariski open subset $U \hookrightarrow \operatorname{rep}_\alpha Q^\bullet$ of points V such that V_a is invertible. By base change in either v_i or v_j we can find a point W in its orbit such that $W_a = \mathbb{1}_e$. If we think of W_a as identifying $\mathbb{C}^{e_i}$ with $\mathbb{C}^{e_j}$ we can view the remaining maps of W as a representation in $\mathbf{rep}_{\alpha_s} Q_s^\bullet$ and denote it by W^s. The map $U \longrightarrow \mathbf{rep}_{\alpha_s} Q_s^\bullet$ is well-defined and maps $GL(\alpha)$-orbits to $GL(\alpha_s)$-orbits. Conversely, given a representation $W' \in \mathbf{rep}_{\alpha_s} Q_s^\bullet$ we can uniquely determine a representation $W \in U$ mapping to W'. Both claims follow immediately from this observation. ⬚

A marked quiver-setting can be uniquely shrinked to its *simplified form* , which has the characteristic property that no arrow-connected vertices can have the same dimension. The shrinking process has a converse operation, which we will call *splitting of a vertex* . However, this splitting operation is usually not uniquely determined.

Before compiling a lists of marked-quiver settings in simplified form for a specific base-dimension d, we bound the components of α.

PROPOSITION 5.8
Let $\alpha = (e_1, \ldots, e_k)$ be the dimension vector of a simple representation of Q and let $1 - \chi_Q(\alpha, \alpha) = d = \dim \mathbf{rep}_\alpha Q/GL(\alpha)$. Then, if $e = \max e_i$, we have that $d \geq e + 1$.

PROOF   By lemma 5.1 we may assume that (Q, α) is brought in its simplified form, that is, no two arrow-connected vertices have the same dimension. Let χ_{ii} denote the number of loops in a vertex v_i, then

$$-\chi_Q(\alpha, \alpha) = \begin{cases} \sum_i e_i \left(\sum_j \chi_{ij} e_j - e_i \right) \\ \sum_i e_i \left(\sum_j \chi_{ji} e_j - e_i \right) \end{cases}$$

and observe that the bracketed terms are positive by the requirement that α is the dimension vector of a simple representation. We call them the incoming in_i, respectively outgoing out_i, contribution of the vertex v_i to d. Let v_m be a vertex with maximal vertex-dimension e.

$$in_m = e\left(\sum_{j \neq m} \chi_{jm} e_j + (\chi_{ii} - 1)e \right) \quad \text{and} \quad out_m = e\left(\sum_{j \neq m} \chi_{ij} e_j + (\chi_{ii} - 1)e \right)$$

If there are loops in v_m, then $in_m \geq 2$ or $out_m \geq 2$ unless the local structure of Q is

in which case $in_m = e = out_m$. Let v_i be the unique incoming vertex of v_m, then we have $out_i \geq e - 1$. But then

$$d = 1 - \chi_Q(\alpha, \alpha) = 1 + \sum_j out_j \geq 2e$$

If v_m has no loops, consider the incoming vertices $\{v_{i_1}, \ldots, v_{i_s}\}$, then

$$in_m = e(\sum_{j=1}^{s} \chi_{i_j m} e_{i_j} - e)$$

which is $\geq e$ unless $\sum \chi_{i_j m} e_{i_j} = e$, but in that case we have

$$\sum_{j=1}^{s} out_{i_j} \geq e^2 - \sum_{j=1}^{s} e_{i_j}^2 \geq e$$

the last inequality because all $e_{i_j} < e$. In either case we have that $d = 1 - \chi_Q(\alpha, \alpha) = 1 + \sum_i out_i = 1 + \sum_i in_i \geq e + 1$. ▯

Example 5.3
In a list of simplified marked quivers we are only interested in $\mathbf{rep}_\alpha Q^\bullet$ as $GL(\alpha)$-module and we call two setting *equivalent* if they determine the same $GL(\alpha)$-module. For example, the marked quiver-settings

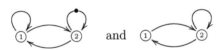

and

determine the same $\mathbb{C}^* \times GL_2$-module, hence are equivalent. ▯

THEOREM 5.6
Let A be a Cayley-Hamilton order of degree n over a central normal variety X of degree d. Then, the local quiver of A in a point $\xi \in X = \mathbf{triss}_n A$ belonging to the smooth locus $Sm_{tr} A$ can be shrunk to one of a finite list of equivalence classes of simplified marked quiver-settings. For $d \leq 4$, the complete list is given in figure 5.2 where the boxed value is the dimension d of X.

An immediate consequence is a noncommutative analog of the fact that commutative smooth varieties have only one type of analytic (or étale) local behavior.

THEOREM 5.7
There are only finitely many types of étale local behavior of smooth Cayley-Hamilton orders of degree n over a central variety of dimension d.

PROOF The foregoing reduction shows that for fixed d there are only a finite number of marked quiver-settings shrinked to their simplified form.

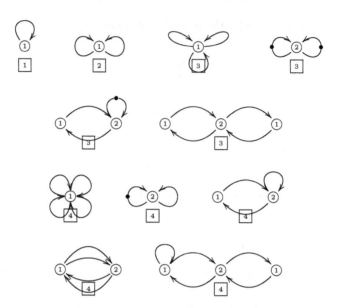

FIGURE 5.2:   The simplified local quivers for $d \leq 4$

As $\sum e_i \leq n$, we can apply the splitting operations on vertices only a finite number of times.   ▯

The second set of reduction steps is due to Raf Bocklandt who found them to prove his theorem, see section 5.7, which is crucial to study the smooth locus and the singularities of $\mathbf{triss}_n \, A$. In essence the reduction steps relate quiver settings that have invariants rings which are isomorphic (up to adding variables).

THEOREM 5.8

We have the following reductions:

1. **b1:** *Let* (Q, α) *be a quiver setting and* v *a vertex without loops such that*

$$\chi_Q(\alpha, \epsilon_v) \geq 0 \text{ or } \chi_Q(\epsilon_v, \alpha) \geq 0.$$

Define the quiver setting (Q', α') *by composing arrows through* v

$$\begin{bmatrix} u_1 & \cdots & u_k \\ & b_k & \\ b_1 & \alpha_v & b_k \\ a_1 & & a_l \\ i_1 & \cdots & i_l \end{bmatrix} \longrightarrow \begin{bmatrix} u_1 & \cdots & u_k \\ c_{11} & \times & c_{lk} \\ i_1 \, c_{1k} \cdots c_{l1} \, i_l \end{bmatrix}.$$

(some of the vertices may be the same). Then

$$\mathbb{C}[\mathbf{iss}_\alpha \, Q] \simeq \mathbb{C}[\mathbf{iss}_{\alpha'} \, Q']$$

2. **b2**: *Let (Q, α) be a quiver setting and v a vertex with k loops such that $\alpha_v = 1$. Let (Q', α) be the quiver setting where Q' is the quiver obtained by removing the loops in v, then*

$$\mathbb{C}[\text{iss}_\alpha \ Q] \simeq \mathbb{C}[\text{iss}_\alpha \ Q'] \otimes \mathbb{C}[X_1, \cdots, X_k]$$

3. **b3**: *Let (Q, α) be a quiver setting and v a vertex with one loop such that $\alpha_v = k \geq 2$ and*

$$\chi_Q(\alpha, \epsilon_v) = -1 \text{ or } \chi_Q(\epsilon_v, \alpha) = -1$$

Define the quiver setting (Q', α) by changing the quiver as below

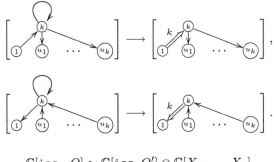

Then,

$$\mathbb{C}[\text{iss}_\alpha \ Q] \simeq \mathbb{C}[\text{iss}_\alpha Q'] \otimes \mathbb{C}[X_1, \ldots, X_k]$$

PROOF (1): $\text{rep}_\alpha \ Q$ can be decomposed as

$$\text{rep}_\alpha \ Q = \underbrace{\bigoplus_{a, \ s(a)=v} M_{\alpha_{t(a)} \times \alpha_{s(a)}}(\mathbb{C})}_{\text{arrows starting in } v} \oplus \underbrace{\bigoplus_{a, \ t(a)=v} M_{\alpha_{t(a)} \times \alpha_{s(a)}}(\mathbb{C})}_{\text{arrows terminating in } v} \oplus \text{rest}$$

$$= M_{\sum_{s(a)=v} \alpha_{t(a)} \times \alpha_v}(\mathbb{C}) \oplus M_{\alpha_v \times \sum_{t(a)=v} \alpha_{s(a)}}(\mathbb{C}) \oplus \text{rest}$$

$$= M_{\alpha_v - \chi(\alpha, \epsilon_v) \times \alpha_v}(\mathbb{C}) \oplus M_{\alpha_v \times \alpha_v - \chi(\epsilon_v, \alpha)}(\mathbb{C}) \oplus \text{rest}$$

$GL_{\alpha_v}(\mathbb{C})$ only acts on the first two terms and not on **rest**. Taking the quotient corresponding to $GL_{\alpha_v}(\mathbb{C})$ involves only the first two terms.

We recall the *first fundamental theorem* for GL_n-invariants , see, for example, [63, II.4.1]. The quotient variety

$$(M_{l \times n}(\mathbb{C}) \oplus M_{n \times m}(\mathbb{C}))/GL_n$$

where GL_n acts in the natural way, is for all $l, n, m \in \mathbb{N}$ isomorphic to the space of all $l \times m$ matrices of rank $\leq n$. The projection map is induced by multiplication

$$M_{l \times n}(\mathbb{C}) \oplus M_{n \times m}(\mathbb{C}) \xrightarrow{\ \pi\ } M_{l \times m}(\mathbb{C}) \qquad (A, B) \mapsto A.B$$

In particular, if $n \geq l$ and $n \geq m$ then π is surjective and the quotient variety is isomorphic to $M_{l \times m}(\mathbb{C})$.

By this fundamental theorem and the fact that either $\chi_Q(\alpha, \epsilon_v) \geq 0$ or $\chi_Q(\epsilon_v, \alpha) \geq 0$, the above quotient variety is isomorphic to

$$M_{\alpha_v - \chi(\alpha, \epsilon_v) \times \alpha_v - \chi(\epsilon_v, \alpha)}(\mathbb{C}) \oplus \mathtt{rest}$$

This space can be decomposed as

$$\bigoplus_{a,\, t(a)=vb,\, s(b)=v} M_{\alpha_{t(b)} \times \alpha_{s(a)}}(\mathbb{C}) \oplus \mathtt{rest} = \mathtt{rep}_{\alpha'} Q'$$

Taking quotients for $GL(\alpha')$ then proves the claim.

(2): Trivial as $GL(\alpha)$ acts trivially on the loop-representations in v.

(3): We only prove this for the first case. Call the loop in the first quiver ℓ and the incoming arrow a. Call the incoming arrows in the second quiver $c_i, i = 0, \ldots, k-1$.

There is a map

$$\pi : \mathtt{rep}_\alpha Q \to \mathtt{rep}_{\alpha'} Q' \times \mathbb{C}^k : V \mapsto (V', Tr(V_\ell), \ldots, Tr(V_{\ell^k})) \text{ with } V'_{c_i} := V_\ell^i V_a$$

Suppose $(V', x_1, \ldots, x_k) \in \mathtt{rep}_{\alpha'} Q' \times \mathbb{C}^k \in$ such that $(x_1, \ldots, x_k)$ correspond to the traces of powers of an invertible diagonal matrix D with k different eigenvalues $(\lambda_i, i = 1, \ldots, k)$ and the matrix A made of the columns $(V_{c_i}, i = 0, \ldots, k-1)$ is invertible. The image of the representation

$$V \in \mathtt{rep}_\alpha Q : V_a = V'_{c_0}, V_\ell = A \begin{pmatrix} \lambda_1^0 & \cdots & \lambda_1^{k-1} \\ \vdots & & \vdots \\ \lambda_k^0 & \cdots & \lambda_k^{k-1} \end{pmatrix}^{-1} D \begin{pmatrix} \lambda_1^0 & \cdots & \lambda_1^{k-1} \\ \vdots & & \vdots \\ \lambda_k^0 & \cdots & \lambda_k^{k-1} \end{pmatrix} A^{-1}$$

under π is $(V', x_1, \ldots, x_k)$ because

$$V_\ell^i V_a = A \begin{pmatrix} \lambda_1^0 & \cdots & \lambda_1^{k-1} \\ \vdots & & \vdots \\ \lambda_k^0 & \cdots & \lambda_k^{k-1} \end{pmatrix}^{-1} D^i \begin{pmatrix} \lambda_1^0 & \cdots & \lambda_1^{k-1} \\ \vdots & & \vdots \\ \lambda_k^0 & \cdots & \lambda_k^{k-1} \end{pmatrix} A^{-1} V'_{c_0}$$

$$= A \begin{pmatrix} \lambda_1^0 & \cdots & \lambda_1^{k-1} \\ \vdots & & \vdots \\ \lambda_k^0 & \cdots & \lambda_k^{k-1} \end{pmatrix}^{-1} \begin{pmatrix} \lambda_1^i \\ \vdots \\ \lambda_k^i \end{pmatrix}$$

$$= V_{c_i}$$

and the traces of V_ℓ are the same as those of D. The conditions on $(V', x_1, \ldots, x_k)$, imply that the image of π, U, is dense, and hence π is a dominant map.

There is a bijection between the generators of $\mathbb{C}[\mathtt{iss}_\alpha Q]$ and $\mathbb{C}[\mathtt{iss}_{\alpha'} Q'] \otimes \mathbb{C}[X_1, \ldots, X_k]$ by identifying

$$f_{\ell^i} \mapsto X_i, i = 1, \ldots, k \ , f_{\ldots a \ell^i \ldots} \mapsto f_{\ldots c_i \ldots}, i = 0, \ldots, k-1$$

Notice that higher orders of ℓ don't occur by the Caley Hamilton identity on V_ℓ. If n is the number of generators of $\mathbb{C}[\mathbf{iss}_\alpha Q]$, we have two maps

$$\phi : \mathbb{C}[Y_1, \cdots Y_n] \to \mathbb{C}[\mathbf{iss}_\alpha Q] \subset \mathbb{C}[\mathbf{rep}_\alpha Q]$$

$$\phi' : \mathbb{C}[Y_1, \cdots Y_n] \to \mathbb{C}[\mathbf{iss}_{\alpha'} Q'] \otimes \mathbb{C}[X_1, \ldots, X_k] \subset \mathbb{C}[\mathbf{rep}_{\alpha'} Q' \times \mathbb{C}^k]$$

Note that $\phi'(f) \circ \pi \equiv \phi(f)$ and $\phi(f) \circ \pi^{-1}|_U \equiv \phi'(f)|_U$. So if $\phi(f) = 0$ then also $\phi'(f)|_U = 0$. Because U is zariski-open and dense in $\mathbf{rep}_{\alpha'} Q' \times \mathbb{C}^2$, $\phi'(f) \equiv 0$. A similar argument holds for the inverse implication whence $Ker(\phi) = Ker(\phi')$. $\qquad\qquad\square$

We have to work with *marked* quiver settings and therefore we need slightly more general reduction steps. The proofs of the claims below follow immediately from the above theorem by separating traces.

With ϵ_v we denote the base vector concentrated in vertex v and α_v will denote the vertex dimension component of α in vertex v. There are three types of reduction moves, each with their own condition and effect on the ring of invariants.

Vertex removal (b1): Let $(Q^\bullet, \alpha)$ be a marked quiver setting and v a vertex satisfying the condition C_V^v, that is, v is without (marked) loops and satisfies

$$\chi_Q(\alpha, \epsilon_v) \geq 0 \quad \text{or} \quad \chi_Q(\epsilon_v, \alpha) \geq 0$$

Define the new quiver setting $(Q^{\bullet'}, \alpha')$ obtained by the operation R_V^v that removes the vertex v and composes all arrows through v, the dimensions of the other vertices are unchanged

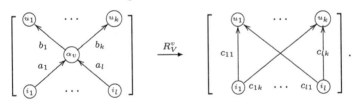

where $c_{ij} = a_i b_j$ (observe that some of the incoming and outgoing vertices may be the same so that one obtains loops in the corresponding vertex). In this case we have

$$\mathbb{C}[\mathbf{rep}_\alpha \ Q^\bullet]^{GL(\alpha)} \simeq \mathbb{C}[\mathbf{rep}_{\alpha'} \ Q^{\bullet'}]^{GL(\alpha')}$$

loop removal (b2): Let $(Q^\bullet, \alpha)$ be a marked quiver setting and v a vertex satisfying the condition C_l^v that the vertex-dimension $\alpha_v = 1$ and there are $k \geq 1$ loops in v. Let $(Q^{\bullet'}, \alpha)$ be the quiver setting obtained by the loop removal operation R_l^v

$$\left[\ \overset{k}{\underset{1}{\circlearrowright}} \ \right] \xrightarrow{\ R_l^v \ } \left[\ \overset{k-1}{\underset{1}{\circlearrowright}} \ \right]$$

removing one loop in v and keeping the dimension vector the same, then

$$\mathbb{C}[\mathrm{rep}_\alpha\ Q^\bullet]^{GL(\alpha)} \simeq \mathbb{C}[\mathrm{rep}_\alpha\ Q^{\bullet'}]^{GL(\alpha)}[x]$$

Loop removal (b3):   Let $(Q^\bullet, \alpha)$ be a marked quiver setting and v a vertex satisfying condition C_L^v, that is, the vertex dimension $\alpha_v \geq 2$, v has precisely one (marked) loop in v and

$$\chi_Q(\epsilon_v, \alpha) = -1 \quad \text{or} \quad \chi_Q(\alpha, \epsilon_v) = -1$$

(that is, there is exactly one other incoming or outgoing arrow from/to a vertex with dimension 1). Let $(Q^{\bullet'}, \alpha)$ be the marked quiver setting obtained by changing the quiver as indicated below (depending on whether the incoming or outgoing condition is satisfied and whether there is a loop or a marked loop in v)

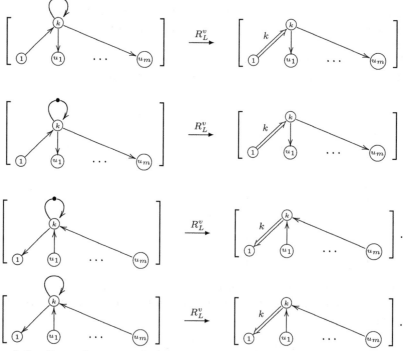

and the dimension vector is left unchanged, then we have

$$\mathbb{C}[\mathrm{rep}_\alpha\ Q^\bullet]^{GL(\alpha)} = \begin{cases} \mathbb{C}[\mathrm{rep}_\alpha\ Q^{\bullet'}]^{GL(\alpha)}[x_1, \ldots, x_k] & \text{(loop)} \\ \mathbb{C}[\mathrm{rep}_\alpha\ Q^{\bullet'}]^{GL(\alpha)}[x_1, \ldots, x_{k-1}] & \text{(marked loop)} \end{cases}$$

DEFINITION 5.2   *A marked quiver $Q^\bullet$ is said to be strongly connected if for every pair of vertices $\{v, w\}$ there is an oriented path from v to w and an oriented path from w to v.*

A marked quiver setting $(Q^\bullet, \alpha)$ *is said to be* reduced *if and only if there is no vertex* v *such that one of the conditions* C_V^v, C_l^v *or* C_L^v *is satisfied.*

LEMMA 5.2

Every marked quiver setting $(Q_1^\bullet, \alpha_1)$ *can be reduced by a sequence of operations* R_V^v, R_l^v *and* R_L^v *to a reduced quiver setting* $(Q_2^\bullet, \alpha_2)$ *such that*

$$\mathbb{C}[\mathbf{rep}_{\alpha_1} Q_1^\bullet]^{GL(\alpha_1)} \simeq \mathbb{C}[\mathbf{rep}_{\alpha_2} Q_2^\bullet]^{GL(\alpha_2)}[x_1, \ldots, x_z]$$

Moreover, the number z *of extra variables is determined by the reduction sequence*

$$(Q_2^\bullet, \alpha_2) = R_{X_u}^{v_{i_u}} \circ \ldots \circ R_{X_1}^{v_{i_1}}(Q_1^\bullet, \alpha_1)$$

where for every $1 \le j \le u$, $X_j \in \{V, l, L\}$. *More precisely*

$$z = \sum_{X_j = l} 1 + \overbrace{\sum_{X_j = L} \alpha_{v_{i_j}}}^{(unmarked)} + \overbrace{\sum_{X_j = L} (\alpha_{v_{i_j}} - 1)}^{(marked)}$$

PROOF   As any reduction step removes a (marked) loop or a vertex, any sequence of reduction steps starting with $(Q_1^\bullet, \alpha_1)$ must eventually end in a reduced marked quiver setting. The statement then follows from the discussion above.   □

As the reduction steps have no uniquely determined inverse, there is no a priori reason why the reduced quiver setting of the previous lemma should be unique. Nevertheless this is true.

We will say that a vertex v is *reducible* if one of the conditions C_V^v (vertex removal), C_l^v (loop removal in vertex dimension one) or C_L^v (one (marked) loop removal) is satisfied. If we let the specific condition unspecified we will say that v satisfies C_X^v and denote R_X^v for the corresponding marked quiver setting reduction. The resulting marked quiver setting will be denoted by

$$R_X^v(Q^\bullet, \alpha)$$

If $w \ne v$ is another vertex in $Q^\bullet$ we will denote the corresponding vertex in $R_X^v(Q^\bullet)$ also with w. The proof of the uniqueness result relies on three claims

1. If $w \ne v$ satisfies R_Y^w in $(Q^\bullet, \alpha)$, then w virtually always satisfies R_Y^w in $R_X^v(Q^\bullet, \alpha)$.

2. If v satisfies R_X^v and w satisfies R_Y^w, then $R_X^v(R_Y^w(Q^\bullet, \alpha)) = R_Y^w(R_X^v(Q^\bullet, \alpha))$.

3. The previous two facts can be used to prove the result by induction on the minimal length of the reduction chain.

By the *neighborhood* of a vertex v in $Q^\bullet$ we mean the (marked) subquiver on the vertices connected to v. A neighborhood of a set of vertices is the union of the vertex-neighborhoods. *Incoming* resp. *outgoing* neighborhoods are defined in the natural manner.

LEMMA 5.3

Let $v \neq w$ be vertices in $(Q^\bullet, \alpha)$.

1. *If v satisfies C_V^v in $(Q^\bullet, \alpha)$ and w satisfies C_X^w, then v satisfies C_V^w in $R_X^w(Q^\bullet, \alpha)$ unless the neighborhood of $\{v, w\}$ looks like*

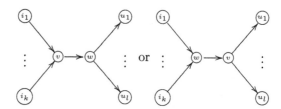

and $\alpha_v = \alpha_w$. Observe that in this case $R_V^v(Q^\bullet, \alpha) = R_V^w(Q^\bullet, \alpha)$

2. *If v satisfies C_l^v and w satisfies C_X^w then then v satisfies C_l^v in $R_X^w(Q^\bullet, \alpha)$*

3. *If v satisfies C_V^v and w satisfies C_X^w then then v satisfies C_V^v in $R_X^w(Q^\bullet, \alpha)$*

PROOF   (1): If $X = l$ then R_X^w does not change the neighborhood of v so C_V^v holds in $R_l^w(Q^\bullet, \alpha)$. If $X = L$ then R_X^w does not change the neighborhood of v unless $\alpha_v = 1$ and $\chi_Q(\epsilon_w, \epsilon_v) = -1$ (resp. $\chi_Q(\epsilon_v, \epsilon_w) = -1$) depending on whether w satisfies the in- or outgoing condition C_L^w. We only consider the first case, the latter is similar. Then v cannot satisfy the outgoing form of C_V^v in $(Q^\bullet, \alpha)$ so the incoming condition is satisfied. Because the R_L^w-move does not change the incoming neighborhood of v, C_V^v still holds for v in $R_L^w(Q^\bullet, \alpha)$.

If $X = V$ and v and w have disjoint neighborhoods then C_V^v trivially remains true in $R_V^w(Q^\bullet, \alpha)$. Hence assume that there is at least one arrow from v to w (the case where there are only arrows from w to v is similar). If $\alpha_v < \alpha_w$ then the incoming condition C_V^v must hold (outgoing is impossible) and hence w does not appear in the incoming neighborhood of v. But then R_V^w preserves the incoming neighborhood of v and C_V^v remains true in the reduction. If $\alpha_v > \alpha_w$ then the outgoing condition C_V^w must hold and hence w does not appear in the incoming neighborhood of v. So if the incoming condition C_V^v holds in $(Q^\bullet, \alpha)$ it will still hold after the application of R_V^w. If the outgoing condition C_V^v holds, the neighborhoods of v and w in $(Q^\bullet, \alpha)$ and v in $R_V^w(Q^\bullet, \alpha)$ are depicted in figure 5.3. Let A be the set of arrows in $Q^\bullet$ and A' the set of arrows in the reduction, then because $\sum_{a \in A, s(a)=w} \alpha_{t(a)} \leq \alpha_w$

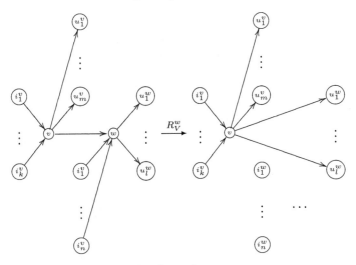

FIGURE 5.3: Neighborhoods of v and w.

(the incoming condition for w) we have

$$
\sum_{a \in A', s(a)=v} \alpha'_{t(a)} = \sum_{\substack{a \in A, \\ s(a)=v, t(a) \neq w}} \alpha_{t(a)} + \sum_{\substack{a \in A \\ t(a)=w, s(a)=v}} \sum_{a \in A, s(a)=w} \alpha_{t(a)}
$$

$$
\leq \sum_{\substack{a \in A, \\ s(a)=v, t(a) \neq w}} \alpha_{t(a)} + \sum_{\substack{a \in A \\ t(a)=w, s(a)=w}} \alpha_w
$$

$$
= \sum_{a \in A, s(a)=v} \alpha_{t(a)} \leq \alpha_v
$$

and therefore the outgoing condition C_V^v also holds in $R_V^w(Q^\bullet, \alpha)$. Finally if $\alpha_v = \alpha_w$, it may be that C_V^v does not hold in $R_V^w(Q^\bullet, \alpha)$. In this case $\chi(\epsilon_v, \alpha) < 0$ and $\chi(\alpha, \epsilon_w) < 0$ (C_V^v is false in $R_V^w(Q^\bullet, \alpha)$). Also $\chi(\alpha, \epsilon_v) \geq 0$ and $\chi(\epsilon_w, \alpha) \geq 0$ (otherwise C_V does not hold for v or w in $(Q^\bullet, \alpha)$). This implies that we are in the situation described in the lemma and the conclusion follows.

(2) : None of the R_X^w-moves removes a loop in v nor changes $\alpha_v = 1$.

(3) : Assume that the incoming condition C_L^v holds in $(Q^\bullet, \alpha)$ but not in $R_V^w(Q^\bullet, \alpha)$, then w must be the unique vertex, which has an arrow to v and $X = V$. Because $\alpha_w = 1 < \alpha_v$, the incoming condition C_L^w holds. This means that there is also only one arrow arriving in w and this arrow is coming from a vertex with dimension 1. Therefore after applying R_V^w, v will still have only one incoming arrow starting in a vertex with dimension 1. A similar argument holds for the outgoing condition C_L^v. □

LEMMA 5.4

Suppose that $v \neq w$ are vertices in $(Q^\bullet, \alpha)$ and that C_X^v and C_Y^w are satisfied. If C_X^v holds in $R_Y^w(Q^\bullet, \alpha)$ and C_Y^w holds in $R_X^v(Q^\bullet, \alpha)$ then

$$R_X^v R_Y^w(Q^\bullet, \alpha) = R_Y^w R_X^v(Q^\bullet, \alpha)$$

PROOF If $X, Y \in \{l, L\}$ this is obvious, so let us assume that $X = V$. If $Y = V$ as well, we can calculate the Euler form $\chi_{R_V^w R_V^v Q}(\epsilon_x, \epsilon_y)$. Because

$$\chi_{R_V^v Q}(\epsilon_x, \epsilon_y) = \chi_Q(\epsilon_x, \epsilon_y) - \chi_Q(\epsilon_x, \epsilon_v)\chi_Q(\epsilon_v, \epsilon_y)$$

it follows that

$$
\begin{aligned}
\chi_{R_V^w R_V^v Q}(\epsilon_x, \epsilon_y) &= \chi_{R_V^v Q}(\epsilon_x, \epsilon_y) - \chi_{R_V^v Q}(\epsilon_x, \epsilon_w)\chi_{R_V^v Q}(\epsilon_v, \epsilon_y) \\
&= \chi_Q(\epsilon_x, \epsilon_y) - \chi_Q(\epsilon_x, \epsilon_v)\chi_Q(\epsilon_v, \epsilon_y) \\
&\quad - (\chi_Q(\epsilon_x, \epsilon_w) - \chi_Q(\epsilon_x, \epsilon_v)\chi_Q(\epsilon_v, \epsilon_w))(\chi_Q(\epsilon_w, \epsilon_y) - \chi_Q(\epsilon_w, \epsilon_v)\chi_Q(\epsilon_v, \epsilon_y)) \\
&= \chi_Q(\epsilon_x, \epsilon_y) - \chi_Q(\epsilon_x, \epsilon_v)\chi_Q(\epsilon_v, \epsilon_y) - \chi_Q(\epsilon_x, \epsilon_w)\chi_Q(\epsilon_w, \epsilon_y) \\
&\quad - \chi_Q(\epsilon_x, \epsilon_v)\chi_Q(\epsilon_v, \epsilon_w)\chi_Q(\epsilon_w, \epsilon_v)\chi_Q(\epsilon_v, \epsilon_y) \\
&\quad + \chi_Q(\epsilon_x, \epsilon_w)\chi_Q(\epsilon_w, \epsilon_v)\chi_Q(\epsilon_v, \epsilon_y) + \chi_Q(\epsilon_x, \epsilon_v)\chi_Q(\epsilon_v, \epsilon_w)\chi_Q(\epsilon_w, \epsilon_y)
\end{aligned}
$$

This is symmetric in v and w and therefore the ordering of R_V^v and R_V^w is irrelevant.
If $Y = l$ we have the following equalities

$$
\begin{aligned}
\chi_{R_l^w R_V^v Q}(\epsilon_x, \epsilon_y) &= \chi_{R_V^v Q}(\epsilon_x, \epsilon_y) - \delta_{wx}\delta_{wy} \\
&= \chi_Q(\epsilon_x, \epsilon_y) - \chi_Q(\epsilon_x, \epsilon_v)\chi_Q(\epsilon_v, \epsilon_y) - \delta_{wx}\delta_{wy} \\
&= \chi_Q(\epsilon_x, \epsilon_y) - \delta_{wx}\delta_{wy} - (\chi_Q(\epsilon_x, \epsilon_v) - \delta_{wx}\delta_{wv})(\chi_Q(\epsilon_v, \epsilon_y) - \delta_{wv}\delta_{wy}) \\
&= \chi_{R_l^w Q}(\epsilon_x, \epsilon_y) - \chi_{R_l^w Q}(\epsilon_x, \epsilon_v)\chi_{R_l^w Q}(\epsilon_v, \epsilon_y) \\
&= \chi_{R_V^v R_l^w Q}.
\end{aligned}
$$

If $Y = L$, an R_L^w-move commutes with the R_V^v move because it does not change the neighborhood of v except when v is the unique vertex of dimension 1 connected to w. In this case the neighborhood of v looks like

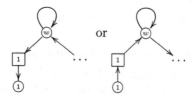

In this case the reduction at v is equivalent to a reduction at v' (i.e. the lower vertex) which certainly commutes with R_L^w. □

We are now in a position to prove the claimed uniqueness result.

THEOREM 5.9

If $(Q^\bullet, \alpha)$ is a strongly connected marked quiver setting and $(Q_1^\bullet, \alpha_1)$ and $(Q_2^\bullet, \alpha_2)$ are two reduced marked quiver setting obtained by applying reduction moves to $(Q^\bullet, \alpha)$ then

$$(Q_1^\bullet, \alpha_1) = (Q_2^\bullet, \alpha_2)$$

PROOF   We do induction on the length l_1 of the reduction chain R_1 reducing $(Q^\bullet, \alpha)$ to $(Q_1^\bullet, \alpha_1)$. If $l_1 = 0$, then $(Q^\bullet, \alpha)$ has no reducible vertices so the result holds trivially. Assume the result holds for all lengths $< l_1$. There are two cases to consider.

There exists a vertex v satisfying a loop removal condition $C_X^v, X = l$ or L. Then, there is a R_X^v-move in both reduction chains R_1 and R_2. This follows from lemma 5.3 and the fact that none of the vertices in $(Q_1^\bullet, \alpha_1)$ and $(Q_2^\bullet, \alpha_2)$ are reducible. By the commutation relations from lemma 5.4, we can bring this reduction to the first position in both chains and use induction.

If there is a vertex v satisfying condition C_V^v, either both chains will contain an R_V^v-move or the neighborhood of v looks like the figure in lemma 5.3 (1). Then, R_1 can contain an R_V^v-move and R_2 an R_V^w-move. But then we change the R_V^w move into a R_V^v move, because they have the same effect. The concluding argument is similar to that above.   ⬜

5.4   Curves and surfaces

W. Schelter has proved in [91] that in dimension one, Cayley-smooth orders are hereditary. We give an alternative proof of this result using the étale local classification. The next result follows also by splitting the dimension 1 case in figure 5.2. We give a direct proof illustrating the type-stratification result of section 5.1.

THEOREM 5.10

Let A be a Cayley-Hamilton order of degree n over an affine curve $X = \mathrm{triss}_n A$. If $\xi \in Sm_{tr} A$, then the étale local structure of A in ξ is determined by a marked quiver-setting, which is an oriented cycle on k vertices with $k \le n$ and an unordered partition $p = (d_1, \ldots, d_k)$ having precisely k parts such that $\sum_i d_i = n$ determining the dimensions of the simple components of V_ξ, see figure 5.4.

PROOF   Let $(Q^\bullet, \alpha)$ be the corresponding local marked quiver-setting. Because $Q^\bullet$ is strongly connected, there exist oriented cycles in $Q^\bullet$. Fix one

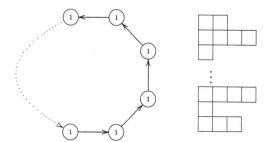

FIGURE 5.4: Cayley-smooth curve types.

such cycle of length $s \leq k$ and renumber the vertices of $Q^{\bullet}$ such that the first s vertices make up the cycle. If $\alpha = (e_1, \ldots, e_k)$, then there exist semisimple representations in $\mathbf{rep}_{\alpha} Q^{\bullet}$ with composition

$$\alpha_1 = \underbrace{(1, \ldots, 1,}_{s} \underbrace{0, \ldots, 0)}_{k-s} \oplus \epsilon_1^{\oplus e_1 - 1} \oplus \ldots \oplus \epsilon_s^{\oplus e_s - 1} \oplus \epsilon_{s+1}^{\oplus e_{s+1}} \oplus \ldots \oplus \epsilon_k^{\oplus e_k}$$

where ϵ_i stands for the simple one-dimensional representation concentrated in vertex v_i. There is a one-dimensional family of simple representations of dimension vector α_1, hence the stratum of semisimple representations in $\mathbf{iss}_{\alpha} Q^{\bullet}$ of representation type $\tau = (1, \alpha_1; e_1 - 1, \epsilon_1; \ldots; e_s - 1, \epsilon_s; e_{s+1}, \epsilon_{s+1}; e_k, \epsilon_k)$ is at least one-dimensional. However, as $dim \ \mathbf{iss}_{\alpha} Q^{\bullet} = 1$ this can only happen if this semisimple representation is actually simple. That is, when $\alpha = \alpha_1$ and $k = s$.   ▯

If V_{ξ} is the semisimple n-dimensional representation of A corresponding to ξ, then

$$V_{\xi} = S_1 \oplus \ldots \oplus S_k \quad \text{with} \quad dim \ S_i = d_i$$

and the stabilizer subgroup is $GL(\alpha) = \mathbb{C}^* \times \ldots \times \mathbb{C}^*$ embedded in GL_n via the diagonal embedding

$$(\lambda_1, \ldots, \lambda_k) \longrightarrow \text{diag}(\underbrace{\lambda_1, \ldots, \lambda_1}_{d_1}, \ldots, \underbrace{\lambda_k, \ldots, \lambda_k}_{d_k})$$

Further, using basechange in $\mathbf{rep}_{\alpha} Q^{\bullet}$ we can bring every simple α-dimensional representation of Q^{α} in standard form

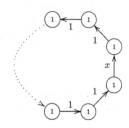

where $x \in \mathbb{C}^*$ is the arrow from v_k to v_1. That is, $\mathbb{C}[\mathbf{rep}_\alpha \ Q^\bullet]^{GL(\alpha)} \simeq \mathbb{C}[x]$ proving that the quotient (or central) variety X must be smooth in ξ by the slice result. Moreover, as $\widehat{A_\xi} \simeq \widehat{\mathbb{T}_\alpha}$ we have, using the numbering conventions of the vertices), the following block decomposition

$$
\widehat{A_\xi} \simeq
\begin{bmatrix}
M_{d_1}(\mathbb{C}[[x]]) & M_{d_1 \times d_2}(\mathbb{C}[[x]]) & \cdots & M_{d_1 \times d_k}(\mathbb{C}[[x]]) \\
M_{d_2 \times d_1}(x\mathbb{C}[[x]]) & M_{d_2}(\mathbb{C}[[x]]) & \cdots & M_{d_2 \times d_k}(\mathbb{C}[[x]]) \\
\vdots & \vdots & \ddots & \vdots \\
M_{d_k \times d_1}(x\mathbb{C}[[x]]) & M_{d_k \times d_2}(x\mathbb{C}[[x]]) & \cdots & M_{d_k}(\mathbb{C}[[x]])
\end{bmatrix}
$$

From the local description of hereditary orders given in [90, Thm. 39.14] we deduce that A_ξ is a hereditary order. That is, we have the following characterization of the smooth locus.

PROPOSITION 5.9
Let A be a Cayley-Hamilton order of degree n over a central affine curve X. Then, $Sm_{tr} \ A$ is the locus of points $\xi \in X$ such that A_ξ is an hereditary order (in particular, ξ must be a smooth point of X).

THEOREM 5.11
Let $\mathcal{A}$ be a Cayley-Hamilton central $\mathcal{O}_X$-order of degree n where X is a projective curve. Equivalent are

1. *$\mathcal{A}$ is a sheaf of Cayley-smooth orders*

2. *X is smooth and $\mathcal{A}$ is a sheaf of hereditary $\mathcal{O}_X$-orders*

We now turn to orders over surfaces. The next result can equally be proved using splitting and the classification of figure 5.2.

THEOREM 5.12
Let A be a Cayley-Hamilton order of degree n over an affine surface $X = \text{triss}_n A$. If $\xi \in Sm_{tr} A$, then the étale local structure of A in ξ is determined by a marked local quiver-setting A_{klm} on $k + l + m \leq n$ vertices and an unordered partition $p = (d_1, \ldots, d_{k+l+m})$ of n with $k + l + m$ nonzero parts determined by the dimensions of the simple components of V_ξ as in figure 5.5.

PROOF Let $(Q^\bullet, \alpha)$ be the marked quiver-setting on r vertices with $\alpha = (e_1, \ldots, e_r)$ corresponding to ξ. As $Q^\bullet$ is strongly connected and the

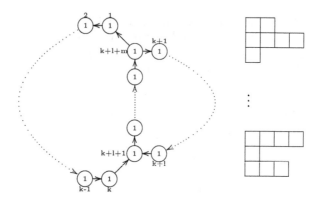

FIGURE 5.5:   Cayley-smooth surface types.

quotient variety is two-dimensional, $Q^\bullet$ must contain more than one oriented cycle, hence it contains a subquiver of type A_{klm}, possibly degenerated with k or l equal to zero. Order the first $k+l+m$ vertices of $Q^\bullet$ as indicated. One verifies that A_{klm} has simple representations of dimension vector $(1,\dots,1)$. Assume that A_{klm} is a proper subquiver and denote $s = k + l + m + 1$ then $Q^\bullet$ has semisimple representations in $\mathbf{rep}_\alpha\, Q^\bullet$ with dimension-vector decomposition

$$\alpha_1 = \underbrace{(1,\dots,1,0,\dots,0)}_{k+l+m} \oplus \epsilon_1^{\oplus e_1 - 1} \oplus \dots \oplus \epsilon_{k+l+m}^{\oplus e_{k+l+m}-1} \oplus \epsilon_s^{\oplus e_s} \oplus \dots \oplus \epsilon_r^{\oplus e_r}$$

Applying the formula for the dimension of the quotient variety shows that $\mathbf{iss}_{(1,\dots,1)}\, A_{klm}$ has dimension 2 so there is a two-dimensional family of such semisimple representation in the two-dimensional quotient variety $\mathbf{iss}_\alpha\, Q^\bullet$. This is only possible if this semisimple representation is actually simple, whence $r = k + l + m$, $Q^\bullet = A_{klm}$ and $\alpha = (1,\dots,1)$. □

If V_ξ is the semisimple n-dimensional representation of A corresponding to ξ, then

$$V_\xi = S_1 \oplus \dots \oplus S_r \quad \text{with} \quad \dim\, S_i = d_i$$

and the stabilizer subgroup $GL(\alpha) = \mathbb{C}^* \times \dots \times \mathbb{C}^*$ embedded diagonally in GL_n

$$(\lambda_1,\dots,\lambda_r) \mapsto \mathrm{diag}(\underbrace{\lambda_1,\dots,\lambda_1}_{d_1},\dots,\underbrace{\lambda_r,\dots,\lambda_r}_{d_r})$$

By base change in $\mathbf{rep}_\alpha\, A_{klm}$ we can bring every simple α-dimensional representation in the following standard form

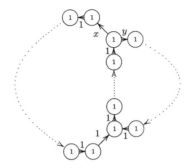

with $x, y \in \mathbb{C}^*$ and as $\mathbb{C}[\mathbf{iss}_\alpha \, A_{klm}] = \mathbb{C}[\mathbf{rep}_\alpha \, A_{klm}]^{GL(\alpha)}$ is the ring generated by traces along oriented cycles in A_{klm}, it is isomorphic to $\mathbb{C}[x, y]$. From the slice result one deduces that ξ must be a smooth point of X and because $\widehat{A_\xi} \simeq \widehat{\mathbb{T}_\alpha}$ we deduce it must have the following block-decomposition

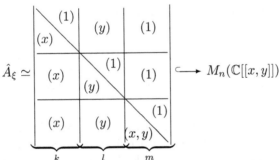

where at spot (i, j) with $1 \leq i, j \leq k + l + m$, there is a block of dimension $d_i \times d_j$ with entries the indicated ideal of $\mathbb{C}[[x, y]]$.

DEFINITION 5.3   *Let A be a Cayley-Hamilton central $\mathbb{C}[X]$-order of degree n in a central simple $\mathbb{C}(X)$- algebra Δ of dimension n^2.*

1. *A is said to be étale locally split in ξ if and only if $\widehat{A_\xi}$ is a central $\hat{\mathcal{O}}_{X,x}$-order in $M_n(\hat{\mathcal{O}}_{X,x} \otimes_{\mathcal{O}_{X,x}} \mathbb{C}(X))$.*

2. *The ramification locus ram_A of A is the locus of points $\xi \in X$ such that*

$$\frac{A}{\mathfrak{m}_\xi A \mathfrak{m}_\xi} \not\simeq M_n(\mathbb{C})$$

The complement $X - ram_A$ is the Azumaya locus X_{az} of A.

THEOREM 5.13
Let $\mathcal{A}$ be a Cayley-smooth central $\mathcal{O}_X$-order of degree n over a projective surface X. Then

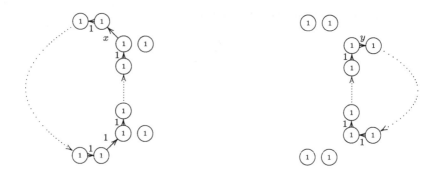

FIGURE 5.6:   Proper semisimples of A_{klm}.

1. *X is smooth.*

2. *A is étale locally split in all points of X.*

3. *The ramification divisor $ram_A \hookrightarrow X$ is either empty or consists of a finite number of isolated (possibly embedded) points and a reduced divisor having as its worst singularities normal crossings.*

PROOF   (1) and (2) follow from the above local description of A. As for (3) we have to compute the local quiver-settings in proper semisimple representations of $\mathbf{rep}_\alpha A_{klm}$. As simples have a strongly connected support, the decomposition types of these proper semisimples are depicted in figure 5.6. with $x, y \in \mathbb{C}^*$. By the description of local quivers given in section 3 we see that they are respectively of the forms in figure 5.7. The associated unordered partitions are defined in the obvious way, that is, to the looped vertex one assigns the sum of the d_i belonging to the loop-contracted circuit and the other components of the partition are preserved. Using the étale local isomorphism between X in a neighborhood of ξ and of $\mathbf{iss}_\alpha A_{klm}$ in a neighborhood of the trivial representation, we see that the local picture of quiver-settings of A in a neighborhood of ξ is described in figure 5.8. The Azumaya points are the points in which the quiver-setting is A_{001} (the two-loop quiver). From this local description the result follows if we take care of possibly degenerated cases.   ☐

An isolated point in ξ can occur if the quiver-setting in ξ is of type A_{00m} with $m \geq 2$. In the case of curves and surfaces, the central variety X of a Cayley-smooth model A had to be smooth and that A is étale locally split in every point $\xi \in X$. Both of these properties are no longer valid in higher dimensions.

A_{0l1} A_{k01}

FIGURE 5.7: Local quivers for A_{klm}.

LEMMA 5.5
For dimension $d \geq 3$, the center Z of a Cayley-smooth order of degree n can have singularities.

PROOF   Consider the marked quiver-setting of figure 5.9 that is allowed for dimension $d = 3$ and degree $n = 2$. The quiver-invariants are generated by the traces along oriented cycles, that is, by ac, ad, bc and bd. The coordinate ring is

$$\mathbb{C}[\text{iss}_\alpha \, Q] \simeq \frac{\mathbb{C}[x, y, z, v]}{(xv - yz)}$$

having a singularity in the origin. This example can be extended to dimensions $d \geq 3$ by adding loops in one of the vertices

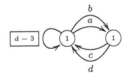

⬚

LEMMA 5.6
For dimension $d \geq 3$, a Cayley-smooth algebra does not have to be locally étale split in every point of its central variety.

PROOF   Consider the following allowable quiver-setting for $d = 3$ and $n = 2$

☐

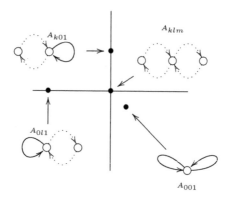

FIGURE 5.8:   Local picture for A_{klm}.

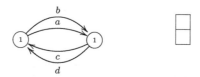

FIGURE 5.9:   Central singularities can arise.

The corresponding Cayley-smooth algebra A is generated by two generic 2×2 trace zero matrices, say A and B. From the description of the trace algebra $\mathbb{T}_2^2$ we see that its center is generated by $A^2 = x$, $B^2 = z$ and $AB + BA = z$. Alternatively, we can identify A with the Clifford-algebra over $\mathbb{C}[x, y, z]$ of the nondegenerate quadratic form

$$\begin{bmatrix} x & y \\ y & z \end{bmatrix}$$

This is a noncommutative domain and remains so over the formal power series $\mathbb{C}[[x, y, z]]$. That is, A cannot be split by an étale extension in the origin. More generally, whenever the local marked quiver contains vertices with dimension ≥ 2, the corresponding Cayley-smooth algebra cannot be split by an étale extension as the local quiver-setting does not change and for a split algebra all vertex-dimensions have to be equal to 1. In particular, the Cayley-smooth algebra of degree 2 corresponding to the quiver-setting

cannot be split by an étale extension in the origin. Its corresponding dimension is

$$d = 3k + 4l - 3$$

whenever $k + l \geq 2$ and all dimensions $d \geq 3$ are obtained.   ▯

Let X be a projective surface. We will characterize the central simple $\mathbb{C}(X)$-algebras Δ allowing a *Cayley-smooth model*. We first need to perform a local calculation. Consider the ring of algebraic functions in two variables $\mathbb{C}\{x,y\}$ and let $X_{loc} = Spec\ \mathbb{C}\{x,y\}$. There is only one codimension two subvariety: $m = (x,y)$. Let us compute the coniveau spectral sequence for X_{loc}. If K is the field of fractions of $\mathbb{C}\{x,y\}$ and if we denote with k_p the field of fractions of $\mathbb{C}\{x,y\}/p$ where p is a height one prime, we have as its first term

0	0	0	0	$\ldots$
$H^2(K,\mu_n)$	$\oplus_p H^1(k_p,\mathbb{Z}_n)$	μ_n^{-1}	0	$\ldots$
$H^1(K,\mu_n)$	$\oplus_p\mathbb{Z}_n$	0	0	$\ldots$
μ_n	0	0	0	$\ldots$

Because $\mathbb{C}\{x,y\}$ is a unique factorization domain, we see that the map

$$H^1_{et}(K,\mu_n) = K^*/(K^*)^n \xrightarrow{\gamma} \oplus_p \mathbb{Z}_n$$

is surjective. Moreover, all fields k_p are isomorphic to the field of fractions of $\mathbb{C}\{z\}$ whose only cyclic extensions are given by adjoining a root of z and hence they are all ramified in m. Therefore, the component maps

$$\mathbb{Z}_n = H^1_{et}(k_p,\mathbb{Z}_n) \xrightarrow{\beta_L} \mu^{-1}$$

are isomorphisms. But then, the second (and limiting) term of the spectral sequence has the form

0	0	0	0	$\ldots$
$Ker\ \alpha$	$Ker\ \beta/Im\ \alpha$	0	0	$\ldots$
$Ker\ \gamma$	0	0	0	$\ldots$
μ_n	0	0	0	$\ldots$

Finally, we use the fact that $\mathbb{C}\{x, y\}$ is strict Henselian whence has no proper étale extensions. But then

$$H^i_{et}(X_{loc}, \mu_n) = 0 \text{ for } i \geq 1$$

and substituting this information in the spectral sequence we obtain that the top sequence of the coniveau spectral sequence

$$0 \longrightarrow Br_n \ K \overset{\alpha}{\longrightarrow} \oplus_p \mathbb{Z}_n \longrightarrow \mathbb{Z}_n \longrightarrow 0$$

is exact. From this sequence we immediately obtain the following

LEMMA 5.7
With notations as before, we have

1. *Let $U = X_{loc} - V(x)$, then $Br_n \ U = 0$*

2. *Let $U = X_{loc} - V(xy)$, then $Br_n \ U = \mathbb{Z}_n$ with generator the quantum-plane algebra*

$$\mathbb{C}_\zeta[u, v] = \frac{\mathbb{C}\langle u, v \rangle}{(vu - \zeta uv)}$$

where ζ is a primitive n-th root of one

Let Δ be a central simple algebra of dimension n^2 over a field L of transcendence degree 2. We want to determine when Δ admits a *Cayley-smooth model* $\mathcal{A}$, that is, a sheaf of Cayley-smooth $\mathcal{O}_X$-algebras where X is a projective surface with functionfield $\mathbb{C}(X) = L$. It follows from theorem 5.13 that, if such a model exists, X must be a smooth projective surface. We may assume that X is a (commutative) smooth model for L. By the Artin-Mumford exact sequence 3.11 the class of Δ in $Br_n \ \mathbb{C}(X)$ is determined by the following geocombinatorial data

- A finite collection $\mathcal{C} = \{C_1, \dots, C_k\}$ of *irreducible curves* in X.

- A finite collection $\mathcal{P} = \{P_1, \dots, P_l\}$ of *points* of X where each P_i is either an intersection point of two or more C_i or a singular point of some C_i.

- For each $P \in \mathcal{P}$ the *branch-data* $b_P = (b_1, \dots, b_{i_P})$ with $b_i \in \mathbb{Z}_n = \mathbb{Z}/n\mathbb{Z}$ and $\{1, \dots, i_P\}$ the different branches of $\mathcal{C}$ in P. These numbers must satisfy the admissibility condition

$$\sum_i b_i = 0 \in \mathbb{Z}_n$$

for every $P \in \mathcal{P}$

- for each $C \in \mathcal{C}$ we fix a cyclic $\mathbb{Z}_n$-cover of smooth curves

$$D \longrightarrow \tilde{C}$$

of the desingularization $\tilde{C}$ of C which is compatible with the branch-data.

If $\mathcal{A}$ is a maximal $\mathcal{O}_X$-order in Δ, then the ramification locus $ram_\mathcal{A}$ coincides with the collection of curves $\mathcal{C}$. We fix such a maximal $\mathcal{O}_X$-order $\mathcal{A}$ and investigate its Cayley-smooth locus.

PROPOSITION 5.10

Let $\mathcal{A}$ be a maximal $\mathcal{O}_X$-order in Δ with X a projective smooth surface and with geocombinatorial data $(\mathcal{C}, \mathcal{P}, b, \mathcal{D})$ determining the class of Δ in $Br_n \mathbb{C}(X)$.

If $\xi \in X$ lies in $X - \mathcal{C}$ or if ξ is a nonsingular point of $\mathcal{C}$, then $\mathcal{A}$ is Cayley-smooth in ξ.

PROOF If $\xi \notin \mathcal{C}$, then $\mathcal{A}_\xi$ is an Azumaya algebra over $\mathcal{O}_{X,x}$. As X is smooth in ξ, $\mathcal{A}$ is Cayley-smooth in ξ. Alternatively, we know that Azumaya algebras are split by étale extensions, whence $\hat{\mathcal{A}}_\xi \simeq M_n(\mathbb{C}[[x,y]])$, which shows that the behavior of $\mathcal{A}$ near ξ is controlled by the local data

and hence $\xi \in Sm_{tr}\, \mathcal{A}$. Next, assume that ξ is a nonsingular point of the ramification divisor $\mathcal{C}$. Consider the pointed spectrum $X_\xi = \operatorname{Spec} \mathcal{O}_{X,\xi} - \{\mathfrak{m}_\xi\}$. The only prime ideals are of height one, corresponding to the curves on X passing through ξ and hence this pointed spectrum is a Dedekind scheme. Further, $\mathcal{A}$ determines a maximal order over X_ξ. But then, tensoring $\mathcal{A}$ with the strict henselization $\mathcal{O}_{X,\xi}^{sh} \simeq \mathbb{C}\{x,y\}$ determines a sheaf of hereditary orders on the pointed spectrum $\hat{X}_\xi = \operatorname{Spec} \mathbb{C}\{x,y\} - \{(x,y)\}$ and we may choose the local variable x such that x is a local parameter of the ramification divisor $\mathcal{C}$ near ξ.

Using the characterization result for hereditary orders over discrete valuation rings, given in [90, Thm. 39.14] we know the structure of this extended sheaf of hereditary orders over any height one prime of $\hat{X}_\xi$. Because $\mathcal{A}_\xi$ is a reflexive (even a projective) $\mathcal{O}_{X,\xi}$-module, this height one information determines $\mathcal{A}_\xi^{sh}$ or $\hat{\mathcal{A}}_\xi$. This proves that $\mathcal{A}_\xi^{sh}$ must be isomorphic to the following block decomposition

$$
\begin{bmatrix}
M_{d_1}(\mathbb{C}\{x,y\}) & M_{d_1\times d_2}(\mathbb{C}\{x,y\}) & \cdots & M_{d_1\times d_k}(\mathbb{C}\{x,y\}) \\
M_{d_2\times d_1}(x\mathbb{C}\{x,y\}) & M_{d_2}(\mathbb{C}\{x,y\}) & \cdots & M_{d_2\times d_k}(\mathbb{C}\{x,y\}) \\
\vdots & \vdots & \ddots & \vdots \\
M_{d_k\times d_1}(x\mathbb{C}\{x,y\}) & M_{d_k\times d_2}(x\mathbb{C}\{x,y\}) & \cdots & M_{d_k}(\mathbb{C}\{x,y\})
\end{bmatrix}
$$

for a certain partition $p = (d_1,\ldots,d_k)$ of n having k parts. In fact, as we started out with a maximal order $\mathcal{A}$ one can even show that all these integers d_i must be equal. This local form corresponds to the following quiver-setting

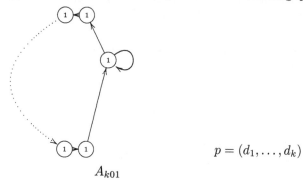

$$p = (d_1,\ldots,d_k)$$

A_{k01}

whence $\xi \in Sm_{tr}\ A$ as this is one of the allowed surface settings.   □

A maximal $\mathcal{O}_X$-order in Δ can have at worst noncommutative singularities in the singular points of the ramification divisor $\mathcal{C}$. Theorem 5.13 a Cayley-smooth order over a surface has as ramification-singularities at worst normal crossings. We are always able to reduce to normal crossings by the following classical result on commutative surfaces, see, for example, [43, V.3.8].

THEOREM 5.14 Embedded resolution of curves in surfaces
Let C be any curve on the surface X. Then, there exists a finite sequence of blow-ups
$$X' = X_s \longrightarrow X_{s-1} \longrightarrow \cdots \longrightarrow X_0 = X$$
and, if $f : X' \longrightarrow X$ is their composition, then the total inverse image $f^{-1}(\mathcal{C})$ is a divisor with normal crossings.

Fix a series of blow-ups $X' \xrightarrow{\ f\ } X$ such that the inverse image $f^{-1}(\mathcal{C})$ is a divisor on X' having as worst singularities normal crossings. We will replace the Cayley-Hamilton $\mathcal{O}_X$-order $\mathcal{A}$ by a Cayley-Hamilton $\mathcal{O}_{X'}$-order $\mathcal{A}'$ where $\mathcal{A}'$ is a sheaf of $\mathcal{O}_{X'}$-maximal orders in Δ. In order to determine the ramification divisor of $\mathcal{A}'$ we need to be able to keep track how the ramification divisor C of Δ changes if we blow up a singular point $p \in \mathcal{P}$.

LEMMA 5.8
Let $\tilde{X} \twoheadrightarrow X$ be the blow-up of X at a singular point p of C, the ramification divisor of Δ on X. Let $\tilde{C}$ be the strict transform of C and E the exceptional line on $\tilde{X}$. Let C' be the ramification divisor of Δ on the smooth model $\tilde{X}$. Then,

1. Assume the local branch data at p distribute in an admissible way on $\tilde{C}$, that is
$$\sum_{i \text{ at } q} b_{i,p} = 0 \text{ for all } q \in E \cap \tilde{C}$$
where the sum is taken only over the branches at q. Then, $C' = \tilde{C}$.

2. Assume the local branch data at p do not distribute in an admissible way, then $C' = \tilde{C} \cup E$.

PROOF   Clearly, $\tilde{C} \hookrightarrow C' \hookrightarrow \tilde{C} \cup E$. By the Artin-Mumford sequence applied to X' we know that the branch data of C' must add up to zero at all points q of $\tilde{C} \cap E$. We investigate the two cases
1.: Assume $E \subset C'$. Then, the E-branch number at q must be zero for all $q \in \tilde{C} \cap E$. But there are no nontrivial étale covers of $\mathbb{P}^1 = E$ so $ram(\Delta)$ gives the trivial element in $H^1_{et}(\mathbb{C}(E), \mu_n)$, a contradiction. Hence $C' = \tilde{C}$.

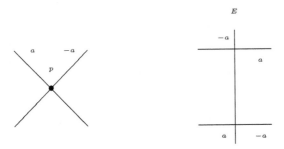

2.: If at some $q \in \tilde{C} \cap E$ the branch numbers do not add up to zero, the only remedy is to include E in the ramification divisor and let the E-branch number be such that the total sum is zero in $\mathbb{Z}_n$. □

THEOREM 5.15
Let Δ be a central simple algebra of dimension n^2 over a field L of transcendence degree two. Then, there exists a smooth projective surface S with

function field $\mathbb{C}(S) = L$ such that any maximal $\mathcal{O}_S$-order $\mathcal{A}_S$ in Δ has at worst a finite number of isolated noncommutative singularities. Each of these singularities is locally étale of quantum-plane type.

PROOF   We take any projective smooth surface X with function field $\mathbb{C}(X) = L$. By the Artin-Mumford exact sequence, the class of Δ determines a geocombinatorial set of data

$$(\mathcal{C}, \mathcal{P}, b, \mathcal{D})$$

as before. In particular, $\mathcal{C}$ is the ramification divisor $ram(\Delta)$ and $\mathcal{P}$ is the set of singular points of $\mathcal{C}$. We can separate $\mathcal{P}$ in two subsets

- $\mathcal{P}_{unr} = \{P \in \mathcal{P}$ where all the branch-data $b_P = (b_1, \ldots, b_{i_P})$ are trivial, that is, all $b_i = 0$ in $\mathbb{Z}_n\}$

- $\mathcal{P}_{ram} = \{P \in \mathcal{P}$ where some of the branch-data $b_P = (b_1, \ldots, b_{i_P})$ are non-trivial, that is, some $b_i \neq 0$ in $\mathbb{Z}_n\}$

After a finite number of blow-ups we get a birational morphism $S_1 \xrightarrow{\pi} X$ such that $\pi^{-1}(\mathcal{C})$ has as its worst singularities normal crossings and all branches in points of $\mathcal{P}$ are separated in S. Let $\mathcal{C}_1$ be the ramification divisor of Δ in S_1. By the foregoing argument we have

- If $P \in \mathcal{P}_{unr}$, then we have that $\mathcal{C}' \cap \pi^{-1}(P)$ consists of smooth points of $\mathcal{C}_1$,

- If $P \in \mathcal{P}_{ram}$, then $\pi^{-1}(P)$ contains at least one singular points Q of $\mathcal{C}_1$ with branch data $b_Q = (a, -a)$ for some $a \neq 0$ in $\mathbb{Z}_n$.

In fact, after blowingup singular points Q' in $\pi^{-1}(P)$ with trivial branchdata we obtain a smooth surface $S \longrightarrow S_1 \longrightarrow X$ such that the only singular points of the ramification divisor $\mathcal{C}'$ of Δ have nontrivial branchdata $(a, -a)$ for some $a \in \mathbb{Z}_n$. Then, take a maximal $\mathcal{O}_S$-order $\mathcal{A}$ in Δ. By the local calculation of $Br_n \ \mathbb{C}\{x, y\}$ performed in the last section we know that locally étale $\mathcal{A}$ is of quantum-plane type in these remaining singularities. As the quantum-plane is not étale locally split, $\mathcal{A}$ is not Cayley-smooth in these finite number of singularities. $\square$

In fact, the above proof gives a complete classification of the central simple algebras admitting a Cayley-smooth model.

THEOREM 5.16
Let Δ be a central simple $\mathbb{C}(X)$-algebra of dimension n^2 determined by the geo-combinatorial data $(\mathcal{C}, \mathcal{P}, b, \mathcal{D})$ given by the Artin-Mumford sequence. Then, Δ admits a Cayley-smooth model if and only if all branchdata are trivial.

PROOF   If all branch-data are trivial, the foregoing proof constructs a Cayley-smooth model of Δ. Conversely, if $\mathcal{A}$ is a Cayley-smooth $\mathcal{O}_S$-order in Δ with S a smooth projective model of $\mathbb{C}(X)$, then $\mathcal{A}$ is locally étale split in every point $s \in S$. But then, so is any maximal $\mathcal{O}_S$-order $\mathcal{A}_{max}$ containing $\mathcal{A}$. By the foregoing arguments this can only happen if all branchdata are trivial.

$\square$

5.5   Complex moment map

We fix a quiver Q on k vertices $\{v_1, \ldots, v_k\}$ and define the *opposite quiver* Q^o the quiver on $\{v_1, \ldots, v_k\}$ obtained by reversing all arrows in Q. That is, there is an arrow $\text{(j)} \xrightarrow{\ a^*\ } \text{(i)}$ in Q^o for each arrow $\text{(j)} \xleftarrow{\ a\ } \text{(i)}$ in the quiver Q. Fix a dimension vector $\alpha = (a_1, \ldots, a_k)$, using the *trace pairings*

$$M_{a_i \times a_j} \times M_{a_j \times a_i} \longrightarrow \mathbb{C} \qquad (V_{a^*}, V_a) \mapsto tr(V_{a^*} V_a)$$

we can identify the representation space $\mathbf{rep}_\alpha\, Q^o$ with the *dual space* $(\mathbf{rep}_\alpha\, Q)^* = Hom_\mathbb{C}(\mathbf{rep}_\alpha\, Q, \mathbb{C})$. Observe that the base change action of $GL(\alpha)$ on $\mathbf{rep}_\alpha\, Q^o$ coincides with the action dual to that of $GL(\alpha)$ on $\mathbf{rep}_\alpha\, Q$.

The *dual quiver* Q^d is the superposition of the quivers Q and Q^o. Clearly, for an dimension vector α we have

$$\mathbf{rep}_\alpha\, Q^d = \mathbf{rep}_\alpha\, Q \oplus \mathbf{rep}_\alpha\, Q^o = \mathbf{rep}_\alpha\, Q \oplus (\mathbf{rep}_\alpha\, Q)^*$$

whence $\mathbf{rep}_\alpha\, Q^d$ can be viewed as the *cotangent bundle* $T^*\mathbf{rep}_\alpha\, Q$ on $\mathbf{rep}_\alpha\, Q$ with structural morphism projection on the first factor. Cotangent bundles are equipped with a canonical *symplectic structure*, see [20, Example 1.1.3] or chapter 8 for more details. The natural action of $GL(\alpha)$ on $\mathbf{rep}_\alpha\, Q$ extends to an action of $GL(\alpha)$ on $T^*\mathbf{rep}_\alpha\, Q$ preserving the symplectic structure and it coincides with the base change action of $GL(\alpha)$ on $\mathbf{rep}_\alpha\, Q^d$. Such an action on the cotangent bundle gives rise to a *complex moment map*

$$T^*\mathbf{rep}_\alpha\, Q \xrightarrow{\ \mu_\mathbb{C}\ } (Lie\ GL(\alpha))^*$$

Recall that $Lie\ GL(\alpha) = M_\alpha(\mathbb{C}) = M_{a_1}(\mathbb{C}) \oplus \ldots \oplus M_{a_k}(\mathbb{C})$. Using the trace pairings on both sides, the complex moment map is the mapping

$$\mathbf{rep}_\alpha\, Q^d \xrightarrow{\ \mu_\mathbb{C}\ } M_\alpha(\mathbb{C})$$

defined by

$$\mu_\mathbb{C}(V)_i = \sum_{\substack{a \in Q_a \\ t(a)=i}} V_a V_{a^*} - \sum_{\substack{a \in Q_a \\ s(a)=i}} V_{a^*} V_a$$

Observe that the image of the complex moment map is contained in $M_\alpha^0(\mathbb{C})$ where

$$M_\alpha^0(\mathbb{C}) = \{(M_1, \ldots, M_k) \in M_\alpha(\mathbb{C}) \mid \sum_i tr(M_i) = 0\} = Lie\ PGL(\alpha)$$

corresponding to the fact that the action of $GL(\alpha)$ on $T^*rep_\alpha\ Q$ is really a $PGL(\alpha) = GL(\alpha)/C^*$ action.

DEFINITION 5.4 *Elements of* $\mathbb{C}^k = \mathbb{C}^{Q_v}$ *are called* weights *. If* λ *is a weight, one defines the* deformed preprojective algebra *of the quiver* Q *to be*

$$\Pi_\lambda(Q) \stackrel{dfn}{=} \Pi_\lambda = \frac{\mathbb{C}Q^d}{c - \lambda}$$

where c *is the* commutator element

$$c = \sum_{a \in Q_a} [a, a^*]$$

in $\mathbb{C}Q^d$ *and where* $\lambda = (\lambda_1, \ldots, \lambda_k)$ *is identified with the element* $\sum_i \lambda_i v_i \in \mathbb{C}Q^d$.
   The algebra $\Pi(Q) = \Pi$ *is known as the* preprojective algebra *of the quiver* Q.

LEMMA 5.9
The ideal $(c - \lambda) \lhd \mathbb{C}Q^d$ *is the same as the ideal with a generator*

$$\sum_{\substack{a \in Q_a \\ t(a)=i}} aa^* - \sum_{\substack{a \in Q_a \\ s(a)=i}} a^*a - \lambda_i v_i$$

for each vertex $v_i \in Q_v$.

PROOF   These elements are of the form $v_j(c - \lambda)v_i$, so they belong to the ideal $(c - \lambda)$. As $c - \lambda$ is also the sum of them, the ideal they generate contains $c - \lambda$. □

That is, α-dimensional representations of the deformed preprojective algebra Π_λ coincide with representations $V \in rep_\alpha\ Q^d$ that satisfy

$$\sum_{\substack{a \in Q_a \\ t(a)=i}} V_a V_{a^*} - \sum_{\substack{a \in Q_a \\ s(a)=i}} V_{a^*} V_a = \lambda_i 1\!\!1_{a_i}$$

for each vertex v_i. That is, we have an isomorphism between the scheme theoretic fiber of the complex moment map and the representation space

$$rep_\alpha\ \Pi_\lambda = \mu_{\mathbb{C}}^{-1}(\lambda)$$

As the image of $\mu_{\mathbb{C}}$ is contained in $M_\alpha^0(\mathbb{C})$ we have in particular the following.

LEMMA 5.10
If $\lambda.\alpha = \sum_i \lambda_i a_i \neq 0$, then there are no α-dimensional representations of Π_λ.

Because we have an embedding $C_k \hookrightarrow \Pi_\lambda$, the n-dimensional representations of the deformed preprojective algebra decompose into disjoint subvarieties

$$\mathbf{rep}_n\, \Pi_\lambda = \bigsqcup_{\alpha : \sum_i a_i = n} GL_n \times^{GL(\alpha)} \mathbf{rep}_\alpha\, \Pi_\lambda$$

Hence, in studying Cayley-smoothness of Π_λ we may reduce to the distinct components and hence to the study of α-*Cayley-smoothness*, that is, smoothness in the category of $C_k(\alpha)$-algebras, which are Cayley-Hamilton algebras of degree $n = \sum_i a_i$. Again, one can characterize this smoothness condition in a geometric way by the property that the restricted representation scheme $\mathbf{rep}_\alpha$ is smooth. In the next section we will investigate this property for the preprojective algebra Π_0, in chapter 8 we will be able to extend these results to arbitrary Π_λ. In this section we will compute the dimension of these representation schemes. First, we will investigate the fibers of the structural map of the cotangent bundle, that is, the projection

$$T^* \mathbf{rep}_\alpha\, Q \simeq \mathbf{rep}_\alpha\, Q^d \longrightarrow \mathbf{rep}_\alpha\, Q$$

PROPOSITION 5.11
If $V \in \mathbf{rep}_\alpha\, Q$, then there is an exact sequence

$$0 \longrightarrow Ext^1_{\mathbb{C}Q}(V,V)^* \longrightarrow \mathbf{rep}_\alpha\, Q^o \xrightarrow{c} M_\alpha(\mathbb{C}) \xrightarrow{t} Hom_{\mathbb{C}Q}(V,V)^* \longrightarrow 0$$

where c maps $W = (W_{a^*})_{a^*} \in \mathbf{rep}_\alpha\, Q^o$ to $\sum_{a \in Q_a}[V_a, W_{a^*}]$ and t maps $M = (M_i)_i \in M_{|alpha}(\mathbb{C})$ to the linear map $Hom_{\mathbb{C}Q}(V,V) \longrightarrow \mathbb{C}$ sending a morphism $N = (N_i)_i$ to $\sum_i tr(M_i N_i)$.

PROOF There is an exact sequence

$$0 \longrightarrow Hom_{\mathbb{C}Q}(V,V) \longrightarrow M_\alpha(\mathbb{C}) \xrightarrow{f} \mathbf{rep}_\alpha\, Q \longrightarrow Ext^1_{\mathbb{C}Q}(V,V) \longrightarrow 0$$

where f sends $M = (M_i)_i \in M_\alpha(\mathbb{C})$ to $V' = (V'_a)_a$ with $V'_a = M_{t(a)} V_a - V_a M_{s(a)}$. By definition, the kernel of f is $Hom_{\mathbb{C}Q}(V,V)$ and by the Euler form interpretation of theorem 4.5 we have

$$dim_{\mathbb{C}} Hom_{\mathbb{C}Q}(V,V) - dim_{\mathbb{C}} Ext^1_{\mathbb{C}Q}(V,V) = \chi_Q(\alpha,\alpha) = dim_{\mathbb{C}} M_\alpha(\mathbb{C}) - dim_{\mathbb{C}} \mathbf{rep}_\alpha\, Q$$

so the cokernel of f has the same dimension as $Ext^1_{\mathbb{C}Q}(V,V)$ and using the standard projective resolution of V one can show that it is naturally isomorphic to it. The required exact sequence follows by dualizing, using the trace pairing to identify $\mathbf{rep}_\alpha\, Q^o$ with $(\mathbf{rep}_\alpha\, Q)^*$ and $M_\alpha(\mathbb{C})$ with its dual. $\square$

This result allows us to give a characterization of the dimension vectors α such that $\mathbf{rep}_\alpha\, Q \neq \emptyset$.

THEOREM 5.17

For a weight $\lambda \in \mathbb{C}^k$ and a representation $V \in \mathbf{rep}_\alpha\, Q$ the following are equivalent

1. *V extends to an α-dimensional representation of the deformed preprojective algebra Π_λ.*

2. *For all dimension vectors β of direct summands W of V we have $\lambda.\beta = 0$.*

Moreover, if $V \in \mathbf{rep}_\alpha\, Q$ does lift, then $\pi^{-1}(V) \simeq (Ext^1_{\mathbb{C}Q}(V,V))^*$.

PROOF   If V lifts to a representation of Π_λ, then there is a representation $W \in \mathbf{rep}_\alpha\, Q^o$ mapping under c of proposition 5.11 to λ. But then, by exactness of the sequence in proposition 5.11 λ must be in the kernel of t. In particular, for any morphism $N = (N_i)_i \in Hom_{\mathbb{C}Q}(V,V)$ we have that $\sum_i \lambda_i tr(N_i) = 0$. In particular, let W be a direct summand of V (as Q-representation) and let $N = (N_i)_i$ be the projection morphism $V \twoheadrightarrow W \hookrightarrow V$, then $\sum_i \lambda_i tr(N_i) = \sum_i \lambda_i b_i$ where $\beta = (b_1,\ldots,b_k)$ is the dimension vector of W.

Conversely, it suffices to prove the lifting of any indecomposable representation W having a dimension vector β satisfying $\lambda.\beta = 0$. Because the endomorphism ring of W is a local algebra, any endomorphism $N = (N_i)_i$ of W is the sum of a nilpotent matrix and a scalar matrix whence $\sum_i \lambda_i tr(N_i) = 0$. But then considering the sequence of proposition 5.11 for β and considering λ as an element of $M_{|\beta|}(\mathbb{C})$, it lies in the kernel of t whence in the image of c and therefore W can be extended to a representation of Π_λ.

The last statement follows again from the exact sequence of proposition 5.11. $\square$

In particular, if α is a root for Q satisfying $\lambda.\alpha = 0$, then there are α-dimensional representations of Π_λ. Recall the definition of the *number of parameters* given in definition 4.8

$$\mu(X) = \max_d\, (dim\, X_{(d)} - d)$$

where $X_{(d)}$ is the union of all orbits of dimension d. We denote $\mu(\mathbf{rep}^{ind}_\alpha\, Q)$ for the $GL(\alpha)$-action on the indecomposables of $\mathbf{rep}_\alpha\, Q$ by $p_Q(\alpha)$. Recall that part of Kac's theorem 4.14 asserts that

$$p_Q(\alpha) = 1 - \chi_Q(\alpha, \alpha)$$

We will apply these facts to the determination of the dimension of the fibers of the complex moment map.

LEMMA 5.11
Let U be a $GL(\alpha)$-stable constructible subset of $\mathbf{rep}_\alpha Q$ contained in the image of the projection map $\mathbf{rep}_\alpha Q^d \xrightarrow{\pi} \mathbf{rep}_\alpha Q$. Then

$$dim \ \pi^{-1}(U) = \mu(U) + \alpha.\alpha - \chi_Q(\alpha, \alpha)$$

If in addition $U = \mathcal{O}(V)$ is a single orbit, then $\pi^{-1}(U)$ is irreducible of dimension $\alpha.\alpha - \chi_Q(\alpha, \alpha)$.

PROOF   Let $V \in U_{(d)}$, then by theorem 5.17, the fiber $\pi^{-1}(V)$ is isomorphic to $(Ext^1_{CQ}(V, V))^*$ and has dimension $dim_\mathbb{C} End(V) - \chi_Q(\alpha, \alpha)$ by theorem 4.5 and

$$dim_\mathbb{C} \ End(V) = dim \ GL(\alpha) - dim \ \mathcal{O}(V) = \alpha.\alpha - d$$

Hence, $dim \ \pi^{-1}(U_{(d)}) = (dim \ U_{(d)} - d) + \alpha.\alpha - \chi_Q(\alpha, \alpha)$. If we now vary d, the result follows.

For the second assertion, suppose that $\pi^{-1}(U) \longleftarrow Z_1 \sqcup Z_2$ with Z_i a $GL(\alpha)$-stable open subset, but then $\pi^{-1}(V) \cap Z_i$ are nonempty disjoint open subsets of the irreducible variety $\pi^{-1}(V)$, a contradiction.   □

THEOREM 5.18
Let λ be a weight and α a dimension vector such that $\lambda.\alpha = 0$. Then

$$dim \ \mathbf{rep}_\alpha \ \Pi_\lambda = dim \ \mu_\mathbb{C}^{-1}(\lambda) = \alpha.\alpha - \chi_Q(\alpha, \alpha) + m$$

where m is the maximum number among all

$$p_Q(\beta_1) + \ldots + p_Q(\beta_r)$$

with $r \geq 1$, all β_i are (positive) roots such that $\lambda.\beta_i = 0$ and $\alpha = \beta_1 + \ldots + \beta_r$.

PROOF   Decompose $\mathbf{rep}_\alpha Q = \bigsqcup_\tau \mathbf{rep}_\alpha(\tau)$ where $\mathbf{rep}_\alpha(\tau)$ are the representations decomposing as a direct sum of indecomposables of dimension vector $\tau = (\beta_1, \ldots, \beta_r)$. By Kac's theorem 4.14 we have that

$$\mu(\mathbf{rep}_\alpha(\tau)) = p_Q(\beta_1) + \ldots + p_Q(\beta_r)$$

If some of the β_i are such that $\lambda.\beta_i \neq 0$, and $\mu_{\mathbb{C}}^{-1}(\lambda) \xrightarrow{\pi} \mathbf{rep}_\alpha\, Q$ is the projection then $\pi^{-1}(\mathbf{rep}_\alpha(\tau)) = \emptyset$ by lemma 5.10. Combining this with lemma 5.11 the result follows. $\qquad\square$

DEFINITION 5.5   *The set of λ-Schur roots S_λ is defined to be the set of $\alpha \in \mathbb{N}^k$ such that $p_Q(\alpha) \geq p_Q(\beta_1) + \ldots + p_Q(\beta_r)$ for all decompositions $\alpha = \beta_1 + \ldots + \beta_r$ with β_i positive roots satisfying $\lambda.\beta_i = 0$.*
   $S_{\underline{0}}$ is the set of $\alpha \in \mathbb{N}^k$ such that $p_Q(\alpha) \geq p_Q(\beta_1) + \ldots + p_Q(\beta_r)$ for all decompositions $\alpha = \beta_1 + \ldots + \beta_r$ with $\beta_i \in \mathbb{N}^k$

Observe that $S_{\underline{0}}$ consists of Schur roots for Q, for if

$$\tau_{can} = (e_1, \beta_1; \ldots; e_s, \beta_s) = (\gamma_1, \ldots, \gamma_t)$$

(the γ_j possibly occurring with multiplicities) is the canonical decomposition of α with $t \geq 2$ we have

$$
\begin{aligned}
p_Q(\alpha) &= 1 - \chi_Q(\alpha, \alpha) \\
&= 1 - \sum_{i,j} \chi_Q(\gamma_i, \gamma_j) \\
&= \sum_i (1 - \chi_Q(\gamma_i, \gamma_i)) - \sum_{i \neq j} \chi_Q(\gamma_i, \gamma_j) - (t-1) \\
&> \sum_i p_Q(\gamma_i)
\end{aligned}
$$

whence $\alpha \notin S_{\underline{0}}$. This argument also shows that in the definition of $S_{\underline{0}}$ we could have taken all decompositions in positive roots, replacing the components β_i by their canonical decompositions.

THEOREM 5.19
For $\alpha \in \mathbb{N}^k$, the following are equivalent

1. *The complex moment map $\mathbf{rep}_\alpha\, Q^d \xrightarrow{\mu_{\mathbb{C}}} \mathbf{rep}_\alpha\, Q$ is flat.*

2. *$\mathbf{rep}_\alpha\, \Pi_0 = \mu_{\mathbb{C}}^{-1}(\underline{0})$ has dimension $\alpha.\alpha - 1 + 2p_Q(\alpha)$.*

3. *$\alpha \in S_{\underline{0}}$.*

PROOF   The dimensions of the relevant representation spaces are

$$
\begin{cases}
dim\ \mathbf{rep}_\alpha\, Q &= \alpha.\alpha - \chi_Q(\alpha, \alpha) = \alpha.\alpha - 1 + p_Q(\alpha) \\
dim\ \mathbf{rep}_\alpha\, Q^d &= 2\alpha.\alpha - 2\chi_Q(\alpha, \alpha) = 2\alpha.\alpha - 2 + 2p_Q(\alpha) \\
dim\ M_\alpha^0(\mathbb{C}) &= \alpha.\alpha - 1
\end{cases}
$$

so the relative dimension of the complex moment map is $d = \alpha.\alpha - 1 + 2p_Q(\alpha)$.

(1) $\Rightarrow$ (2): Because $\mu_{\mathbb{C}}$ os flat, its image U is an open subset of $M_\alpha^0(\mathbb{C})$ which obviously contains $\underline{0}$, but then the dimension of $\mu_{\mathbb{C}}^{-1}(\underline{0})$ is equal to the relative dimension d.

(2) $\Rightarrow$ (3): Assume $p_Q(\alpha) < \sum_i p_Q(\beta_i)$ for some decomposition $\alpha = \beta_1 + \ldots + \beta_s$ with $\beta_i \in \mathbb{N}^k$. Replacing each β_i by its canonical decomposition, we may assume that the β_i are actually positive roots. But then, theorem 5.18 implies that $\mu_{\mathbb{C}}^{-1}(\underline{0})$ has dimension greater than d.

(3) $\Rightarrow$ (1): We have that α is a Schur root. We claim that $\mathbf{rep}_\alpha Q^d \xrightarrow{\mu_{\mathbb{C}}} M_\alpha^0(\mathbb{C})$ is surjective. Let $V \in \mathbf{rep}_\alpha Q$ be a general representation, then $Hom_{\mathbb{C}Q}(V,V) = \mathbb{C}$. But then, the map c in proposition 5.11 has a one-dimensional cokernel. But as the image of c is contained in $M_\alpha^0(\mathbb{C})$, this shows that

$$\mathbf{rep}_\alpha Q^0 \xrightarrow{c} M_\alpha^0(\mathbb{C})$$

is surjective from which the claim follows. Let $M = (M_i)_i \in M_\alpha^0(\mathbb{C})$ and consider the projection

$$\mu_{\mathbb{C}}^{-1}(M) \xrightarrow{\tilde{\pi}} \mathbf{rep}_\alpha Q$$

If U is a constructible $GL(\alpha)$-stable subset of $\mathbf{rep}_\alpha Q$, then by an argument as in lemma 5.11 we have that

$$dim \; \tilde{\pi}^{-1}(U) \leq \mu(U) + \alpha.\alpha - \chi_Q(\alpha, \alpha)$$

But then, decomposing $\mathbf{rep}_\alpha Q$ into types τ of direct sums of indecomposables, it follows from the assumption that $\mu_{\mathbb{C}}^{-1}(M)$ has dimension at most d. But then by the dimension formula it must be equidimensional of dimension d whence flat. $\qquad\square$

5.6 Preprojective algebras

In this section we will determine the n-smooth locus of the preprojective algebra Π_0. By the étale local description of section 4.2 it is clear that we need to control the Ext^1-spaces of representations of Π_0.

PROPOSITION 5.12

Let V and W be representations of Π_0 of dimension vectors α and β, then we have

$$dim_{\mathbb{C}} \; Ext_{\Pi_0}^1(V,W) = dim_{\mathbb{C}} \; Hom_{\Pi_0}(V,W) + dim_{\mathbb{C}} \; Hom_{\Pi_0}(W,V) - T_Q(\alpha,\beta)$$

PROOF   It is easy to verify by direct computation that V has a projective resolution as Π_0-module, which starts as

$$\cdots \longrightarrow \bigoplus_{i \in Q_v} \Pi_0 v_i \otimes v_i V \xrightarrow{\ f\ } \bigoplus_{\substack{\text{\textcircled{\scriptsize j}} \xleftarrow{\ a\ } \text{\textcircled{\scriptsize i}} \\ a \in Q_a^d}} \Pi_0 v_j \otimes v_i V \xrightarrow{\ g\ } \bigoplus_{i \in Q_v} \Pi_0 v_i \otimes v_i V \xrightarrow{\ h\ } V \longrightarrow 0$$

where f is defined by

$$f(\sum_i p_i \otimes m_i) = \sum_{\substack{\text{\textcircled{\scriptsize j}} \xleftarrow{\ a\ } \text{\textcircled{\scriptsize i}} \\ a \in Q_a}} (p_i a^* \otimes m_i - p_j \otimes a^* m_j)_a - (p_j a \otimes m_j - p_i \otimes a m_i)_{a^*}$$

where $p_i \in \Pi_0 v_i$ and $m_i \in v_i V$. The map g is defined on the summand corresponding to an arrow $\text{\textcircled{\scriptsize j}} \xleftarrow{\ a\ } \text{\textcircled{\scriptsize i}}$ in Q^d by

$$g(pa \otimes m) = (pa \otimes m)_i - (p \otimes am)_j$$

for $p \in \Pi_0 v_j$ and $m \in v_i V$. the map h is the multiplication map. If we compute homomorphisms to W and use the identification

$$Hom_{\Pi_0}(\Pi_0 v_j \otimes v_i V, W) = Hom_{\mathbb{C}}(v_i V, v_j W)$$

we obtain a complex

$$\bigoplus_{i \in Q_v} Hom_{\mathbb{C}}(v_i V, v_i W) \lhook\joinrel\longrightarrow \bigoplus_{\substack{\text{\textcircled{\scriptsize j}} \xleftarrow{\ a\ } \text{\textcircled{\scriptsize i}} \\ a \in Q_a^d}} Hom_{\mathbb{C}}(v_i V, v_j W) \longrightarrow \bigoplus_{i \in Q_v} Hom_{\mathbb{C}}(v_i V, v_i W)$$

in which the left-hand cohomology is $Hom_{\Pi_0}(V, W)$ and the middle cohomology is $Ext^1_{\Pi_0}(V, W)$. Moreover, the alternating sum of the dimensions of the terms is $T_Q(\alpha, \beta)$. It remains to prove that the cokernel of the right-hand side map has the same dimension as $Hom_{\Pi_0}(W, V)$. But using the trace pairing to identify

$$Hom_{\mathbb{C}}(M, N)^* = Hom_{\mathbb{C}}(N, M)$$

we obtain that the dual of this complex is

$$\bigoplus_{i \in Q_v} Hom_{\mathbb{C}}(v_i W, v_i V) \longrightarrow \bigoplus_{\substack{\text{\textcircled{\scriptsize j}} \xleftarrow{\ a\ } \text{\textcircled{\scriptsize i}} \\ a \in Q_a^d}} Hom_{\mathbb{C}}(v_i W, v_j V) \longrightarrow\joinrel\rhook \bigoplus_{i \in Q_v} Hom_{\mathbb{C}}(v_i W, v_i V)$$

and, up to changing the sign of components in the second direct sum corresponding to arrows that are not in Q, this is the same complex as the complex arising with V and W interchanged. From this the result follows. $\qquad \Box$

In order to determine the n-smooth locus we observe that the representation space decomposes into a disjoint union and we have quotient morphisms

$$\mathbf{rep}_n \, \Pi_0 \xrightarrow{=} \bigsqcup_{\substack{\alpha=(a_1,\ldots,a_k) \\ a_1+\ldots+a_k=n}} GL_n \times^{GL(\alpha)} \mathbf{rep}_\alpha \, \Pi_0$$

$$\pi_n \downarrow \qquad\qquad \sqcup\pi_\alpha \downarrow$$

$$\mathbf{iss}_n \, \Pi_0 \xrightarrow{=} \bigsqcup_{\substack{\alpha=(a_1,\ldots,a_k) \\ a_1+\ldots+a_k=n}} \mathbf{iss}_\alpha \, \Pi_0$$

Hence if $\xi \in \mathbf{iss}_\alpha \, \Pi_0$ for $\xi \in Sm_{tr} \, \Pi_0$ it is necessary and sufficient that $\mathbf{rep}_\alpha \, \Pi_0$ is smooth along $\mathcal{O}(M_\xi)$ where M_ξ is the semisimple α-dimensional representation of Π_0 corresponding to ξ. Assume that ξ is of type $\tau = (e_1, \alpha_1; \ldots; e_z, \alpha_z)$, that is

$$M_\xi = S_1^{\oplus e_1} \oplus \ldots \oplus S_z^{\oplus e_z}$$

with S_i a simple Π_0-representation of dimension vector α_i. Again, the normal space to the orbit $\mathcal{O}(M_\xi)$ is determined by $Ext_{\Pi_o}^1(M_\xi, M_\xi)$ and can be depicted by a local quiver setting (Q_ξ, α_ξ) where Q_ξ is a quiver on z vertices and where $\alpha_\xi = \alpha_\tau = (e_1, \ldots, e_z)$. Repeating the arguments of section 4.2 we have the following.

LEMMA 5.12
With notations as above, $\xi \in Sm_n \, \Pi_0$ if and only if

$$dim \, GL(\alpha) \times^{GL(\alpha_\xi)} Ext_{\Pi_0}^1(M_\xi, M_\xi) = dim_{M_\xi} \, \mathbf{rep}_\alpha \, \Pi_0$$

As we have enough information to compute both sides, we can prove the following.

THEOREM 5.20
If $\xi \in \mathbf{iss}_\alpha \, \Pi_0$ with $\alpha = (a_1, \ldots, a_k) \in S_{\underline{0}}$ and $\sum_i a_i = n$, then $\xi \in Sm_n \, \Pi_0$ if and only if M_ξ is a simple n-dimensional representation of Π_0.

PROOF   Assume that ξ is a point of semisimple representation type $\tau = (e_1, \alpha_1; \ldots; e_z, \alpha_z)$, that is

$$M_\xi = S_1^{\oplus e_1} \oplus \ldots \oplus S_z^{\oplus e_z} \qquad \text{with} \qquad dim(S_i) = \alpha_i$$

and S_i a simple Π_0-representation. Then, by proposition 5.12 we have

$$\begin{cases} dim_{\mathbb{C}} \, Ext_{\Pi_0}^1(S_i, S_j) & = -T_Q(\alpha_i, \alpha_j) \qquad i \neq j \\ dim_{\mathbb{C}} \, Ext_{\Pi_0}^1(S_i, S_i) & = 2 - T_Q(\alpha_i, \alpha_i) \end{cases}$$

But then, the dimension of $Ext^1_{\Pi_0}(M_\xi, M_\xi)$ is equal to

$$\sum_{i=1}^{z}(2 - T_Q(\alpha_i, \alpha_i))e_i^2 + \sum_{i \neq j} e_i e_j(-T_Q(\alpha_i, \alpha_j)) = 2\sum_{i=1}^{z} e_i - T_Q(\alpha, \alpha)$$

from which it follows immediately that

$$dim \; GL(\alpha) \times^{GL(\alpha_\xi)} Ext^1_{\Pi_0}(M_\xi, M_\xi) = \alpha.\alpha + \sum_{i=1}^{z} e_i^2 - T_Q(\alpha, \alpha)$$

On the other hand, as $\alpha \in S_{\underline{0}}$ we know from theorem 5.19 that

$$dim \; \mathbf{rep}_\alpha \; \Pi_0 = \alpha.\alpha - 1 + 2p_Q(\alpha) = \alpha.\alpha - 1 + 2 + 2\chi_Q(\alpha, \alpha) = \alpha.\alpha + 1 - T_Q(\alpha, \alpha)$$

But then, equality occurs if and only if $\sum_i e_i^2 = 1$, that is, $\tau = (1, \alpha)$ or M_ξ is a simple n-dimensional representation of Π_0. $\square$

In particular it follows that the preprojective algebra Π_0 is *never* Quillen-smooth. Further, as $\vec{v}_i = (0, \ldots, 1, 0, \ldots, 0)$ are dimension vectors of simple representations of Π_0 it follows that Π_0 is α-smooth if and only if $\alpha = \vec{v}_i$ for some i. In chapter 8 we will determine the dimension vectors of simple representations of the (deformed) preprojective algebras.

Example 5.4
Let Q be an extended Dynkin diagram and δ_Q the corresponding dimension vector. Then, we will show that δ_Q is the dimension vector of a simple representation and $\delta_Q \in S_{\underline{0}}$. Then, the dimension of the quotient variety

$$dim \; \mathbf{iss}_{\delta_Q} \; \Pi_0 = dim \; \mathbf{rep}_{\delta_Q} \; \Pi_0 - \delta_Q.\delta_Q + 1$$
$$= 2p_Q(\delta_Q) = 2$$

so it is a surface. The only other semisimple δ_Q-dimensional representation of Π_0 is the trivial representation. By the theorem, this must be an isolated singular point of $\mathbf{iss}_{\delta_Q} \; Q$. In fact, one can show that $\mathbf{iss}_{\delta_Q} \; \Pi_0$ is the Kleinian singularity corresponding to the extended Dynkin diagram Q. $\square$

5.7 Central smooth locus

In this section we will prove the characterization, due to Raf Bocklandt, of (marked) quiver settings such that the ring of invariants is smooth. Remark that as the ring of invariants is a positively graded algebra, this is equivalent to being a polynomial algebra.

DEFINITION 5.6 *A quiver setting* (Q, α) *is said to be* final *iff none of the reduction steps* **b1**, **b2** *or* **b3** *of theorem 5.8 can be applied. Every quiver setting can be reduced to a final quiver setting that we denote as* $(Q, \alpha) \rightsquigarrow (Q_f, \alpha_f)$.

THEOREM 5.21
For a quiver setting (Q, α) *with* $Q = \mathrm{supp}\,\alpha$ *strongly connected, the following are equivalent :*

1. $\mathbb{C}[\mathrm{iss}_\alpha\, Q] = \mathbb{C}[\mathrm{rep}_\alpha\, Q]^{GL(\alpha)}$ *is* commalg-*smooth.*

2. $(Q_f, \alpha_f) \rightsquigarrow (Q_f, \alpha_f)$ *with* (Q_f, α_f) *one of the following quiver settings*

PROOF (2) $\Rightarrow$ (1): Follows from the foregoing theorem and the fact that the rings of invariants of the three quiver settings are resp. $\mathbb{C}$, $\mathbb{C}[tr(X), tr(X^2), \ldots, tr(X^k)]$ and $\mathbb{C}[tr(X), tr(Y), tr(X^2), tr(Y^2), tr(XY)]$.

(1) $\Rightarrow$ (2): Take a final reduction $(Q, \alpha) \rightsquigarrow (Q_f, \alpha_f)$ and to avoid subscripts rename $(Q_f, \alpha_f) = (Q, \alpha)$ (observe that the condition of the theorem as well as (1) is preserved under the reduction steps by the foregoing theorem). That is, we will assume that (Q, α) is final whence, in particular as **b1** cannot be applied,

$$\chi_Q(\alpha, \epsilon_v) < 0 \qquad \chi_Q(\epsilon_v, \alpha) < 0$$

for all vertices v of Q. With **1** we denote the dimension vector $(1, \ldots, 1)$.

claim 1 : Either $(Q, \alpha) = \textcircled{k}$ or Q has loops. Assume neither, then if $\alpha \neq 1$ we can choose a vertex v with maximal α_v. By the above inequalities and theorem 4.10 we have that

$$\tau = (1, \alpha - \epsilon_v; 1, \epsilon_v) \in \mathbf{types}_\alpha\, Q$$

As there are no loops in v, we have

$$\begin{cases} \chi_Q(\alpha - \epsilon_v, \epsilon_v) &= \chi(\alpha, \epsilon_v) - 1 < -1 \\ \chi_Q(\epsilon_v, \alpha - \epsilon_v) &= \chi(\epsilon_v, \alpha) - 1 < -1 \end{cases}$$

and the local quiver setting (Q_τ, α_τ) contains the subquiver

$$\textcircled{1} \underset{l}{\overset{k}{\rightleftarrows}} \textcircled{1} \qquad \text{with } k, l \geq 2$$

The invariant ring of the local quiver setting cannot be a polynomial ring as it contains the subalgebra

$$\frac{\mathbb{C}[a,b,c,d]}{(ab-cd)}$$

where $a = x_1 y_1$, $b = x_2 y_2$, $c = x_1 y_2$ and $d = x_2 y_1$ are necklaces of length 2 with x_i arrows from w_1 to w_2 and y_i arrows from w_2 to w_1. This contradicts the assumption (1) by the étale local structure result.

Hence, $\alpha = 1$ and because (Q,α) is final, every vertex must have least have two incoming and two outgoing arrows. Because Q has no loops

$$dim \, \texttt{iss}_1 \, Q = 1 - \chi_Q(1,1) = \#\texttt{arrows} - \#\texttt{vertices} + 1$$

On the other hand, a minimal generating set for $\mathbb{C}[\texttt{iss}_1 \, Q]$ is the set of *Eulerian necklaces* , that is, those necklaces in Q not re-entering any vertex. By (1) both numbers must be equal, so we will reach a contradiction by showing that #\texttt{euler}, the number of Eulerian necklaces is strictly larger than $\chi(Q) = \#\texttt{arrows} - \#\texttt{vertices} + 1$. We will do this by induction on the number of vertices.

If #\texttt{vertices} $= 2$, the statement is true because

whence $\#\texttt{euler} = kl > \chi(Q) = k + l - 1$

as both k and l are at least 2.

Assume #\texttt{vertices} > 2 and that there is a subquiver of the form

If $k > 1$ and $l > 1$ we have seen before that this subquiver and hence Q cannot have a polynomial ring of invariants.

If $k = 1$ and $l = 1$ then substitute this subquiver by one vertex

The new quiver Q' is again final without loops because there are at least four incoming arrows in the vertices of the subquiver and we only deleted two (the same holds for the outgoing arrows). Q' has one Eulerian necklace less than Q. By induction, we have that

$$\begin{aligned} \#\texttt{euler} &= \#\texttt{euler}' + 1 \\ &> \chi(Q') + 1 \\ &= \chi(Q) \end{aligned}$$

If $k > 1$ then one can look at the subquiver Q' of Q obtained by deleting $k-1$ of these arrows. If Q' is final, we are in the previous situation and obtain the inequality as before. If Q' is not final, then Q contains a subquiver of the form

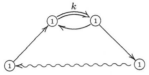

which cannot have a polynomial ring of invariants, as it is reducible to `basic` with both k and l at least equal to 2.

Finally, if `#vertices` > 2 and there is no `basic`-subquiver, take an arbitrary vertex v. Construct a new quiver Q' bypassing v

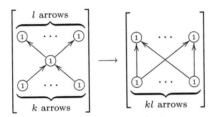

Q' is again final without loops and has the same number of Eulerian necklaces. By induction

$$\begin{aligned}
\#\texttt{euler} = \#\texttt{euler}' \\
> \#\texttt{arrows}' - \#\texttt{vertices}' + 1 \\
= \#\texttt{arrows} + (kl - k - l) - \#\texttt{vertices} + 1 + 1 \\
> \#\texttt{arrows} - \#\texttt{vertices} + 1
\end{aligned}$$

In all cases, we obtain a contradiction with (1) and hence have proved **claim 1**. So we may assume from now on that Q has loops.

claim 2: If Q has loops in v, then there is at most one loop in v or (Q, α) is

$$\texttt{2twobytwo} =$$

Because (Q, α) is final, we have $\alpha_v \geq 2$. If $\alpha_v = a \geq 3$ then there is only one loop in v. If not, there is a subquiver of the form

and its ring of invariants cannot be a polynomial algebra. Indeed, consider its representation type $\tau = (1, k - 1; 1, 1)$ then the local quiver is of type

basic with $k = l = a - 1 \geq 2$ and we know already that this cannot have a polynomial algebra as invariant ring. If $\alpha_v = 2$ then either we are in the **2twobytwo** case or there is at most one loop in v. If not, we either have at least three loops in v or two loops and a cyclic path through v, but then we can use the reductions

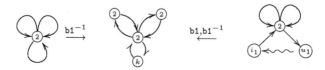

The middle quiver cannot have a polynomial ring as invariants because we consider the type

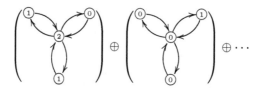

The number of arrows between the first and the second simple component equals

$$- (2\ 1\ 1\ 0) \begin{pmatrix} 1 & -1 & -1 & -1 \\ -1 & 1 & 0 & 0 \\ -1 & 0 & 1 & 0 \\ -1 & 0 & 0 & 1 \end{pmatrix} \begin{pmatrix} 0 \\ 0 \\ 0 \\ 1 \end{pmatrix} = 2$$

whence the corresponding local quiver contains **basic** with $k = l = 2$ as subquiver. This proves **claim 2**. From now on we will assume that the quiver setting (Q, α) is such that there is precisely one loop in v and that $k = \alpha_v \geq 2$. Let

$$\tau = (1, 1; 1, \epsilon_v; \alpha_{v_1} - 1, \epsilon_{v_1}; \ldots; \ldots; \alpha_v - 2, \epsilon_v; \ldots; \alpha_{v_l} - 1, \epsilon_{v_l}) \in \text{types}_\alpha Q$$

Here, the second simple representation, concentrated in v has nonzero trace in the loop whereas the remaining $\alpha_v - 2$ simple representations concentrated in v have zero trace. Further, $1 \in \text{simp}\mathbb{C}Q$ as Q is strongly connected by theorem 4.10. We work out the local quiver setting (Q_τ, α_τ). The number of arrows between the vertices in Q_τ corresponding to simple components concentrated in a vertex is equal to the number of arrows in Q between these vertices. We will denote the vertex (and multiplicity) in Q_τ corresponding to the simple component of dimension vector 1 by $\boxed{1}$.

The number of arrows between the vertex in Q_τ corresponding to a simple concentrated in vertex w in Q to $\boxed{1}$ is $-\chi_Q(\epsilon_w, 1)$ and hence is one less than the number of outgoing arrows from w in Q. Similarly, the number of arrows from the vertex $\boxed{1}$ to that of the simple concentrated in w is $-\chi_Q(1, \epsilon_w)$ and is equal to one less than the number of incoming arrows in w in Q. But then we must have for all vertices w in Q that

$$\chi_Q(\epsilon_w, 1) = -1 \quad \text{or} \quad \chi_Q(1, \epsilon_w) = -1$$

Indeed, because (Q, α) is final we know that these numbers must be strictly negative, but they cannot be both ≤ -2 for then the local quiver Q_τ will contain a subquiver of type

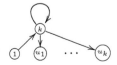

contradicting that the ring of invariants is a polynomial ring. Similarly, we must have

$$\chi_Q(\epsilon_w, \epsilon_v) \geq -1 \quad \text{or} \quad \chi_Q(\epsilon_v, \epsilon_v)$$

for all vertices w in Q for which $\alpha_w \geq 2$. Let us assume that $\chi_Q(\epsilon_v, 1) = -1$.

claim 3: If w_1 is the unique vertex in Q such that $\chi_Q(\epsilon_v, \epsilon_{w_1}) = -1$, then $\alpha_{w_1} = 1$. If this was not the case there is a vertex corresponding to a simple representation concentrated in w_1 in the local quiver Q_τ. If $\chi_Q(1, \epsilon_{w_1}) = 0$ then the dimension of the unique vertex w_2 with an arrow to w_1 has strictly bigger dimension than w_1, otherwise $\chi_Q(\alpha, \epsilon_{w_1}) \geq 0$ contradicting finality of (Q, α). The vertex w_2 corresponds again to a vertex in the local quiver. If $\chi_Q(1, \epsilon_{w_2}) = 0$, the unique vertex w_3 with an arrow to w_2 has strictly bigger dimension than w_2. Proceeding this way one can find a sequence of vertices with increasing dimension, which attains a maximum in vertex w_k. Therefore $\chi_Q(1, \epsilon_{w_k}) \leq -1$. This last vertex is in the local quiver connected with W, so one has a path from 1 to ϵ_v.

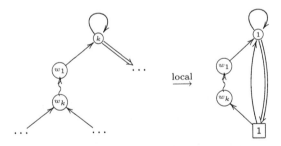

The subquiver of the local quiver Q_τ consisting of the vertices corresponding to the simple representation of dimension vector 1 and the simples concentrated in vertex v resp. w_k is reducible via b1 to ①⟲①, at least if $\chi_Q(1, \epsilon_v) \leq -2$, a contradiction finishing the proof of the claim. But then, the quiver setting (Q, α) has the following shape in the neighborhood of v

contradicting finality of (Q, α) for we can apply b3. In a similar way one proves that the quiver setting (Q, α) has the form

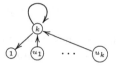

in a neighborhood of v if $\chi_Q(1, \epsilon_v) = -1$ and $\chi_Q(\epsilon_v, 1) \leq -2$, again contradicting finality.

There remains one case to consider: $\chi_Q(1, \epsilon_v) = -1$ *and* $\chi_Q(\epsilon_v, 1) = -1$. Suppose w_1 is the unique vertex in Q such that $\chi_Q(\epsilon_v, \epsilon_{w_1}) = -1$ and w_k is the unique vertex in Q such that $\chi_Q(\epsilon_{w_k}, \epsilon_v) = -1$, then we claim the following.

claim 4: Either $\alpha_{w_1} = 1$ or $\alpha_{w_k} = 1$. If not, consider the path connecting w_k and w_1 and call the intermediate vertices w_i, $1 < i < k$. Starting from w_1 we go back the path until α_{w_i} reaches a maximum. at that point we know that $\chi_Q(1, \epsilon_{w_k}) \leq -1$, otherwise $\chi_Q(\alpha, \epsilon_{w_k}) \geq 0$. In the local quiver there is a path from the vertex corresponding to the 1-dimensional simple over the ones corresponding to the simples concentrated in w_i to v. Repeating the argument, starting from w_k we also have a path from the vertex of the simple v-representation over the vertices of the w_j-simples to the vertex of the 1-dimensional simple

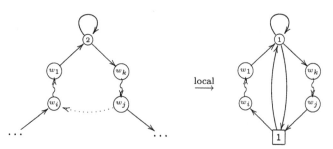

The subquiver consisting of 1, ϵ_v and the two paths through the ϵ_{w_i} is reducible to ⓵⇄1 and we again obtain a contradiction.

The only way out of these dilemmas is that the final quiver setting (Q, α) is of the form

finishing the proof.   ⬜

DEFINITION 5.7 *Let (Q, α) and (Q', α') be two quiver settings such that there is a vertex v in Q and a vertex v' in Q' with $\alpha_v = 1 = \alpha'_{v'}$. We* *define the* connected sum *of the two settings to be the quiver setting*

$$(\; Q \overset{v}{\underset{v'}{\#}} Q' \; , \; \alpha \overset{v}{\underset{v'}{\#}} \alpha' \;)$$

where $Q \overset{v}{\underset{v'}{\#}} Q$ is the quiver obtained by identifying the two vertices v and v'

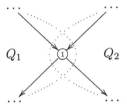

*and where $\alpha \overset{v}{\underset{v'}{\#}} \alpha'$ is the dimension vector that restricts to α (resp. α') on Q
(resp. Q').*

Example 5.5
With this notation we have

$$\mathbb{C}[\mathrm{iss}_{\alpha \overset{v}{\underset{v'}{\#}} \alpha'} Q \overset{v}{\underset{v'}{\#}} Q'] \simeq \mathbb{C}[\mathrm{iss}_\alpha Q] \otimes \mathbb{C}[\mathrm{iss}_{\alpha'} Q']$$

Because traces of necklaces passing more than once through a vertex where the dimension vector is equal to 1 can be split as a product of traces of necklaces which pass through this vertex only one time, we see that the invariant ring of the connected sum is generated by Eulerian necklaces fully contained in Q or in Q'. ▫

Theorem 5.21 gives a procedure to decide whether a given quiver setting (Q, α) has a regular ring of invariants. However, is is not feasible to give a graph theoretic description of all such settings in general. Still, in the special (but important) case of *symmetric* quivers, there is a nice graph theoretic characterization.

THEOREM 5.22
Let (Q, α) be a symmetric quiver setting such that Q is connected and has no loops. Then, the ring of polynomial invariants

$$\mathbb{C}[\mathrm{iss}_\alpha Q] = \mathbb{C}[\mathrm{rep}_\alpha Q]^{GL(\alpha)}$$

is a polynomial ring if and only if the following conditions are satisfied.

1. *Q is tree-like, that is, if we draw an edge between vertices of Q whenever there is at least one arrow between them in Q, the graph obtained in a tree.*

2. *α is such that in every branching vertex v of the tree we have $\alpha_v = 1$.*

3. *The quiver subsetting corresponding to branches of the tree are connected sums of the following atomic pieces*

I (n) ⟶ (m)

II (1) (n), $k \leq n$

III (1) (n) (m)

IV (n) (2) (m)

PROOF Using theorem 5.21 any of the atomic quiver settings has a polynomial ring of invariants. Type **I** reduces via **b1** to

where $k = min(m, n)$, type **II** reduces via **b1** and **b2** to ①, type **III** reduces via **b1, b3, b1** and **b2** to ① and finally, type **IV** reduces via **b1** to

By the previous example, any connected sum constructed out of these atomic quiver settings has a regular ring of invariants. Observe that such connected sums satisfy the first two requirements. Therefore, any quiver setting satisfying the requirements has indeed a polynomial ring of invariants.

Conversely, assume that the ring of invariants $\mathbb{C}[\text{iss}_\alpha Q]$ is a polynomial ring, then there can be no quiver subsetting of the form

#vertices ≥ 3

For we could look at a semisimple representation type τ with decomposition

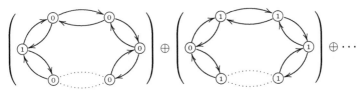

The local quiver contains a subquiver (corresponding to the first two components) of type `basic` with k and $l \geq 2$ whence cannot give a polynomial ring. That is, Q is tree-like.

Further, the dimension vector α cannot have components ≥ 2 at a branching vertex v. For we could consider the semisimple representation type with decomposition

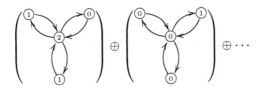

and again the local quiver contains a subquiver setting of type `basic` with $k = 2 = l$ (the one corresponding to the first two components). Hence, α satisfies the second requirement.

It remains to be shown that the branches do not contain other subquiver settings than those made of the atomic components. That is, we have to rule out the following subquiver settings

with $a_2 \geq 2$ and $a_3 \geq 2$

with $a_2 \geq 3$ and $a_1 \geq 2$, $a_3 \geq 2$ and

whenever $a_2 \geq 2$. These situations are easily ruled out by theorem 5.21 and we leave this as a pleasant exercise. ◻

Example 5.6
The quiver setting

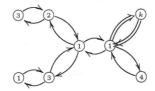

has a polynomial ring of invariants if and only if $k \geq 2$.   ⬚

Example 5.7

Let $(Q^\bullet, \alpha)$ be a *marked* quiver setting and assume that $\{l_1, \ldots, l_u\}$ are the marked loops in $Q^\bullet$. If Q is the underlying quiver forgetting the markings we have by separating traces that

$$\mathbb{C}[\text{iss}_\alpha Q] \simeq \mathbb{C}[\text{iss}_\alpha Q^\bullet][tr(l_1), \ldots, tr(l_u)]$$

Hence, we do not have to do extra work in the case of marked quivers:

A marked quiver setting $(Q^\bullet, \alpha)$ has a regular ring of invariants if and only if (Q, α) can be reduced to a one of the three final quiver settings of theorem 5.21.
⬚

5.8   Central singularities

Surprisingly, the reduction steps of section 5.3 allow us to classify all central singularities of a Cayley-smooth algebra $A \in \text{alg}@n$ up to *smooth equivalence*. Recall that two commutative *local* rings $C_\mathfrak{m}$ and $D_\mathfrak{n}$ are said to be smooth equivalent if there are numbers k and l such that

$$\hat{C}_\mathfrak{m}[[x_1, \ldots, x_k]] \simeq \hat{D}_\mathfrak{n}[[y_1, \ldots, y_l]]$$

By theorem 5.8 (and its extension to marked quivers) and the étale local classification of Cayley-smooth orders it is enough to classify the rings of invariants of *reduced* marked quiver settings up to smooth equivalence. We can always assume that the quiver Q is strongly connected (if not, the ring of invariants is the tensor product of the rings of invariants of the maximal strongly connected subquivers). Our aim is to classify the reduced quiver singularities up to equivalence, so we need to determine the Krull dimension of the rings of invariants.

LEMMA 5.13

Let $(Q^\bullet, \alpha)$ be a reduced marked quiver setting and Q strongly connected. Then,

$$dim\ \text{iss}_\alpha\ Q^\bullet = 1 - \chi_Q(\alpha, \alpha) - m$$

where m is the total number of marked loops in $Q^\bullet$.

PROOF   Because $(Q^\bullet, \alpha)$ is reduced, none of the vertices satisfies condition C_V^v, whence

$$\chi_Q(\epsilon_v, \alpha) \leq -1 \quad \text{and} \quad \chi_Q(\alpha, \epsilon_v) \leq -1$$

for all vertices v. In particular it follows (because Q is strongly connected) from section 4.3 that α is the dimension vector of a simple representation of Q and that the dimension of the quotient variety

$$dim \; iss_\alpha \; Q = 1 - \chi_Q(\alpha, \alpha)$$

Finally, separating traces of the loops to be marked gives the required formula.
$\square$

Extending theorem 5.21 to the setting of marked quivers, we can classify all smooth points of $triss_n \; A$ for a Cayley-smooth order A.

THEOREM 5.23
Let $(Q^\bullet, \alpha)$ be a marked quiver setting such that Q is strongly connected. Then $iss_\alpha \; Q^\bullet$ is smooth if and only if the unique reduced marked quiver setting to which $(Q^\bullet, \alpha)$ can be reduced is one of the following five types

The next step is to classify for a given dimension d all reduced marked quiver settings $(Q^\bullet, \alpha)$ such that $dim \; iss_\alpha \; Q^\bullet = d$. The following result limits the possible cases drastically in low dimensions.

LEMMA 5.14
Let $(Q^\bullet, \alpha)$ be a reduced marked quiver setting on $k \geq 2$ vertices. Then,

$$dim \; iss_\alpha \; Q^\bullet \geq 1 + \overset{a \geq 1}{\underset{\textcircled{a}}{\sum}} a + \overset{a > 1}{\underset{\textcircled{a}}{\sum}} (2a - 1) + \overset{a > 1}{\underset{\textcircled{a}}{\sum}} (2a) + \overset{a > 1}{\underset{\textcircled{a}}{\sum}} (a^2 + a - 2)) +$$

$$\overset{a > 1}{\underset{\textcircled{a}}{\sum}} (a^2 + a - 1) + \overset{a > 1}{\underset{\textcircled{a}}{\sum}} (a^2 + a) + \ldots + \overset{a > 1}{\underset{k \, \textcircled{a} \, l}{\sum}} ((k + l - 1)a^2 + a - k) + \ldots$$

In this sum the contribution of a vertex v with $\alpha_v = a$ is determined by the number of (marked) loops in v. By the reduction steps (marked) loops only occur at vertices where $\alpha_v > 1$.

PROOF We know that the dimension of $iss_\alpha \; Q^\bullet$ is equal to

$$1 - \chi_Q(\alpha, \alpha) - m = 1 - \sum_v \chi_Q(\epsilon_v, \alpha)\alpha_v - m$$

If there are no (marked) loops at v, then $\chi_Q(\epsilon_v, \alpha) \leq -1$ (if not we would reduce further) which explains the first sum. If there is exactly one (marked) loop at v then $\chi_Q(\epsilon_v, \alpha) \leq -2$ for if $\chi_Q(\epsilon_v, \alpha) = -1$ then there is just one outgoing arrow to a vertex w with $\alpha_w = 1$ but then we can reduce the quiver setting further. This explains the second and third sums. If there are k marked loops and l ordinary loops in v (and Q has at least two vertices), then

$$-\chi_Q(\epsilon_v, \alpha)\alpha_v - k \geq ((k+l)\alpha_v - \alpha_v + 1)\alpha_v - k$$

which explains all other sums.   ▯

Observe that the dimension of the quotient variety of the one vertex marked quivers

$$k \bullet \; \overset{a}{\circlearrowleft\circlearrowright} \; l$$

is equal to $(k+l-1)a^2 + 1 - k$ and is singular (for $a \geq 2$) unless $k+l = 2$. We will now classify the reduced singular settings when there are at least two vertices in low dimensions. By the previous lemma it follows immediately that

1. the maximal number of vertices in a reduced marked quiver setting $(Q^\bullet, \alpha)$ of dimension d is $d-1$ (in which case all vertex dimensions must be equal to one)

2. if a vertex dimension in a reduced marked quiver setting is $a \geq 2$, then the dimension $d \geq 2a$.

LEMMA 5.15

Let $(Q^\bullet, \alpha)$ be a reduced marked quiver setting such that $\mathbf{iss}_\alpha \, Q^\bullet$ is singular of dimension $d \leq 5$, then $\alpha = (1, \dots, 1)$. Moreover, each vertex must have at least two incoming and two outgoing arrows and no loops.

PROOF   From the lower bound of the sum formula it follows that if some $\alpha_v > 1$ it must be equal to 2 and must have a unique marked loop and there can only be one other vertex w with $\alpha_w = 1$. If there are x arrows from w to v and y arrows from v to w, then

$$dim \; \mathbf{iss}_\alpha \, Q^\bullet = 2(x+y) - 1$$

whence x or y must be equal to 1 contradicting reducedness. The second statement follows as otherwise we could perform extra reductions.   ▯

PROPOSITION 5.13

The only reduced marked quiver singularity in dimension 3 is

$$3_{con} \; : \;$$

The reduced marked quiver singularities in dimension 4 are

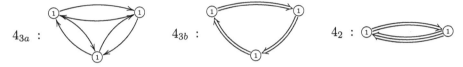

4_{3a} : $\qquad$ 4_{3b} : $\qquad$ 4_2 :

PROOF All one vertex marked quiver settings with quotient dimension ≤ 5 are smooth, so we are in the situation of lemma 5.15. If the dimension is 3 there must be two vertices each having exactly two incoming and two outgoing arrows, whence the indicated type is the only one. The resulting singularity is the *conifold singularity*

$$\frac{\mathbb{C}[[x, y, u, v]]}{(xy - uv)}$$

In dimension 4 we can have three or two vertices. In the first case, each vertex must have exactly two incoming and two outgoing arrows whence the first two cases. If there are two vertices, then just one of them has three incoming arrows and one has three outgoing arrows. ⬜

Assume that all vertex dimensions are equal to one, then one can write any (trace of an) oriented cycle as a product of (traces of) *primitive* oriented cycles (that is, those that cannot be decomposed further). From this one deduces immediately the following.

LEMMA 5.16
Let $(Q^{\bullet}, \alpha)$ *be a reduced marked quiver setting such that all* $\alpha_v = 1$. *Let* m *be the maximal graded ideal of* $\mathbb{C}[\mathrm{rep}_\alpha Q^{\bullet}]^{GL(\alpha)}$, *then a vectorspace basis of*

$$\frac{m^i}{m^{i+1}}$$

is given by the oriented cycles in Q *which can be written as a product of* i *primitive cycles, but not as a product of* $i + 1$ *such cycles.*

Clearly, the dimensions of the quotients m^i/m^{i+1} are (étale) isomorphism invariants. Recall that the first of these numbers m/m^2 is the embedding dimension of the singularity. Hence, for $d \leq 5$ this simple-minded counting method can be used to separate quiver singularities.

THEOREM 5.24
There are precisely three reduced quiver singularities in dimension $d = 4$.

PROOF The number of primitive oriented cycles of the three types of reduced marked quiver settings in dimension four

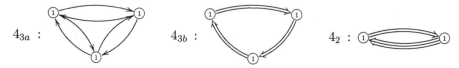

is 5, respectively 8 and 6. Hence, they give nonisomorphic rings of invariants.
▯

If some of the vertex dimensions are ≥ 2 we have no easy description of the vectorspaces m^i/m^{i+1} and we need a more refined argument. The idea is to answer the question "what other singularities can the reduced singularity see?" An α-*representation type* is a datum

$$\tau = (e_1, \beta_1; \ldots; e_l, \beta_l)$$

where the e_i are natural numbers ≥ 1, the β_i are dimension vectors of simple representations of Q such that $\alpha = \sum_i e_i \beta_i$. Any neighborhood of the trivial representation contains semisimple representations of Q of type τ for any α-representation type. Let $(Q^\bullet_\tau, \alpha_\tau)$ be the associated (marked) local quiver setting. Assume that $\text{iss}_{\alpha_\tau} Q_\tau$ has a singularity, then the couple

(dimension of strata, type of singularity)

is a characteristic feature of the singularity of $\text{iss}_\alpha Q^\bullet$ and one can often distinguish types by these couples. The *fingerprint* of a reduced quiver singularity will be the Hasse diagram of those α-representation types τ such that the local marked quiver setting $(Q^\bullet_\tau, \alpha_\tau)$ can be reduced to a reduced quiver singularity (necessarily occurring in lower dimension and the difference between the two dimensions gives the dimension of the stratum). Clearly, this method fails in case the marked quiver singularity is an *isolated singularity*. Fortunately, we have a complete characterization of these.

THEOREM 5.25

[13] The only reduced marked quiver settings $(Q^\bullet, \alpha)$ such that the quotient variety is an isolated singularity are of the form

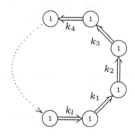

where Q has l vertices and all $k_i \geq 2$. *The dimension of the corresponding quotient is*

$$d = \sum_i k_i + l - 1$$

and the unordered *l-tuple* $\{k_1, \ldots, k_l\}$ *is an (étale) isomorphism invariant of the ring of invariants.*

Not only does this result distinguish among isolated reduced quiver singularities, but it also shows that in all other marked quiver settings we will have additional families of singularities. We will illustrate the method in some detail to separate the reduced marked quiver settings in dimension 6 having one vertex of dimension two.

PROPOSITION 5.14
The reduced singularities of dimension 6 such that α contains a component equal to 2 are pairwise non-equivalent.

PROOF   One can show that the reduced marked quiver setting for $d = 6$ with at least one component ≥ 2 are

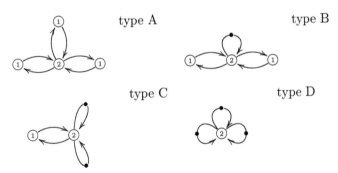

type A   type B

type C   type D

We will order the vertices such that $\alpha_1 = 2$.

type A: There are three different representation types $\tau_1 = (1, (2; 1, 1, 0); 1, (0; 0, 0, 1))$ (and permutations of the 1-vertices). The local quiver setting has the form

because for $\beta_1 = (2; 1, 1, 0)$ and $\beta_2 = (0; 0, 0, 1)$ we have that $\chi_Q(\beta_1, \beta_1) = -2$, $\chi_Q(\beta_1, \beta_2) = -2$, $\chi_Q(\beta_2, \beta_1) = -2$ and $\chi(\beta_2, \beta_2) = 1$. These three representation types each give a three-dimensional family of conifold (type 3_{con}) singularities.

Further, there are three different representation types $\tau_2 = (1, (1; 1, 1, 0); 1, (1; 0, 0, 1))$ (and permutations) of which the local quiver setting is of the form

as with $\beta_1 = (1; 1, 1, 0)$ and $\beta_2 = (1; 0, 0, 1)$ we have $\chi_Q(\beta_1, \beta_1) = -1$, $\chi_Q(\beta_1, \beta_2) = -2$, $\chi_Q(\beta_2, \beta_1) = -2$ and $\chi_Q(\beta_2, \beta_2) = 0$. These three representation types each give a three-dimensional family of conifold singularities.

Finally, there are the three representation types

$$\tau_3 = (1, (1; 1, 0, 0); 1, (1; 0, 1, 0); 1, (0; 0, 0, 1))$$

(and permutations) with local quiver setting

These three types each give a two-dimensional family of reduced singularities of type 4_{3a}.

The degeneration order on representation types gives $\tau_1 < \tau_3$ and $\tau_2 < \tau_3$ (but for different permutations) and the *fingerprint* of this reduced singularity can be depicted as

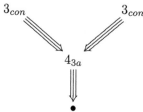

type B: There is one representation type $\tau_1 = (1, (1; 1, 0); 1, (1; 0, 1))$ giving as above a three-dimensional family of conifold singularities, one representation type $\tau_2 = (1, (1; 1, 1); 1, (1; 0, 0))$ giving a three-dimensional family of conifolds and finally one representation type

$$\tau_3 = (1, (1; 0, 0); 1, (1; 0, 0); 1, (0; 1, 1); 1, (0; 0, 1))$$

of which the local quiver setting has the form

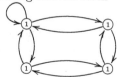

(the loop in the downright corner is removed to compensate for the marking) giving rise to a one-dimensional family of five-dimensional singularities of type 5_{4a}. This gives the fingerprint

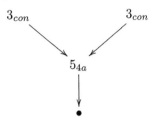

type C: We have a three-dimensional family of conifold singularities coming from the representation type $(1, (1; 1); 1, (1; 0))$ and a two-dimensional family of type 4_{3a} singularities corresponding to the representation type $(1, (1; 0); 1, (1, 0); 1, (0; 1))$. Therefore, the fingerprint is depicted as

$$3_{con} \longrightarrow 4_{3a} \longrightarrow \bullet$$

type D: We have just one three-dimensional family of conifold singularities determined by the representation type $(1, (1); 1, (1))$ so the fingerprint is $3_{con} \longrightarrow \bullet$. As fingerprints are isomorphism invariants of the singularity, this finishes the proof.

We claim that the minimal number of generators for these invariant rings is 7. The structure of the invariant ring of three 2×2 matrices upto simultaneous conjugation was determined by Ed Formanek [33] who showed that it is generated by 10 elements

$$\{tr(X_1), tr(X_2), tr(X_3), det(X_1), det(X_2), det(X_3), tr(X_1X_2), tr(X_1X_3),$$
$$tr(X_2X_3), tr(X_1X_2X_3)\}$$

and even gave the explicit quadratic polynomial satisfied by $tr(X_1X_2X_3)$ with coefficients in the remaining generators. The rings of invariants of the four cases of interest to us are quotients of this algebra by the ideal generated by three of its generators : for type A it is $(det(X_1), det(X_2), det(X_3))$, for type B: $(det(X_1), tr(X_2), det(X_3))$, for type C: $(det(X_1), tr(X_2), tr(X_3))$ and for type D: $(tr(X_1), tr(X_2), tr(X_3))$. $\qquad\Box$

These two tricks (counting cycles and fingerprinting) are sufficient to classify all central singularities of Cayley-smooth orders for central dimension $d \leq 6$. We will give the details for $d = 5$, the remaining cases for $d = 6$ can be found in the paper [14].

PROPOSITION 5.15
The reduced marked quiver settings for $d = 5$ are

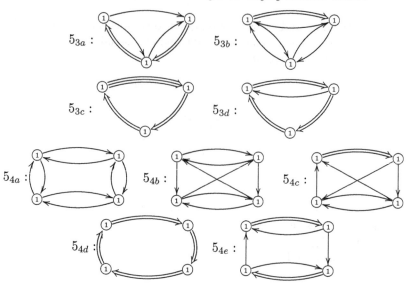

PROOF   We are in the situation of lemma 5.15 and hence know that all vertex-dimensions are equal to one, every vertex has at least two incoming and two outgoing arrows and the total number of arrows is equal to $5 - 1 + k$ where k is the number of arrows, which can be at most 4.

$k = 2$:   There are 6 arrows and as there must be at least two incoming arrows in each vertex, the only possibilities are types 5_{2a} and 5_{2b}.

$k = 3$:   There are seven arrows. Hence every two vertices are connected, otherwise one needs at least 8 arrows

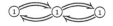

There is one vertex with 3 incoming arrows and one vertex with 3 outgoing arrows. If these vertices are equal ($= v$), there are no triple arrows. Call x the vertex with 2 arrows coming from v and y the other one. Because there are already two incoming arrows in x, $\chi_Q(\epsilon_y, \epsilon_x) = 0$. This also implies that $\chi_Q(\epsilon_y, \epsilon_v) = -2$ and $\chi_Q(\epsilon_x, \epsilon_v) = \chi_Q(\epsilon_x, \epsilon_y) = -1$. This gives us setting 5_{3a}. If the two vertices are different, we can delete one arrow between them, which leaves us with a singularity of dimension $d = 4$ (because now all vertices have 2 incoming and 2 outgoing vertices). So starting from the types 4_{3a-b} and adding one extra arrow we obtain three new types 5_{3b-d}.

$k = 4$:   There are 8 arrows so each vertex must have exactly two incoming and two outgoing arrows. First consider the cases having no double arrows. Fix a vertex v, there is at least one vertex connected to v in both directions. This is because there are 3 remaining vertices and four arrows connected to v (two incoming and two outgoing). If there are two such vertices, w_1 and w_2, the remaining vertex w_3 is not connected to v. Because there are no double arrows we must be in case 5_{4a}. If there is only one such vertex, the quiver contains two disjoint cycles of length 2. This leads to type 5_{4b}.

If there is precisely one double arrow (from v to w), the two remaining vertices must be contained in a cycle of length 2 (if not, there would be 3 arrows leaving v). This leads to type 5_{4c}.

If there are two double arrows, they can be consecutive or disjoint. In the first case, all arrows must be double (if not, there are three arrows leaving one vertex), so this is type 5_{4d}. In the latter case, let v_1 and v_2 be the starting vertices of the double arrows and w_1 and w_2 the end points. As there are no consecutive double arrows, the two arrows leaving w_1 must go to different vertices not equal to w_2. An analogous condition holds for the arrows leaving w_2 and therefore we are in type 5_{4e}. $\Box$

Next, we have to separate the corresponding rings of invariants up to isomorphism.

THEOREM 5.26

There are exactly ten reduced marked quiver singularities in dimension $d = 5$. Only the types 5_{3a} and 5_{4e} have an isomorphic ring of invariants.

PROOF Recall that the dimension of m/m^2 is given by the number of primitive cycles in Q. These numbers are

type	$dim\ m/m^2$	type	$dim\ m/m^2$
5_{2a}	8	5_{4a}	6
5_{2b}	9	5_{4b}	6
5_{3a}	8	5_{4c}	9
5_{3b}	7	5_{4d}	16
5_{3c}	12	5_{4e}	8
5_{3d}	10		

Type 5_{4a} can be separated from type 5_{4b} because 5_{4a} contains $2 + 4$ two-dimensional families of conifold singularities corresponding to representation types of the form

$$\begin{cases} {\tiny\begin{matrix}1&1\\0&0\end{matrix}} \oplus {\tiny\begin{matrix}0&0\\1&1\end{matrix}} \\ {\tiny\begin{matrix}1&0\\1&0\end{matrix}} \oplus {\tiny\begin{matrix}0&1\\0&1\end{matrix}} \end{cases} \quad \text{and} \quad 4 \times {\tiny\begin{matrix}1&1\\1&0\end{matrix}} \oplus {\tiny\begin{matrix}0&0\\0&1\end{matrix}}$$

whereas type 5_{4b} has only $1 + 4$ such families as the decomposition

$$\begin{matrix}0&1\\0&1\end{matrix} \oplus \begin{matrix}1&0\\1&0\end{matrix}$$

is not a valid representation type.

Type 5_{2a} and 5_{2b} are both isolated singularities because we have no non-trivial representation types, whereas types 5_{4c}, and 5_{4e} are not as they have representation types of the form

$$\begin{smallmatrix} 0 & 1 \\ 0 & 0 \end{smallmatrix} \oplus \begin{smallmatrix} 1 & 0 \\ 0 & 0 \end{smallmatrix} \oplus \begin{smallmatrix} 0 & 0 \\ 1 & 1 \end{smallmatrix}$$

giving local quivers smooth equivalent to type 4_{3b} (in the case of type 5_{4c}) and to type 3_a (in the case of 5_{3e}).

Finally, as we know the algebra generators of the rings of invariants (the primitive cycles) it is not difficult to compute these rings explicitly. Type 5_{3a} and type 5_{4e} have a ring of invariants isomorphic to

$$\frac{\mathbb{C}[X_i, Y_i, Z_{ij} : 1 \leq i, j \leq 2]}{(Z_{11}Z_{22} = Z_{12}Z_{21}, X_1 Y_1 Z_{22} = X_1 Y_2 Z_{21} = X_2 Y_1 Z_{12} = X_2 Y_2 Z_{11})}$$

$\square$

References

The results of section 5.1 are due to L. Le Bruyn and C. Procesi [76]. The results of sections 5.2, 5.3, 5.4 are due to L. Le Bruyn, [70], [74]. Lemma 5.7 and proposition 5.10 are due to M. Artin [4], [5]. The results of section 5.5 are due to W. Crawley-Boevey and M. Holland [26], [28]. Proposition 5.12 is due to W. Crawley-Boevey [25] and theorem 5.20 is due to R. Bockland and L. Le Bruyn [12]. The results of section 5.7 are due to R. Bocklandt [10], [11] and the classification of central singularities is due to R. Bocklandt, L. Le Bruyn and G. Van de Weyer [14].

Chapter 6

Nilpotent Representations

Having obtained some control over the quotient variety $\mathrm{triss}_n \, A$ of a Cayley-smooth algebra A we turn to the study of the *fibers* of the quotient map

$$\mathrm{trep}_n \, A \xrightarrow{\pi} \mathrm{triss}_n \, A$$

If $(Q^\bullet, \alpha)$ is the local marked quiver setting of a point $\xi \in \mathrm{triss}_n \, A$ then the GL_n-structure of the fiber $\pi^{-1}(\xi)$ is isomorphic to the $GL(\alpha)$-structure of the *nullcone* $Null_\alpha \, Q^\bullet$ consisting of all nilpotent α-dimensional representations of $Q^\bullet$. In geometric invariant theory, nullcones are investigated by a refinement of the Hilbert criterion: *Hesselink's stratification*.

The main aim of the present chapter is to prove that the different strata in the Hesselink stratification of the nullcone of quiver-representations can be studied via *moduli spaces* of semistable quiver-representations. We will illustrate the method first by considering nilpotent m-tuples of $n \times n$ matrices and generalize the results later to quivers and Cayley-smooth orders. The methods allow us to begin to attack the "hopeless" problem of studying simultaneous conjugacy classes of matrices. We then turn to the description of representation fibers, which can be studied quite explicitly for low-dimensional Cayley-smooth orders, and investigate the fibers of the Brauer-Severi fibration. Before reading the last two sections on Brauer-Severi varieties, it may be helpful to glance through the final chapter where similar, but easier, constructions are studied.

6.1 Cornering matrices

In this section we will outline the main idea of the Hesselink stratification of the nullcone [45] in the generic case, that is, the action of GL_n by simultaneous conjugation on m-tuples of matrices $M_n^m = M_n \oplus \ldots \oplus M_n$. With $Null_n^m$ we denote the nullcone of this action

$$Null_n^m = \{x = (A_1, \ldots, A_m) \in M_n^m \mid \underline{0} = (0, \ldots, 0) \in \overline{\mathcal{O}(x)}\}$$

It follows from the Hilbert criterium 2.2 that $x = (A_1, \ldots, A_m)$ belongs to the nullcone if and only if there is a one-parameter subgroup $\mathbb{C}^* \xrightarrow{\lambda} GL_n$ such that

$$\lim_{t \to 0} \lambda(t).(A_1, \ldots, A_m) = (0, \ldots, 0)$$

We recall from proposition 2.5 that any one-parameter subgroup of GL_n is conjugated to one determined by an integral n-tuple $(r_1, \ldots, r_n) \in \mathbb{Z}^n$ by

$$\lambda(t) = \begin{bmatrix} t^{r_1} & & 0 \\ & \ddots & \\ 0 & & t^{r_n} \end{bmatrix}$$

Moreover, permuting the basis if necessary, we can conjugate this λ to one where the n-tuple if *dominant*, that is, $r_1 \geq r_2 \geq \ldots \geq r_n$. By applying *permutation Jordan-moves*, that is, by simultaneously interchanging certain rows and columns in all A_i, we may therefore assume that the limit-formula holds for a dominant one-parameter subgroup λ of the maximal torus

$$T_n \simeq \underbrace{\mathbb{C}^* \times \ldots \times \mathbb{C}^*}_{n} = \{ \begin{bmatrix} c_1 & & 0 \\ & \ddots & \\ 0 & & c_n \end{bmatrix} \mid c_i \in \mathbb{C}^* \} \hookrightarrow GL_n$$

of GL_n. Computing its action on an $n \times n$ matrix A we obtain

$$\begin{bmatrix} t^{r_1} & & 0 \\ & \ddots & \\ 0 & & t^{r_n} \end{bmatrix} \begin{bmatrix} a_{11} & \ldots & a_{1n} \\ \vdots & & \vdots \\ a_{n1} & \ldots & a_{nn} \end{bmatrix} \begin{bmatrix} t^{-r_1} & & 0 \\ & \ddots & \\ 0 & & r^{-r_n} \end{bmatrix} = \begin{bmatrix} t^{r_1-r_1}a_{11} & \ldots & t^{r_1-r_n}a_{1n} \\ \vdots & & \vdots \\ t^{r_n-r_1}a_{n1} & \ldots & t^{r_n-r_n}a_{nn} \end{bmatrix}$$

But then, using dominance $r_i \leq r_j$ for $i \geq j$, we see that the limit is only defined if $a_{ij} = 0$ for $i \geq j$, that is, when A is a strictly upper triangular matrix. We have proved the first "cornering" result.

LEMMA 6.1

Any m-tuple $x = (A_1, \ldots, A_m) \in Null_n^m$ has a point in its orbit $\mathcal{O}(x)$ under simultaneous conjugation $x' = (A_1', \ldots, A_m')$ with all A_i' strictly upper triangular matrices. In fact permutation Jordan-moves suffice to arrive at x'.

For specific m-tuples $x = (A_1, \ldots, A_m)$ it might be possible to improve on this result. That is, we want to determine the smallest "corner" C in the upper right-hand corner of the matrix, such that all the component matrices A_i can be conjugated simultaneously to matrices A_i' having only nonzero entries in

the corner C

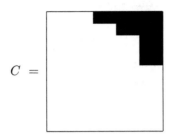

$$C =$$

and no strictly smaller corner C' can be found with this property. Our first task will be to compile a list of the relevant corners and to define an order relation on this set. Consider the *weight space decomposition* of M_n^m for the action by simultaneous conjugation of the maximal torus T_n

$$M_n^m = \oplus_{1 \leq i,j \leq n} M_n^m(\pi_i - \pi_j) = \oplus_{1 \leq i,j \leq n} \mathbb{C}_{\pi_i - \pi_j}^{\oplus m}$$

where $c = diag(c_1, \ldots, c_n) \in T_m$ acts on any element of $M_n^m(\pi_i - \pi_j)$ by multiplication with $c_i c_j^{-1}$, that is, the eigenspace $M_n^m(\pi_i - \pi_j)$ is the space of the (i,j)-entries of the m-matrices. We call

$$\mathcal{W} = \{\pi_i - \pi_j \mid 1 \leq i,j \leq n\}$$

the set of T_n-*weights* of M_n^m. Let $x = (A_1, \ldots, A_m) \in Null_n^m$ and consider the subset $E_x \subset \mathcal{W}$ consisting of the elements $\pi_i - \pi_j$ such that for at least one of the matrix components A_k the (i,j)-entry is nonzero. Repeating the argument above, we see that if λ is a one-parameter subgroup of T_n determined by the integral n-tuple $(r_1, \ldots, r_n) \in \mathbb{Z}^n$ such that $lim\ \lambda(t).x = \underline{0}$ we have

$$\forall\ \pi_i - \pi_j \in E_x \quad \text{we have} \quad r_i - r_j \geq 1$$

Conversely, let $E \subset \mathcal{W}$ be a subset of weights, we want to determine the subset

$$\{s = (s_1, \ldots, s_n) \in \mathbb{R}^n \mid s_i - s_j \geq 1 \ \forall\ \pi_i - \pi_j \in E \}$$

and determine a point in this set, minimal with respect to the usual norm

$$\| s \| = \sqrt{s_1^2 + \ldots + s_n^2}$$

Let $s = (s_1, \ldots, s_n)$ attain such a minimum. We can partition the entries of s in a disjoint union of *strings*

$$\{p_i, p_i + 1, \ldots, p_i + k_i\}$$

with $k_i \in \mathbb{N}$ and subject to the condition that all the numbers $p_{ij} \overset{def}{=} p_i + j$ with $0 \leq j \leq k_i$ occur as components of s, possibly with a multiplicity that we denote by a_{ij}. We call a string $string_i = \{p_i, p_i + 1, \ldots, p_i + k_i\}$ of s *balanced* if and only if

$$\sum_{s_k \in string_i} s_j = \sum_{j=0}^{k_i} a_{ij}(p_i + j) = 0$$

In particular, all balanced strings consists entirely of rational numbers. We have

LEMMA 6.2
Let $E \subset W$, then the subset of $\mathbb{R}^n$ determined by

$$\mathbb{R}_E^n = \{ (r_1, \ldots, r_n) \mid r_i - r_j \geq 1 \, \forall \, \pi_i - \pi_j \in E\}$$

has a unique point $s_E = (s_1, \ldots, s_n)$ of minimal norm $\| s_E \|$. This point is determined by the characteristic feature that all its strings are balanced. In particular, $s_E \in \mathbb{Q}^n$.

PROOF   Let s be a minimal point for the norm in $\mathbb{R}_E^n$ and consider a string of s and denote with S the indices $k \in \{1, \ldots, n\}$ such that $s_k \in string$. Let $\pi_i - \pi_j \in E$, then if only one of i or j belongs to S we have a strictly positive number a_{ij}

$$s_i - s_j = 1 + r_{ij} \quad \text{with} \quad r_{ij} > 0$$

Take $\epsilon_0 > 0$ smaller than all r_{ij} and consider the n-tuple

$$s_\epsilon = s + \epsilon(\delta_{1S}, \ldots, \delta_{nS}) \quad \text{with} \quad \delta_{kS} = 1 \text{ if } k \in S \text{ and } 0 \text{ otherwise}$$

with $\mid \epsilon \mid \leq \epsilon_0$. Then, $s_\epsilon \in \mathbb{R}_E^n$ for if $\pi_i - \pi_j \in E$ and i and j both belong to S or both do not belong to S then $(s_\epsilon)_i - (s_\epsilon)_j = s_i - s_j \geq 1$ and if one of i or j belong to S, then

$$(s_\epsilon)_i - (s_\epsilon)_j = 1 + r_{ij} \pm \epsilon \geq 1$$

by the choice of ϵ_0. However, the norm of s_ϵ is

$$\| s_\epsilon \| = \sqrt{\| s \| + 2\epsilon \sum_{k \in S} s_k + \epsilon^2 \#S}$$

Hence, if the string would not be balanced, $\sum_{k \in S} s_k \neq 0$ and we can choose ϵ small enough such that $\| s_\epsilon \| < \| s \|$, contradicting minimality of s. $\square$

For given n we have the following **algorithm** to compile the list $\mathcal{S}_n$ of all dominant n-tuples $(s_1, \ldots, s_n)$ (that is, $s_i \leq s_j$ whenever $i \geq j$) having all its strings balanced.

- List all Young-diagrams $\mathcal{Y}_n = \{Y_1, \ldots\}$ having $\leq n$ boxes.

- For every diagram Y_l fill the boxes with strictly positive integers subject to the rules

  1. the total sum is equal to n

  2. no two rows are filled identically

  3. at most one row has length 1

This gives a list $\mathcal{T}_n = \{T_1, \ldots\}$ of tableaux.

- For every tableau $T_l \in \mathcal{T}_n$, for each of its rows $(a_1, a_2, \ldots, a_k)$ find a solution p to the linear equation

$$a_1 x + a_2 (x+1) + \ldots + a_k (x+k) = 0$$

and define the $\sum a_i$-tuple of rational numbers

$$(\underbrace{p, \ldots, p}_{a_1}, \underbrace{p+1, \ldots, p+1}_{a_2}, \ldots \underbrace{p+k, \ldots, p+k}_{a_k})$$

Repeating this process for every row of T_l we obtain an n-tuple, which we then order.

The list $\mathcal{S}_n$ will be the combinatorial object underlying the relevant corners and the stratification of the nullcone.

Example 6.1 $\mathcal{S}_n$ for small n

For $n = 2$, we have $\boxed{1\ 1}$ giving $(\frac{1}{2}, -\frac{1}{2})$ and $\boxed{2}$ giving $(0,0)$. For $n = 3$ we have five types

tableau	s_1	s_2	s_3	$\|s\|^2$
$\boxed{1\ 1\ 1}$	1	0	-1	2
$\boxed{1\ 2}$	$\frac{1}{3}$	$\frac{1}{3}$	$-\frac{2}{3}$	$\frac{2}{3}$
$\boxed{2\ 1}$	$\frac{2}{3}$	$-\frac{1}{3}$	$-\frac{1}{3}$	$\frac{2}{3}$
$\boxed{1\ 1}\ \boxed{1}$	$\frac{1}{2}$	0	$-\frac{1}{2}$	$\frac{1}{2}$
$\boxed{3}$	0	0	0	0

$$\mathcal{S}_3 =$$

$\mathcal{S}_4$ has eleven types

$$\mathcal{S}_4 =$$

tableau	s_1	s_2	s_3	s_4	$\|\| s \|\|^2$
$\boxed{1\,1\,1\,1}$	$\frac{3}{2}$	$\frac{1}{2}$	$-\frac{1}{2}$	$-\frac{3}{2}$	5
$\boxed{2\,1\,1}$	$\frac{5}{4}$	$\frac{1}{4}$	$-\frac{3}{4}$	$-\frac{3}{4}$	$\frac{11}{4}$
$\boxed{1\,1\,2}$	$\frac{3}{4}$	$\frac{3}{4}$	$-\frac{1}{4}$	$-\frac{5}{4}$	$\frac{11}{4}$
$\boxed{1\,2\,1}$	1	0	0	-1	2
$\boxed{2\,2}$	$\frac{1}{2}$	$\frac{1}{2}$	$-\frac{1}{2}$	$-\frac{1}{2}$	1
$\boxed{3\,1}$	$\frac{3}{4}$	$-\frac{1}{4}$	$-\frac{1}{4}$	$-\frac{1}{4}$	$\frac{3}{4}$
$\boxed{1\,3}$	$\frac{1}{4}$	$\frac{1}{4}$	$\frac{1}{4}$	$-\frac{3}{4}$	$\frac{3}{4}$
$\boxed{1\,2}\;\boxed{1}$	$\frac{1}{3}$	$\frac{1}{3}$	0	$-\frac{2}{3}$	$\frac{2}{3}$
$\boxed{2\,1}\;\boxed{1}$	$\frac{2}{3}$	0	$-\frac{1}{3}$	$-\frac{1}{3}$	$\frac{2}{3}$
$\boxed{1\,1}\;\boxed{2}$	$\frac{1}{2}$	0	0	$-\frac{1}{2}$	$\frac{1}{2}$
$\boxed{4}$	0	0	0	0	0

Observe that we ordered the elements in $\mathcal{S}_n$ according to $\|\| s \|\|$. The reader is invited to verify that $\mathcal{S}_5$ has 28 different types.   ▯

To every $s = (s_1, \ldots, s_n) \in \mathcal{S}_n$ we associate the following data

- the *corner* C_s is the subspace of M_n^m consisting of those m tuples of $n \times n$ matrices with zero entries except perhaps at position (i,j) where $s_i - s_j \geq 1$. A partial ordering is defined on these corners by the rule

$$C_{s'} < C_s \iff \|\| s' \|\| < \|\| s \|\|$$

- the *parabolic subgroup* P_s which is the subgroup of GL_n consisting of matrices with zero entries except perhaps at entry (i,j) when $s_i - s_j \geq 0$

- the *Levi subgroup* L_s which is the subgroup of GL_n consisting of matrices with zero entries except perhaps at entry (i,j) when $s_i - s_j = 0$. Observe that $L_s = \prod GL_{a_{ij}}$ where the a_{ij} are the multiplicities of $p_i + j$

Example 6.2

Using the sequence of types in the previous example, we have that the relevant corners and subgroup for 3×3 matrices are

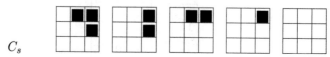

C_s

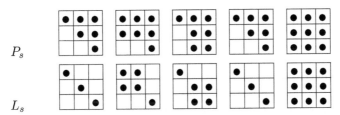

P_s

L_s

For 4×4 matrices the relevant corners are

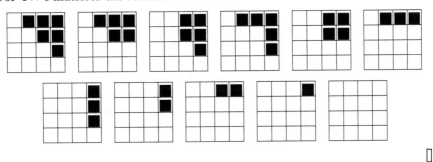

□

Returning to the corner-type of an m-tuple $x = (A_1, \ldots, A_m) \in Null_n^m$, we have seen that $E_x \subset W$ determines a unique $s_{E_x} \in \mathbb{Q}^n$ which is unique up to permuting the entries an element s of $\mathcal{S}_n$. As permuting the entries of s translates into permuting rows and columns in $M_n(\mathbb{C})$ we have the following.

THEOREM 6.1
Every $x = (A_1, \ldots, A_m) \in Null_n^m$ can be brought by permutation Jordan-moves to an m-tuple $x' = (A'_1, \ldots, A'_m) \in C_s$. Here, s is the dominant re-ordering of s_{E_x} with $E_x \subset W$ the subset $\pi_i - \pi_j$ determined by the nonzero entries at place (i, j) of one of the components A_k. The permutation of rows and columns is determined by the dominant reordering.

The m-tuple s (or s_{E_x}) determines a one-parameter subgroup λ_s of T_n where λ corresponds to the unique n-tuple of integers

$$(r_1, \ldots, r_n) \in \mathbb{N}_+ s \cap \mathbb{Z}^n \quad \text{with} \quad gcd(r_i) = 1$$

For any one-parameter subgroup μ of T_n determined by an integral n-tuple $\mu = (a_1, \ldots, a_n) \in \mathbb{Z}^n$ and any $x = (A_1, \ldots, A_n) \in Null_n^m$ we define the integer

$$m(x, \mu) = min \{a_i - a_j \mid x \text{ contains a nonzero entry in } M_n^m(\pi_i - \pi_j) \}$$

From the definition of $\mathbb{R}_E^n$ it follows that the minimal value s_E and λ_{s_E} is

$$s_{E_x} = \frac{\lambda_{s_{E_x}}}{m(x, \lambda_{s_{E_x}})} \quad \text{and} \quad s = \frac{\lambda_s}{m(x, \lambda_s)}$$

We can now state to what extent λ_s is an optimal one-parameter subgroup of T_n.

THEOREM 6.2
Let $x = (A_1, \ldots, A_m) \in Null_n^m$ and let μ be a one-parameter subgroup contained in T_n such that $\lim\limits_{t \to 0} \lambda(t).x = \underline{0}$, then

$$\frac{\| \lambda_{s_{E_x}} \|}{m(x, \lambda_{s_{E_x}})} \leq \frac{\| \mu \|}{m(x, \mu)}$$

The proof follows immediately from the observation that $\frac{\mu}{m(x,\mu)} \in \mathbb{R}_{E_x}^n$ and the minimality of s_{E_x}. Phrased differently, there is no simultaneous reordering of rows and columns that admit an m-tuple $x" = (A"_1, \ldots, A"_m) \in C_{s'}$ for a corner $C_{s'} < C_s$. In the next section we will improve on this result.

6.2   Optimal corners

We have seen that one can transform an m-tuple $x = (A_1, \ldots, A_m) \in Null_n^m$ by interchanging rows and columns to an m-tuple in cornerform C_s. However, it is possible that another point in the orbit $\mathcal{O}(x)$, say, $y = g.x = (B_1, \ldots, B_m)$ can be transformed by permutation Jordan moves in a strictly smaller corner.

Example 6.3
Consider one 3×3 nilpotent matrix of the form

$$x = \begin{bmatrix} 0 & a & b \\ 0 & 0 & 0 \\ 0 & 0 & 0 \end{bmatrix} \quad \text{with} \quad ab \neq 0$$

Then, $E_x = \{\pi_1 - \pi_2, \pi_1 - \pi_3\}$ and the corresponding $s = s_{E_x} = (\frac{2}{3}, -\frac{1}{3}, -\frac{1}{3})$ so x is clearly of corner type

$$C_s = \quad \blacksquare\blacksquare$$

However, x is a nilpotent matrix of rank 1 and by the Jordan-normalform we can conjugate it in standard form, that is, there is some $g \in GL_3$ such that

$$y = g.x = gxg^{-1} = \begin{bmatrix} 0 & 1 & 0 \\ 0 & 0 & 0 \\ 0 & 0 & 0 \end{bmatrix}$$

For this y we have $E_y = \{\pi_1 - \pi_2\}$ and the corresponding $s_{E_y} = (\frac{1}{2}, -\frac{1}{2}, 0)$, which can be brought into standard dominant form $s' = (\frac{1}{2}, 0, -\frac{1}{2})$ by interchanging the two last entries. Hence, by interchanging the last two rows and columns, y is indeed of corner type

$$C_{s'} = \quad \text{}$$

and we have that $C_{s'} < C_s$. ▯

We have used the Jordannormalform to produce this example. As there are no known canonical forms for m tuples of $n \times n$ matrices, it is a difficult problem to determine the optimal corner type in general.

DEFINITION 6.1 *We say that $x = (A_1, \ldots, A_m) \in Null_n^m$ is of optimal corner type C_s if after reordering rows and columns, x is of corner type C_s and there is no point $y = g.x$ in the orbit that is of corner type $C_{s'}$ with $C_{s'} < C_s$.*

We can give an elegant solution to the problem of determining the optimal corner type of an m-tuple in $Null_n^m$ by using results on θ-semistable representations. We assume that $x = (A_1, \ldots, A_m)$ is brought into corner type C_s with $s = (s_1, \ldots, s_n) \in S_n$. We will associate a quiver-representation to x. As we are interested in checking whether we can transform x to a smaller corner-type, it is intuitively clear that the *border* region of C_s will be important.

- the *border* B_s is the subspace of C_s consisting of those m-tuples of $n \times n$ matrices with zero entries except perhaps at entries (i, j) where $s_i - s_j = 1$.

Example 6.4
For 3×3 matrices we have the following corner-types C_s having border-regions B_s and associated Levi-subgroups L_s

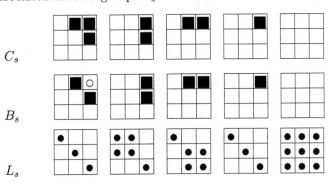

For 4×4 matrices the relevant data are

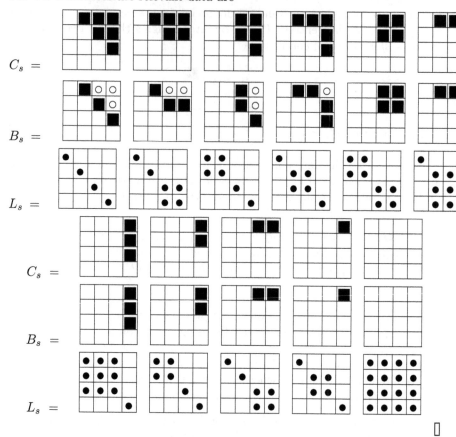

From these examples, it is clear that the action of the Levi-subgroup L_s on the border B_s is a quiver-setting. In general, let $s \in S_n$ be determined by the tableau T_s, then the *associated quiver-setting* (Q_s, α_s) is

- Q_s is the quiver having as many connected components as there are rows in the tableau T_s. If the i-th row in T_s is

$$(a_{i0}, a_{i1}, \ldots, a_{ik_i})$$

then the corresponding string of entries in s is of the form

$$\{\underbrace{p_i, \ldots, p_i}_{a_{i0}}, \underbrace{p_i + 1, \ldots, p_i + 1}_{a_{i1}}, \ldots, \underbrace{p_i + k_i, \ldots, p_i + k_i}_{a_{ik_i}}\}$$

and the i-th component of Q_s is defined to be the quiver Q_i on $k_i + 1$ vertices having m arrows between the consecutive vertices, that is, Q_i is

$$\textcircled{0}\overset{m}{\Longrightarrow}\textcircled{1}\overset{m}{\Longrightarrow}\textcircled{2}\overset{m}{\Longrightarrow} \cdots \overset{m}{\Longrightarrow}\textcircled{k_i}$$

- the dimension vector α_i for the i-th component quiver Q_i is equal to the i-th row of the tableau T_s, that is

$$\alpha_i = (a_{i0}, a_{i1}, \ldots, a_{ik_i})$$

and the total dimension vector α_s is the collection of these component dimension vectors.

- the character $GL(\alpha_s) \xrightarrow{\chi_s} \mathbb{C}^*$ is determined by the integral n-tuple $\theta_s = (t_1, \ldots, t_n) \in \mathbb{Z}^n$ where if entry k corresponds to the j-th vertex of the i-th component of Q_s we have

$$t_k = n_{ij} \overset{\text{def}}{=} d.(p_i + j)$$

where d is the least common multiple of the numerators of the p_i's for all i. Equivalently, the n_{ij} are the integers appearing in the description of the one-parameter subgroup $\lambda_s = (r_1, \ldots, r_n)$ grouped together according to the ordering of vertices in the quiver Q_s. Recall that the character χ_s is then defined to be

$$\chi_s(g_1 \ldots, g_n) = \prod_{i=1}^{n} det(g_i)^{t_i}$$

or in terms of $GL(\alpha_s)$ it sends an element $g_{ij} \in GL(\alpha_s)$ to $\prod_{i,j} det(g_{ij})^{n_{ij}}$.

PROPOSITION 6.1

The action of the Levi-subgroup $L_s = \prod_{i,j} GL_{a_{ij}}$ on the border B_s coincides with the base-change action of $GL(\alpha_s)$ on the representation space $\mathrm{rep}_{\alpha_s} Q_s$. The isomorphism

$$B_s \longrightarrow \mathrm{rep}_{\alpha_s} Q_s$$

is given by sending an m-tuple of border B_s-matrices $(A_1, \ldots, A_m)$ to the representation in $\mathrm{rep}_{\alpha_s} Q_s$ where the j-th arrow between the vertices v_a and v_{a+1} of the i-th component quiver Q_i is given by the relevant block in the matrix A_j.

Some examples are depicted in figure 6.1 Using these conventions we can now state the main result of this section, giving a solution to the problem of optimal corners.

THEOREM 6.3

Let $x = (A_1, \ldots, A_m) \in Null_n^m$ be of corner type C_s. Then, x is of optimal corner type C_s if and only if under the natural maps

$$C_s \longrightarrow B_s \xrightarrow{\simeq} \mathrm{rep}_{\alpha_s} Q_s$$

tableau	L_s	B_s	θ_s	$(Q_s, \alpha_s, \theta_s)$

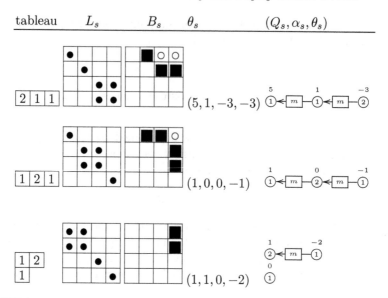

FIGURE 6.1:   Some examples for 4×4 matrices.

(the first map forgets the nonborder entries) x is mapped to a θ_s-semistable representation in $\mathbf{rep}_{\alpha_s} Q_s$.

6.3   Hesselink stratification

Every orbit in $Null_n^m$ has a representative $x = (A_1, \ldots, A_m)$ with all A_i strictly upper triangular matrices. That is, if $N \subset M_n$ is the subspace of strictly upper triangular matrices, then the action map determines a surjection

$$GL_n \times N^m \xrightarrow{\ ac\ } Null_n^m$$

Recall that the standard *Borel subgroup* B is the subgroup of GL_n consisting of all upper triangular matrices and consider the action of B on $GL_n \times M_n^m$ determined by

$$b.(g, x) = (gb^{-1}, b.x)$$

Then, B-orbits in $GL_n \times N^m$ are mapped under the action map ac to the same point in the nullcone $Null_n^m$. Consider the morphisms

$$GL_n \times M_n^m \xrightarrow{\ \pi\ } GL_n/B \times M_n^m$$

which sends a point (g, x) to $(gB, g.x)$. The quotient GL_n/B is called a *flag variety* and is a projective manifold. Its points are easily seen to correspond to complete *flags*

$$\mathcal{F}: \ 0 \subset F_1 \subset F_2 \subset \ldots \subset F_n = \mathbb{C}^n \quad \text{with} \quad dim_{\mathbb{C}} \ F_i = i$$

of subspaces of $\mathbb{C}^n$. For example, if $n = 2$ then $GL_2/B \simeq \mathbb{P}^1$. Consider the fiber π^{-1} of a point $(\overline{g}, (B_1, \ldots, B_m)) \in GL_n/B \times M_n^m$. These are the points

$$(h, (A_1, \ldots, A_m)) \quad \text{such that} \quad \begin{cases} g^{-1}h & = b \in B \\ bA_ib^{-1} & = g^{-1}B_ig \quad \text{for all } 1 \leq i \leq m. \end{cases}$$

Therefore, the fibers of π are precisely the B-orbits in $GL_n \times M_n^m$. That is, there exists a quotient variety for the B-action on $GL_n \times M_n^m$ which is the trivial vectorbundle of rank mn^2

$$\mathcal{T} = GL_n/B \times M_n^m \xrightarrow{\ p\ } GL_n/B$$

over the flag variety GL_n/B. We will denote with $GL_n \times^B N^m$ the image of the subvariety $GL_n \times N^m$ of $GL_n \times M_n^m$ under this quotient map. That is, we have a commuting diagram

$$
\begin{array}{ccc}
GL_n \times N^m & \hookrightarrow & GL_n \times M_n^m \\
\downarrow & & \downarrow \\
GL_n \times^B N^m & \hookrightarrow & GL_n/B \ \times M_n^m
\end{array}
$$

Hence, $\mathcal{V} = GL_n \times^B N^m$ is a subbundle of rank $m.\frac{n(n-1)}{2}$ of the trivial bundle $\mathcal{T}$ over the flag variety. Note however that $\mathcal{V}$ itself is not trivial as the action of GL_n does not map N^m to itself.

THEOREM 6.4
Let U be the open subvariety of m-tuples of strictly upper triangular matrices N^m consisting of those tuples such that one of the component matrices has rank $n - 1$. The action map ac induces the commuting diagram of figure 6.2. The upper map is an isomorphism of GL_n-varieties for the action on fiber bundles to be left multiplication in the first component.

Therefore, there is a natural one-to-one correspondence between GL_n-orbits in $GL_n.U$ and B-orbits in U. Further, ac is a desingularization of the nullcone and $Null_n^m$ is irreducible of dimension

$$(m + 1)\frac{n(n-1)}{2}.$$

PROOF Let $A \in N$ be a strictly upper triangular matrix of rank $n - 1$ and $g \in GL_n$ such that $gAg^{-1} \in N$, then $g \in B$ as one verifies by first

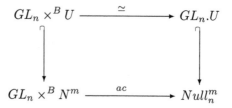

FIGURE 6.2:   Resolution of the nullcone.

bringing A into Jordan-normal form $J_n(0)$. This implies that over a point $x = (A_1, \ldots, A_m) \in U$ the fiber of the action map

$$GL_n \times N^m \xrightarrow{ac} Null_n^m$$

has dimension $\frac{n(n-1)}{2} = dim\ B$. Over all other points the fiber has at least dimension $\frac{n(n-1)}{2}$. But then, by the dimension formula we have

$$dim\ Null_n^m = dim\ GL_n + dim\ N^m - dim\ B = (m+1)\frac{n(n-1)}{2}$$

Over $GL_n.U$ this map is an isomorphism of GL_n-varieties. Irreducibility of $Null_n^m$ follows from surjectivity of ac as $\mathbb{C}[Null_n^m] \hookrightarrow \mathbb{C}[GL_n] \otimes \mathbb{C}[N^m]$ and the latter is a domain. These facts imply that the induced action map

$$GL_n \times^B N^m \xrightarrow{ac} Null_n^m$$

is birational and as the former is a smooth variety (being a vector bundle over the flag manifold), this is a desingularization.   ∎

Example 6.5
Let $n = 2$ and $m = 1$. We have seen in chapter 3 that $Null_2^1$ is a cone in 3-space with the singular top the orbit of the zero-matrix and the open complement the orbit of

$$\begin{bmatrix} 0 & 1 \\ 0 & 0 \end{bmatrix}$$

In this case the flag variety is $\mathbb{P}^1$ and the fiber bundle $GL_2 \times^B N$ has rank one. The action map is depicted in figure 6.3 and is a GL_2-isomorphism over the complement of the fiber of the top.   ∎

Theorem 6.4 gives us a complexity-reduction, both in the dimension of the acting group and in the dimension of the space acted upon, from

- GL_n-orbits in the nullcone $Null_n^m$, to

- B-orbits in N^m.

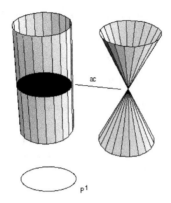

FIGURE 6.3: Resolution of $Null_2^1$.

at least on the stratum $GL_n.U$ described before. The aim of the *Hesselink stratification* of the nullcone is to extend this reduction also to the complement.

Let $s \in \mathcal{S}_n$ and let C_s be the vector space of all m-tuples in M_n^m which are of cornertype C_s. We have seen that there is a Zariski open subset (but, possibly empty) U_s of C_s consisting of m-tuples of optimal corner type C_s. Observe that the action of conjugation of GL_n on M_n^m induces an action of the associated parabolic subgroup P_s on C_s.

DEFINITION 6.2 *The* Hesselink stratum S_s *associated to s is the subvariety $GL_n.U_s$ where U_s is the open subset of C_s consisting of the optimal C_s-type tuples.*

THEOREM 6.5
With notations as before we have a commuting diagram

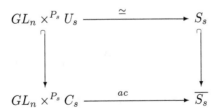

where ac is the action map, $\overline{S_s}$ is the Zariski closure of S_s in $Null_n^m$ and the upper map is an isomorphism of GL_n-varieties.

Here, GL_n/P_s is the flag variety associated to the parabolic subgroup P_s and is a projective manifold. The variety $GL_n \times^{P_s} C_s$ is a vector bundle over the flag variety GL_n/P_s and is a subbundle of the trivial bundle $GL_n \times^{P_s} M_n^m$.

Therefore, the Hesselink stratum S_s is an irreducible smooth variety of dimension

$$dim \ S_s = dim \ GL_n/P_s + rk \ GL_n \times^{P_s} C_s$$
$$= n^2 - dim \ P_s + dim_{\mathbb{C}} \ C_s$$

and there is a natural one-to-one correspondence between the GL_n-orbits in S_s and the P_s-orbits in U_s.

Moreover, the vector bundle $GL_n \times^{P_s} C_s$ is a desingularization of $\overline{S_s}$ hence "feels" the gluing of S_s to the remaining strata. Finally, the ordering of corners has the geometric interpretation

$$\overline{S_s} \subset \bigcup_{\|s'\| \leq \|s\|} S_{s'}$$

We have seen that $U_s = p^{-1} \ \mathbf{rep}^{ss}_{\alpha_s}(Q_s, \theta_s)$ where $C_s \xrightarrow{p} B_s$ is the canonical projection forgetting the nonborder entries. As the action of the parabolic subgroup P_s restricts to the action of its Levi-part L_s on $B_s = \mathbf{rep}_{\alpha_s} Q$ we have a canonical projection

$$U_s/P_s \xrightarrow{p} M^{ss}_{\alpha_s}(Q_s, \theta_s)$$

to the moduli space of θ_s-semistable representations in $\mathbf{rep}_{\alpha_s} \ Q_s$. As none of the components of Q_s admits cycles, these moduli spaces are projective varieties. For small values of m and n these moduli spaces give good approximations to the study of the orbits in the nullcone.

Example 6.6 Nullcone of m-tuples of 2×2 matrices
In the first volume we have seen by a brute force method that the orbits in $Null_2^2$ correspond to points on $\mathbb{P}^1$ together with one extra orbit, the zero representation. For arbitrary m, the relevant strata-information for $Null_2^m$ is contained in the following table

tableau	s	$B_s = C_s$	P_s	$(Q_s, \alpha_s, \theta_s)$
$\boxed{1\,1}$	$(\frac{1}{2}, -\frac{1}{2})$			①$\xleftarrow[m]{1 \quad -1}$①
$\boxed{2}$	$(0,0)$			②^0

Because $B_s = C_s$ we have that the orbit space $U_s/P_s \simeq M^{ss}_{\alpha_s}(Q_s, \theta_s)$. For the first stratum, every representation in $\mathbf{rep}_{\alpha_s} \ Q_s$ is θ_s-semistable except the zero-representation (as it contains a subrepresentation of dimension $\beta = (1,0)$ and $\theta_s(\beta) = -1 < 0$. The action of $L_s = \mathbb{C}^* \times \mathbb{C}^*$ on $\mathbb{C}^m - \underline{0}$ has as orbit space

$\mathbb{P}^{m-1}$, classifying the orbits in the maximal stratum. The second stratum consists of one point, the zero representation. ∎

Example 6.7

A more interesting application, illustrating all of the general phenomena, is the description of orbits in the nullcone of two 3×3 matrices. H. Kraft described them in [62, p. 202] by brute force. The orbit space decomposes as a disjoint union of tori and can be represented by the picture

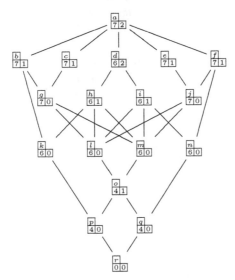

Here, each node corresponds to a torus of dimension the right-hand side number in the bottom row. A point in this torus represents an orbit with dimension the left-hand side number. The top letter is included for classification purposes. That is, every orbit has a unique representant in the following list of couples of 3×3 matrices (A, B). The top letter gives the torus, the first 2 rows give the first two rows of A and the last two rows give the first two rows of B, $x, y \in \mathbb{C}^*$

a

0	1	0
0	0	1
0	x	0
0	0	y

b

0	1	0
0	0	1
0	0	0
0	0	x

c

0	1	0
0	0	1
0	x	0
0	0	0

d

0	1	0
0	0	1
0	x	y
0	0	x

e

0	1	0
0	0	1
0	x	0
0	0	0

f

0	0	0
0	0	1
0	1	0
0	0	x

g

0	1	0
0	0	0
0	0	0
0	0	1

h

0	1	0
0	0	1
0	0	x
0	0	0

i

0	0	x
0	0	0
0	1	0
0	0	1

j

0	0	0
0	0	1
0	1	0
0	0	0

k

0	0	1
0	0	0
0	1	0
0	0	0

l

0	0	0
0	0	1
0	0	1
0	0	0

m

0	0	1
0	0	0
0	1	0
0	0	0

n

0	0	0
0	0	0
0	1	0
0	0	1

o

0	1	0
0	0	0
0	x	0
0	0	0

p

0	1	0
0	0	0
0	0	0
0	0	0

q

0	0	0
0	0	0
0	1	0
0	0	0

r

0	0	0
0	0	0
0	0	0
0	0	0

We will now derive this result from the above description of the Hesselink stratification. To begin, the relevant data concerning $\mathcal{S}_3$ is summarized in the

following table

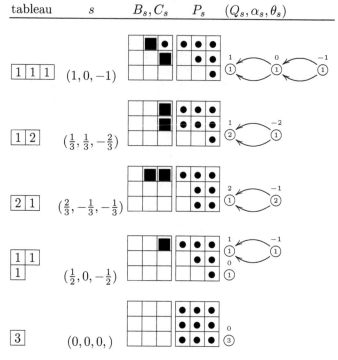

tableau	s	B_s, C_s	P_s	$(Q_s, \alpha_s, \theta_s)$

For the last four corner types, $B_s = C_s$ whence the orbit space U_s/P_s is isomorphic to the moduli space $M^{ss}_{\alpha_s}(Q_s, \theta_s)$. Consider the quiver-setting

If the two arrows are not linearly independent, then the representation contains a proper subrepresentation of dimension-vector $\beta = (1,1)$ or $(1,0)$ and in both cases $\theta_s(\beta) < 0$ whence the representation is not θ_s-semistable. If the two arrows are linearly independent, we can use the GL_2-component to bring

them in the form $(\begin{bmatrix} 0 \\ 1 \end{bmatrix}, \begin{bmatrix} 1 \\ 0 \end{bmatrix})$, whence $M^{ss}_{\alpha_s}(Q_s, \alpha_s)$ is reduced to one point, corresponding to the matrix-couple of type l

$$(\begin{bmatrix} 0 & 0 & 0 \\ 0 & 0 & 1 \\ 0 & 0 & 0 \end{bmatrix}, \begin{bmatrix} 0 & 0 & 1 \\ 0 & 0 & 0 \\ 0 & 0 & 0 \end{bmatrix})$$

A similar argument, replacing linear independence by common zero-vector shows that also the quiver-setting corresponding to the tableau $\boxed{2\,1}$ has one point as its moduli space, the matrix-tuple of type k. Incidentally, this shows that the corners corresponding to the tableaux $\boxed{2}\boxed{1}$ or $\boxed{1}\boxed{2}$ cannot be optimal when $m = 1$ as then the row or column vector always has a kernel or cokernel whence cannot be θ_s-semistable. This of course corresponds to the fact that the only orbits in $Null_3^1$ are those corresponding to the Jordan-matrixes

$$\begin{bmatrix} 0 & 1 & 0 \\ 0 & 0 & 1 \\ 0 & 0 & 0 \end{bmatrix} \quad \begin{bmatrix} 0 & 1 & 0 \\ 0 & 0 & 0 \\ 0 & 0 & 0 \end{bmatrix} \quad \begin{bmatrix} 0 & 0 & 0 \\ 0 & 0 & 0 \\ 0 & 0 & 0 \end{bmatrix}$$

which are, respectively, of corner type $\boxed{1}\boxed{1}\boxed{1}$, $\boxed{\begin{smallmatrix}1&1\\1&\end{smallmatrix}}$ and $\boxed{3}$, whence the two other types do not occur. Next, consider the quiver setting

A representation in $\mathbf{rep}_{\alpha_s} Q_s$ is θ_s-semistable if and only if the two maps are not both zero (otherwise, there is a subrepresentation of dimension $\beta = (1, 0)$ with $\theta_s(\beta) < 0$). The action of $GL(\alpha_s) = \mathbb{C}^* \times \mathbb{C}^*$ on $\mathbb{C}^2 - \underline{0}$ has as orbit space $\mathbb{P}^1$ and they are represented by matrix-couples

$$\left(\begin{bmatrix} 0 & 0 & a \\ 0 & 0 & 0 \\ 0 & 0 & 0 \end{bmatrix}, \begin{bmatrix} 0 & 0 & b \\ 0 & 0 & 0 \\ 0 & 0 & 0 \end{bmatrix} \right)$$

with $[a : b] \in \mathbb{P}^1$ giving the types o, p and q. Clearly, the stratum $\boxed{3}$ consists just of the zero-matrix, which is type r. Remains to investigate the quiver-setting

Again, one easily verifies that a representation in $\mathbf{rep}_{\alpha_s} Q_s$ is θ_s-semistable if and only if $(a, b) \neq (0, 0) \neq (c, d)$ (for otherwise one would have subrepresentations of dimensions $(1, 1, 0)$ or $(1, 0, 0)$). The corresponding $GL(\alpha_s)$-orbits are classified by

$$M_{\alpha_s}^{ss}(Q_s.\theta_s) \simeq \mathbb{P}^1 \times \mathbb{P}^1$$

corresponding to the matrix-couples of types a, b, c, e, f, g, j, k and n

$$\left(\begin{bmatrix} 0 & c & 0 \\ 0 & 0 & a \\ 0 & 0 & 0 \end{bmatrix}, \begin{bmatrix} 0 & d & 0 \\ 0 & 0 & b \\ 0 & 0 & 0 \end{bmatrix} \right)$$

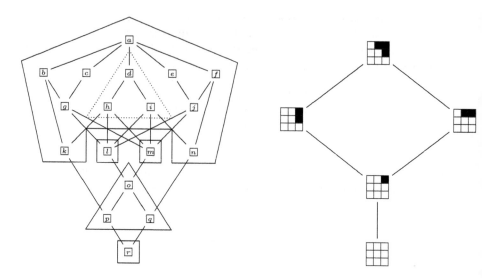

FIGURE 6.4: Nullcone of couples of 3×3 matrices.

where $[a : b]$ and $[c : d]$ are points in $\mathbb{P}^1$. In this case, however, $C_s \neq B_s$ and we need to investigate the fibers of the projection

$$U_s/P_s \xrightarrow{\ p\ } M_{\alpha_s}^{ss}(Q_s, \alpha_s)$$

Now, P_s is the Borel subgroup of upper triangular matrices and one verifies that the following two couples

$$\left(\ \begin{bmatrix} 0 & c & 0 \\ 0 & 0 & a \\ 0 & 0 & 0 \end{bmatrix} ,\ \begin{bmatrix} 0 & d & 0 \\ 0 & 0 & b \\ 0 & 0 & 0 \end{bmatrix}\ \right) \quad \text{and} \quad \left(\ \begin{bmatrix} 0 & c & x \\ 0 & 0 & a \\ 0 & 0 & 0 \end{bmatrix} ,\ \begin{bmatrix} 0 & d & y \\ 0 & 0 & b \\ 0 & 0 & 0 \end{bmatrix}\ \right)$$

lie in the same B-orbit if and only if $det \begin{bmatrix} a & c \\ b & d \end{bmatrix} \neq 0$, that is, if and only if $[a : b] \neq [c : d]$ in $\mathbb{P}^1$. Hence, away from the diagonal p is an isomorphism. On the diagonal one can again verify by direct computation that the fibers of p are isomorphic to $\mathbb{C}$, giving rise to the cases d, h and i in the classification.

The connection between this approach and Kraft's result is depicted in figure 6.4. The picture on the left is Kraft's toric degeneration picture where we enclosed all orbits belonging to the same Hesselink strata, that is, having the same optimal corner type. The dashed region enclosed the orbits which do not come from the moduli spaces $M_{\alpha_s}^{ss}(Q_s, \theta_s)$, that is, those coming from the projection $U_s/P_s \longrightarrow M_{\alpha_s}^{ss}(Q_s, \theta_s)$). The picture on the right gives the ordering of the relevant corners.   ⬜

Example 6.8

We see that we get most orbits in the nullcone from the moduli spaces $M_{\alpha_s}^{ss}(Q_s,\theta_s)$. The reader is invited to work out the orbits in $Null_4^2$. We list here the moduli spaces of the relevant corners

corner	$M_{\alpha_s}^{ss}(Q_s,\theta_s)$	corner	$M_{\alpha_s}^{ss}(Q_s,\theta_s)$	corner	$M_{\alpha_s}^{ss}(Q_s,\theta_s)$
	$\mathbb{P}^1 \times \mathbb{P}^1 \times \mathbb{P}^1$		$\mathbb{P}^1$		$\mathbb{P}^1$
	$\mathbb{P}^3 \sqcup \mathbb{P}^1 \times \mathbb{P}^1 \sqcup \mathbb{P}^1 \times \mathbb{P}^1$		$\mathbb{P}^1 \sqcup S^2(\mathbb{P}^1)$		$\mathbb{P}^0$
	$\mathbb{P}^1$		$\mathbb{P}^1$		$\mathbb{P}^0$

Observe that two potential corners are missing in this list. This is because we have the following quiver setting for the corner

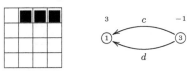

and there are no θ_s-semistable representations as the two maps have a common kernel, whence a subrepresentation of dimension $\beta = (1,0)$ and $\theta_s(\beta) < 0$. A similar argument holds for the other missing corner. □

For general n, a similar argument proves that the corners associated to the tableaux $\boxed{1\,|\,n}$ and $\boxed{n\,|\,1}$ are not optimal for tuples in $Null_{n+1}^m$ unless $m \geq n$. It is also easy to see that with $m \geq n$ all relevant corners appear in $Null_{n+1}^m$, that is all potential Hesselink strata are non-empty.

6.4 Cornering quiver representations

In this section we generalize the results on matrices to representation of arbitrary quivers. Let Q be a quiver on k vertices $\{v_1,\ldots,v_k\}$ and fix a dimension vector $\alpha = (a_1, \ldots, a_k)$ and denote the total dimension $\sum_{i=1}^k a_i$ by a. A representation $V \in \mathbf{rep}_\alpha Q$ is said to belong to the nullcone $Null_\alpha Q$ if the trivial representation $\underline{0} \in \overline{\mathcal{O}(V)}$. Equivalently, all polynomial invariants are zero when evaluated in V, that is, the traces of all oriented cycles in Q are zero in V. By the Hilbert criterium 2.2 for $GL(\alpha)$, $V \in Null_\alpha Q$ if and only if there is a one-parameter subgroup

$$\mathbb{C}^* \xrightarrow{\lambda} GL(\alpha) = \begin{bmatrix} GL_{a_1} & & \\ & \ddots & \\ & & GL_{a_k} \end{bmatrix} \hookrightarrow GL_a$$

such that $\underset{\to}{lim}\ \lambda(t).V = \underline{0}$. Up to conjugation in $GL(\alpha)$, or equivalently, replacing V by another point in the orbit $\mathcal{O}(V)$, we may assume that λ lies in the maximal torus T_a of $GL(\alpha)$ (and of GL_a) and can be represented by an integral a-tuple $(r_1, \ldots, r_a) \in \mathbb{Z}^a$ such that

$$\lambda(t) = \begin{bmatrix} t^{r_1} & & \\ & \ddots & \\ & & t^{r_a} \end{bmatrix}$$

We have to take the vertices into account, so we decompose the integer interval $[1, 2, \ldots, a]$ into *vertex intervals* I_{v_i} such that

$$[1, 2, \ldots, a] = \sqcup_{i=1}^k I_{v_i} \quad \text{with} \quad I_{v_i} = [\sum_{j=1}^{i-1} a_j + 1, \ldots, \sum_{j=1}^i a_j]$$

If we recall that the weights of T_a are isomorphic to $\mathbb{Z}^a$ having canonical generators π_p for $1 \le p \le a$ we can decompose the representation space into weight spaces

$$\mathbf{rep}_\alpha Q = \bigoplus_{\pi_{pq} = \pi_q - \pi_p} \mathbf{rep}_\alpha Q(\pi_{pq})$$

where the eigenspace of π_{pq} is nonzero if and only if for $p \in I_{v_i}$ and $q \in I_{v_j}$, there is an arrow

$$\textcircled{j} \longleftarrow \textcircled{i}$$

in the quiver Q. Call $\pi_\alpha Q$ the set of weights π_{pq} which have nonzero eigenspace in $\mathbf{rep}_\alpha Q$. Using this weight space decomposition we can write every representation as $V = \sum_{p,q} V_{pq}$ where V_{pq} is a vector of the (p, q)-entries of the maps $V(a)$ for all arrows a in Q from v_i to v_j. Using the fact that the action of T_a on $\mathbf{rep}_\alpha Q$ is induced by conjugation, we deduce as before that for λ determined by $(r_1, \ldots, r_a)$

$$\underset{t \to 0}{lim}\ \lambda(t).V = \underline{0} \Leftrightarrow r_q - r_p \ge 1 \text{ whenever } V_{pq} \ne 0$$

Again, we define the corner type C of the representation V by defining the subset of real a-tuples

$$E_V = \{(x_1, \ldots, x_a) \in \mathbb{R}^a \mid x_q - x_p \geq 1 \; \forall \; V_{pq} \neq 0\}$$

and determine a minimal element s_V in it, minimal with respect to the usual norm on $\mathbb{R}^a$. Similar to the case of matrices considered before, it follows that s_V is a uniquely determined point in $\mathbb{Q}^a$, having the characteristic property that its entries can be partitioned into strings

$$\{\underbrace{p_l, \ldots, p_l}_{a_{l0}}, \underbrace{p_l + 1, \ldots, p_l + 1}_{a_{l1}}, \ldots, \underbrace{p_l + k_l, \ldots, p_l + k_l}_{a_{lk_l}}\} \quad \text{with all } a_{lm} \geq 1$$

which are balanced, that is, $\sum_{m=0}^{k_l} a_{lm}(p_l + m) = 0$.

Note however that this time we are not allowed to bring s_V into dominant form, as we can only permute base-vectors of the vertex-spaces. That is, we can only use the action of the *vertex-symmetric groups*

$$S_{a_1} \times \ldots \times S_{a_k} \hookrightarrow S_a$$

to bring s_V into *vertex dominant form* , that is if $s_V = (s_1, \ldots, s_a)$ then

$$s_q \leq s_p \quad \text{whenever} \quad p, q \in I_{v_i} \text{ for some } i \text{ and } p < q$$

We compile a list $\mathcal{S}_\alpha$ of such rational a-tuples by the following `algorithm`

- Start with the list $\mathcal{S}_a$ of matrix corner types.

- For every $s \in \mathcal{S}_a$ consider all permutations $\sigma \in S_a/(S_{a_1} \times \ldots \times S_{a_k})$ such that $\sigma.s = (s_{\sigma(1)}, \ldots, s_{\sigma(a)})$ is vertex dominant.

- Take $\mathcal{H}_\alpha$ to be the list of the distinct a-tuples $\sigma.s$ which are vertex dominant.

- Remove $s \in \mathcal{H}_\alpha$ whenever there is an $s' \in \mathcal{H}_\alpha$ such that

$$\pi_s \, Q = \{\pi_{pq} \in \pi_\alpha \, Q \mid s_q - s_p \geq 1\} \subset \pi_{s'} \, Q = \{\pi_{pq} \in \pi_\alpha \, Q \mid s'_q - s'_p \geq 1\}$$

and $\| s \| > \| s' \|$.

- The list $\mathcal{S}_\alpha$ are the remaining entries s from $\mathcal{H}_\alpha$.

For $s \in \mathcal{S}_\alpha$, we define associated data similar to the case of matrices

- The *corner* C_s is the subspace of $\mathbf{rep}_\alpha \, Q$ such that all arrow matrices V_b, when viewed as $a \times a$ matrices using the partitioning in vertex-entries, have only nonzero entries at spot (p, q) when $s_q - s_p \geq 1$.

- The *border* B_s is the subspace of $\mathbf{rep}_\alpha \, Q$ such that all arrow matrices V_b, when viewed as $a \times a$ matrices using the partitioning in vertex-entries, have only nonzero entries at spot (p, q) when $s_q - s_p = 1$.

- The *parabolic subgroup* $P_s(\alpha)$ is the intersection of $P_s \subset GL_a$ with $GL(\alpha)$ embedded along the diagonal. $P_s(\alpha)$ is a parabolic subgroup of $GL(\alpha)$, that is, contains the product of the Borels $B(\alpha) = B_{a_1} \times \ldots \times B_{a_k}$.

- The *Levi-subgroup* $L_s(\alpha)$ is the intersection of $L_s \subset GL_a$ with $GL(\alpha)$ embedded along the diagonal.

We say that a representation $V \in \mathbf{rep}_\alpha Q$ is of *corner type* C_s whenever $V \in C_s$.

THEOREM 6.6
By permuting the vertex-bases, every representation $V \in \mathbf{rep}_\alpha Q$ can be brought to a corner type C_s for a uniquely determined s, which is a vertex-dominant reordering of s_V.

Example 6.9
Consider the following quiver setting

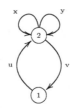

Then, the relevant corners have the following block decomposition

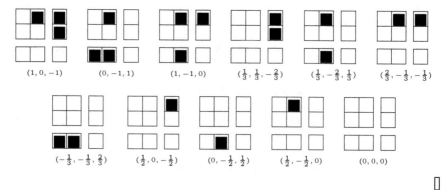

$$(1,0,-1) \quad (0,-1,1) \quad (1,-1,0) \quad (\tfrac{1}{3},\tfrac{1}{3},-\tfrac{2}{3}) \quad (\tfrac{1}{3},-\tfrac{2}{3},\tfrac{1}{3}) \quad (\tfrac{2}{3},-\tfrac{1}{3},-\tfrac{1}{3})$$

$$(-\tfrac{1}{3},-\tfrac{1}{3},\tfrac{2}{3}) \quad (\tfrac{1}{2},0,-\tfrac{1}{2}) \quad (0,-\tfrac{1}{2},\tfrac{1}{2}) \quad (\tfrac{1}{2},-\tfrac{1}{2},0) \quad (0,0,0)$$

□

Again, we solve the problem of *optimal corner representations* by introducing a new quiver setting.

Fix a type $s \in \mathcal{S}_\alpha Q$ and let $J_1, \ldots, J_u$ be the distinct strings partitioning the entries of s, say with

$$J_l = \{\underbrace{p_l, \ldots, p_l}_{\sum_{i=1}^k b_{i,l0}}, \underbrace{p_l+1, \ldots, p_l+1}_{\sum_{i=1}^k b_{i,l1}}, \ldots, \underbrace{p_l+k_l, \ldots, p_l+k_l}_{\sum_{i=1}^k b_{i,lk_l}}\}$$

where $b_{i,lm}$ is the number of entries $p \in I_{v_i}$ such that $s_p = p_l + m$. To every string l we will associate a quiver $Q_{s,l}$ and dimension vector $\alpha_{s,l}$ as follows

- $Q_{s,l}$ has $k.(k_l+1)$ vertices labeled (v_i, m) with $1 \leq i \leq k$ and $0 \leq m \leq k_l$.

- In $Q_{s,l}$ there are as many arrows from vertex (v_i, m) to vertex $(v_j, m+1)$ as there are arrows in Q from vertex v_i to vertex v_j. There are no arrows between (v_i, m) and (v_j, m') if $m' - m \neq 1$.

- The dimension-component of $\alpha_{s,l}$ in vertex (v_i, m) is equal to $b_{i,lm}$.

Example 6.10

For the above quiver, all component quivers $Q_{s,l}$ are pieces of the quiver below

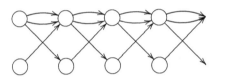

Clearly, we only need to consider that part of the quiver $Q_{s,l}$ where the dimensions of the vertex spaces are nonzero.   ⬚

The quiver-setting (Q_s, α_s) associated to a type $s \in S_\alpha Q$ will be the disjoint union of the string quiver-settings $(Q_{s,l}, \alpha_{s,l})$ for $1 \leq l \leq u$.

THEOREM 6.7

With notations as before, for $s \in S_\alpha Q$ we have isomorphisms

$$\begin{cases} B_s & \simeq \mathbf{rep}_{\alpha_s} Q_s \\ L_s(\alpha) & \simeq GL(\alpha_s) \end{cases}$$

Moreover, the base-change action of $GL(\alpha_s)$ on $\mathbf{rep}_{\alpha_s} Q_s$ coincides under the isomorphisms with the action of the Levi-subgroup $L_s(\alpha)$ on the border B_s.

In order to determine the representations in $\mathbf{rep}_{\alpha_s} Q_s$ which have *optimal corner type C_s* we define the following character on the Levi-subgroup

$$L_s(\alpha) = \prod_{l=1}^{u} \times_{i=1}^{k} \times_{m=0}^{k_l} GL_{b_{i,lm}} \xrightarrow{\chi_{\theta_s}} \mathbb{C}^*$$

determined by sending a tuple $(g_{i,lm})_{ilm} \longrightarrow \prod_{ilm} \det g_{i,lm}^{m_{i,lm}}$ where the exponents are determined by

$$\theta_s = (m_{i,lm})_{ilm} \quad \text{where} \quad m_{i,lm} = d(p_l + m)$$

with d the least common multiple of the numerators of the rational numbers p_l for all $1 \leq l \leq u$.

THEOREM 6.8

Consider a representation $V \in Null_\alpha \, Q$ of corner type C_s. Then, V is of optimal corner type C_s if and only if under the natural maps

$$C_s \xrightarrow{\pi} B_s \xrightarrow{\simeq} rep_{\alpha_s} \, Q_s$$

V is mapped to a θ_s-semistable representation in $rep_{\alpha_s} \, Q_s$. If U_s is the open subvariety of C_s consisting of all representations of optimal corner type C_s, then

$$U_s = \pi^{-1} \, rep_{\alpha_s}^{ss}(Q_s, \theta_s)$$

For the corresponding Hesselink stratum $S_s = GL(\alpha).U_s$ we have the commuting diagram

$$
\begin{array}{ccc}
GL(\alpha) \times^{P_s(\alpha)} U_s & \xrightarrow{\quad \simeq \quad} & S_s \\[2mm]
\Big\uparrow & & \Big\uparrow \\[2mm]
\Big\downarrow & & \Big\downarrow \\[2mm]
GL(\alpha) \times^{P_s(\alpha)} C_s & \xrightarrow{\quad ac \quad} & \overline{S_s}
\end{array}
$$

where ac is the action map, $\overline{S_s}$ is the Zariski closure of S_s in $Null_\alpha \, Q$ and the upper map is an isomorphism as $GL(\alpha)$-varieties.

Here, $GL(\alpha)/P_s(\alpha)$ is the flag variety associated to the parabolic subgroup $P_s(\alpha)$ and is a projective manifold. The variety $GL(\alpha) \times^{P_s(\alpha)} C_s$ is a vectorbundle over the flag variety $GL(\alpha)/P_s(\alpha)$ and is a subbundle of the trivial bundle $GL(\alpha) \times^{P_s(\alpha)} rep_\alpha \, Q$.

Hence, the Hesselink stratum S_s is an irreducible smooth variety of dimension

$$dim \, S_s = dim \, GL(\alpha)/P_s(\alpha) + rk \, GL(\alpha) \times^{P_s(\alpha)} C_s$$

$$= \sum_{i=1}^{k} a_i^2 - dim \, P_s(\alpha) + dim_{\mathbb{C}} \, C_s$$

and there is a natural one-to-one correspondence between the $GL(\alpha)$-orbits in S_s and the $P_s(\alpha)$-orbits in U_s.

Moreover, the vector bundle $GL(\alpha) \times^{P_s(\alpha)} C_s$ is a desingularization of $\overline{S_s}$ hence "feels" the gluing of S_s to the remaining strata. The ordering of corners has the geometric interpretation

$$\overline{S_s} \subset \bigcup_{\|s'\| \leq \|s\|} S_{s'}$$

Finally, because $P_s(\alpha)$ acts on B_s by the restriction to its subgroup $L_s(\alpha) = GL(\alpha_s)$ we have a projection from the orbit space

$$U_s/P_s \xrightarrow{\ p\ } M^{ss}_{\alpha_s}(Q_s, \theta_s)$$

to the moduli space of θ_s-semistable quiver representations.

Example 6.11
Above we have listed the relevant corner-types for the nullcone of the quiver-setting

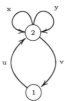

In the table below we list the data of the three irreducible components of $Null_\alpha\, Q/GL(\alpha)$ corresponding to the three maximal Hesselink strata

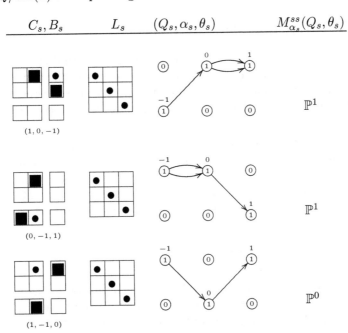

C_s, B_s	L_s	$(Q_s, \alpha_s, \theta_s)$	$M^{ss}_{\alpha_s}(Q_s, \theta_s)$
$(1, 0, -1)$			$\mathbb{P}^1$
$(0, -1, 1)$			$\mathbb{P}^1$
$(1, -1, 0)$			$\mathbb{P}^0$

There are 6 other Hesselink strata consisting of precisely one orbit. Finally, two possible corner-types do not appear as there are no θ_s-semistable representations for the corresponding quiver setting as depicted in figure 6.5

☐

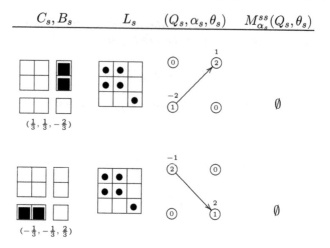

FIGURE 6.5: Missing strata.

6.5   Simultaneous conjugacy classes

We have come a long way from our bare-hands description of the simultaneous conjugacy classes of couples of 2×2 matrices in the first chapter of volume 1. In this section we will summarize what we have learned so far to approach the hopeless problem of classifying conjugacy classes of m tuples of $n \times n$ matrices.

First, we show how one can reduce the study of representations of a Quillen-smooth algebra to that of studying nullcones of quiver representations. Let A be an affine $\mathbb{C}$-algebra and M_ξ is a semisimple n-dimensional module such that the representation variety $\mathbf{rep}_n \int_n A$ is smooth in M_ξ, that is $\xi \in Sm_n A$. Let M_ξ be of representation type $\tau = (e_1, d_1; \ldots; e_k, d_k)$, that is

$$M_\xi = S_1^{\oplus e_1} \oplus \ldots \oplus S_k^{\oplus e_k}$$

with distinct simple components S_i of dimension d_i and occurring in M_ξ with multiplicity e_i, then the $GL(\alpha) = Stab\ M_\xi$-structure on the normal space N_ξ to the orbit $\mathcal{O}(M_\xi)$ is isomorphic to that of the representation space

$$\mathbf{rep}_\alpha\ Q^\bullet$$

of a certain marked quiver on k vertices. The slice theorem asserts the existence of a slice $S_\xi \xrightarrow{\phi} N_\xi$ and a commuting diagram

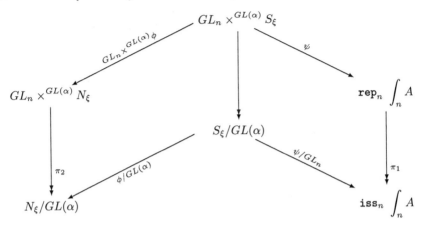

in a neighborhood of $\xi \in \mathtt{iss}_n \int_n A$ on the right and a neighborhood of the image $\underline{0}$ of the trivial representation in $N_\xi/GL(\alpha)$ on the left. In this diagram, the vertical maps are the quotient maps, all diagonal maps are étale and the upper ones are GL_n-equivariant. In particular, there is a GL_n-isomorphism between the fibers

$$\pi_2^{-1}(\underline{0}) \simeq \pi_1^{-1}(\xi)$$

Because $\pi_2^{-1}(\underline{0}) \simeq GL_n \times^{GL(\alpha)} \pi^{-1}(\underline{0})$ with π is the quotient morphism for the marked quiver representations $N_\xi = \mathtt{rep}_\alpha Q^\bullet \xrightarrow{\pi} \mathtt{iss}_\alpha Q^\bullet = N_x/GL(\alpha)$ we have a GL_n-isomorphism

$$\pi_1^{-1}(\xi) \simeq GL_n \times^{GL(\alpha)} \pi^{-1}(\underline{0})$$

That is, there is a natural one-to-one correspondence between

- GL_n-orbits in the fiber $\pi_1^{-1}(\zeta)$, that is, isomorphism classes of n-dimensional representations of A with Jordan-Hölder decomposition M_ξ, and

- $GL(\alpha)$-orbits in $\pi^{-1}(\underline{0})$, that is, the nullcone of the marked quiver $Null_\alpha Q^\bullet$.

A summary follows.

THEOREM 6.9
Let A be an affine Quillen-smooth $\mathbb{C}$-algebra and M_ξ a semisimple n-dimensional representation of A. Then, the isomorphism classes of n-dimensional representations of A with Jordan-Hölder decomposition isomorphic to M_ξ are given by the $GL(\alpha)$-orbits in the nullcone $Null_\alpha Q^\bullet$ of the local marked quiver setting.

The problem of classifying simultaneous conjugacy classes of m-tuples of $n \times n$ matrices is the same as n-dimensional representations of the Quillen-smooth algebra $\mathbb{C}\langle x_1, \ldots, x_m \rangle$. To study semisimple representations, one considers the quotient map

$$M_n^m = \mathbf{rep}_n \mathbb{C}\langle x_1, \ldots, x_m \rangle \xrightarrow{\ \pi\ } \mathbf{iss}_n \, \mathbb{C}\langle x_1, \ldots, x_m \rangle = \mathbf{iss}_n^m$$

Fix a point $\xi \in \mathbf{iss}_n^m$ and assume that the corresponding semi-simple n-dimensional representation M_ξ is of representation type $\tau = (e_1, d_1; \ldots; e_k, d_k)$.

We have shown that the coordinate ring $\mathbb{C}[\mathbf{iss}_n^m] = \mathbb{N}_n^m$ is the *necklace algebra*, that is, is generated by traces of monomials in the generic $n \times n$ matrices $X_1, \ldots, X_m$ of length bounded by $n^2 + 1$. Further, if we collect all M_ξ with representation type τ in the subset $\mathbf{iss}_n^m(\tau)$, then

$$\mathbf{iss}_n = \bigsqcup_\tau \mathbf{iss}_n^m(\tau)$$

is a finite stratification of $\mathbf{iss}_n^m$ into locally closed smooth algebraic subvarieties.

We have an ordering on the representation types $\tau' < \tau$ indicating that the stratum $\mathbf{iss}_n^m(\tau')$ is contained in the Zariski closure of $\mathbf{iss}_n^m(\tau)$. This order relation is induced by the *direct ordering*

$$\tau' = (e_1', d_1'; \ldots; e_{k'}', d_{k'}') <^{dir} \tau = (e_1, d_1; \ldots; e_k, d_k)$$

if there is a permutation σ of $[1, 2, \ldots, k']$ and there are numbers

$$1 = j_0 < j_1 < j_2 \ldots < j_k = k'$$

such that for every $1 \le i \le k$ we have the following relations

$$\begin{cases} e_i d_i &= \sum_{j=j_{i-1}+1}^{j_i} e_{\sigma(j)}' d_{\sigma(j)}' \\ e_i &\le e_{\sigma(j)}' \text{ for all } j_{i-1} < j \le j_i \end{cases}$$

Because $\mathbf{iss}_n^m$ is irreducible, there is an open stratum corresponding to the simple representations, that is, type $(1, n)$. The subgeneric strata are all of the form

$$\tau = (1, m_1; 1, m_2) \quad \text{with} \quad m_1 + m_2 = n$$

The (in)equalities describing the locally closed subvarieties $\mathbf{iss}_n^m(\tau)$ can (in principle) be deduced from the theory of trace identities. Remains to study the local structure of the quotient variety $\mathbf{iss}_n^m$ near ξ and the description of the fibers $\pi^{-1}(\xi)$.

Both problems can be tackled by studying the local quiver setting (Q_ξ, α_ξ) corresponding to ξ, which describes the $GL(\alpha_\xi) = Stab(M_\xi)$-module structure of the normal space to the orbit of M_ξ. If ξ is of representation type

$\tau = (e_1, d_1; \ldots; e_k, d_k)$ then the local quiver Q_ξ has k-vertices $\{v_1, \ldots, v_k\}$ corresponding to the k distinct simple components $S_1, \ldots, S_k$ of M_ξ and the number of arrows (resp. loops) from v_i to v_j (resp. in v_i) are given by the dimensions

$$dim_\mathbb{C} Ext^1(S_i, S_j) \quad \text{resp.} \quad dim_\mathbb{C} Ext^1(S_i, S_i)$$

and these numbers can be computed from the dimensions of the simple components

$$\begin{cases} \# \; \circled{j} \xleftarrow{\;a\;} \circled{i} & = (m-1)d_i d_j \\ \# \; \circled{i} \circlearrowright & = (m-1)d_i^2 + 1 \end{cases}$$

Further, the local dimension vector α_ξ is given by the multiplicities $(e_1, \ldots, e_k)$. The étale local structure of $\mathbf{iss}_n^m$ in a neighborhood of ξ is the same as that of the quotient variety $\mathbf{iss}_{\alpha_\xi} Q_\xi$ in a neighborhood of $\underline{0}$. The local algebra of the latter is generated by traces along oriented cycles in the quiver Q_ξ. A direct application is discussed below.

PROPOSITION 6.2

For $m \geq 2$, ξ is a smooth point of $\mathbf{iss}_n^m$ if and only if M_ξ is a simple representation, unless $(m, n) = (2, 2)$ in which case $\mathbf{iss}_2^2 \simeq \mathbb{C}^5$ is a smooth variety.

PROOF If ξ is of representation type $(1, n)$, the local quiver setting (Q_ξ, α_ξ) is

where $d = (m-1)n^2 + 1$, whence the local algebra is the formal power series ring in d variables and so $\mathbf{iss}_n^m$ is smooth in ξ. Because the singularities form a Zariski closed subvariety of $\mathbf{iss}_n^m$, the result follows if we prove that all points ξ lying in subgeneric strata, say, of type $(1, m_1; 1, m_2)$ are singular. In this case the local quiver setting is equal to

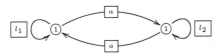

where $a = (m-1)m_1 m_2$ and $l_i = (m-1)m_i^2 + 1$. Let us denote the arrows from v_1 to v_2 by $x_1, \ldots, x_a$ and those from v_2 to v_1 by $y_1, \ldots, y_a$. If $(m, n) \neq (2, 2)$ then $a \geq 2$, but then we have traces along cycles

$$\{x_i y_j \mid 1 \leq i, j \leq a\}$$

that is, the polynomial ring of invariants is the polynomial algebra in $l_1 + l_2$ variables (the traces of the loops) over the homogeneous coordinate ring of the Segre embedding

$$\mathbb{P}^{a-1} \times \mathbb{P}^{a-1} \lhook\joinrel\longrightarrow \mathbb{P}^{a^2-1}$$

which has a singularity at the top (for example, we have equations of the form $(x_1y_2)(x_2y_1) - (x_1y_1)(x_2y_2)$). Thus, the local algebra of iss_n^m cannot be a formal power series ring in ξ whence iss_n^m is singular in ξ. We have seen in section 1.2 that for the exceptional case $\text{iss}_2^2 \simeq \mathbb{C}^5$.   ☐

To determine the fibers of the quotient map $M_n^m \xrightarrow{\ \pi\ } \text{iss}_n^m$ we have to study the nullcone of this local quiver setting, $Null_{\alpha_\xi} Q_\xi$. Observe that the quiver Q_ξ has loops in every vertex and arrows connecting each ordered pair of vertices, whence we do not have to worry about potential corner-type removals. Denote $\sum e_i = z \leq n$ and let $\mathcal{C}_z$ be the set of all $s = (s_1, \ldots, s_z) \in \mathbb{Q}^z$, which are disjoint unions of strings of the form

$$\{p_i, p_i + 1, \ldots, p_i + k_i\}$$

where $l_i \in \mathbb{N}$, all intermediate numbers $p_i + j$ with $j \leq k_i$ do occur as components in s with multiplicity $a_{ij} \geq 1$ and p_i satisfies the balance-condition

$$\sum_{j=0}^{k_i} a_{ij}(p_i + j) = 0$$

for every string in s. For fixed $s \in \mathcal{C}_z$ we can distribute the components s_i over the vertices of Q_ξ (e_j of them to vertex v_j) in all possible ways modulo the action of the small Weyl group $S_{e_1} \times \ldots S_{e_k} \lhook\joinrel\longrightarrow S_z$. That is, we can rearrange the s_i's belonging to a fixed vertex such that they are in decreasing order. This gives us the list $\mathcal{S}_{\alpha_\xi}$ or $\mathcal{S}_\tau$ of all corner-types in $Null_{\alpha_\xi} Q_\xi$. For each $s \in \mathcal{S}_{\alpha_\xi}$ we then construct the corner-quiver setting

$$(Q_{\xi\ s}, \alpha_{\xi\ s}, \theta_{\xi\ s})$$

and study the Hesselink strata S_s that actually do appear, which is equivalent to verifying whether there are $\theta_{\xi s}$-semistable representations in $\mathbf{rep}_{\alpha_{\xi s}} Q_{\xi s}$. We have given a purely combinatorial way to settle this (in general quite hard) problem of optimal corner-types.

That is, we can determine which Hesselink strata S_s actually occur in $\pi^{-1}(\xi) \simeq Null_{\alpha_{xi}} Q_\xi$. The $GL(\alpha_{\xi\ s})$-orbits in the stratum S_s are in natural one-to-one correspondence with the orbits under the associated parabolic subgroup P_s acting on the semistable representations

$$U_s = \pi^{-1} \mathbf{rep}_{\alpha_{\xi\ s}}^{ss}(Q_{\xi\ s}, \theta_{\xi\ s})$$

and there is a natural projection morphism from the corresponding orbit-space

$$U_s / P_s \xrightarrow{\ p_s\ } M_{\alpha_{\xi\ s}}^{ss}(Q_{\xi\ s}, \theta_{\xi\ s})$$

type	τ	(Q_τ, α_τ)
2_a	$(1,2)$	
2_b	$(1,1;1,1)$	
2_c	$(2,1)$	

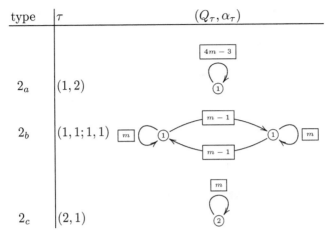

FIGURE 6.6: Local quiver settings for 2×2 matrices.

to the moduli space of $\theta_{\xi\ s}$-semistable representations. The remaining (hard) problem in the classification of m-tuples of $n \times n$ matrices under simultaneous conjugation is the description of the fibers of this projection map p_s.

Example 6.12 m-tuples of 2×2 matrices
There are three different representation types τ of 2-dimensional representations of $\mathbb{C}\langle x_1, \ldots, x_m \rangle$ with corresponding local quiver settings (Q_τ, α_τ) given in figure 6.6 The defining (in)equalities of the strata $\mathrm{iss}_2^m(\tau)$ are given by $k \times k$ minors (with $k \leq 4$ of the symmetric $m \times m$ matrix

$$\begin{bmatrix} tr(x_1^0 x_1^0) & \cdots & tr(x_1^0 x_m^0) \\ \vdots & & \vdots \\ tr(x_m^0 x_1^0) & \cdots & tr(x_m^0 x_m^0) \end{bmatrix}$$

where $x_i^0 = x_i - \frac{1}{2} tr(x_i)$ is the generic trace zero matrix. These facts follow from the description of the trace algebras $\mathbb{T}_2^m$ as polynomial algebras over the generic Clifford algebras of rank ≤ 4 (determined by the above symmetric matrix) and the classical matrix decomposition of Clifford algebras over $\mathbb{C}$. For more details we refer to [67].

To study the fibers $M_2^m \longrightarrow \mathrm{iss}_2^m$ we need to investigate the different Hesselink strata in the nullcones of these local quiver settings. Type 2_a has just one potential corner type corresponding to $s = (0) \in \mathcal{S}_1$ and with corresponding corner-quiver setting

which obviously has $\mathbb{P}^0$ (one point) as corresponding moduli (and orbit) space. This corresponds to the fact that for $\xi \in \mathrm{iss}_2^m(1,2)$, M_ξ is simple and hence

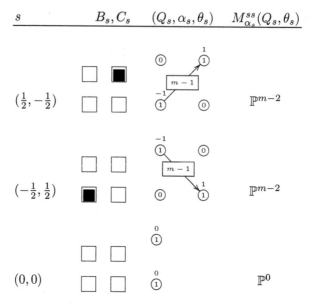

FIGURE 6.7: Moduli spaces for type 2_b.

the fiber $\pi^{-1}(\xi)$ consists of the closed orbit $\mathcal{O}(M_\xi)$.

For type 2_b the list of figure 6.7 gives the potential corner-types C_s together with their associated corner-quiver settings and moduli spaces (note that as $B_s = C_s$ in all cases, these moduli spaces describe the full fiber) That is, for $\xi \in \mathrm{iss}_2^m(1,1;1,1)$, the fiber $\pi^{-1}(\xi)$ consists of the unique closed orbit $\mathcal{O}(M_\xi)$ (corresponding to the $\mathbb{P}^0$) and two families $\mathbb{P}^{m-2}$ of nonclosed orbits. Observe that in the special case $m = 2$ we recover the two nonclosed orbits found in section 1.2.

Finally, for type 2_c, the fibers are isomorphic to the nullcones of m-tuples of 2×2 matrices. We have the following list of corner-types, corner-quiver settings and moduli spaces. Again, as $B_s = C_s$ in all cases, these moduli spaces describe the full fiber.

s	B_s, C_s	$(Q_s, \alpha_s, \theta_s)$	$M_{\alpha_s}^{ss}(Q_s, \theta_s)$
$(\frac{1}{2}, -\frac{1}{2})$		$\overset{-1}{①} \!-\! \boxed{m} \!\to\! \overset{1}{①}$	$\mathbb{P}^{m-1}$
$(0,0)$		$\overset{0}{②}$	$\mathbb{P}^0$

whence the fiber $\pi^{-1}(\xi)$ consists of the closed orbit, together with a $\mathbb{P}^{m-1}$-family of non-closed orbits. Again, in the special case $m = 2$, we recover the $\mathbb{P}^1$-family found in section 1.2. □

type	τ	(Q_τ, α_τ)

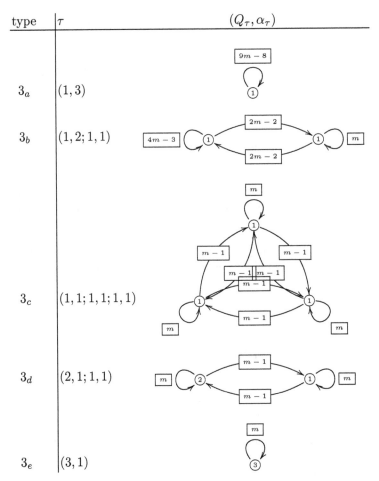

3_a $(1,3)$

3_b $(1,2;1,1)$

3_c $(1,1;1,1;1,1)$

3_d $(2,1;1,1)$

3_e $(3,1)$

FIGURE 6.8: Local quiver settings for 3×3 matrices.

Example 6.13 *m-tuples of 3×3 matrices*

There are 5 different representation-types for 3-dimensional representations. Their associated local quiver settings are given in figure 6.8 For each of these types we can perform an analysis of the nullcones as before. We leave the details to the interested reader and mention only the end-result

- For type 3_a the fiber is one closed orbit

- For type 3_b the fiber consists of the closed orbit together with two $\mathbb{P}^{2m-3}$-families of nonclosed orbits

- For type 3_c the fiber consists of the closed orbit together with twelve $\mathbb{P}^{m-2} \times \mathbb{P}^{m-2}$-families and one $\mathbb{P}^{m-2}$-family of nonclosed orbits

- For type 3_d the fiber consists of the closed orbit together with four $\mathbb{P}^{m-1} \times \mathbb{P}^{m-2}$-families, one $\mathbb{P}^{m-2} \times \mathbb{P}^{m-2}$-family, two $\mathbb{P}^{m-2}$-families, one $\mathbb{P}^{m-1}$-family and two M-families of nonclosed orbits determined by moduli spaces of quivers, where M is the moduli space of the following quiver setting

together with some additional orbits coming from the projection maps p_s.

- For type 3_e we have to study the nullcone of m-tuples of 3×3 matrices, which can be done as in the case of couples but for $m \geq 3$ the two extra strata do occur.

We see that in this case the only representation-types where the fiber is not fully determined by moduli spaces of quivers are 3_d and 3_e.   □

6.6   Representation fibers

Let A be a Cayley-Hamilton algebra of degree n and consider the algebraic quotient map

$$\mathbf{trep}_n\, A \xrightarrow{\ \pi\ } \mathbf{triss}_n\, A$$

from the variety of n-dimensional trace preserving representations to the variety classifying isomorphism classes of trace preserving n-dimensional semi-simple representations. Assume $\xi \in Sm_{tr}\, A \hookrightarrow \mathbf{triss}_n\, A$. That is, the representation variety $\mathbf{trep}_n\, A$ is smooth along the GL_n-orbit of M_ξ where M_ξ is the semi-simple representation determined by $\xi \in \mathbf{triss}_n\, A$. We have seen that the local structure of A and $\mathbf{trep}_n\, A$ near ξ is fully determined by a local marked quiver setting $(Q_\xi^\bullet, \alpha_\xi)$. That is, we have a GL_n-isomorphism between the fiber of the quotient map, that is, the n-dimensional trace preserving representation degenerating to M_ξ

$$\pi^{-1}(\xi) \simeq GL_n \times^{GL(\alpha_\xi)} Null_{\alpha_\xi}\, Q_\xi$$

and the nullcone of the marked quiver-setting. In this section we will apply the results on nullcones to the study of these representation fibers $\pi^{-1}(\xi)$.

Observe that all the facts on nullcones of quivers extend verbatim to marked quivers $Q^\bullet$ using the underlying quiver Q with the proviso that we drop all loops in vertices with vertex-dimension 1 that get a marking in $Q^\bullet$. This is clear as nilpotent quiver representations obviously have zero trace along each oriented cycle, in particular in each loop.

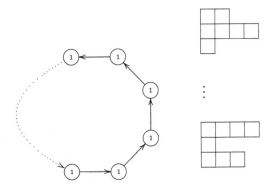

FIGURE 6.9: Local quiver settings for curve orders.

The examples given before illustrate that a complete description of the nullcone is rather cumbersome. For this reason we restrict ourselves here to the determination of the number of irreducible components and their dimensions in the representation fibers. Modulo the GL_n-isomorphism above this study amounts to describing the irreducible components of $Null_{\alpha_\xi} Q_\xi$, which are determined by the maximal corner-types C_s, that is, such that the set of weights in C_s is maximal among subsets of $\pi_{\alpha_{xi}} Q_\xi$ (and hence $\| s \|$ is maximal among $S_{\alpha_\xi} Q_\xi$).

To illustrate our strategy, consider the case of curve orders. In section 5.4 we proved that if A is a Cayley-Hamilton order of degree n over an affine curve $X = \mathtt{triss}_n A$ and if $\xi \in Sm_n A$, then the local quiver setting (Q, α) is determined by an oriented cycle Q on k vertices with $k \leq n$ being the number of distinct simple components of M_ξ, the dimension vector $\alpha = (1, \dots, 1)$ as in figure 6.9 and an unordered partition $p = (d_1, \dots, d_k)$ having precisely k parts such that $\sum_i d_i = n$, determining the dimensions of the simple components of M_ξ. Fixing a cyclic ordering of the k-vertices $\{v_1, \dots, v_k\}$ we have that the set of weights of the maximal torus $T_k = \mathbb{C}^* \times \dots \times \mathbb{C}^* = GL(\alpha)$ occurring in $\mathtt{rep}_\alpha Q$ is the set

$$\pi_\alpha Q = \{\pi_{k1}, \pi_{12}, \pi_{23}, \dots, \pi_{k-1k}\}$$

Denote $K = \sum_{i=0}^{k-1} i = \frac{k(k-1)}{2}$ and consider the one string vector

$$s = (\dots, k - 2 - \frac{K}{k}, k - 1 - \frac{K}{k}, \underbrace{-\frac{K}{k}}_{i}, 1 - \frac{K}{k}, 2 - \frac{K}{k}, \dots)$$

then s is balanced and vertex-dominant, $s \in S_\alpha Q$ and $\pi_s Q = \Pi$. To check whether the corresponding Hesselink strata in $Null_\alpha Q$ is nonempty we have to consider the associated quiver-setting $(Q_s, \alpha_s, \theta_s)$, which is

It is well known and easy to verify that $\mathbf{rep}_{\alpha_s} Q_s$ has an open orbit with all representative arrows equal to 1. For this representation all proper subrepresentations have dimension vector $\beta = (0, \ldots, 0, 1, \ldots, 1)$ and hence $\theta_s(\beta) > 0$. That is, the representation is θ_s-stable and hence the corresponding Hesselink stratum $S_s \neq \emptyset$. Finally, because the dimension of $\mathbf{rep}_{\alpha_s} Q_s$ is $k - 1$ we have that the dimension of this component in the representation fiber $\pi^{-1}(x)$ is equal to

$$dim \ GL_n - dim \ GL(\alpha) + dim \ \mathbf{rep}_{\alpha_s} Q_s = n^2 - k + k - 1 = n^2 - 1$$

which completes the proof of the following.

THEOREM 6.10
Let A be a Cayley-Hamilton order of degree n over an affine curve X such that A is smooth in $\xi \in X$. Then, the representation fiber $\pi^{-1}(\xi)$ has exactly k irreducible components of dimension $n^2 - 1$, each the closure of one orbit. In particular, if A is Cayley-smooth over X, then the quotient map

$$\mathbf{trep}_n A \xrightarrow{\ \pi\ } \mathbf{triss}_n A = X$$

is flat, that is, all fibers have the same dimension $n^2 - 1$.

For Cayley-Hamilton orders over surfaces, the situation is slightly more complicated. From section 5.4 we recall that if A is a Cayley-Hamilton order of degree n over an affine surface $S = \mathbf{triss}_n A$ and if A is smooth in $\xi \in X$, then the local structure of A is determined by a quiver setting (Q, α) where $\alpha = (1, \ldots, 1)$ and Q is a two-circuit quiver on $k + l + m \leq n$ vertices, corresponding to the distinct simple components of M_ξ as in figure 6.10 and an unordered partition $p = (d_1, \ldots, d_{k+l+m})$ of n with $k + l + m$ nonzero parts determined by the dimensions of the simple components of M_ξ. With the indicated ordering of the vertices we have that

$$\pi_\alpha Q = \{\pi_{i \ i+1} \ | \ \begin{cases} 1 & \leq i \leq k - 1 \\ k+1 & \leq i \leq k+l-1 \\ k+l+1 & \leq i \leq k+l+m-1 \end{cases} \}$$

$$\cup \{\pi_{k \ k+l+1}, \pi_{k+l \ k+l+1}, \pi_{k+l+m \ 1}, \pi_{k+l+m \ k+1}\}$$

As the weights of a corner cannot contain all weights of an oriented cycle in Q we have to consider the following two types of potential corner-weights Π of maximal cardinality

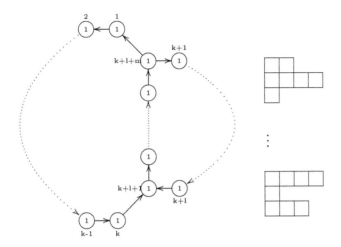

FIGURE 6.10: Local quiver settings for surface orders.

- (outer type): $\Pi = \pi_\alpha\, Q - \{\pi_a, \pi_b\}$ where a is an edge in the interval $[v_1, \ldots, v_k]$ and b is an edge in the interval $[v_{k+1}, \ldots, v_{k+l}]$

- (inner type): $\Pi = \pi_\alpha\, Q - \{\pi_c\}$ where c is an edge in the interval $[v_{k+l+1}, v_{k+l+m}]$

There are $2 + (k-1)(l-1)$ different subsets Π of outer type, each occurring as the set of weights of a corner C_s, that is, $\Pi = \pi_s\, Q$ for some $s \in S_\alpha\, Q$. The two exceptional cases correspond to

$$\begin{cases} \Pi_1 &= \pi_\alpha\, Q - \{\pi_{k+l+m\;1}, \pi_{k+l\;k+l+1}\} \\ \Pi_2 &= \pi_\alpha\, Q - \{\pi_{k+l+m\;k+1}, \pi_{k\;k+l+1}\} \end{cases}$$

which are of the form $\pi_{s_i}\, Q$ with associated border quiver-setting $(Q_{s_i}, \alpha_{s_i}, \theta_{s_i})$ where $\alpha_{s_i} = (1, \ldots, 1)$, Q_{s_i} are the full line subquivers of Q given in figure 6.11 with starting point v_1 (resp. v_{k+1}). The corresponding $s_i \in S_\alpha\, Q$ is a single string with minimal entry

$$-\frac{\sum_{i=0}^{k+l+m-1} i}{k+l+m} = -\frac{k+l+m-1}{2} \quad \text{at place} \quad \begin{cases} 1 \\ k+1 \end{cases}$$

and going with increments equal to one along the unique path. Again, one verifies that $\mathbf{rep}_{\alpha_s}\, Q_s$ has a unique open and θ_s-stable orbit, whence these Hesselink strata do occur and the border B_s is the full corner C_s. The corresponding irreducible component in $\pi^{-1}(\xi)$ has therefore a dimension equal to $n^2 - 1$ and is the closure of a unique orbit. The remaining $(k-1)(l-1)$ subsets Π of outer type are of the form

$$\Pi_{ij} = \pi_\alpha\, Q - \{\pi_{i\;i+1}, \pi_{j\;j+1}\}$$

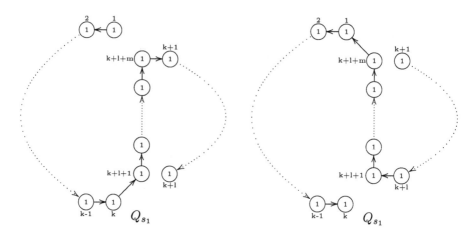

FIGURE 6.11:   Border quiver settings.

with $1 \leq i \leq k - 1$ and $k + 1 \leq j \leq k + l - 1$. We will see in a moment that they are again of type $\pi_s Q$ for some $s \in \mathcal{S}_\alpha Q$ with associated border quiver-setting $(Q_s, \alpha_s, \theta_s)$ where $\alpha_s = (1, \ldots, 1)$ and Q_s is the full subquiver of Q

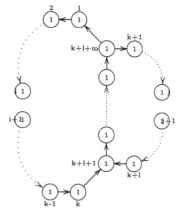

If we denote with A_l the directed line quiver on $l + 1$ vertices, then Q_s can be decomposes into full line subquivers

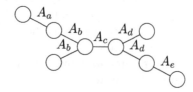

but then we consider the one string $s \in \mathcal{S}_\alpha Q$ with minimal entry equal to

$-\frac{x}{k+l+m}$ where with notations as above

$$x = \sum_{i=1}^{a} i + 2\sum_{i=1}^{b}(a+i) + \sum_{i=1}^{c}(a+b+i)$$

$$+ 2\sum_{i=1}^{d}(a+b+c+i) + \sum_{i=1}^{e}(a+b+c+d+i)$$

where the components of s are given to the relevant vertex-indices. Again, there is a unique open orbit in $\mathbf{rep}_{\alpha_s} Q_s$ which is a θ_s-stable representation and the border B_s coincides with the corner C_s. That is, the corresponding Hesselink stratum occurs and the irreducible component of $\pi^{-1}(\xi)$ it determines had dimension equal to

$$dim\ GL_n - dim\ GL(\alpha) + dim\ \mathbf{rep}_{\alpha_s} Q_s = n^2 - (k+l+m) + (k+l+m-1)$$
$$= n^2 - 1$$

There are $m-1$ different subsets Π_u of inner type, where for $k+l+1 \le u < k+l+m$ we define $\Pi_u = \pi_\alpha Q - \{\pi_{u\ u+1}\}$, that is, dropping an edge in the middle

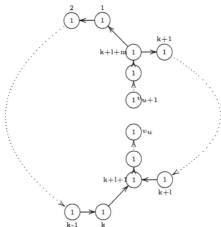

First assume that $k = l$. In this case we can walk through the quiver (with notations as before)

and hence the full subquiver of Q is part of a corner quiver-setting $(Q_s, \alpha_s, \theta_s)$ where $\alpha = (1, \ldots, 1)$ and where s has as its minimal entry $-\frac{x}{k+l+m}$ where

$$x = \sum_{i=1}^{a} i + 2\sum_{i=1}^{b}(a+i) + \sum_{i=1}^{c}(a+b+i)$$

In this case we see that $\mathbf{rep}_{\alpha_s} Q_s$ has θ_s-stable representations, in fact, there is a $\mathbb{P}^1$-family of such orbits. The corresponding Hesselink stratum is nonempty and the irreducible component of $\pi^{-1}(\xi)$ determined by it has dimension

$$dim \ GL_n - dim \ GL(\alpha) + dim \ \mathbf{rep}_{\alpha_s} Q_s = n^2 - (k+l+m) + (k+l+m) = n^2$$

If $l < k$, then $\Pi_u = \pi_s Q$ for some $s \in \mathcal{S}_\alpha Q$ but this time the border quiver-setting $(Q_s, \alpha_s, \theta_s)$ is determined by $\alpha_s = (1, \ldots, 1)$ and Q_s the full subquiver of Q by also dropping the arrow corresponding to $\pi_{k+l+1 \ k+l}$, that is

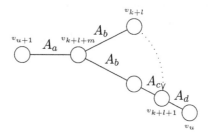

If Q_s is this quiver (without the dashed arrow) then $B_s = \mathbf{rep}_{\alpha_s} Q_s$ and it contains an open orbit of a θ_s-stable representation. Observe that s is determined as the one string vector with minimal entry $-\frac{x}{k+l+m}$ where

$$x = \sum_{i=1}^{a} i + 2 \sum_{i=1}^{b}(a+i) + \sum_{i=1}^{c}(a+b+i) + \sum_{i=1}^{d}(a+b+c+i)$$

However, in this case $B_s \neq C_s$ and we can identify C_s with $\mathbf{rep}_{\alpha_s} Q'_s$ where Q'_s is Q_s together with the dashed arrow. There is an $\mathbb{A}^1$-family of orbits in C_s mapping to the θ_s-stable representation. In particular, the Hesselink stratum exists and the corresponding irreducible component in $\pi^{-1}(\xi)$ has a dimension equal to

$$dim \ GL_n - dim \ GL(\alpha) + dim \ C_s = n^2 - (k+l+m) + (k+l+m) = n^2$$

This concludes the proof of the description of the representation fibers of smooth orders over surfaces, summarized in the following result.

THEOREM 6.11
Let A be a Cayley-Hamilton order of degree n over an affine surface $X = \mathbf{triss}_n A$ and assume that A is smooth in $\xi \in X$ of local type (A_{klm}, α). Then, the representation fiber $\pi^{-1}(\xi)$ has exactly $2 + (k-1)(l-1) + (m-1)$ irreducible components of which $2 + (k-1)(l-1)$ are of dimension $n^2 - 1$ and are closures of one orbit and the remaining $m - 1$ have dimension n^2 and are closures of a one-dimensional family of orbits. In particular, if A is Cayley-smooth, then the algebraic quotient map

$$\mathbf{trep}_n A \xrightarrow{\ \pi\ } \mathbf{triss}_n A = X$$

is flat if and only if all local quiver settings of A have quiver A_{klm} with $m = 1$.

The final example will determine the fibers over *smooth* points in the quotient varieties (or moduli spaces) provided the local quiver is *symmetric*. This computation is due to Geert Van de Weyer.

Example 6.14 Smooth symmetric settings
Recall from theorem 5.22 that a smooth symmetric quiver setting (**sss**) if and only if it is a tree constructed as a connected sum of three different types of quivers

- $(m) \rightleftarrows (n)$

- $(1) \overset{m}{\underset{m}{\rightrightarrows}} (n)$, with $m \leq n$

- $(1) \rightleftarrows (m) \rightleftarrows (n)$

- $(m) \rightleftarrows (2) \rightleftarrows (n)$

where the connected sum is taken in the vertex with dimension 1. We call the vertices where the connected sum is taken *connecting vertices* and depict them by a square vertex $\square$. We want to study the nullcone of connected sums composed of more than one of these quivers so we will focus on instances of these four quivers having at least one vertex with dimension 1

I $(1) \overset{m}{\underset{m}{\rightrightarrows}} (n)$, with $m \leq n$

II(1) $(1) \rightleftarrows (m) \rightleftarrows (1)$

II(2) $(1) \rightleftarrows (m) \rightleftarrows (n)$

We will call the quiver settings of type **I** and **II** forming an **sss** (Q, α) the *terms of Q.*

claim 1: *Let (Q, α) be an* **sss** *and Q_μ a type quiver for Q, then any string quiver of Q_μ is either a connected sum of string quivers of type quivers for terms of Q or a string quiver of type quivers of*

$(m) \rightleftarrows (n)$, (n)

Consider a string quiver $Q_{\mu(i)}$ of Q_μ. By definition vertices in a type quiver are only connected if they originate from the same term in Q. This means we may divide the string quiver $Q_{\mu(i)}$ into segments, each segment either a string quiver of a type quiver of a term of Q (if it contains the connecting vertex) or a level quiver of a type quiver of the quivers listed above (if it does not contain the connecting vertex).

The only vertices these segments may have in common are instances of the connecting vertices. Now note that there is only one instance of each connecting vertex in Q_μ because the dimension of each connecting vertex is 1. Moreover, two segments cannot have more than one connecting vertex in common as this would mean that in the original quiver there is a cycle, proving the claim.

Hence, constructing a type quiver for an **sss** boils down to patching together string quivers of its terms. These string quivers are subquivers of the following two quivers

Observe that the second quiver has two components. So a string quiver will either be a tree (possible from all components) or a quiver containing a square. We will distinguish two different types of squares; S_1 corresponding to a term of type **II**(1) and S_2 corresponding to a term of type **II**(2)

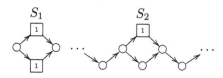

These squares are the only polygons that can appear in our type quiver. Indeed, consider a possible polygon

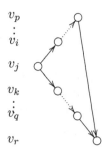

This polygon corresponds to the following subquiver of Q

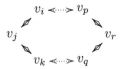

But Q is a tree, so this is only a subquiver if it collapses to $v_i \longleftrightarrow v_j \longleftrightarrow v_k$.

claim 2: *Let (Q, α) be an* **sss** *and Q_μ a type quiver containing (connected) squares. If Q_μ determines a nonempty Hesselink stratum then*

(i) *the 0-axis in Q_μ lies between the axes containing the outer vertices of the squares of type S_1*

(ii) *squares of type S_1 are connected through paths of maximum length 2*

(iii) *squares of type S_1 that are connected through a path of length 2 are connected to other quivers in top and bottom vertex (and hence originate from type* **II(1)** *terms that are connected to other terms in both their connecting vertices)*

(iv) *the string $\mu(i)$ containing squares of type S_1 connected through a path of length two equals $(\ldots, -2, -1, 0, 1, 2, \ldots)$*

(v) *for a square of type S_2*

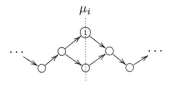

with p vertices on its left branch and q vertices on its right branch we have

$$-\frac{q}{2} \le \mu_i \le \frac{p}{2}$$

Let us call the string quiver of Q_μ containing the squares $Q_{\mu(i)}$ and let $\theta \in \mu(i)\mathbb{N}_0$ be the character determining this string quiver. Consider the subrepresentation

$$\theta_i \qquad \theta_{i+1} \qquad \theta_{i+2}$$

This subrepresentation has character $\theta(\alpha_{\mu(i)}) - \alpha_{\mu(i)}(\mathbf{v})\theta_i \geq 0$ where $\mathbf{v}$ is the vertex whose dimension we reduced to 0, so $\theta_i \leq 0$. But then the subrepresentation

$$\theta_i \quad \theta_{i+1} \quad \theta_{i+2}$$

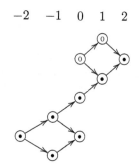

gives $\theta_{i+2} \geq 0$, whence (i). Note that the left vertex of one square can never lie on an axis right of the right vertex of another square. At most it can lie on the same axis as the right vertex, in which case this axis is the 0-axis and the squares are connected by a path of length 2. In order to prove (iii) look at the subrepresentation

$$-2 \quad -1 \quad 0 \quad 1 \quad 2$$

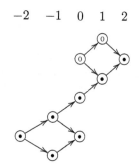

This subrepresentation has negative character and hence the original representation was not semistable. Finally, for (v) we look at the subrepresentation obtained by reducing the dimension of all dotted vertices by 1

$$\mu_i$$

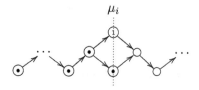

having character $-((p+1)\mu_i - \sum_{j=1}^{p} j) \geq 0$. So $\mu_i \leq \frac{p}{2}$. Mirroring this argument yields the other inequality $\mu_i \geq -\frac{q}{2}$.

claim 3:   *Let (Q, α) be an* **sss** *and Q_μ be a type quiver determining a nonempty stratum and let $Q_{\mu(i)}$ be a string quiver determined by a segment $\mu(i)$ not containing 0. Then the only possible dimension vectors for squares of type S_1 in $Q_{\mu(i)}$ are those of figure 6.12.*

Top and bottom vertex of the square are constructed from the connecting vertices so can only be one-dimensional. Left and right vertex of the square are constructed from a vertex of dimension n. Claim 2 asserts that the leftmost

$$\alpha_1 = \begin{pmatrix} & 1 & \\ 1 & & 1 \\ & 1 & \end{pmatrix}$$

$$\alpha_2 = \begin{pmatrix} & 1 & \\ 1 & & 2 \\ & 1 & \end{pmatrix}$$

$$\alpha_3 = \begin{pmatrix} & 1 & \\ 2 & & 2 \\ & 1 & \end{pmatrix}$$

FIGURE 6.12: Possible dimension vectors for squares.

vertex lies on a negative axis while the rightmost vertex lies on a positive axis. If the left dimension is > 2 then the representation splits

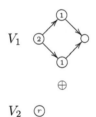

with $r = m - 2$. By semistability the character of V_2 must be zero. A similar argument applies to the right vertex.

claim 4: *Let μ be a type determining a non-empty stratum.*

(i) *When a vertex (v, i) in Q_μ determined by a term of type **II(1)** has $\alpha(v, i) > 2$ then $\mu_i = 0$*

(ii) *When a vertex (v, i) in Q_μ determined by a term of type **I** with m arrows has $\alpha(v, i) > m$ then $\mu_i = 0$*

Suppose we have a vertex v with dimension $\alpha_{\mu(i)}(v) > 2$, then the number of paths running through this vertex is at most 2: would there be at least three paths arriving or departing in the vertex, it would be a connecting vertex, which is not possible because of its dimension. If there are two paths arriving and at least one path departing, it must be a central vertex of a type **II(2)** term. But then the only possible subtrees generated from type **II(1)** terms with vertices of dimension at least three are (modulo reversing all arrows)

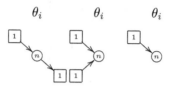

In the last tree there are no other arrows from the vertex with dimension n. For each of these trees we have a subrepresentation whence $\theta_i \geq 0$. But if $\theta_i > 0$, reducing the dimension of the vertex with dimension ≥ 3 gives a subrepresentation with negative character, so $\theta_i = 0$. The second part is proved similarly.

Let (Q, α) be an sss and μ a type determining a nonempty stratum in $\mathbf{null}_\alpha \, Q$. Let Q_μ be the corresponding type quiver and α_μ the corresponding dimension vector, then

(i) *every connected component $Q_{\mu(i)}$ of Q_μ is a connected sum of string quivers of either terms of Q or quivers generated from terms of Q by removing the connecting vertex. The connected sum is taken in the instances of the connecting vertices and results in a connected sum of trees and quivers of the form*

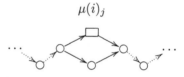

(ii) *For a square of type S_1 we have $\mu(i)_{j-1} \leq 0 \leq \mu(i)_{j+1}$. Moreover, such squares cannot be connected by paths longer than two arrows and can only be connected by paths of this length if $\mu(i)_{j+1} = 0$.*

(iii) *For vertices (v, j) constructed from type **II(1)** terms we have $\alpha_{\mu_i}(v, j) \leq 2$ when $\mu_i \neq 0$.*

(iv) *For a vertex (v, j) constructed from a type **I** term with m arrows we have $\alpha_{\mu_i}(v, j) \leq m$ when $\mu_i \neq 0$.*

$\square$

6.7   Brauer-Severi varieties

In this section we will reconsider the Brauer-Severi scheme $BS_n(A)$ of an algebra A. In the generic case, that is when A is the free algebra $\mathbb{C}\langle x_1, \ldots, x_m \rangle$, we will show that it is a moduli space of a certain quiver situation. This then allows us to give the étale local description of $BS_n(A)$ whenever A is a Cayley-smooth algebra. Again, this local description will be a moduli space.

The generic Brauer-Severi scheme of degree n for m-generators, $BS_n^m(gen)$ is defined as follows. Consider the free algebra on m generators $\mathbb{C}\langle x_1, \ldots, x_m \rangle$ and consider the GL_n-action on $\mathbf{rep}_n \, \mathbb{C}\langle x_1, \ldots, x_m \rangle \times \mathbb{C}^n = M_n^m \oplus \mathbb{C}^n$ given by

$$g.(A_1, \ldots, A_m, v) = (gA_1g^{-1}, \ldots, gA_mg^{-1}, gv)$$

and consider the open subset $Brauer^s(gen)$ consisting of those points $(A_1, \ldots, A_m, v)$ where v is a cyclic vector, that is, there is no proper subspace of $\mathbb{C}^n$ containing v and invariant under left multiplication by the matrices A_i. The GL_n-stabilizer is trivial in every point of $Brauer^s(gen)$ whence we can define the orbit space

$$BS_n^m(gen) = Brauer^s(gen)/GL_n$$

Consider the following quiver situation

on two vertices $\{v_1, v_2\}$ such that there are m loops in v_2 and consider the dimension vector $\alpha = (1, n)$. Then, clearly

$$\mathbf{rep}_\alpha \, Q = \mathbb{C}^n \oplus M_n^m \simeq \mathbf{rep}_n \, \mathbb{C}\langle x_1, \ldots, x_m \rangle \oplus \mathbb{C}^n$$

where the isomorphism is as GL_n-module. On $\mathbf{rep}_\alpha \, Q$ we consider the action of the larger group $GL(\alpha) = \mathbb{C}^* \times GL_n$ acting as

$$(\lambda, g).(v, A_1, \ldots, A_m) = (gv\lambda^{-1}, gA_1g^{-1}, \ldots, gA_mg^{-1})$$

Consider the character χ_θ where $\theta = (-n, 1)$, then $\theta(\alpha) = 0$ and consider the open subset of θ-semistable representations in $\mathbf{rep}_\alpha \, Q$.

LEMMA 6.3
The following are equivalent for $V = (v, A_1, \ldots, A_m) \in \mathbf{rep}_\alpha \, Q$

1. V *is* θ-*semistable*

2. V *is* θ-*stable*

3. $V \in Brauer^s(gen)$

Consequently

$$M_\alpha^{ss}(Q, \alpha) \simeq BS_n^m(gen)$$

PROOF   1. $\Rightarrow$ 2.: If V is θ-semistable it must contain a largest θ-stable subrepresentation W (the first term in the Jordan-Hölder filtration for θ-semistables). In particular, if the dimension vector of W is $\beta = (a, b) < (1, n)$, then $\theta(\beta) = 0$, which is impossible unless $\beta = \alpha$ whence $W = V$ is θ-stable.

2. $\Rightarrow$ 3.: Observe that $v \neq 0$, for otherwise V would contain a subrepresentation of dimension vector $\beta = (1, 0)$ but $\theta(\beta) = -n$ is impossible. Assume that v is noncyclic and let $U \hookrightarrow \mathbb{C}^n$ be a proper subspace, say, of dimension $l < n$ containing v and stable under left multiplication by the A_i, then V has a subrepresentation of dimension vector $\beta' = (1, l)$ and again $\theta(\beta') = l - n < 0$ is impossible.

3. $\Rightarrow$ 1.: By cyclicity of v, the only proper subrepresentations of V have dimension vector $\beta = (0, l)$ for some $0 < l \leq n$, but they satisfy $\theta(\beta) > 0$, whence V is θ-(semi)stable.

As for the last statement, recall that geometric points of $M_\alpha^{ss}(Q, \alpha)$ classify isomorphism classes of direct sums of θ-stable representations. As there are no proper θ-stable subrepresentations, $M_\alpha^{ss}(Q, \alpha)$ classifies the $GL(\alpha)$-orbits in $Brauer^s(gen)$. Finally, as in chapter 1, there is a one-to-one correspondence between the GL_n-orbits as described in the definition of the Brauer-Severi variety and the $GL(\alpha)$-orbits on $\mathbf{rep}_\alpha Q$. $\square$

By definition, $M_\alpha^{ss}(Q, \theta) = \mathtt{proj}\ \oplus_{n=0}^\infty\ \mathbb{C}[\mathbf{rep}_\alpha\ Q]^{GL(\alpha), \chi^n \theta}$ and we can either use the results of section 3 or the previous section to show that these semi-invariants f are generated by brackets, that is

$$f(V) = det\ \big[w_1(A_1, \ldots, A_m)v\ \ldots\ w_n(A_1, \ldots, A_m)v\big]$$

where the w_i are words in the noncommuting variables $x_1, \ldots, x_m$. As before we can restrict these n-tuples of words $\{w_1, \ldots, w_n\}$ to sequences arising from multicolored Hilbert n-stairs. That is, the lower triangular part of a square $n \times n$ array

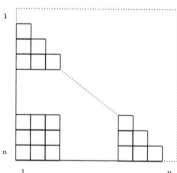

this time filled with colored stones $\odot$ where $1 \leq i \leq m$ subject to the two coloring rules

- each row contains exactly one stone

- each column contains at most one stone of each color

The relevant sequences $W(\sigma) = \{1, w_2, \ldots, w_n\}$ of words are then constructed by placing the identity element 1 at the top of the stair, and descend according to the rule

- Every go-stone has a *top word* T that we may assume we have constructed before and a *side word* S and they are related as indicated below

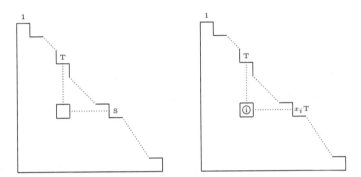

In a similar way to the argument in chapter 1 we can cover $M_\alpha^{ss}(Q, \alpha) = BS_n^m(gen)$ by open sets determined by Hilbert stairs and find representatives of the orbits in σ-standard form, that is, replacing every i-colored stone in σ by a 1 at the same spot in A_i and fill the remaining spots in the same column of A_i by zeroes

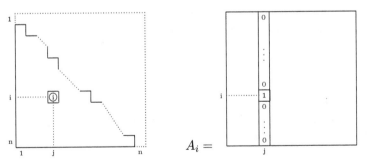

As this fixes $(n-1)n$ entries of the $mn^2 + n$ entries of V, one recovers the following result of M. Van den Bergh [101].

THEOREM 6.12

The generic Brauer-Severi variety $BS_n^m(gen)$ of degree n in m generators is a smooth variety, which can be covered by affine open subsets each isomorphic to $\mathbb{C}^{(m-1)n^2+n}$.

For an arbitrary affine $\mathbb{C}$-algebra A, one defines the Brauer stable points to be the open subset of $\mathbf{rep}_n A \times \mathbb{C}^n$

$$Brauer_n^s(A) = \{(\phi, v) \in \mathbf{rep}_n A \times \mathbb{C}^n \mid \phi(A)v = \mathbb{C}^n\}$$

As Brauer stable points have trivial stabilizer in GL_n all orbits are closed and we can define the Brauer-Severi variety of A of degree n to be the orbit space

$$BS_n(A) = Brauer_n^s(A)/GL_n$$

We claim that Quillen-smooth algebras have smooth Brauer-Severi varieties. Indeed, as the quotient morphism

$$Brauer_n^s(A) \longrightarrow BS_n(A)$$

is a principal GL_n-fibration, the base is smooth whenever the total space is smooth. The total space is an open subvariety of $\mathbf{rep}_n A \times \mathbb{C}^n$, which is smooth whenever A is Quillen-smooth.

PROPOSITION 6.3
If A is Quillen-smooth, then for every n we have that the Brauer-Severi variety of A at degree n is smooth.

Next, we bring in the approximation at level n. Observe that for every affine $\mathbb{C}$-algebra A we have a GL_n-equivariant isomorphism

$$\mathbf{rep}_n A \simeq \mathbf{trep}_n \int_n A$$

More generally, we can define for every Cayley-Hamilton algebra A of degree n the trace preserving Brauer-Severi variety to be the orbit space of the Brauer stable points in $\mathbf{trep}_n A \times \mathbb{C}^n$. We denote this variety with $BS_n^{tr}(A)$. Again, the same argument applies.

PROPOSITION 6.4
If A is Cayley-smooth of degree n, then the trace preserving Brauer-Severi variety $BS_n^{tr}(A)$ is smooth.

We have seen that the moduli spaces are projective fiber bundles over the variety determined by the invariants

$$M_\alpha^{ss}(Q, \theta) \longrightarrow \mathbf{iss}_\alpha Q$$

Similarly, the (trace preserving) Brauer-Severi variety is a projective fiber bundle over the quotient variety of $\mathbf{rep}_n A$, that is, there is a proper map

$$BS_n(A) \xrightarrow{\pi} \mathbf{iss}_n A$$

and we would like to study the fibers of this map. Recall that when A is an order in a central simple algebra of degree n, then the general fiber will be isomorphic to the projective space $\mathbb{P}^{n-1}$ embedded in a higher dimensional $\mathbb{P}^N$. Over non-Azumaya points we expect this $\mathbb{P}^{n-1}$ to degenerate to more complex projective varieties that we would like to describe. To perform this study we need to control the étale local structure of the fiber bundle π in a neighborhood of $\xi \in \mathtt{iss}_n \, A$. Again, it is helpful to consider first the generic case, that is, when $A = \mathbb{C}\langle x_1, \ldots, x_m \rangle$ or $\mathbb{T}_n^m$. In this case, we have seen that the following two fiber bundles are isomorphic

$$BS_n^m(gen) \longrightarrow \mathtt{iss}_n \, \mathbb{T}_n^m \quad \text{and} \quad M_\alpha^{ss}(Q, \theta) \longrightarrow \mathtt{iss}_\alpha \, Q$$

where $\alpha = (1, n)$, $\theta = (-n, 1)$ and the quiver

①——————⟶ⓝ $\overset{m}{\circlearrowright}$ has Euler form $\quad \chi_Q = \begin{bmatrix} 1 & -1 \\ 0 & 1 - m \end{bmatrix}$

A semi-simple α-dimensional representation V_ζ of Q has representation type

$$(1, 0) \oplus (0, d_1)^{\oplus e_1} \oplus \ldots \oplus (0, d_k)^{\oplus e_k} \quad \text{with} \quad \sum_i d_i e_i = n$$

and hence corresponds uniquely to a point $\xi \in \mathtt{iss}_n \, \mathbb{T}_n^m$ of representation type $\tau = (e_1, d_1; \ldots; e_k, d_k)$. The étale local structure of $\mathtt{rep}_\alpha \, Q$ and of $\mathtt{iss}_\alpha \, Q$ near ζ is determined by the local quiver Q_ζ on $k + 1$-vertices, say $\{v_0, v_1, \ldots, v_k\}$, with dimension vector $\alpha_\zeta = (1, e_1, \ldots, e_k)$ and where Q_ζ has the following local form for every triple (v_0, v_i, v_j) as can be verified from the Euler-form

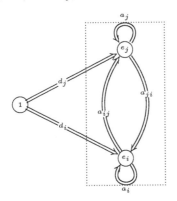

where $a_{ij} = (m - 1)d_i d_j = a_{ji}$ and $a_i = (m - 1)d_i^2 + 1$, $a_j = (m - 1)d_j^2 + 1$. The dashed part of Q_ζ is the same as the local quiver Q_ξ describing the étale local structure of $\mathtt{iss}_n \, \mathbb{T}_n^m$ near ξ. Hence, we see that the fibration $BS_n^m(gen) \longrightarrow \mathtt{iss}_n \, \mathbb{T}_n^m$ is étale isomorphic in a neighborhood of ξ to the fibration of the moduli space

$$M_{\alpha_\zeta}^{ss}(Q_\zeta, \theta_\zeta) \longrightarrow \mathtt{iss}_{\alpha_\zeta} \, Q_\zeta \simeq \mathtt{iss}_{\alpha_\xi} \, Q_\xi$$

in a neighborhood of the trivial representation and where $\theta_\zeta =$ $(-n, d_1, \ldots, d_k)$. Another application of the Luna slice results gives the following.

THEOREM 6.13
Let A be a Cayley-smooth algebra of degree n. Let $\xi \in \mathrm{triss}_n\ A$ correspond to the trace preserving n-dimensional semisimple representation

$$M_\xi = S_1^{\oplus e_1} \oplus \ldots \oplus S_k^{\oplus e_k}$$

where the S_i are distinct simple representations of dimension d_i and occurring with multiplicity e_i. Then, the projective fibration

$$BS_n^{tr}(A) \overset{\pi}{\longrightarrow\!\!\!\triangleright} \mathrm{triss}_n\ A$$

is étale isomorphic in a neighborhood of ξ to the fibration of the moduli space

$$M_{\alpha_\zeta}^{ss}(Q_\zeta^\bullet, \theta_\zeta) \longrightarrow\!\!\!\triangleright \mathrm{iss}_{\alpha_\zeta}\ Q_\zeta^\bullet \simeq \mathrm{iss}_{\alpha_\xi}\ Q_\xi^\bullet$$

in a neighborhood of the trivial representation. Here, $Q_\xi^\bullet$ is the local marked quiver describing the étale local structure of $\mathrm{trep}_n\ A$ near ξ, where $Q_\zeta^\bullet$ is the extended marked quiver situation, which locally for every triple (v_0, v_i, v_j) has the following shape where the dashed region is the local marked quiver $Q_\xi^\bullet$ describing $Ext_A^{tr}(M_\xi, M_\xi)$ and where $\alpha_\zeta = (1, e_1, \ldots, e_k)$ and $\theta_\zeta = (-n, d_1, \ldots, d_k)$

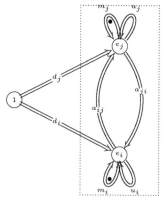

6.8 Brauer-Severi fibers

In the foregoing section we have given a description of the generic Brauer-Severi variety $BS_n^m(gen)$ as a moduli space of quiver representation as well as a local description of the fibration

$$BS_n^m(gen) \xrightarrow{\psi} \texttt{iss}_n^m$$

in an étale neighborhood of a point $\xi \in \texttt{iss}_n^m$ of representation type $\tau = (e_1, d_1; \ldots; e_k, d_k)$. We proved that it is étale locally isomorphic to the fibration

$$M_{\alpha_\zeta}^{ss}(Q_\zeta, \theta_\zeta) \longrightarrow \texttt{iss}_{\alpha_\zeta} Q_\zeta$$

in a neighborhood of the trivial representation. That is, we can obtain the generic Brauer-Severi fiber $\psi^{-1}(\xi)$ from the description of the nullcone $Null_{\alpha_\zeta} Q_\zeta$ provided we can keep track of θ_ζ-semistable representations. Let us briefly recall the description of the quiver-setting $(Q_\zeta, \alpha_\zeta, \theta_\zeta)$.

- The quiver Q_ζ has $k + 1$ vertices $\{v_0, v_1, \ldots, v_k\}$ such that there are d_i arrows from v_0 to v_i for $1 \le i \le k$. For $1 \le i, j \le k$ there are $a_{ij} = (m - 1)d_i d_j + \delta_{ij}$ directed arrows from v_i to v_j

- The dimension vector $\alpha_\zeta = (1, e_1, \ldots, e_k)$

- The character θ_ζ is determined by the integral $k + 1$-tuple $(-n, d_1, \ldots, d_k)$

That is, for any triple (v_0, v_i, v_j) of vertices, the full subquiver of Q_ζ on these three vertices has the following form

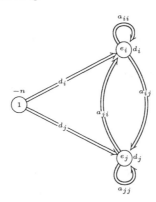

Let $E = \sum_{i=1}^k e_i$ and T the usual (diagonal) maximal torus of dimension $1 + E$ in $GL(\alpha_\zeta) \longrightarrow GL_E$ and let $\{\pi_0, \pi_1, \ldots, \pi_E\}$ be the obvious basis for the weights of T.. As there are loops in every v_i for $i \ge 1$ and there are arrows from v_i to v_j for all $i, j \ge 1$ we see that the set of weights of $\texttt{rep}_{\alpha_\zeta} Q_\zeta$ is

$$\pi_{\alpha_\zeta} Q_\zeta = \{\pi_{ij} = \pi_j - \pi_i \mid 0 \le i \le E, 1 \le j \le E\}$$

The maximal sets $\pi_s Q_\zeta$ for $s \in \mathcal{S}_{\alpha_\zeta} Q_\zeta$ are of the form

$$\pi_s Q_\zeta \stackrel{dfn}{=} \pi_\sigma = \{\pi_{ij} \mid i = 0 \text{ or } \sigma(i) < \sigma(j)\}$$

for some fixed permutation $\sigma \in S_E$ of the last E entries. To begin, there can be no larger subset as this would imply that for some $1 \leq i, j \leq E$ both π_{ij} and π_{ji} would belong to it, which cannot be the case for a subset $\pi_{s'} Q_\zeta$. Next, $\pi_\sigma = \pi_s Q_\zeta$ where

$$s = (p, p + \sigma(1), p + \sigma(2), \ldots, p + \sigma(E)) \quad \text{where} \quad p = -\frac{E}{2}$$

If we now make s vertex-dominant, or equivalently if we only take a σ in the factor $S_E/(S_{e_1} \times S_{e_2} \times \ldots \times S_{e_k})$, then s belongs to $S_{\alpha_\zeta} Q_\zeta$. For example, if $E = 3$ and $\sigma = id \in S_3$, then the corresponding border and corner regions for π_s are

$$C_s = \qquad \text{and} \quad B_s =$$

We have to show that the corresponding Hesselink stratum is nonempty in $Null_{\alpha_\zeta} Q_\zeta$ and that it contains θ_ζ-semistable representations. For s corresponding to a fixed $\sigma \in S_E$ the border quiver-setting $(Q_s, \alpha_s, \theta_s)$ is equal to

$$\overset{-E}{\underset{}{\textcircled{1}}} =^{z_0} \Rightarrow \overset{-E+2}{\textcircled{1}} =^{z_1} \Rightarrow \overset{-E+4}{\textcircled{1}} =^{z_2} \Rightarrow \cdots \cdots =^{z_{E-1}} \overset{E-2}{\textcircled{1}} =^{z_E} \Rightarrow \overset{E}{\textcircled{1}}$$

where the number of arrows z_i are determined by

$$\begin{cases} z_0 &= p_u \text{ if } \sigma(1) \in I_{v_u} \\ z_i = a_{uv} \text{ if } \sigma(i) \in I_{v_u} \text{ and } \sigma(i+1) \in I_{v_v} \end{cases}$$

where we recall that I_{v_i} is the interval of entries in $[1, \ldots, E]$ belonging to vertex v_i. As all the $z_i \geq 1$ it follows that $\mathbf{rep}_{\alpha_s} Q_s$ contains θ_s-stable representations, so the stratum in $Null_{\alpha_\zeta} Q_\zeta$ determined by the corner-type C_s is nonempty. We can depict the $L_s = T$-action on the corner as a representation space of the extended quiver-setting

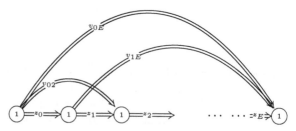

Translating representations of this extended quiver back to the original quiver-setting (Q_ζ, α_ζ) we see that the corner C_s indeed contains θ_ζ-semistable representations and hence that this stratum in the nullcone determines an irreducible component in the Brauer-Severi fiber $\psi(\xi)$ of the generic Brauer-Severi variety.

THEOREM 6.14
Let $\xi \in \text{iss}_n^m$ be of representation type $\tau = (e_1, d_1; \ldots; e_k, d_k)$ and let $E = \sum_{i=1}^k e_i$. Then, the fiber $\pi^{-1}(\xi)$ of the Brauer-Severi fibration

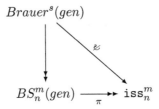

$Brauer^s(gen)$

$BS_n^m(gen) \xrightarrow{\quad \pi \quad} \text{iss}_n^m$

has exactly $\frac{E!}{e_1! e_2! \ldots e_k!}$ irreducible components, all of dimension

$$n + (m-1) \sum_{i<j} e_i e_j d_i d_j + (m-1) \sum_i \frac{e_i(e_i - 1)}{2} - \sum_i e_i$$

PROOF In view of the foregoing remarks we only have to compute the dimension of the irreducible components. For a corner type C_s as above we have that the corresponding irreducible component in $Null_{\alpha_\zeta} Q_\zeta$ has dimension

$$dim \ GL(\alpha_\zeta) - dim \ P_s + dim \ C_s$$

and from the foregoing description of C_s as a quiver-representation space we see that

- $dim \ P_s = 1 + \frac{e_i(e_i+1)}{2}$

- $dim \ C_s = n + \sum_i \frac{e_i(e_i-1)}{2}((m-1)d_i^2 + 1) + \sum_{i<j}(m-1)e_i e_j d_i d_j$

as we can identify $P_s \simeq \mathbb{C}^* \times B_{e_1} \times \ldots \times B_{e_k}$ where B_e is the Borel subgroup of GL_e. Moreover, as $\psi^{-1}(\xi)$ is a Zariski open subset of

$$(\mathbb{C}^* \times GL_n) \times^{GL(\alpha_\zeta)} Null_{\alpha_\zeta} Q_\zeta$$

we see that the corresponding irreducible component of $\psi^{-1}(\xi)$ has dimension

$$1 + dim \ GL_n - dim \ P_s + dim \ C_s$$

As the quotient morphism $\psi^{-1}(\xi) \longrightarrow \pi^{-1}(\xi)$ is surjective, we have that the Brauer-Severi fiber $\pi^{-1}(\xi)$ has the same number of irreducible components of $\psi^{-1}(\xi)$. As the quotient

$$\psi^{-1}(\xi) \longrightarrow \pi^{-1}(\xi)$$

is by Brauer-stability of all point a principal $PGL(1, n)$-fibration, substituting the obtained dimensions finishes the proof. □

In particular, we deduce that the Brauer-Severi fibration $BS_n^m(gen) \xrightarrow{\pi} \mathtt{iss}_n^m$ is a flat morphism if and only if $(m,n) = (2,2)$ in which case all Brauer-Severi fibers have dimension one.

As a final application, let us compute the Brauer-Severi fibers in a point $\xi \in X = \mathtt{triss}_n A$ of the smooth locus $Sm_n A$ of a Cayley-Hamilton order of degree n, which is of local quiver type (Q, α) where $\alpha = (1, \ldots, 1)$ and Q is the quiver

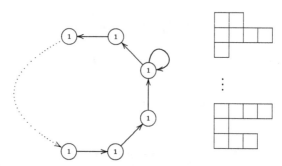

where the cycle has k vertices and $p = (p_1, \ldots, p_k)$ is an unordered partition of n having exactly k parts. That is, A is a local Cayley-smooth order over a surface of type A_{k-101}. These are the only types that can occur for smooth surface orders which are maximal orders and have a nonsingular ramification divisor. Observe also that in the description of nullcones, the extra loop will play no role, so the discussion below also gives the Brauer-Severi fibers of smooth curve orders. The Brauer-Severi fibration is étale locally isomorphic to the fibration

$$M_{\alpha'}^{ss}(Q', \theta') \xrightarrow{\pi} \mathtt{iss}_\alpha Q = \mathtt{iss}_{\alpha'} Q'$$

in a neighborhood of the trivial representation. Here, Q' is the extended quiver by one vertex v_0

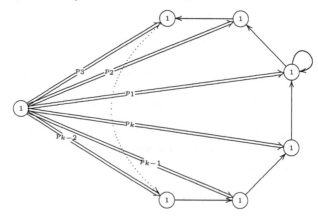

the extended dimension vector is $\alpha' = (1, 1, \ldots, 1)$ and the character is determined by the integral $k + 1$-tuple $(-n, p_1, p_2, \ldots, p_k)$. The weights of the maximal torus $T = GL(\alpha')$ of dimension $k + 1$ that occur in representations in the nullcone are

$$\pi_{\alpha'} \, Q' = \{\pi_{0 \ i}, \pi_{i \ i+1}, 1 \le i \le k\}$$

Therefore, maximal corners C_s are associated to $s \in \mathcal{S}_{\alpha'} \, Q'$ where

$$\pi_s \, Q' = \{\pi_{0 \ j}, 1 \le j \le k\} \cup \{\pi_{i \ i+1}, \pi_{i+1 \ i+2}, \dots, \pi_{i-2 \ i-1}\}$$

for some fixed i. For such a subset the corresponding s is a one string $k + 1$-tuple having as minimal value $-\frac{k}{2}$ at entry 0, $-\frac{k}{2} + 1$ at entry i, $-\frac{k}{2} + 2$ at entry $i + 1$ and so on. To verify that this corner-type occurs in $Null_{\alpha'} \, Q'$ we have to consider the corresponding border quiver-setting $(Q'_s, \alpha'_s, \theta'_s)$, which is

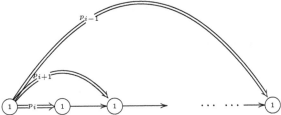

which clearly has θ'_s-semistable representations, in fact, the corresponding moduli space $M^{ss}_{\alpha'_s}(Q'_s, \theta'_s) \simeq \mathbb{P}^{p_1 - 1}$. In this case we have that $L_s = P_s = GL(\alpha'_s)$ and therefore we can also interpret the corner as an open subset of the representation space

$$C_s \hookrightarrow \mathbf{rep}_{\alpha'_s} \, Q"_s$$

where the embedding is $P_s = GL(\alpha'_s)$-equivariant and the extended quiver $Q"_s$ is

Translating corner representations back to $\mathbf{rep}_{\alpha'} \, Q'$ we see that C_s contains θ'-semistable representations, so will determine an irreducible component in the Brauer-Severi fiber $\pi^{-1}(\xi)$. Let us calculate its dimension. The irreducible component N_s of $Null_{\alpha'} \, Q'$ determined by the corner C_s has dimension

$$dim \ GL(\alpha') - dim \ P_s + dim \ C_s = (k + 1) - (k + 1) + \sum_i p_i + (k - 1)$$

$$= n + k - 1$$

But then, the corresponding component in the Brauer-stable is an open subvariety of $(\mathbb{C}^* \times GL_n) \times^{GL(\alpha')} N_s$ and therefore has dimension

$$dim \ \mathbb{C}^* \times GL_n - dim \ GL(\alpha') + dim \ N_s = 1 + n^2 - (k + 1) + n + k - 1$$

$$= n^2 + n - 1$$

But then, as the stabilizer subgroup of all Brauer-stable points is one dimensional in $\mathbb{C}^* \times GL_n$ the corresponding irreducible component in the Brauer-Severi fiber $\pi^{-1}(\xi)$ has dimension $n - 1$. The following completes the proof.

THEOREM 6.15
Let A be a Cayley-Hamilton order of degree n over a surface $X = \text{triss}_n A$ and let A be Cayley-smooth in $\xi \in X$ of type A_{k-101} and p as before. Then, the fiber of the Brauer-Severi fibration

$$BS_n^t(A) \longrightarrow X$$

in ξ has exactly k irreducible components, each of dimension $n - 1$. In particular, if A is a Cayley-smooth order over the surface X such that all local types are $(A_{k-101}.p)$ for some $k \geq 1$ and partition p of n in having k-parts, then the Brauer-Severi fibration is a flat morphism.

In fact, one can give a nice geometric interpretation to the different components. Consider the component corresponding to the corner C_s with notations as before. Consider the sequence of $k - 1$ rational maps

$$\mathbb{P}^{n-1} \longrightarrow \mathbb{P}^{n-1-p_{i-1}} \longrightarrow \mathbb{P}^{n-1-p_{i-1}-p_{i-2}} \longrightarrow \ldots \longrightarrow \mathbb{P}^{p_i-1}$$

defined by killing the right-hand coordinates

$$[x_1 : \ldots : x_n] \mapsto [x_1 : \ldots : x_{n-p_{i-1}} : \underbrace{0 : \ldots : 0}_{p_{i-1}}] \mapsto \ldots \mapsto [x_1 : \ldots : x_{p_i} : \underbrace{0 : \ldots : 0}_{n-p_i}]$$

that is in the extended corner-quiver setting

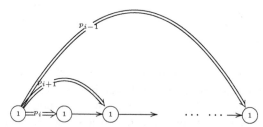

we subsequently set all entries of the arrows from v_0 to v_{i-j} zero for $j \geq 1$, the extreme projection $\mathbb{P}^{n-1} \longrightarrow \mathbb{P}^{p_i-1}$ corresponds to the projection $C_s/P_s \longrightarrow B_s/L_s = M_{\alpha'_s}^{ss}(Q'_s, \theta'_s)$. Let V_i be the subvariety in $\times_{j=1}^k \mathbb{P}^{n-1}$ be the closure of the graph of this sequence of rational maps. If we label the coordinates in the $k - j$-th component $\mathbb{P}^{n-1}$ as $x(j) = [x_1(j) : \ldots : x_n(j)]$, then the multihomogeneous equations defining V_i are

$$\begin{cases} x_a(j) & = 0 \text{ if } a > p_i + p_{i+1} + \ldots + p_{i+j} \\ x_a(j)x_b(j-1) & = x_b(j)x_a(j-1) \text{ if } 1 \leq a < b \leq p_i + \ldots + p_{i+l-1} \end{cases}$$

One verifies that V_i is a smooth variety of dimension $n - 1$. If we would have the patience to work out the whole nullcone (restricting to the θ'-semistable representations) rather than just the irreducible components, we would see that the Brauer-Severi fiber $\pi^{-1}(\xi)$ consists of the varieties $V_1, \ldots, V_k$ intersecting transversally. The reader is invited to compare our description of the Brauer-Severi fibers with that of M. Artin [3] in the case of Cayley-smooth maximal curve orders.

References

The stratification of the nullcone used in this chapter is due to W. Hesselink [45]. The proof of theorem 6.3 uses the results of F. Kirwan [55]. Theorem 6.4 is due to H.P. Kraft [62], theorem 6.5 and theorem 6.8 are special cases of Hesselinks result [45]. Example 6.14 is due to G. Van de Weyer . The definition of Brauer-Severi varieties of orders used here is due to M. Van den Bergh [101]. All other results are due to L. Le Bruyn [71], [72] and [75].

Chapter 7

Noncommutative Manifolds

By now we have developed enough machinery to study the representation varieties $\text{trep}_n\ A$ and $\text{triss}_n\ A$ of a Cayley-smooth algebra $A \in \text{alg@n}$. In particular, we now understand the varieties

$$\text{rep}_n\ A = \text{trep}_n\ \int_n A \quad \text{and} \quad \text{iss}_n\ A = \text{triss}_n\ \int_n A$$

for the level n approximation $\int_n A$ of a Quillen-smooth algebra A, for all n. In this chapter we begin to study noncommutative manifolds, that is, *families* $(X_n)_n$ of commutative varieties that are locally controlled by Quillen-smooth algebras. Observe that for every $\mathbb{C}$-algebra A, the direct sum of representations induces *sum maps*

$$\text{rep}_n\ A \times \text{rep}_m\ A \longrightarrow \text{rep}_{n+m}\ A \quad \text{and} \quad \text{iss}_n\ A \times \text{iss}_m\ A \longrightarrow \text{iss}_{n+m}\ A$$

The characteristic feature of a family $(X_n)_n$ of varieties defining a noncommutative variety is that they are connected by sum-maps

$$X_n \times X_m \longrightarrow X_{n+m}$$

and that these morphisms are locally of the form $\text{iss}_n\ A \times \text{iss}_m\ A \longrightarrow \text{iss}_{n+m}\ A$ for a Quillen-smooth algebra A. An important class of examples of such noncommutative manifolds is given by moduli spaces of quiver representations. In order to prove that they are indeed of the above type, we have to recall results on semi-invariants of quiver representations and universal localization.

7.1 Formal structure

Objects in noncommutative geometry@n are families of varieties $(X_i)_i$, which are locally controlled by a set of noncommutative algebras $\mathcal{A}$. That is, X_i is locally the quotient variety of a representation variety $\text{rep}_n\ A$ for some n and some $\mathbb{C}$-algebra $A \in \mathcal{A}$. In section 2.7 we have seen that the representation varieties form a somewhat mysterious subclass of the category of all (affine) GL_n-varieties. For this reason it is important to equip them with additional structures that make them stand out among the GL_n-varieties. In this section we define the *formal structure* on representation varieties, extending in a natural way the formal structure introduced by M. Kapranov on smooth affine varieties. Let us give an illustrative example of this structure.

Example 7.1 Formal structure on $\mathbb{A}^d$

Consider the affine space $\mathbb{A}^d$ with coordinate ring $\mathbb{C}[x_1, \dots, x_d]$ and order the coordinate functions $x_1 < x_2 < \dots < x_d$. Let $\mathfrak{f}_d$ be the free Lie algebra on $\mathbb{C}x_1 \oplus \dots \oplus \mathbb{C}x_d$, which has an ordered basis $B = \cup_{k \geq 1} B_k$ defined as follows. B_1 is the ordered set $\{x_1, \dots, x_d\}$ and $B_2 = \{[x_i, x_j] \mid j < i\}$, ordered such that $B_1 < B_2$ and $[x_i, x_j] < [x_k, x_l]$ iff $j < l$ or $j = l$ and $i < k$. Having constructed the ordered sets B_l for $l < k$ we define

$$B_k = \{[t, w] \mid t = [u, v] \in B_l, w \in B_{k-l} \text{ such that } v \leq w < t \text{ for } l < k\}$$

For $l < k$ we let $B_l < B_k$ and B_k is ordered by $[t, w] < [t'.w']$ iff $w < w'$ or $w = w'$ and $t < t'$.

It is well known that B is an ordered $\mathbb{C}$-basis of the Lie algebra $\mathfrak{f}_d$ and that its enveloping algebra

$$U(\mathfrak{f}_d) = \mathbb{C}\langle x_1, \dots, x_d \rangle$$

is the free associative algebra on the x_i. We number the elements of $\cup_{k \geq 2} B_k$ according to the order $\{b_1, b_2, \dots\}$ and for $b_i \in B_k$ we define $ord(b_i) = k - 1$ (the number of brackets needed to define b_i). Let Λ be the set of all functions with finite support $\lambda : \cup_{k \geq 2} B_k \longrightarrow \mathbb{N}$ and define $ord(\lambda) = \sum \lambda(b_i) ord(b_i)$. Rephrasing the *Poincaré-Birkhoff-Witt* result for $U(\mathfrak{f}_d)$ we have that any noncommutative polynomial $p \in \mathbb{C}\langle x_1, \dots, x_d \rangle$ can be written uniquely as a finite sum

$$p = \sum_{\lambda \in \Lambda} [\![f_\lambda]\!] \, M_\lambda$$

where $[\![f_\lambda]\!] \in \mathbb{C}[x_1, \dots, x_d] = S(B_1)$ and $M_\lambda = \prod_i b_i^{\lambda(b_i)}$. In particular, for every $\lambda, \mu, \nu \in \Lambda$, there is a unique bilinear differential operator with polynomial coefficients

$$C_{\lambda\mu}^{\nu} : \mathbb{C}[x_1, \dots, x_d] \otimes_{\mathbb{C}} \mathbb{C}[x_1, \dots, x_d] \longrightarrow \mathbb{C}[x_1, \dots, x_d]$$

defined by expressing the product $[\![f]\!] \, M_\lambda \cdot [\![g]\!] \, M_\mu$ in $\mathbb{C}\langle x_1, \dots, x_d \rangle$ uniquely as $\sum_{\nu \in \Lambda} [\![C_{\lambda\mu}^{\nu}(f, g)]\!] \, M_\nu$.

By associativity of $\mathbb{C}\langle x_1, \dots, x_d \rangle$ the $C_{\lambda\mu}^{\nu}$ satisfy the *associativity constraint*, that is, we have equality of the trilinear differential operators

$$\sum_{\mu_1} C_{\mu_1\lambda_3}^{\nu} \circ (C_{\lambda_1\lambda_2}^{\mu_1} \otimes id) = \sum_{\mu_2} C_{\lambda_1\mu_2}^{\nu} \circ (id \otimes C_{\lambda_2\lambda_3}^{\mu_2})$$

for all $\lambda_1, \lambda_2, \lambda_3, \nu \in \Lambda$. That is, one can define the algebra $\mathbb{C}\langle x_1, \ldots, x_d \rangle_{[[ab]]}$ to be the $\mathbb{C}$-vector space of possibly *infinite formal sums* $\sum_{\lambda \in \Lambda} [\![f_\lambda]\!] \, M_\lambda$ with multiplication defined by the operators $C_{\lambda\mu}^\nu$.

Let $A_d(\mathbb{C})$ be the d-th *Weyl algebra*, that is, the ring of differential operators with polynomial coefficients on $\mathbb{A}^d$. Let $\mathcal{O}_{\mathbb{A}^d}$ be the structure sheaf on $\mathbb{A}^d$, then it is well-known that the ring of sections $\mathcal{O}_{\mathbb{A}^d}(U)$ on any Zariski open subset $U \hookrightarrow \mathbb{A}^d$ is a left $A_d(\mathbb{C})$-module. Define a sheaf $\mathcal{O}_{\mathbb{A}^d}^f$ of noncommutative algebras on $\mathbb{A}^d$ by taking as its sections over U the algebra

$$\mathcal{O}_{\mathbb{A}^d}^f(U) = \mathbb{C}\langle x_1, \ldots, x_d \rangle_{[[ab]]} \underset{\mathbb{C}[x_1, \ldots, x_d]}{\otimes} \mathcal{O}_{\mathbb{A}^d}(U)$$

that is, the $\mathbb{C}$-vector space of possibly infinite formal sums $\sum_{\lambda \in \Lambda} [\![f_\lambda]\!] \, M_\lambda$ with $f_\lambda \in \mathcal{O}_{\mathbb{A}^d}(U)$ and the multiplication is given as before by the action of the bilinear differential operators $C_{\lambda\mu}^\nu$ on the left $A_d(\mathbb{C})$-module $\mathcal{O}_{\mathbb{A}^d}(U)$, that is, for all $f, g \in \mathcal{O}_{\mathbb{A}^d}(U)$ we have

$$[\![f]\!] \, M_\lambda . [\![g]\!] \, M_\mu = \sum_\nu [\![C_{\lambda\mu}^\nu(f, g)]\!] \, M_\nu$$

This sheaf of noncommutative algebras $\mathcal{O}_{\mathbb{A}^d}^f$ is called *the formal structure* on $\mathbb{A}^d$. ⬚

We will now define formal structures on arbitrary affine smooth varieties. Let R be an associative $\mathbb{C}$-algebra, R^{Lie} its Lie structure and R_m^{Lie} the subspace spanned by the expressions $[r_1, [r_2, \ldots, [r_{m-1}, r_m] \ldots]]$ containing $m - 1$ instances of Lie brackets. The *commutator filtration* of R is the (increasing) filtration by ideals $(F^k R)_{k \in \mathbb{Z}}$ with $F^k R = R$ for $d \in \mathbb{N}$ and

$$F^{-k} R = \sum_m \sum_{i_1 + \ldots + i_m = k} R R_{i_1}^{Lie} R \ldots R R_{i_m}^{Lie} R$$

Observe that all $\mathbb{C}$-algebra morphisms preserve the commutator filtration. The *associated graded algebra* $\mathrm{gr}_F R$ is a (negatively) graded commutative *Poisson algebra* with part of degree zero, the *abelianization* $R_{ab} = \frac{R}{[R,R]}$. If $R = \mathbb{C}\langle x_1, \ldots, x_d \rangle$, then the commutator filtration has components

$$F^{-k} \mathbb{C}\langle x_1, \ldots, x_d \rangle = \{ \sum_\lambda [\![f_\lambda]\!] \, M_\lambda, \forall \lambda : ord(\lambda) \geq k \}$$

DEFINITION 7.1 *Denote with* $\mathtt{nil}_k$ *the category of associative $\mathbb{C}$-algebras R such that $F^{-k}R = 0$ (in particular, $\mathtt{nil}_1 = \mathtt{commalg}$ the category of commutative $\mathbb{C}$-algebras). An algebra $A \in Ob(\mathtt{nil}_k)$ is said to be k-smooth if and only if for all $T \in Ob(\mathtt{nil}_k)$, all nilpotent two-sided ideals $I \triangleleft T$ and all*

$\mathbb{C}$-algebra morphisms $A \xrightarrow{\phi} \frac{T}{I}$ there exist a lifted $\mathbb{C}$-algebra morphism

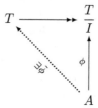

making the diagram commutative. Alternatively, an algebra is k-smooth if and only if it is $\mathtt{nil}_k$-smooth.

For example, the quotient $\frac{\mathbb{C}\langle x_1,...,x_d\rangle}{F^{-k}\,\mathbb{C}\langle x_1,...,x_d\rangle}$ is k-smooth using the lifting property of free algebras and the fact that algebra morphisms preserve the commutator filtration. Generalizing this, if A is Quillen-smooth then the quotient

$$A_{(k)} = \frac{A}{F^{-k}\,A}$$

is k-smooth.

Kapranov proves [50, Thm 1.6.1] that an affine commutative *Grothendieck-smooth* algebra C has a *unique* (upto $\mathbb{C}$-algebra isomorphism identical on C) k-smooth *thickening* $C^{(k)}$ with $C_{ab}^{(k)} \simeq C$. The inverse limit (connecting morphisms are given by the unicity result)

$$C^f = \lim_{\leftarrow} C^{(k)}$$

is then called the *formal completion* of C. Clearly, one has $C_{ab}^f = C$. For example

$$\mathbb{C}[x_1,\ldots,x_d]^f = \lim_{\leftarrow} \frac{\mathbb{C}\langle x_1,\ldots,x_d\rangle}{F^{-k}\,\mathbb{C}\langle x_1,\ldots,x_d\rangle} \simeq \mathbb{C}\langle x_1,\ldots,x_d\rangle_{[[\mathtt{ab}]]}$$

If X is an affine smooth (commutative) variety, one can use the formal completion $\mathbb{C}[X]^f$ to define a sheaf of noncommutative algebras $\mathcal{O}_X^f$ defining the *formal structure* on X.

The fact that C is Grothendieck-smooth is essential to construct and prove uniqueness of the formal completion. At present, no sufficiently functorial extension of formal completion is known for arbitrary commutative $\mathbb{C}$-algebras. It is not true that any (nonaffine) smooth variety can be equipped with a formal structure. In fact, the obstruction gives important new invariants of a smooth variety related to *Atiyah classes*. We refer to [50, §4] for more details.

We recall briefly the algebraic construction of *microlocalization*. Let R be a filtered algebra with a *separated filtration* $\{F_n\}_{n\in\mathbb{Z}}$ and let S be a multiplicatively closed subset of R containing 1 but not 0. For any $r \in F_n - F_{n-1}$ we denote its *principal character* $\sigma(r)$ to be the image of r in the associated graded algebra $gr(R)$. We assume that the set $\sigma(S)$ is a multiplicatively closed subset of $gr(R)$. We define the *Rees ring* $\tilde{R}$ to be the graded algebra

$$\tilde{R} = \oplus_{n \in Z} F_n t^n \hookrightarrow R[t, t^{-1}]$$

where t is an extra central variable. If $\sigma(s) \in gr(R)_n$ then we define the element $\tilde{s} = st^n \in \tilde{R}_n$. The set $\tilde{S} = \{\tilde{s}, s \in S\}$ is a multiplicatively closed subset of homogeneous elements in $\tilde{R}$.

Assume that $\sigma(S)$ is an Ore set in $gr(R) = \frac{\tilde{R}}{(t)}$, then for every $n \in \mathbb{N}_0$ the image $\pi_n(\tilde{S})$ is an Ore set in $\frac{\tilde{R}}{(t^n)}$ where $\tilde{R} \twoheadrightarrow \frac{\tilde{R}}{(t^n)}$ is the quotient morphism. Hence, we have an inverse system of graded localizations and can form the inverse limit in the graded sense

$$Q^\mu_{\tilde{S}}(\tilde{R}) = \lim^g_{\leftarrow} \pi_n(\tilde{S})^{-1} \frac{\tilde{R}}{(t^n)}$$

The element t acts torsion-free on this limit and hence we can form the filtered algebra

$$Q^\mu_S(R) = \frac{Q^\mu_{\tilde{S}}(\tilde{R})}{(t-1)Q^\mu_{\tilde{S}}(\tilde{R})}$$

which is the *microlocalization* of R at the multiplicatively closed subset S. We recall that the associated graded algebra of the microlocalization can be identified with the graded localization

$$gr(Q^\mu_S(R)) = \sigma(S)^{-1} gr(R)$$

Let R be a $\mathbb{C}$-algebra with $R_{ab} = \frac{R}{[R,R]} = C$. We assume that the commutator filtration $(F^k)_{k \in \mathbb{Z}}$ is a separated filtration on R. Observe that this is not always the case (for example, consider $U(\mathfrak{g})$ for $\mathfrak{g}$ a semisimple Lie algebra) but often one can repeat the argument below replacing R with $\frac{R}{\cap F^n}$.

Observe that $gr(R)$ is a negatively graded commutative algebra with part of degree zero C. Take a multiplicatively closed subset S_c of C, then $S = S_c + [R,R]$ is a multiplicatively closed subset of R with the property that $\sigma(S) = S_c$ and clearly S_c is an Ore set in $gr(R)$. Therefore, $\tilde{S}$ is a multiplicatively closed set of the Rees ring $\tilde{R}$ consisting of homogeneous elements of degree zero. Observing that $(t^n)_0 = F^{-n} t^n$ for all $n \in \mathbb{N}_0$ we see that

$$Q^\mu_S(R) = \lim_{\leftarrow} \pi_n(S)^{-1} \frac{R}{F^{-n}}$$

where $R \xrightarrow{\pi_n} \frac{R}{F^n}$ is the quotient morphism and Q^μ_S is filtered again by the commutator filtration and has as associated graded algebra

$$gr(Q^\mu_S(R)) = S_c^{-1} gr(R).$$

One can define a *microstructure sheaf* $\mathcal{O}_R^\mu$ on the affine scheme X of C by taking as its sections over the affine Zariski open set $X(f)$

$$\Gamma(X(f), \mathcal{O}_R^\mu) = Q_{S_f}^\mu(R)$$

where $S = \{1, f, f^2, \ldots\} + [R, R]$. For C a Grothendieck-smooth affine commutative algebra this sheaf of noncommutative algebras is the *formal structure* on X introduced by M. Kapranov.

An important remark to make is that one really needs microlocalization to construct a sheaf of noncommutative algebras on X. If by some fluke we would have that all the S_f are already Ore sets in R, we might optimistically assume that taking as sections over $X(f)$ the Ore localization $S_f^{-1}R$ we would define a sheaf $\mathcal{O}_R$ over X. This is in general *not* the case as the Ore set S_g need no longer be Ore in a localization $S_f^{-1}R$!

Still one can remedy this by defining a *noncommutative Zariski topology* on X using *words* in the Ore sets S_f, see [104, §1.3]. Whereas we do not need this to define formal structures it seems to me inevitable that at a later stage in the development of noncommutative geometry we will need to resort to such *noncommutative Grothendieck topologies* on usual commutative schemes.

Having defined a formal structure on affine smooth varieties, we will now extend it to arbitrary representation varieties. The starting point is that for every associative algebra A the functor

$$\mathtt{alg} \xrightarrow{\ Hom_{\mathtt{alg}}(A, M_n(-))\ } \mathtt{sets}$$

is *representable* in $\mathtt{alg}$. That is, there exists an associative $\mathbb{C}$-algebra $\sqrt[n]{A}$ such that there is a natural equivalence between the functors

$$Hom_{\mathtt{alg}}(A, M_n(-)) \underset{n.e.}{\sim} Hom_{\mathtt{alg}}(\sqrt[n]{A}, -)$$

In other words, for every associative $\mathbb{C}$-algebra B, there is a functorial one-to-one correspondence between the sets

$$\begin{cases} \text{algebra maps} & A \longrightarrow M_n(B) \\ \text{algebra maps} & \sqrt[n]{A} \longrightarrow B \end{cases}$$

We call $\sqrt[n]{A}$ the *n-th root algebra* of A .

Example 7.2

If $A = \mathbb{C}\langle x_1, \ldots, x_d \rangle$, then it is easy to see that $\sqrt[n]{A}$ is the free algebra $\mathbb{C}\langle x_{11,1}, \ldots, x_{nn,d} \rangle$ on dn^2 variables. For, given an algebra map $A \xrightarrow{\phi} M_n(B)$ we obtain an algebra map $\sqrt[n]{A} \longrightarrow B$ by sending the free variable $x_{ij,k}$ to the (i,j)-entry of the matrix $\phi(x_k) \in M_n(B)$. Conversely, to an algebra map $\sqrt[n]{A} \xrightarrow{\psi} B$ we assign the algebra map $A \longrightarrow M_n(B)$ by sending x_k to the matrix $(\psi(x_{ij,k}))_{i,j} \in M_n(B)$. Clearly, these operations are each others inverses. □

To define $\sqrt[n]{A}$ in general, consider the free algebra product $A * M_n(\mathbb{C})$ and consider the subalgebra

$$\sqrt[n]{A} = A * M_n(\mathbb{C})^{M_n(\mathbb{C})} = \{p \in A * M_n(\mathbb{C}) \mid p.(1*m) = (1*m).p \; \forall m \in M_n(\mathbb{C})\}$$

Before we can prove the universal property of $\sqrt[n]{A}$ we need to recall a property that $M_n(\mathbb{C})$ shares with any Azumaya algebra : if $M_n(\mathbb{C}) \xrightarrow{\phi} R$ is an algebra morphism and if $R^{M_n(\mathbb{C})} = \{r \in R \mid r.\phi(m) = \phi(m).r \; \forall m \in M_n(\mathbb{C})\}$, then we have $R \simeq M_n(\mathbb{C}) \otimes_{\mathbb{C}} R^{M_n(\mathbb{C})}$.

In particular, if we apply this to $R = A * M_n(\mathbb{C})$ and the canonical map $M_n(\mathbb{C}) \xrightarrow{\phi} A * M_n(\mathbb{C})$ where $\phi(m) = 1 * m$ we obtain that $M_n(\sqrt[n]{A}) = M_n(\mathbb{C}) \otimes_{\mathbb{C}} \sqrt[n]{A} = A * M_n(\mathbb{C})$.

Hence, if $\sqrt[n]{A} \xrightarrow{f} B$ is an algebra map we can consider the composition

$$A \xrightarrow{id_A * 1} A * M_n(\mathbb{C}) \simeq M_n(\sqrt[n]{A}) \xrightarrow{M_n(f)} M_n(B)$$

to obtain an algebra map $A \longrightarrow M_n(B)$. Conversely, consider an algebra map $A \xrightarrow{g} M_n(B)$ and the canonical map $M_n(\mathbb{C}) \xrightarrow{i} M_n(B)$, which centralizes B in $M_n(B)$. Then, by the universal property of free algebra products we have an algebra map $A * M_n(\mathbb{C}) \xrightarrow{g*i} M_n(B)$ and restricting to $\sqrt[n]{A}$ we see that this maps factors

$$A * M_n(\mathbb{C}) \xrightarrow{g*i} M_n(B)$$

$$\sqrt[n]{A} \cdots\cdots\cdots\rightarrow B$$

and one verifies that these two operations are each other's inverses.

It follows from the functoriality of the $\sqrt[n]{\cdot}$ construction that $\mathbb{C}\langle x_1, \ldots, x_d \rangle \longrightarrow A$ implies that $\sqrt[n]{\mathbb{C}\langle x_1, \ldots, x_d \rangle} \longrightarrow \sqrt[n]{A}$. Therefore, if A is affine and generated by $\leq d$ elements, then $\sqrt[n]{A}$ is also affine and generated by $\leq dn^2$ elements.

These properties allow us define a *formal completion* of $\mathbb{C}[\mathbf{rep}_n \; A]$ in a functorial way for any associative algebra A. Equip $\sqrt[n]{A}$ with the commutator filtration

$$\cdots \hookrightarrow F_{-2} \sqrt[n]{A} \hookrightarrow F_{-1} \sqrt[n]{A} \hookrightarrow \sqrt[n]{A} \quad = \quad \sqrt[n]{A} \quad = \quad \cdots$$

Because algebra morphisms are commutator filtration preserving, it follows from the universal property of $\sqrt[n]{A}$ that $\frac{\sqrt[n]{A}}{F_{-k} \sqrt[n]{A}}$ is the object in $\mathbf{nil}_k$ representing the functor

$$\mathbf{nil}_k \xrightarrow{\; Hom_{\mathbf{alg}}(A, M_n(-)) \;} \mathbf{sets}$$

In particular, because the categories `commalg` and `nil`$_1$ are naturally equivalent, we deduce that

$$\sqrt[n]{A}_{ab} = \frac{\sqrt[n]{A}}{[\sqrt[n]{A}, \sqrt[n]{A}]} = \frac{\sqrt[n]{A}}{F_{-1}\sqrt[n]{A}} \simeq \mathbb{C}[\mathbf{rep}_n\, A]$$

because both algebras represent the same functor. We now define

$$\sqrt[n]{A}_{[[ab]]} = \lim_{\leftarrow} \frac{\sqrt[n]{A}}{F_{-k}\,\sqrt[n]{A}}.$$

Assume that A is Quillen-smooth, then so is $\sqrt[n]{A}$ because we have seen before that

$$M_n(\sqrt[n]{A}) \simeq A * M_n(\mathbb{C})$$

and the class of Quillen-smooth algebras is easily seen to be closed under free products and matrix algebras.

As a consequence, we have for every $k \in \mathbb{N}$ that the quotient $\frac{\sqrt[n]{A}}{F_{-k}\,\sqrt[n]{A}}$ is k-smooth. Moreover, we have that

$$(\frac{\sqrt[n]{A}}{F_{-k}\,\sqrt[n]{A}})_{ab} \simeq \frac{\sqrt[n]{A}}{[\sqrt[n]{A}, \sqrt[n]{A}]} \simeq \mathbb{C}[\mathbf{rep}_n\, A]$$

Because $\mathbb{C}[\mathbf{rep}_n\, A]$ is an affine commutative Grothendieck-smooth algebra, we deduce from the uniqueness of k-smooth thickenings that

$$\mathbb{C}[\mathbf{rep}_n\, A]^{(k)} \simeq \frac{\sqrt[n]{A}}{F_{-k}\,\sqrt[n]{A}}$$

and consequently that the formal completion of $\mathbb{C}[\mathbf{rep}_n\, A]$ can be identified with

$$\mathbb{C}[\mathbf{rep}_n\, A]^f \simeq \sqrt[n]{A}_{[[ab]]}$$

Therefore, if we define for an *arbitrary* $\mathbb{C}$-algebra A the *formal completion* of $\mathbb{C}[\mathbf{rep}_n\, A]$ to be $\sqrt[n]{A}_{[[ab]]}$ we have a canonical extension of the formal structure on affine Grothendieck-smooth commutative algebras to the class of coordinate rings of representation spaces on which it is functorial in the algebras.

There is a natural action of GL_n by algebra automorphisms on $\sqrt[n]{A}$. Let u_A denote the universal morphism $A \xrightarrow{u_A} M_n(\sqrt[n]{A})$ corresponding to the identity map on $\sqrt[n]{A}$. For $g \in GL_n$ we can consider the composed algebra map

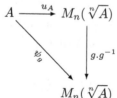

Then g acts on $\sqrt[n]{A}$ via the automorphism $\sqrt[n]{A} \xrightarrow{\phi_g} \sqrt[n]{A}$ corresponding to the the composition ψ_g. It is easy to verify that this indeed defines a GL_n-action on $\sqrt[n]{A}$.

The *formal structure sheaf* $\mathcal{O}^f_{\mathtt{rep}_n\ A}$ defined over $\mathtt{rep}_n\ A$ constructed from $\sqrt[n]{A}$ will be denoted by $\mathcal{O}^f_{\sqrt[n]{A}}$. We see that it actually has a GL_n-structure that is compatible with the GL_n-action on $\mathtt{rep}_n\ A$.

7.2 Semi-invariants

An important class of examples of noncommutative varieties are *moduli spaces* of θ-semistable representations of quivers. Because the moduli space $M^{ss}_\alpha(Q, \theta)$ is by definition the projective scheme of the graded *algebra of semi-invariants* of weight χ^n_θ for some n

$$M^{ss}_\alpha(Q, \theta) = \mathtt{proj}\ \oplus_{n=0}^\infty \mathbb{C}[\mathtt{rep}_\alpha\ Q]^{GL(\alpha), \chi^n \theta}$$

we need some control on these semi-invariants of quivers.

In this section we will give a generating set of semi-invariants. The strategy of proof should be clear by now. First, we will describe a large set of semi-invariants. Then we use classical invariant theory to describe all multilinear semi-invariants of $GL(\alpha)$, or equivalently, all multilinear invariants of $SL(\alpha) = SL_{a_1} \times \ldots \times SL_{a_k}$ and describe them in terms of these determinantal semi-invariants. Finally, we show by polarization and restitution that these semi-invariants do indeed generate all semi-invariants.

Let Q be a quiver on k vertices $\{v_1, \ldots, v_k\}$. We introduce the *additive* $\mathbb{C}$-*category* $\mathtt{add}\ Q$ generated by the quiver. For every vertex v_i we introduce an indecomposable object, which we denote by $\textcircled{i}$. An arbitrary object in $\mathtt{add}\ Q$ is then a sum of these

$$\textcircled{1}^{\oplus e_1} \oplus \ldots \oplus \textcircled{k}^{\oplus e_k}$$

That is, we can identify $\mathtt{add}\ Q$ with $\mathbb{N}^k$. Morphisms in the category $\mathtt{add}\ Q$ are defined by the rules

$$\begin{cases} Hom_{\mathtt{add}\ Q}(\,\textcircled{j}\,,\,\textcircled{i}\,) = \textcircled{j} \overset{\diagup}{} \textcircled{i} \\[1em] Hom_{\mathtt{add}\ Q}(\,\textcircled{i}\,,\,\textcircled{i}\,) = \textcircled{i} \end{cases}$$

where the right-hand sides are the $\mathbb{C}$-vector spaces spanned by all oriented paths from v_i to v_j in the quiver Q, including the idempotent (trivial) path e_i when $i = j$.

Clearly, for any k-tuples of positive integers $\alpha = (u_1, \ldots, u_k)$ and $\beta = (v_1, \ldots, v_k)$

$$Hom_{\mathsf{add}\ Q}(\ \textcircled{\scriptsize 1}^{\oplus u_1} \oplus \ldots \oplus \textcircled{\scriptsize k}^{\oplus u_k}\ ,\ \textcircled{\scriptsize 1}^{\oplus v_1} \oplus \ldots \oplus \textcircled{\scriptsize k}^{\oplus v_k}\ )$$

is defined by matrices and composition arises via matrix multiplication

$$\begin{bmatrix} M_{v_1 \times u_1}(\ \textcircled{\scriptsize 1}\ ) & \ldots & M_{v_1 \times u_k}(\ \textcircled{\scriptsize 1} \qquad \textcircled{\scriptsize k}\ ) \\ \vdots & \ddots & \vdots \\ M_{v_k \times u_1}(\ \textcircled{\scriptsize k} \qquad \textcircled{\scriptsize 1}\ ) & \ldots & M_{v_k \times u_k}(\ \textcircled{\scriptsize k}\ ) \end{bmatrix}$$

Fix a dimension vector $\alpha = (a_1, \ldots, a_k)$ and a morphism ϕ in $\mathsf{add}\ Q$

$$\textcircled{\scriptsize 1}^{\oplus u_1} \oplus \ldots \oplus \textcircled{\scriptsize k}^{\oplus u_k} \xrightarrow{\ \phi\ } \textcircled{\scriptsize 1}^{\oplus v_1} \oplus \ldots \oplus \textcircled{\scriptsize k}^{\oplus v_k}$$

For any representation $V \in \mathbf{rep}_\alpha\ Q$ we can replace each occurrence of an arrow $\textcircled{\scriptsize j} \xleftarrow{\ a\ } \textcircled{\scriptsize i}$ of Q in ϕ by the $a_j \times a_i$-matrix V_a. This way we obtain a rectangular matrix

$$V(\phi) \in M_{\sum_{i=1}^k a_i v_i \times \sum_{i=1}^k a_i u_i}(\mathbb{C})$$

If we are in a situation where $\sum a_i v_i = \sum a_i u_i$, then we can define a *semi-invariant polynomial function* on $\mathbf{rep}_\alpha\ Q$ by

$$P_{\alpha,\phi}(V) = det\ V(\phi)$$

We call such semi-invariants *determinantal semi-invariants* . One verifies that $P_{\phi,\alpha}$ is a semi-invariant of weight χ_θ where $\theta = (u_1 - v_1, \ldots, u_k - v_k)$. We will show that such determinantal semi-invariant together with traces along oriented cycles in the quiver Q generate all semi-invariants.

Because semi-invariants for the $GL(\alpha)$-action on $\mathbf{rep}_\alpha\ Q$ are the same as invariants for the restricted action of $SL(\alpha) = SL_{a_1} \times \ldots \times SL_{a_k}$, we will describe the multilinear $SL(\alpha)$-invariants from classical invariant theory. Because

$$\mathbf{rep}_\alpha\ Q = \bigoplus_{\textcircled{\scriptsize j} \xleftarrow{\ a\ } \textcircled{\scriptsize i}} M_{a_j \times a_i}(\mathbb{C})$$

$$= \bigoplus_{\textcircled{\scriptsize j} \xleftarrow{\ a\ } \textcircled{\scriptsize i}} \mathbb{C}^{a_i} \otimes \mathbb{C}^{*a_j}$$

we have to consider multilinear $SL(\alpha)$-invariants of

$$\bigotimes_{\textcircled{j}\leftarrow\textcircled{i}} \mathbb{C}^{a_i} \otimes \mathbb{C}^{*a_j} = \bigotimes_{\textcircled{i}} \left[\bigotimes_{\textcircled{\scriptsize$\ast$}\leftarrow\textcircled{i}} \mathbb{C}^{a_i} \otimes \bigotimes_{\textcircled{i}\leftarrow\textcircled{\scriptsize$\circ$}} \mathbb{C}^{*a_i} \right]$$

Hence, any multilinear $SL(\alpha)$-invariant can be written as $f = \prod_{i=1}^{k} f_i$ where f_i is a SL_{a_i}-invariant of

$$\bigotimes_{\textcircled{\scriptsize$\ast$}\leftarrow\textcircled{i}} \mathbb{C}^{a_i} \otimes \bigotimes_{\textcircled{i}\leftarrow\textcircled{\scriptsize$\circ$}} \mathbb{C}^{*a_i}$$

Let us recall the classical description of multilinear SL_n-invariants on $M_n^{\oplus i} \oplus V_n^{\oplus j} \oplus V_n^{*\oplus z}$, that is, the SL_n-invariant linear maps

$$\underbrace{M_n \otimes \ldots \otimes M_n}_{i} \otimes \underbrace{V_n \otimes \ldots \otimes V_n}_{j} \otimes \underbrace{V_n^* \otimes \ldots \otimes V_n^*}_{z} \xrightarrow{f} \mathbb{C}$$

By the identification $M_n = V_n \otimes V_n^*$ we have to determine the SL_n-invariant linear maps

$$V_n^{\otimes i+j} \otimes V_n^{*\otimes i+z} \xrightarrow{f} \mathbb{C}$$

The description of such invariants is given by classical invariant theory, see [107, II.5,Thm. 2.5.A].

THEOREM 7.1
The multilinear SL_n-invariants f are linear combinations of invariants of one of the following two types

1. *For $(i_1, \ldots, i_n, h_1, \ldots, h_n, \ldots, t_1, \ldots, t_n, s_1, \ldots, s_r)$ a permutation of the $i + j$ vector indices and $(u_1, \ldots, u_r)$ a permutation of the $i + z$ covector indices, consider the SL_n-invariant*

$$[v_{i_1}, \ldots, v_{i_n}] \, [v_{h_1}, \ldots, v_{h_n}] \, \ldots \, [v_{t_1}, \ldots, v_{t_n}] \, \phi_{u_1}(v_{s_1}) \ldots \phi_{u_r}(v_{s_r})$$

where the brackets are the determinantal invariants

$$[v_{a_1}, \ldots, v_{a_n}] = \det \begin{bmatrix} v_{a_1} & v_{a_2} & \cdots & v_{a_n} \end{bmatrix}$$

2. *For $(i_1, \ldots, i_n, h_1, \ldots, h_n, \ldots, t_1, \ldots, t_n, s_1, \ldots, s_r)$ a permutation of the $i + z$ covector indices and $(u_1, \ldots, u_r)$ a permutation of the $i + j$ vector indices, consider the SL_n-invariant*

$$[\phi_{i_1}, \ldots, \phi_{i_n}]^* \, [\phi_{h_1}, \ldots, \phi_{h_n}]^* \, \ldots \, [\phi_{t_1}, \ldots, \phi_{t_n}]^* \, \phi_{u_1}(v_{s_1}) \ldots \phi_{u_r}(v_{s_r})$$

where the cobrackets are the determinantal invariants

$$[\phi_{a_1}, \ldots, \phi_{a_n}]^* = \det \begin{bmatrix} \phi_{a_1} \\ \vdots \\ \phi_{a_n} \end{bmatrix}$$

Observe that we do not have at the same time brackets and cobrackets, due to the relation

$$[v_1,\ldots,v_n]\,[\phi_1,\ldots,\phi_n] = \det \begin{bmatrix} \phi_1(v_1) & \cdots & \phi_1(v_n) \\ \vdots & & \vdots \\ \phi_n(v_1) & \cdots & \phi_n(v_n) \end{bmatrix}$$

We can give a matrix-interpretation of these basic invariants. Let us consider the case of a bracket of vectors (the case of cobrackets is similar)

$$[v_{i_1},\ldots,v_{i_n}]$$

If all the indices $\{i_1,\ldots,i_n\}$ are original vector-indices (and so do not come from the matrix-terms) we save this term and go to the next factor. Otherwise, if, say, i_1 is one of the matrix indices, $A_{i_1} = \phi_{i_1} \otimes v_{i_1}$, then the covector ϕ_{i_1} must be paired up in a scalar product $\phi_{i_1}(v_{u_1})$ with a vector v_{u_1}. Again, two cases can occur. If u_1 is a vector index, we have that

$$\phi_{i_1}(v_{u_1})[v_{i_1},\ldots,v_{i_n}] = [A_{i_1}v_{u_1}, v_{i_2},\ldots,v_{i_n}] = [v'_{i_1}, v_{i_2},\ldots,v_{i_n}]$$

Otherwise, we can keep on matching the matrix indices and get an expression

$$\phi_{i_1}(v_{u_1})\,\phi_{u_1}(v_{u_2})\,\phi_{u_2}(v_{u_3}) \,\cdots$$

until we finally hit again a vector index, say, u_l, but then we have the expression

$$\phi_{i_1}(v_{u_1})\,\phi_{u_1}(v_{z_1}) \,\cdots\, \phi_{u_{l-1}}(v_{u_l})\,[v_{i_1},\ldots,v_{i_n}] = [Mv_{u_l}, v_{i_2},\ldots,v_{i_n}]$$

where $M = A_{i_1} A_{u_1} \cdots A_{u_{l-1}}$. One repeats the same argument for all vectors in the brackets. As for the remaining scalar product terms, we have a similar procedure of matching up the matrix indices and one verifies that in doing so one obtains factors of the type

$$\phi(Mv) \quad \text{and} \quad tr(M)$$

where M is a monomial in the matrices. As we mentioned, the case of covector-brackets is similar except that in matching the matrix indices with a covector ϕ, one obtains a monomial in the transposed matrices.

Having found these interpretations of the basic SL_n-invariant linear terms, we can proceed by polarization and restitution processes to prove the following.

THEOREM 7.2

The SL_n-invariants of $W = \mathbf{rep}_\alpha Q'$ where Q' is the quiver

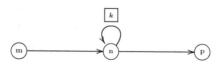

are generated by the following four types of functions, where we write a typical element in W as

$$\underbrace{(A_1, \ldots, A_k}_{k}, \underbrace{v_1, \ldots, v_m}_{m}, \underbrace{\phi_1, \ldots, \phi_p)}_{p}$$

with the A_i the matrices corresponding to the loops, the v_j making up the rows of the $n \times m$ matrix and the ϕ_j the columns of the $p \times n$ matrix.

- *$tr(M)$ where M is a monomial in the matrices A_i*

- *scalar products $\phi_j(Mv_i)$ where M is a monomial in the matrices A_i*

- *brackets $[M_1 v_{i_1}, M_2 v_{i_2}, \ldots, M_n v_{i_n}]$ where the M_j are monomials in the matrices A_i*

- *cobrackets $[M_1 \phi_{i_1}^\tau, \ldots, M_n \phi_{i_n}^\tau]$ where the M_j are monomials in the matrices A_i*

Returning to the special case under consideration, that is, of SL_m-invariants on $\otimes_B \mathbb{C}^m \otimes \otimes_C \mathbb{C}^{*m}$, it follows from this that the linear SL_m-invariants are determined by the following three sets

- *traces*, that is, for each pair (b, c) we have $\mathbb{C}^m \otimes \mathbb{C}^{*m} = M_m(\mathbb{C}) \xrightarrow{Tr} \mathbb{C}$

- *brackets*, that is, for each m-tuple $(b_1, \ldots, b_m)$ we have an invariant $\otimes_{b_j} \mathbb{C}^m \longrightarrow \mathbb{C}$ defined by

$$v_{b_1} \otimes \ldots \otimes v_{b_m} \mapsto det \begin{bmatrix} v_{b_1} & \cdots & v_{b_m} \end{bmatrix}$$

- *cobrackets*, that is, for each m-tuple $(c_1, \ldots, c_m)$ we have an invariant $\otimes_{c_i} \mathbb{C}^{*m} \longrightarrow \mathbb{C}$ defined by

$$\phi_{c_1} \otimes \ldots \otimes \phi_{c_m} \mapsto det \begin{bmatrix} \phi_{c_1} \\ \vdots \\ \phi_{c_m} \end{bmatrix}$$

Multilinear SL_m-invariants of $\otimes_B \mathbb{C}^m \otimes \otimes_C \mathbb{C}^{*m}$ are then spanned by invariants constructed from the following data. Take three disjoint index-sets I, J and K and consider surjective maps

$$\begin{cases} B \xrightarrow{\mu} I \sqcup K \\ C \xrightarrow{\nu} J \sqcup K \end{cases}$$

subject to the following conditions

$$\begin{cases} \# \mu^{-1}(k) = 1 = \# \nu^{-1}(k) & \text{for all } k \in K. \\ \# \mu^{-1}(i) = m = \# \nu^{-1}(j) & \text{for all } i \in I \text{ and } j \in J \end{cases}$$

To this data $\gamma = (\mu, \nu, I, J, K)$ we can associate a multilinear SL_m-invariant $f_\gamma(\otimes_B v_b \otimes \otimes_C \phi_c)$ defined by

$$
\prod_{k \in K} \phi_{\nu^{-1}(k)}(v_{\mu^{-1}(k)}) \prod_{i \in I} \det \begin{bmatrix} v_{b_1} & \cdots & v_{b_m} \end{bmatrix} \prod_{j \in J} \det \begin{bmatrix} \phi_{c_1} \\ \vdots \\ \phi_{c_m} \end{bmatrix}
$$

where $\mu^{-1}(i) = \{b_1, \ldots, b_m\}$ and $\nu^{-1}(j) = \{c_1, \ldots, c_m\}$. Observe that f_γ is determined only up to a sign by the data γ.

But then, we also have a spanning set for the multilinear $SL(\alpha)$-invariants on

$$
\mathbf{rep}_\alpha Q = \bigotimes_{\textcircled{v}} [\bigotimes_{\textcircled{v}\!-\!\textcircled{v}} \mathbb{C}^{a_v} \otimes \bigotimes_{\textcircled{v}\!\leftarrow\!\textcircled{v}} \mathbb{C}^{*a_v}]
$$

determined by quintuples $\Gamma = (\mu, \nu, I, J, K)$ where we have disjoint index-sets partitioned over the vertices $v \in \{v_1, \ldots, v_k\}$ of Q

$$
\begin{cases} I & = \bigsqcup_v I_v \\ J & = \bigsqcup_v J_v \\ K & = \bigsqcup_v K_v \end{cases}
$$

together with surjective maps from the set of all arrows A of Q

$$
\begin{cases} A & \xrightarrow{\;\mu\;} I \sqcup K \\ A & \xrightarrow{\;\nu\;} J \sqcup K \end{cases}
$$

where we have for every arrow $\textcircled{w}\xleftarrow{\;\;a\;\;}\textcircled{v}$ that

$$
\begin{cases} \mu(a) & \in I_v \sqcup K_v \\ \nu(a) & \in J_w \sqcup K_w \end{cases}
$$

and these maps μ and ν are subject to the numerical restrictions

$$
\begin{cases} \#\,\mu^{-1}(k) = 1 = \#\,\nu^{-1}(k) & \text{for all } k \in K. \\ \#\,\mu^{-1}(i) = a_v = \#\,\nu^{-1}(j) & \text{for all } i \in I_v \text{ and all } j \in J_v \end{cases}
$$

Such a quintuple $\Gamma = (\mu, \nu, I, J, K)$ determines for every vertex v a quintuple

$$
\gamma_v = (\mu_v = \mu \mid \{\textcircled{v}\xrightarrow{\;a\;}\textcircled{v}\}, \; \nu_v = \nu \mid \{\textcircled{v}\xleftarrow{\;a\;}\textcircled{v}\}, I_v, J_v, K_v)
$$

satisfying the necessary numerical restrictions to define the SL_{a_v}-invariant f_{γ_v} described before. Then, the multilinear $SL(\alpha)$-invariant on $\mathbf{rep}_\alpha Q$ determined by Γ is defined to be

$$
f_\gamma = \prod_v f_{\gamma_v}
$$

and we have to show that these semi-invariants lie in the linear span of the determinantal semi-invariants.

First, consider the case where the index set K is empty. If we denote the total number of arrows in Q by n, then the numerical restrictions imposed give us two expressions for n

$$\sum_v a_v . \# I_v = n = \sum_v a_v . \# J_v$$

Every arrow $\underset{w}{\bigcirc}\xleftarrow{\quad a \quad}\underset{v}{\bigcirc}$ determines a pair of indices $\mu(a) \in I_v$ and $\nu(a) \in J_w$. To the quintuple Γ we assign a map Φ_Γ in add Q

$$\underset{\textstyle 1}{\bigcirc}^{\oplus I_1} \oplus \ldots \oplus \underset{\textstyle k}{\bigcirc}^{\oplus I_k} \xrightarrow{\Phi_\Gamma} \underset{\textstyle 1}{\bigcirc}^{\oplus J_1} \oplus \ldots \oplus \underset{\textstyle k}{\bigcirc}^{\oplus J_k}$$

which decomposes as a block-matrix in blocks $M_{v,w} \in Hom(\underset{v}{\bigcirc}^{\oplus I_v}, \underset{w}{\bigcirc}^{\oplus J_w})$ of which the (i,j) entry is given by the sum of arrows

$$\sum_{\substack{\mu(a)=i \\ \nu(a)=j}} \underset{w}{\bigcirc}\xleftarrow{\quad a \quad}\underset{v}{\bigcirc}$$

For a representation $V \in \mathbf{rep}_\alpha Q$, $V(\Phi_\Gamma)$ is an $n \times n$ matrix and the determinant defines the determinantal semi-invariant $P_{\Phi_{\alpha,\Gamma}}$, which we claim to be equal to the basic invariant f_Γ possibly up to a sign.

We introduce a new quiver situation. Let Q' be the quiver with vertices the elements of $I \sqcup J$ and with arrows the set A of arrows of Q, but this time we take the starting point of the arrow $\bigcirc\xrightarrow{\quad a \quad}\bigcirc$ in Q to be $\mu(a) \in I$ and the terminating vertex to be $\nu(a) \in J$. That is, Q' is a bipartite quiver

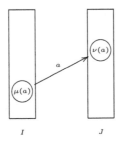

$$I \qquad\qquad J$$

On Q' we have the quintuple $\Gamma' = (\mu', \nu', I', J', K')$ where $K' = \emptyset$

$$I' = \bigsqcup_{i \in I} I'_i = \bigsqcup_{i \in I} \{i\} \qquad J' = \bigsqcup_{j \in J} J'_j = \bigsqcup_{j \in J} \{j\}$$

and $\mu' = \mu$, $\nu' = \nu$. We define an additive functor $\mathbf{add}\ Q' \xrightarrow{\ s\ } \mathbf{add}\ Q$ by

$$\underset{i}{\bigcirc} \xrightarrow{\ s\ } \underset{v}{\bigcirc} \qquad \underset{j}{\bigcirc} \xrightarrow{\ s\ } \underset{w}{\bigcirc} \qquad \bigcirc\xleftarrow{\quad a \quad}\bigcirc \xrightarrow{\ s\ } \bigcirc\xleftarrow{\quad a \quad}\bigcirc$$

for all $i \in I_v$ and all $j \in J_w$. The functor s induces a functor $rep\, Q \xrightarrow{s} rep\, Q'$ defined by $V \xrightarrow{s} V \circ s$. If $V \in \mathbf{rep}_\alpha Q$ then $s(V) \in \mathbf{rep}_{\alpha'} Q'$ where

$$\alpha' = (\underbrace{c_1,\ldots,c_p}_{\#\ I}, \underbrace{d_1,\ldots,d_q}_{\#\ J}) \quad \text{with} \quad \begin{cases} c_i = a_v & \text{if } i \in I_v \\ d_j = a_w & \text{if } j \in J_w \end{cases}$$

That is, the characteristic feature of Q' is that every vertex $i \in I$ is the source of exactly c_i arrows (follows from the numerical condition on μ) and that every vertex $j \in J$ is the sink of exactly d_j arrows in Q'. That is, locally Q' has the following form

There are induced maps

$$\mathbf{rep}_\alpha Q \xrightarrow{s} \mathbf{rep}_{\alpha'} Q' \qquad GL(\alpha) \xrightarrow{s} GL(\alpha')$$

where the latter follows from functoriality by considering $GL(\alpha)$ as the automorphism group of the trivial representation in $\mathbf{rep}_\alpha Q$. These maps are compatible with the actions as one checks that $s(g.V) = s(g).s(V)$. Also s induces a map on the coordinate rings $\mathbb{C}[\mathbf{rep}_\alpha Q] \xrightarrow{s} \mathbb{C}[\mathbf{rep}_{\alpha'} Q']$ by $s(f) = f \circ s$. In particular, for the determinantal semi-invariants we have

$$s(P_{\alpha',\phi'}) = P_{\alpha,s(\phi')}$$

and from the compatibility of the action it follows that when f is a semi-invariant the $GL(\alpha')$ action on $\mathbf{rep}_{\alpha'} Q'$ with character χ', then $s(f)$ is a semi-invariant for the $GL(\alpha)$-action on $\mathbf{rep}_\alpha Q$ with character $s(\chi) = \chi' \circ s$. In particular we have that

$$s(P_{\alpha',\Phi_{\Gamma'}}) = P_{\alpha,s(\Phi_{\Gamma'})} = P_{\alpha,\Phi_\Gamma} \quad \text{and} \quad s(f_{\Gamma'}) = f_\Gamma$$

Hence in order to prove our claim, we may replace the triple (Q,α,Γ) by the triple (Q',α',Γ'). We will do this and forget the dashes from here on.

In order to verify that $f_\Gamma = \pm P_{\alpha,\Phi_\Gamma}$ it suffices to check this equality on the image of

$$W = \bigoplus_{\underset{i \xrightarrow{a} j}{}} \mathbb{C}^{c_i} \oplus \mathbb{C}^{*d_j} \quad \text{in} \quad \bigotimes_{\underset{i \xrightarrow{a} j}{}} \mathbb{C}^{c_i} \otimes \mathbb{C}^{*d_j}$$

One verifies that both f_Γ and P_{α,Φ_Γ} are $GL(\alpha)$-semi-invariants on W of weight χ_θ where

$$\theta = (\underbrace{1,\ldots,1}_{\#\ I}, \underbrace{-1,\ldots,-1}_{\#\ J})$$

Using the characteristic local form of $Q = Q'$, we see that W is isomorphic to the $GL(\alpha)$- module

$$\bigoplus_{i \in I} \underbrace{(\mathbb{C}^{c_i} \oplus \ldots \oplus \mathbb{C}^{c_i})}_{c_i} \oplus \bigoplus_{j \in J} \underbrace{(\mathbb{C}^{*d_j} \oplus \ldots \oplus \mathbb{C}^{*d_j})}_{d_j} \simeq \bigoplus_{i \in I} M_{c_i}(\mathbb{C}) \oplus \bigoplus_{j \in J} M_{d_j}(\mathbb{C})$$

and the i factors of $GL(\alpha)$ act by inverse right-multiplication on the component M_{c_i} (and trivially on all others) and the j factors act by left-multiplication on the component M_{d_j} (and trivially on the others). That is, $GL(\alpha)$ acts on W with an open orbit, say, that of the element

$$w = (\mathbb{1}_{c_1}, \ldots, \mathbb{1}_{c_p}, \mathbb{1}_{d_1}, \ldots, \mathbb{1}_{d_q}) \in W$$

One verifies immediately from the definitions that that both f_Γ and P_{α, Φ_Γ} evaluate to ± 1 in w. Hence, indeed, f_Γ can be expressed as a determinantal semi-invariant.

Remains to consider the case when K is nonempty. For $k \in K$ two situations can occur

- $\mu^{-1}(k) = a$ and $\nu^{-1}(k) = b$ are distinct, then k corresponds to replacing the arrows a and b by their concatenation

$$\xrightarrow{\quad} \underset{a}{\bullet} \xrightarrow{\,(k)\,} \underset{b}{\bullet} \xleftarrow{\quad}$$

- $\mu^{-1}(k) = a = \nu^{-1}(k)$ then a is a loop in Q and k corresponds

to taking the trace of a.

This time we construct a new quiver $\tilde{Q}$ with vertices $\{w_1, \ldots, w_n\}$ corresponding to the set A of arrows in Q. The arrows in $\tilde{Q}$ will correspond to elements of K, that is, if $k \in K$ we have the arrow (or loop) in $\tilde{Q}$ with notations as before

$$\underset{b}{\bigcirc} \xleftarrow{\quad k \quad} \underset{a}{\bigcirc} \quad \text{or} \quad \underset{a}{\bigcirc}$$

We consider the connected components of $\tilde{Q}$. They are of the following three types

- (oriented cycle): To an oriented cycle C in $\tilde{Q}$ corresponds an oriented cycle C'_C in the original quiver Q. We associate to it the trace $tr(C'_C)$ of this cycle.

- (open paths): An open path P in $\tilde{Q}$ corresponds to an oriented path P_P' in Q which may be a cycle. To P we associate the corresponding path P_P' in Q.

- (isolated points): They correspond to arrows in Q.

We will now construct a new quiver Q' having the same vertex set $\{v_1, \ldots, v_k\}$ as Q but with arrows corresponding to the set of paths P_P' described above. The starting and ending vertex of the arrow corresponding to P_P' are of course the starting and ending vertex of the path P_P in Q. Again, we define an additive functor $\mathbf{add}\, Q' \xrightarrow{\;s\;} \mathbf{add}\, Q$ by the rules

$$
\underset{v}{\textcircled{}} \xrightarrow{\;s\;} \underset{v}{\textcircled{}} \quad \text{and} \quad \underset{j}{\textcircled{}} \xleftarrow{\;P_P'\;} \underset{i}{\textcircled{}} \xrightarrow{\;s\;} \underset{j}{\textcircled{}} \overset{P_P'}{\cdots\cdots} \underset{i}{\textcircled{}}
$$

If the path P_P' is the concatenation of the arrows $a_d \circ \ldots \circ a_1$ in Q, we define the maps

$$
\begin{cases} \mu'(P_P') &= \mu(a_1) \\ \nu'(P_P') &= \nu(a_d) \end{cases} \quad \text{whence} \quad \begin{cases} \{P_P'\} &\xrightarrow{\;\mu\;} I' \\ \{P_P'\} &\xrightarrow{\;\nu\;} J' \end{cases}
$$

that is, a quintuple $\Gamma' = (\mu', \nu', I', J', K' = \emptyset)$ for the quiver Q'. One then verifies that

$$
f_\Gamma = s(f_{\Gamma'}) \prod_C tr(C_C') = s(P_{\alpha, \Phi_{\Gamma'}}) \prod_C tr(C_C')
$$

$$
= P_{\alpha, s(\Phi_{\Gamma'})} \prod_C tr(C_C')
$$

finishing the proof of the fact that multilinear semi-invariants lie in the linear span of determinantal semi-invariants (and traces of oriented cycles).

   The arguments above can be reformulated in a more combinatorial form that is often useful in constructing semi-invariants of a specific weight, as is necessary in the study of the moduli spaces $M_\alpha^{ss}(Q, \theta)$. Let Q be a quiver on the vertices $\{v_1, \ldots, v_k\}$, fix a dimension vector $\alpha = (a_1, \ldots, a_k)$ and a character χ_θ where $\theta = (t_1, \ldots, t_k)$ such that $\theta(\alpha) = 0$. We will call a bipartite quiver Q' as in figure 7.1 on left vertex-set $L = \{l_1, \ldots, l_p\}$ and right vertex-set $R = \{r_1, \ldots, r_q\}$ and a dimension vector $\beta = (c_1, \ldots, c_p; d_1, \ldots, d_q)$ to be of type (Q, α, θ) if the following conditions are met

- All left and right vertices correspond to vertices of Q, that is, there are maps

$$
\begin{cases} L &\xrightarrow{\;l\;} \{v_1, \ldots, v_k\} \\ R &\xrightarrow{\;r\;} \{v_1, \ldots, v_k\} \end{cases}
$$

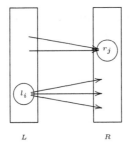

L R

FIGURE 7.1:   Left-right bipartite quiver.

possibly occurring with multiplicities, that is there is a map

$$L \cup R \xrightarrow{\ m\ } \mathbb{N}_+$$

such that $c_i = m(l_i)a_z$ if $l(l_i) = v_z$ and $d_j = m(r_j)a_z$ if $r(r_j) = v_z$.

- There can only be an arrow $\textcircled{\scriptsize l_i} \longrightarrow \textcircled{\scriptsize r_j}$ if for $v_k = l(l_i)$ and $v_l = r(r_i)$ there is an oriented path

in Q allowing the trivial path and loops if $v_k = v_l$.

- Every left vertex l_i is the source of exactly c_i arrows in Q' and every right-vertex r_j is the sink of precisely d_j arrows in Q'.

- Consider the $u \times u$ matrix where $u = \sum_i c_i = \sum_j d_j$ (both numbers are equal to the total number of arrows in Q') where the i-th row contains the entries of the i-th arrow in Q' with respect to the obvious left and right bases. Observe that this is a $GL(\beta)$ semi-invariant on $\mathbf{rep}_\beta Q'$ with weight determined by the integral $k + l$-tuple $(-1, \ldots, -1; 1, \ldots, 1)$. If we fix for every arrow a from l_i to r_j in Q' an $m(r_j) \times m(l_i)$ matrix p_a of linear combinations of paths in Q from $l(l_i)$ to $r(r_j)$, we obtain a morphism

$$\mathbf{rep}_\alpha Q \longrightarrow \mathbf{rep}_\beta Q'$$

sending a representation $V \in \mathbf{rep}_\alpha Q$ to the representation W of Q' defined by $W_a = p_a(V)$. Composing this map with the above semi-invariant we obtain a $GL(\alpha)$ semi-invariant of $\mathbf{rep}_\alpha Q$ with weight determined by the k-tuple $\theta = (t_1, \ldots, t_k)$ where

$$t_i = \sum_{j \in r^{-1}(v_i)} m(r_j) - \sum_{j \in l^{-1}(v_i)} m(l_j)$$

.

We call such semi-invariants *standard determinantal*. Summarizing the arguments of this section we have proved after applying polarization and restitution processes

THEOREM 7.3
The semi-invariants of the $GL(\alpha)$-action on $\mathrm{rep}_\alpha\, Q$ are generated by traces of oriented cycles and by standard determinantal semi-invariants.

7.3   Universal localization

In order to prove that the moduli spaces $M_\alpha^{ss}(Q, \theta)$ are locally controlled by Quillen-smooth algebras, we need to recall the notion of *universal localization* and refer to the monograph by A. Schofield [92] for full details.

Let A be a $\mathbb{C}$-algebra and $\mathrm{projmod}\, A$ the category of finitely generated projective left A-modules. Let Σ be some class of maps in this category (that is, some left A-module morphisms between certain projective modules). Then, there exists an algebra map $A \xrightarrow{\ j_\Sigma\ } A_\Sigma$ with *the universal property* that the maps $A_\Sigma \otimes_A \sigma$ have an inverse for all $\sigma \in \Sigma$. A_Σ is called the *universal localization* of A with respect to the set of maps Σ.

PROPOSITION 7.1
When A is Quillen-smooth, then so is A_Σ.

PROOF   Consider a test-object (T, I) in alg, then we have the following diagram

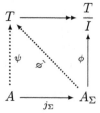

where ψ exists by Quillen-smoothness of A. By *Nakayama's lemma* all maps $\sigma \in \Sigma$ become isomorphisms under tensoring with ψ. Then, $\tilde{\phi}$ exists by the universal property of A_Σ.   □

Consider the special case when A is the path algebra $\mathbb{C}Q$ of a quiver on k vertices. Then, we can identify the isomorphism classes in $\mathrm{projmod}\, \mathbb{C}Q$ with the opposite category of $\mathbf{add}\, Q$ introduced in the foregoing section. To each vertex v_i corresponds an *indecomposable projective* left $\mathbb{C}Q$-ideal $P_i =$

$\mathbb{C}Qe_i$ having as $\mathbb{C}$-vector space basis all paths in Q *starting at* v_i. For the homomorphisms we have

$$Hom_{\mathbb{C}Q}(P_i, P_j) = \bigoplus_{\substack{\text{(i)}\xleftarrow{\;p\;}\text{(j)}}} \mathbb{C}p = Hom_{\text{add }Q}(\,\text{(j)}\,,\,\text{(i)}\,)$$

where p is an oriented path in Q starting at v_j and ending at v_i. Therefore, any A-module morphism σ between two projective left modules

$$P_{i_1} \oplus \ldots \oplus P_{i_u} \xrightarrow{\;\sigma\;} P_{j_1} \oplus \ldots \oplus P_{j_v}$$

can be represented by an $u \times v$ matrix M_σ whose (p,q)-entry m_{pq} is a linear combination of oriented paths in Q starting at v_{j_q} and ending at v_{i_p}.

Now, form an $v \times u$ matrix N_σ of free variables y_{pq} and consider the algebra $\mathbb{C}Q_\sigma$, which is the quotient of the free product $\mathbb{C}Q * \mathbb{C}\langle y_{11}, \ldots, y_{uv}\rangle$ modulo the ideal of relations determined by the matrix equations

$$M_\sigma.N_\sigma = \begin{bmatrix} v_{i_1} & & 0 \\ & \ddots & \\ 0 & & v_{i_u} \end{bmatrix} \qquad N_\sigma.M_\sigma = \begin{bmatrix} v_{j_1} & & 0 \\ & \ddots & \\ 0 & & v_{j_v} \end{bmatrix}$$

Equivalently, $\mathbb{C}Q_\sigma$ is the path algebra of a quiver *with relations* where the quiver is Q extended with arrows y_{pq} from v_{i_p} to v_{j_q} for all $1 \le p \le u$ and $1 \le q \le v$ and the relations are the above matrix entry relations.

Repeating this procedure for every $\sigma \in \Sigma$ we obtain the universal localization $\mathbb{C}Q_\Sigma$. This proves the following.

PROPOSITION 7.2
If Σ is a finite set of maps, then the universal localization $\mathbb{C}Q_\Sigma$ is an affine $\mathbb{C}$-algebra.

It is easy to verify that the representation space $\mathbf{rep}_n \, \mathbb{C}Q_\sigma$ is an affine Zariski open subscheme (but possibly empty) of $\mathbf{rep}_n \, \mathbb{C}Q$. Indeed, if $V = (V_a)_a \in \mathbf{rep}_\alpha \, Q$, then V determines a point in $\mathbf{rep}_n \, \mathbb{C}Q_\Sigma$ if and only if the matrices $M_\sigma(V)$ in which the arrows are all replaced by the matrices V_a are invertible for all $\sigma \in \Sigma$.

In particular, this induces numerical conditions on the dimension vectors α such that $\mathbf{rep}_\alpha \, Q_\Sigma \ne \emptyset$. Let $\alpha = (a_1, \ldots, a_k)$ be a dimension vector such that $\sum a_i = n$ then every $\sigma \in \Sigma$, say, with

$$P_1^{\oplus e_1} \oplus \ldots \oplus P_k^{\oplus e_k} \xrightarrow{\;\sigma\;} P_1^{\oplus f_1} \oplus \ldots \oplus P_k^{\oplus f_k}$$

gives the numerical condition

$$e_1 a_1 + \ldots + e_k a_k = f_1 a_1 + \ldots + f_k a_k$$

These numerical restrictions will be used to relate θ-stable representations of Q to simple representations of universal localizations of $\mathbb{C}Q$.

Fix a character $\theta = (t_1, \ldots, t_k) \in \mathbb{Z}^k$ and divide the set of indices $1 \leq i \leq k$ into the *left set* $L = \{i_1, \ldots, i_u\}$ consisting of those i such that $t_i \leq 0$ and the *right set* $R = \{j_1, \ldots, j_v\}$ consisting of those j such that $t_j \geq 0$. Consider a dimension vector α such that $\theta.\alpha = 0$, then θ determines the *character*

$$GL(\alpha) \xrightarrow{\chi_\theta} \mathbb{C}^* \qquad (g_1, \ldots, g_k) \mapsto \prod_i det(g_i)^{t_i}$$

Next, consider the sets of morphisms

$$\Sigma_\theta = \bigcap_{z \in \mathbb{N}_+} \Sigma_\theta(z)$$

where $\Sigma_\theta(z)$ is the set of all morphisms

$$P_{i_1}^{\oplus -zt_{i_1}} \oplus \ldots \oplus P_{i_u}^{\oplus -zt_{i_u}} \xrightarrow{\sigma} P_{j_1}^{\oplus zt_{j_1}} \oplus \ldots \oplus P_{j_v}^{\oplus zt_{j_v}}$$

With notation as before, it follows that

$$d_\sigma(V) = det\, M_\sigma(V) \qquad V \in \mathbf{rep}_\alpha\, Q$$

is a semi-invariant on $\mathbf{rep}_\alpha\, Q$ of weight $z\chi_\theta$. This semi-invariant determines the Zariski open subset of $\mathbf{rep}_\alpha\, Q$

$$X_\sigma(\alpha) = \{V \in \mathbf{rep}_\alpha\, Q \mid d_\sigma(V) \neq 0\}$$

It is clear from the results of section 4.8 that $X_\sigma(\alpha)$ consists of θ-semistable representations. We can characterize the θ-stable representations in this open set.

LEMMA 7.1
For $V \in X_\sigma(\alpha)$ the following are equivalent

1. *V is a θ-stable representation*

2. *V is a simple α-dimensional representation of the universal localization $\mathbb{C}Q_\sigma$*

PROOF   Let W be a β-dimensional subrepresentation of V with $\beta = (b_1, \ldots, b_k)$, then for W to be a β-dimensional representation of the universal localization $\mathbb{C}Q_\sigma$ it must satisfy the numerical restriction

$$-t_{i_1} b_{i_1} - \ldots - t_{i_u} b_{i_u} = t_{j_1} b_{j_1} + \ldots + t_{j_v} b_{j_v} \quad \text{that is,} \quad \theta.\beta = 0$$

Hence, if V is θ-stable, there are no proper subrepresentations of V as a $\mathbb{C}Q_\sigma$-representation. Conversely, if V is an α-dimensional subrepresentation of $\mathbb{C}Q_\sigma$ we must have that $d_\sigma(V) \neq 0$. But then, if W is a β-dimensional Q-subrepresentation of V we must have that $\sum_a -t_{i_a} b_{i_a} \leq \sum_b t_{j_b} b_{i_b}$ (if not, $\sigma(V)$ would have a kernel) whence $\theta.\beta \geq 0$. If W is a subrepresentation such that $\theta.\beta = 0$, then W would be a proper $\mathbb{C}Q_\sigma$ subrepresentation of V, a contradiction. Therefore, V is θ-stable. ▯

THEOREM 7.4
The moduli space of θ-semistable representations of the quiver Q

$$M_\alpha^{ss}(Q, \theta)$$

is locally controlled by the set of Quillen-smooth algebras $\{\mathbb{C}Q_\sigma \mid \sigma \in \Sigma_\theta\}$.

PROOF By the results of the foregoing section we know that the quotient varieties of the Zariski open affine subsets $X_\sigma(\alpha)$ cover the moduli space $M_\alpha^{ss}(Q, \theta)$. Further, by lemma 7.1 we have a canonical isomorphism

$$X_\sigma(\alpha)/GL(\alpha) \simeq \mathrm{iss}_\alpha \mathbb{C}Q_\sigma$$

Finally, because

$$\mathrm{rep}_n \mathbb{C}Q_\sigma = \sqcup_\alpha GL_n \times^{GL(\alpha)} \mathrm{rep}_\alpha \mathbb{C}Q_\sigma$$

where the disjoint union is taken over all $\alpha = (a_1, \ldots, a_k)$ such that $\sum_i a_i = n$, we have that $\mathrm{iss}_\alpha \mathbb{C}Q_\sigma$ is an irreducible component of $\mathrm{iss}_n \mathbb{C}Q_\sigma$ finishing the proof. ▯

Lemma 7.1 also allows us to study the moduli spaces $M_\alpha^{ss}(Q, \theta)$ locally by the local quiver settings associated to semisimple representations. That is, let $\xi \in M_\alpha^{ss}(Q, \theta)$ be the point corresponding to

$$M_\xi = S_1^{\oplus e_1} \oplus \ldots \oplus S_z^{\oplus e_z}$$

where S_i is a θ-stable representation of dimension vector β_i occurring in M_ξ with multiplicity e_i.

THEOREM 7.5
With notations as above, the étale local structure of the moduli space $M_\alpha^{ss}(Q, \theta)$ near ξ is that of the quotient variety $\mathrm{iss}_\beta Q_\xi$ where $\beta = (e_1, \ldots, e_z)$ and Q_ξ is the quiver on z vertices such that

$$\begin{cases} \# \; \text{(j)} \xleftarrow{\;\;a\;\;} \text{(i)} & = \; -\chi_Q(\beta_i, \beta_j) \\[2em] \# \; \text{(i)} \circlearrowleft & = 1 - \chi_Q(\beta_i, \beta_i) \end{cases}$$

near the trivial representation.

PROOF   In view of the above results and the slice theorems, we only have to compute the ext-spaces $Ext^1_{\mathbb{C}Q_\sigma}(S_i, S_j)$. From [92, Thm. 4.7] we recall that the category of $\mathbb{C}Q_\sigma$ representations is closed under extensions in the category of representations of Q. Therefore, we have for all $\mathbb{C}Q_\sigma$-representations V and W that

$$Ext^1_{\mathbb{C}Q}(V, W) \simeq Ext^1_{\mathbb{C}Q_\sigma}(V, W)$$

from which the result follows using theorem 4.5. $\square$

In the following section we will give some applications of this result. Universal localizations can also be used to determine the *formal structure* on representation spaces of quivers.

Let Q be a quiver on k vertices and consider the extended quiver $Q^{(n)}$

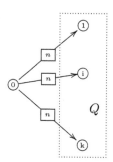

That is, we add to the vertices and arrows of Q one extra vertex v_0 and for every vertex v_i in Q we add n directed arrows from v_0 to v_i. We will denote the j-th arrow $1 \leq j \leq n$ from v_0 to v_i by x_{ij}.

Consider the morphism between projective left $\mathbb{C}Q^{(n)}$-modules

$$P_1 \oplus P_2 \oplus \ldots \oplus P_k \xrightarrow{\ \sigma\ } \underbrace{P_0 \oplus \ldots \oplus P_0}_{n}$$

determined by the matrix

$$M_\sigma = \begin{bmatrix} x_{11} \cdots \cdots x_{1n} \\ \vdots \qquad\quad \vdots \\ x_{k1} \cdots \cdots x_{kn} \end{bmatrix}$$

We consider the universal localization $\mathbb{C}Q_\sigma^{(n)}$, that is, we add for each vertex v_i in Q another n arrows y_{ij} with $1 \leq j \leq n$ from v_i to v_0.

With these arrows y_{ij} one forms the $n \times k$ matrix

$$N_\sigma = \begin{bmatrix} y_{11} & \cdots & y_{k1} \\ \vdots & & \vdots \\ \vdots & & \vdots \\ y_{1n} & \cdots & y_{kn} \end{bmatrix}$$

and the universal localization $\mathbb{C}Q_\sigma^{(n)}$ is described by the relations

$$M_\sigma . N_\sigma = \begin{bmatrix} v_1 & & 0 \\ & \ddots & \\ 0 & & v_k \end{bmatrix} \quad \text{and} \quad N_\sigma . M_\sigma = \begin{bmatrix} v_0 & & & 0 \\ & \ddots & & \\ & & \ddots & \\ 0 & & & v_1 \end{bmatrix}$$

We will depict this quiver with relations by the picture $Q_\sigma^{(n)}$

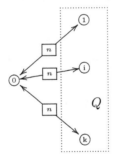

From the discussion above it follows that there is a canonical isomorphism

$$\mathbf{rep}_m \sqrt[n]{\mathbb{C}Q} \simeq \mathbf{rep}_m \mathbb{C}Q_\sigma^{(n)}$$

In fact we can even identify

$$\sqrt[n]{\mathbb{C}Q} = v_0 \, \mathbb{C}Q_\sigma^{(n)} \, v_0$$

Indeed, the right-hand side is generated by all the oriented cycles in $Q_\sigma^{(n)}$ starting and ending at v_0 and is therefore generated by the $y_{ip}x_{iq}$ and the $y_{ip}ax_{jq}$ where a is an arrow in Q starting in v_j and ending in v_i. If we have an algebra morphism

$$\mathbb{C}Q \xrightarrow{\phi} M_n(B)$$

then we have an associated algebra morphism

$$v_0 \, \mathbb{C}Q_\sigma^{(n)} \, v_0 \xrightarrow{\psi} B$$

defined by sending $y_{ip}ax_{jq}$ to the (p,q)-entry of the $n \times n$ matrix $\phi(a)$ and $y_{ip}x_{iq}$ to the (p,q)-entry of $\phi(v_i)$. The defining relations among the x_{ip} and y_{iq} introduced before imply that ψ is indeed an algebra morphism.

Example 7.3

Let $A = \mathbb{C}\langle a, b\rangle$, that is, A is the path algebra of the quiver

In order to describe $\sqrt[n]{A}$ we consider the quiver with relations

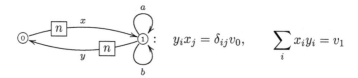

$$y_i x_j = \delta_{ij} v_0, \qquad \sum_i x_i y_i = v_1$$

We see that the algebra of oriented cycles in v_0 in this quiver with relations is isomorphic to the free algebra in $2n^2$ free variables

$$\mathbb{C}\langle y_1 a x_1, \ldots, y_n a x_n, y_1 b x_1, \ldots, y_n b x_n\rangle$$

which coincides with our knowledge of $\sqrt[n]{\mathbb{C}\langle a, b\rangle}$.   □

There is some elementary calculus among the n-th roots of algebras. For example, it follows from the universal property of $\sqrt[n]{A}$ that there is a natural morphism

$$\sqrt[k_1]{\sqrt[k_2]{\ldots \sqrt[k_z]{A}}} \longleftarrow \sqrt[k]{A}$$

where $k = \prod k_i$. When $A = \mathbb{C}Q$ we can represent this morphism graphically by the picture

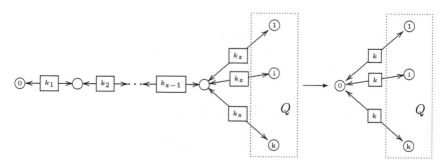

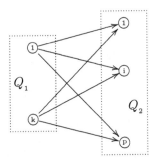

FIGURE 7.2: Free product of quivers.

where the map is given by composing paths from v_0 to v_i. Also observe that we used the isomorphisms in the rightmost part of the left quiver to remove additional arrows from the extra vertices to v_i at each stage.

Probably more important are the *connecting morphisms*

$$\sqrt[k_1]{A} * \sqrt[k_2]{A} * \ldots * \sqrt[k_z]{A} \xleftarrow{\;c_{(k_1,\ldots,k_z)}\;} \sqrt[k]{A}$$

with $k = \sum k_i$ obtained from the universal property of $\sqrt[n]{A}$ by composing algebra morphisms $A \xrightarrow{\phi_i} M_{k_i}(B)$ to an algebra morphism

$$A \xrightarrow{\begin{bmatrix} \phi_1 & & 0 \\ & \ddots & \\ 0 & & \phi_z \end{bmatrix}} M_k(B)$$

Observe that the ordering of the factors is important (but only up to isomorphism of the representations).

We need to have a quiver interpretation of the *free product* $\mathbb{C}Q_1 * \mathbb{C}Q_2$ of two path algebras (at least as far as finite dimensional representations are concerned). Let Q_1 be a quiver on k vertices $\{v_1, \ldots, v_k\}$ and Q_2 a quiver on p vertices $\{w_1, \ldots, w_p\}$ and consider the extended quiver $Q_1 * Q_2$ of figure 7.2. That is, we add one extra arrow from each vertex of Q_1 to each arrow of Q_2.

Let $\{P_1, \ldots, P_k\}$ be the projective left $\mathbb{C}Q_1 * Q_2$-modules corresponding to the vertices of Q_1 and $\{P'_1, \ldots, P'_p\}$ those corresponding to the vertices of Q_2 and consider the morphism

$$P'_1 \oplus \ldots \oplus P'_p \xrightarrow{\;\sigma\;} P_1 \oplus \ldots \oplus P_k$$

determined by the $p \times k$ matrix

$$M_\sigma = \begin{bmatrix} x_{11} & \ldots & x_{1k} \\ \vdots & & \vdots \\ x_{p1} & \ldots & x_{pk} \end{bmatrix}$$

where x_{ij} denotes the extra arrow from vertex v_j to vertex w_i.

Let $Q_1 * Q_{2_\sigma}$ denote the quiver with relations one obtains by inverting this map (as above). Then, it is fairly easy to see that

$$\mathtt{rep}_n \ \mathbb{C}Q_1 * Q_2 \simeq \mathtt{rep}_n \ Q_1 * Q_{2_\sigma}$$

where the right-hand side denote the subscheme of n-dimensional representations of the quiver Q_1 times the n-dimensional representations of Q_2 where the extra arrows determine an isomorphism of the representations.

Using this interpretation of the free product one can now give a graphical interpretation of the connecting morphisms in the case of the two-loop quiver (the general case is similar).

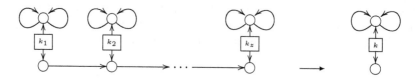

obtained by "grafting" the bottom tree. Observe that we used the isomorphisms given by the k_i bundles to eliminate adding extra arrows in the free products.

7.4   Compact manifolds

`noncommutative geometry@n` is the study of *families* of algebraic varieties (with specified connecting morphisms) that are controlled locally by a set of noncommutative algebras. If this set of algebras consists of Quillen-smooth algebras we say that the family of varieties is a *noncommutative manifold* . If all varieties in the family are in addition projective (possibly with singularities) we say that the family is a *compact noncommutative manifold.*

So far, we have not specified the properties of the connecting morphisms. In this section we present a first class of examples, the *sum families*. In the next chapter we will encounter another possibility coming from the theory of completely integrable dynamical systems.

DEFINITION   7.2   *A   sum   family   is   an   object $(X_n)_n$ in* `noncommutative geometry@n` *indexed   over   the   positive   integers   such that for each n there is a GL_n-variety Y_n and a quotient morphism*

$$Y_n \longrightarrow Y_n/GL_n \simeq X_n$$

and Y_n is locally of the form $\mathbf{rep}_n A$ *for an affine $\mathbb{C}$-algebra A belonging to a set $\mathcal{A}$ of algebras. Moreover, there are* equivariant connecting sum-maps

$$Y_m \times Y_n \xrightarrow{\ \oplus\ } Y_{m+n}$$

for all $m, n \in \mathbb{N}_+$ where equivariance means with respect to the group $GL_m \times GL_n$ embedded diagonally in GL_{m+n}. If the set $\mathcal{A}$ consists of Quillen-smooth algebras, we call $(X_n)_n$ a sum manifold .

THEOREM 7.6
For a quiver Q on k vertices and a fixed character $\theta \in \mathbb{Z}^k$, the family of varieties

$$\left(\ \bigsqcup_{\substack{\alpha=(a_1,\ldots,a_k) \\ \sum_i a_i = n}} M_\alpha^{ss}(Q,\theta)\ \right)_n$$

is a sum manifold in noncommutative geometry@n. *If Q has no oriented cycles, then this family is a compact sum manifold.*

PROOF   In view of theorem 7.5 we only need to construct equivariant-sum maps. They are induced from the direct sums of representations

$$\mathbf{rep}_\alpha Q \times \mathbf{rep}_\beta Q \xrightarrow{\ \oplus\ } \mathbf{rep}_{\alpha+\beta} Q \qquad (V,W) \mapsto V \oplus W$$

and the required properties are clearly satisfied.   ⬚

Example 7.4
Let $M_{\mathbb{P}^2}(n;0,n)$ be the moduli space of semistable vector bundles of rank n over the projective plane $\mathbb{P}^2$ with *Chern numbers* $c_1 = 0$ and $c_2 = n$. Using results of K. Hulek [46] one can identify this moduli space with

$$M_{\mathbb{P}^2}(n;0,n) \simeq M_{(n,n)}^{ss}(Q,\theta)$$

where Q and θ are the following quiver-setting

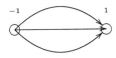

Therefore, the family of moduli spaces $(M_{\mathbb{P}^2}(n;0,n))_n$ is a compact noncommutative manifold.

Let C be a smooth projective curve of genus g and let $M_C(n,0)$ be the moduli space of semistable vector bundles of rank n and degree 0 over C. We expect that the family of moduli spaces $(M_C(n,0))_n$ is a compact noncommutative manifold.   ⬚

Next, we will investigate another class of examples: representations of the *torus knot groups* . Consider a solid cylinder C with m line segments on its curved face, equally spaced and parallel to the axis. If the ends of C are identified with a twist of $2\pi\frac{n}{m}$ where n is an integer relatively prime to m, we obtain a single curve $K_{m,n}$ on the surface of a solid torus T. If we assume that the torus T lies in $\mathbb{R}^3$ in the standard way, the curve $K_{m,n}$ is called the (m,n) *torus knot* .

Computing the *fundamental group* of the complement $\mathbb{R}^3 - K_{m,n}$ one obtains the (m,n)-*torus knot group*

$$\pi_1(\mathbb{R}^3 - K_{m,n}) = G_{m,n} \simeq \langle\, a,b \mid a^m = b^n \,\rangle$$

An important example is the *three-string braid group*.

Example 7.5

Consider Artin's braid group B_3 on three strings. B_3 has the presentation

$$B_3 \simeq \langle L, R \mid LR^{-1}L = R^{-1}LR^{-1}\rangle$$

where L and R are the fundamental 3-braids

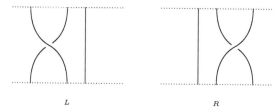

$$L \qquad\qquad\qquad\qquad R$$

If we let $S = LR^{-1}L$ and $T = R^{-1}L$, an algebraic manipulation shows that

$$B_3 = \langle S,T \mid T^3 = S^2\rangle$$

is an equivalent presentation for B_3. The center of B_3 is the infinite cyclic group generated by the braid

$$Z = S^2 = (LR^{-1}L)^2 = (R^{-1}L)^3 = T^3$$

It follows from the second presentation of B_3 that the quotient group modulo the center is isomorphic to

$$\frac{B_3}{\langle Z\rangle} \simeq \langle s,t \mid s^2 = 1 = t^3\rangle \simeq \mathbb{Z}_2 * \mathbb{Z}_3$$

the free product of the cyclic group of order 2 (with generator s) and the cyclic group of order 3 (with generator t). This group is isomorphic to the modular group $PSL_2(\mathbb{Z})$ via

$$\overline{L} \longrightarrow \begin{bmatrix} 1 & 1 \\ 0 & 1 \end{bmatrix} \quad \text{and} \quad \overline{R} \longrightarrow \begin{bmatrix} 1 & 0 \\ 1 & 1 \end{bmatrix}$$

It is well known that the modular group $PSL_2(\mathbb{Z})$ acts on the upper half-plane H^2 by left multiplication in the usual way, that is

$$\begin{bmatrix} a & b \\ c & d \end{bmatrix} : H^2 \longrightarrow H^2 \quad \text{given by} \quad z \longrightarrow \frac{az + b}{cz + d}$$

The fundamental domain $H^2/PSL_2(\mathbb{Z})$ for this action is the hyperbolic triangle bounded by the thick geodesics below, and the action defines a quilt-tiling on the hyperbolic plane, indexed by elements of $PSL_2(\mathbb{Z}) = \mathbb{Z}_2 * \mathbb{Z}_3$

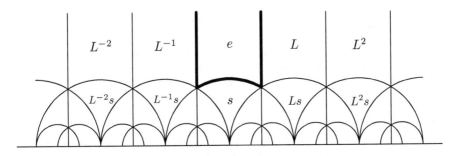

We want to study the irreducible representations of the torus knot group $G_{m,n}$. Recall that the *center* of $G_{m,n}$ is generated by a^m and that the quotient group is the *free product group*

$$\overline{G}_{m,n} = \frac{G_{m,n}}{\langle a^m \rangle} = \langle\, x, y \mid x^m = 1 = y^n \,\rangle = \mathbb{Z}_m * \mathbb{Z}_n$$

of the cyclic groups of order m and n. As the center acts by scalar multiplication on an irreducible representation by *Schur's lemma* the representation theory of $G_{m,n}$ essentially reduces to that of the quotient $\overline{G}_{m,n}$. The latter can be studied by noncommutative geometry as the *group algebra* $\mathbb{C}\overline{G}_{m,n}$ is Quillen-smooth. This follows from

$$\mathbb{C}\overline{G}_{m,n} = \mathbb{C}\mathbb{Z}_m * \mathbb{Z}_n \simeq \mathbb{C}\mathbb{Z}_m * \mathbb{C}\mathbb{Z}_n \simeq \underbrace{\mathbb{C} \times \ldots \times \mathbb{C}}_{m} * \underbrace{\mathbb{C} \times \ldots \times \mathbb{C}}_{n}$$

and as both factors of the *free algebra product* on the right are Quillen-smooth (in fact, semisimple) so is the product by the universal property. Further, as both factors are the path algebras of quivers on m resp. n vertices without arrows, we know that the representation theory of the free algebra product, and hence of $\mathbb{C}\overline{G}_{m,n}$ can be reduced to θ-semistable representations the quiver

$Q_{m,n}$

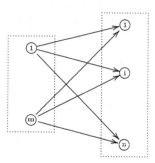

where $\theta = (\underbrace{-1,\ldots,-1}_{m},\underbrace{1,\ldots,1}_{n})$, by the results of the foregoing section. The left vertex spaces S_i, $1 \le i \le m$ for a $\overline{G}_{m,n}$-representation are the eigenspaces for the restricted $\mathbb{Z}_m$-action and the left vertex spaces T_j, $1 \le j \le n$ are the eigenspaces for the restricted $\mathbb{Z}_n$-action.

Example 7.6

Consider the modular group $PSL_2(\mathbb{Z}) \simeq \mathbb{Z}_2 * \mathbb{Z}_3$, the free product of the cyclic groups of orders two and three with generators σ resp. τ. Let S be an n-dimensional simple representation of $PSL_2(\mathbb{Z})$. Let ξ be a 3rd root of unity, then restricting S to these finite Abelian subgroups we have

$$\begin{cases} S\downarrow_{\mathbb{Z}_2} \simeq S_1^{\oplus a_1} \oplus S_{-1}^{\oplus a_2} \\ S\downarrow_{\mathbb{Z}_3} \simeq T_1^{\oplus b_1} \oplus T_{\xi}^{\oplus b_2} \oplus T_{\xi^2}^{\oplus b_3} \end{cases}$$

where S_x resp. T_x are the one-dimensional representations on which σ resp. τ acts via multiplication with x. Observe that $a_1 + a_2 = b_1 + b_2 + b_3 = n$ and we associate to S a representation V of the quiver situation

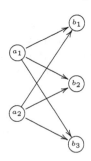

with $V_{1i} = S_i^{\oplus a_i}$ and $V_{2j} = T_j^{\oplus b_j}$ and where the linear map corresponding to an arrow $\overset{a_{ij}}{(b_j)\longrightarrow(a_i)}$ is the composition of

$$V_{a_{ij}} : \quad S_i^{\oplus a_i} \hookrightarrow S\downarrow_{\mathbb{Z}_2} = V\downarrow_{\mathbb{Z}_3} \twoheadrightarrow T_j^{\oplus b_j}$$

of the canonical injections and projections. If $\alpha = (a_1, a_2, b_1, b_2, b_3)$ then we take $\theta = (-1, -1, +1, +1, +1)$. Observe that $\oplus_{i,j} V_{a_{ij}} : \mathbb{C}^n \longrightarrow \mathbb{C}^n$ is a linear isomorphism. If $W \hookrightarrow V$ is a subrepresentation, then $\theta(W) \geq 0$. Indeed, if the dimension vector of W is $\beta = (c_1, c_2, d_1, d_2, d_3)$ and if we assume that $\theta(W) < 0$, then $k = c_1 + c_2 > l = d_1 + d_2 + d_3$, but then the restriction of $\oplus V_{a_{ij}}$ to W gives a linear map $\mathbb{C}^k \longrightarrow \mathbb{C}^l$ having a kernel, which is impossible. Hence, V is a θ-semistable representation of the quiver. In fact, V is even θ-stable, for consider a subrepresentation $W \hookrightarrow V$ with dimension vector β as before and $\theta(W) = 0$, that is, $c_1 + c_2 = d_1 + d_2 + d_3 = m$, then the isomorphism $\oplus_{i,j} V_{a_{ij}} \mid W$ and the decomposition into eigenspaces of $\mathbb{C}^m$ with respect to the $\mathbb{Z}_2$ and $\mathbb{Z}_3$-action, makes $\mathbb{C}^m$ into an m-dimensional representation of $PSL_2(\mathbb{Z})$, which is a subrepresentation of S. S being simple then implies that $W = V$ or $W = 0$, whence V is θ-stable. The underlying reason is that the group algebra $\mathbb{C}PSL_2(\mathbb{Z})$ is a universal localization of the path algebra $\mathbb{C}Q$ of the above quiver. $\quad\square$

As irreducible $\overline{G}_{m,n}$-representations correspond to θ-stable representations of the quiver $Q_{m,n}$ we need to determine the dimension vectors α of θ-stables. In section 4.8 we have given an inductive algorithm to determine them. However, using the fact that the moduli spaces are locally controlled and hence are determined locally by local quivers we can apply the easier classification of simple roots given in section 4.4 so solve this problem.

Example 7.7
With S_{ij} we denote the simple 1-dimensional representation of $PSL_2(\mathbb{Z})$ determined by

$$S_{ij} \downarrow_{\mathbb{Z}_2} = S_i \quad \text{and} \quad S_{ij} \downarrow_{\mathbb{Z}_3}$$

Let $n = x_1 + \ldots + x_6$ and we aim to study the local structure of $\mathbf{rep}_n \, \mathbb{C}PSL_2(\mathbb{Z})$ in a neighborhood of the semisimple n-dimensional representation

$$V_\xi = S_{11}^{\oplus x_1} \oplus S_{12}^{\oplus x_2} \oplus S_{13}^{\oplus x_3} \oplus S_{21}^{\oplus x_4} \oplus S_{22}^{\oplus x_5} \oplus S_{23}^{\oplus x_6}$$

To determine the structure of Q_ξ we have to compute $dim \, Ext^1(S_{ij}, S_{kl})$. To do this we view the S_{ij} as representations of the quiver $Q_{2,3}$ in the example above. For example S_{12} is the representation

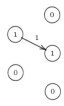

of dimension vector $(1, 0; 0, 1, 0)$. For representations of $Q_{2,3}$, the dimensions

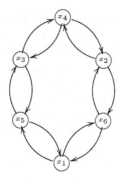

FIGURE 7.3:   The local quiver.

of *Hom* and *Ext*-groups are determined by the bilinear form

$$\chi_Q = \begin{bmatrix} 1 & 0 & -1 & -1 & -1 \\ 0 & 1 & -1 & -1 & -1 \\ 0 & 0 & 1 & 0 & 0 \\ 0 & 0 & 0 & 1 & 0 \\ 0 & 0 & 0 & 0 & 1 \end{bmatrix}$$

If $V \in \mathbf{rep}_\alpha Q$ and $W \in \mathbf{rep}_\beta Q$ where $\alpha = (a_1, a_2; b_1, b_2, b_3)$ with $a_1 + a_2 = b_1 + b_2 + b_3 = k$ and $\beta = (c_1, c_2; d_1, d_2, d_3)$ with $c_1 + c_2 = d_1 + d_2 + d_3 = l$ we have $dim\ Hom(V, W) - dim\ Ext^1(V, W) = \chi_Q(\alpha, \beta)$

$$= kl - (a_1 c_1 + a_2 c_2 + b_1 d_1 + b_2 d_2 + b_3 d_3)$$

As $Hom(S_{ij}, S_{kl}) = \mathbb{C}^{\oplus \delta_{ik}\delta_{jl}}$ we have that

$$dim\ Ext^1(S_{ij}, S_{kl}) = \begin{cases} 1 & \text{if } i \neq k \text{ and } j \neq l \\ 0 & \text{otherwise} \end{cases}$$

The local quiver setting (Q_ξ, α_ξ) is depicted in figure 7.3. We want to determine whether the irreducible component of $\mathbf{rep}_n\ \mathbb{C}PSL_2(\mathbb{Z})$ containing V_ξ contains simple $PSL_2(\mathbb{Z})$-representations, or equivalently, whether α_ξ is the dimension vector of a simple representation of Q_ξ, that is

$$\chi_{Q_\xi}(\alpha_\xi, \epsilon_j) \leq 0 \quad \text{and} \quad \chi_{Q_\xi}(\epsilon_j, \alpha_\xi) \quad \text{for all } 1 \leq j \leq 6$$

The Euler-form of Q_ξ is determined by the matrix where we number the vertices cyclically

$$\chi_{Q_\xi^\bullet} = \begin{bmatrix} 1 & -1 & 0 & 0 & 0 & -1 \\ -1 & 1 & -1 & 0 & 0 & 0 \\ 0 & -1 & 1 & -1 & 0 & 0 \\ 0 & 0 & -1 & 1 & -1 & 0 \\ 0 & 0 & 0 & -1 & 1 & -1 \\ -1 & 0 & 0 & 0 & -1 & 1 \end{bmatrix}$$

leading to the following set of inequalities

$$
\begin{cases}
x_1 & \leq x_5 + x_6 \\
x_2 & \leq x_4 + x_6 \\
x_3 & \leq x_4 + x_5
\end{cases}
\qquad
\begin{cases}
x_4 & \leq x_2 + x_3 \\
x_5 & \leq x_1 + x_3 \\
x_6 & \leq x_1 + x_2
\end{cases}
$$

Finally, observe that V_ξ corresponds to a $Q_{2,3}$-representation of dimension vector $(x_1 + x_2 + x_3, x_4 + x_5 + x_6; x_1 + x_4, x_2 + x_5, x_3 + x_6)$. If we write this dimension vector as $(a_1, a_2; b_1, b_2, b_3)$ then the inequalities are equivalent to the conditions

$$
a_i \geq b_j \quad \text{for all } 1 \leq i \leq 2 \text{ and } 1 \leq j \leq 3
$$

which gives us the desired restriction on the quintuples

at least when $a_i \geq 3$ and $b_j \geq 2$. The remaining cases are handled similarly.
☐

We can use a similar strategy to determine the restrictions on irreducible representations of any torus knot group quotient $\overline{G}_{m,n} \simeq \mathbb{Z}_m * \mathbb{Z}_n$. Having the classification of the dimension vectors α of θ-semistable representations of $Q_{m,n}$ we can use the local quiver settings to study these projective varieties $M_\alpha^{ss}(Q_{m,n}, \theta)$, in particular to determine the α for which this moduli space is a projective smooth variety. For example, $\mathtt{iss}_4 \ PSL_2(\mathbb{Z})$ has several components of dimensions 3 and 2. For one of the three 3-dimensional components, the one corresponding to $\alpha = (2, 2; 2, 1, 1)$, the different types of semisimples M_ξ and corresponding local quivers Q_ξ are all give a smooth ring of invariants. For example, consider a point $\xi \in \mathtt{iss}_4 \ PSL_2(\mathbb{Z})$ of type

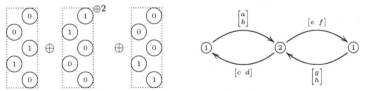

Then, the traces along oriented cycles in Q_ξ are generated by the following three algebraic independent polynomials

$$
\begin{cases}
x & = ac + bd \\
y & = eg + fh \\
z & = (cg + dh)(ea + fb)
\end{cases}
$$

and hence $iss_4\ PSL_2(\mathbb{Z})$ is smooth in ξ. The other cases being easier, we see that this component of $iss_4\ PSL_2(\mathbb{Z})$ is a smooth compact manifold.

A further application of our local quiver-settings (Q_ξ, α_ξ) is that one can often describe large families of irreducible $G_{m,n}$-representations, starting from knowing only rather trivial ones.

Example 7.8

Consider the semisimple $PSL_2(\mathbb{Z})$-representation ξ of type

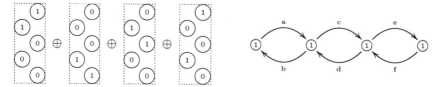

Then, M_ξ is determined by the following matrices

$$\left(\begin{bmatrix} 1 & 0 & 0 & 0 \\ 0 & -1 & 0 & 0 \\ 0 & 0 & 1 & 0 \\ 0 & 0 & 0 & -1 \end{bmatrix}\ ,\ \begin{bmatrix} 1 & 0 & 0 & 0 \\ 0 & \zeta^2 & 0 & 0 \\ 0 & 0 & \zeta & 0 \\ 0 & 0 & 0 & 1 \end{bmatrix}\right)$$

The quiver-setting (Q_ξ, α_ξ) implies that any nearby orbit is determined by a matrix-couple

$$\left(\begin{bmatrix} 1 & b_1 & 0 & 0 \\ a_1 & -1 & d_1 & 0 \\ 0 & c_1 & 1 & f_1 \\ 0 & 0 & e_1 & -1 \end{bmatrix}\ ,\ \begin{bmatrix} 1 & b_2 & 0 & 0 \\ a_2 & \zeta^2 & d_2 & 0 \\ 0 & c_2 & \zeta & f_2 \\ 0 & 0 & e_2 & 1 \end{bmatrix}\right)$$

and as there is just one arrow in each direction these entries must satisfy

$$0 = a_1 a_2 = b_1 b_2 = c_1 c_2 = d_1 d_2 = e_1 e_2 = f_1 f_2$$

As the square of the first matrix must be the identity matrix $\mathbb{1}_4$, we have in addition that

$$0 = a_1 b_1 = c_1 d_1 = e_1 f_1$$

Hence, we get several sheets of 3-dimensional families of representations (possibly, matrix-couples lying on different sheets give isomorphic $PSL_2(\mathbb{Z})$-representations, as the isomorphism holds in the étale topology and not necessarily in the Zariski topology). One of the sheets has representatives

$$\left(\begin{bmatrix} 1 & 0 & 0 & 0 \\ a & -1 & d & 0 \\ 0 & 0 & 1 & 0 \\ 0 & 0 & e & -1 \end{bmatrix}\ ,\ \begin{bmatrix} 1 & b & 0 & 0 \\ 0 & \zeta^2 & 0 & 0 \\ 0 & c & \zeta & f \\ 0 & 0 & 0 & 1 \end{bmatrix}\right)$$

From the description of dimension vectors of semisimple quiver representations it follows that such a representation is simple if and only if

$$ab \neq 0 \quad cd \neq 0 \quad \text{and} \quad ef \neq 0$$

Moreover, these simples are not isomorphic unless their traces ab, cd and ef evaluate to the same numbers. $\square$

Finally, one can use the local quiver-settings (Q_ξ, α_ξ) to determine the isomorphism classes of $G_{m,n}$-representations having a specified Jordan-Hölder sequence. For this we apply the theory on nullcones developed in the foregoing chapter.

Example 7.9
In the above example, this nullcone problem is quite trivial. A representation has M_ξ as Jordan-Hölder sum if and only if all traces vanish, that is

$$ab = cd = ef = 0$$

Under the action of the group $GL(\alpha_\xi) = \mathbb{C}^* \times \mathbb{C}^* \times \mathbb{C}^* \times \mathbb{C}^*$, these orbits are easily seen to be classified by the arrays

a	c	e
b	d	f

filled with zeroes and ones subject to the rule that no column can have two 1's, giving $27 = 3^3$-orbits. $\square$

7.5 Differential forms

In this section we will define the complex of *noncommutative differential forms* of an arbitrary $\mathbb{C}$-algebra A and deduce some extra features in case A is Quillen-smooth. In the following section we will compute the *noncommutative deRham cohomology* spaces that will be of crucial importance in the final chapter.

Let us recall briefly the classical (commutative) case. When A is a commutative $\mathbb{C}$-algebra, the A-module of *Kähler differentials* Ω_A^1 is generated by the $\mathbb{C}$-linear symbols da for $a \in A$ satisfying the relations

$$d(ab) = adb + bda \qquad \forall a, b \in A$$

and the map $A \xrightarrow{\ d\ } \Omega_A^1$ is the universal derivation. By convention we define

$$\begin{cases} \Omega_A^0 = A \\ \Omega_A^n = \wedge_A^n \ \Omega_A^1 \end{cases}$$

where the *exterior product* is taken over A (not over $\mathbb{C}$). Observe that it is spanned by the elements $a_0 da_1 \wedge \ldots \wedge da_n$ that we usually write $a_0 da_1 \ldots da_n$.

The *exterior differential operator*

$$\Omega_A^n \xrightarrow{\ d\ } \Omega_A^{n+1}$$

is defined by

$$d(a_0 da_1 \ldots da_n) = da_0 da_1 \ldots da_n$$

and gives rise to a sequence

$$A = \Omega_A^0 \xrightarrow{\ d\ } \Omega_A^1 \xrightarrow{\ d\ } \ldots \xrightarrow{\ d\ } \Omega_A^n \xrightarrow{\ d\ } \Omega_A^{n+1} \xrightarrow{\ d\ } \ldots$$

which is a complex (that is, $d \circ d = 0$) called the *deRham complex* . The *homology groups* of this complex

$$\mathrm{H}_{dR}^n \ \mathbf{A} = \frac{\mathrm{Ker}\ \Omega_A^n \xrightarrow{\ d\ } \Omega_A^{n+1}}{\mathrm{Im}\ \Omega_A^{n-1} \xrightarrow{\ d\ } \Omega_A^n}$$

are called the *de Rham cohomology groups* of A (over $\mathbb{C}$).

We will extend this to noncommutative $\mathbb{C}$-algebras. We denote by $\mathtt{dgalg}$ the category of *differential graded $\mathbb{C}$-algebras* , that is, an object $R \in \mathtt{dgalg}$ is a $\mathbb{Z}$-graded $\mathbb{C}$-algebra

$$R = \oplus_{i \in \mathbb{Z}} R_i$$

endowed with a differential d of degree one

$$\ldots \xrightarrow{\ d\ } R_{i-1} \xrightarrow{\ d\ } R_i \xrightarrow{\ d\ } R_{i+1} \xrightarrow{\ d\ } \ldots$$

such that $d \circ d = 0$ and for all $r \in R_i$ and $s \in R$ we have

$$d(rs) = (dr)s + (-1)^i r(ds)$$

Clearly, morphisms in $\mathtt{dgalg}$ are $\mathbb{C}$-algebra morphisms $R \xrightarrow{\ \phi\ } S$ which are graded and commute with the differentials.

To a $\mathbb{C}$-algebra A we will now associate the differential graded algebra $\Omega\ A$ of *noncommutative differential forms* . Denote the quotient vector space $A/\mathbb{C}.1$ with $\overline{A}$ and define

$$\Omega^n\ A = A \otimes \underbrace{\overline{A} \otimes \ldots \otimes \overline{A}}_{n}$$

for $n \geq 0$ and $\Omega^n\ A = 0$ for $n < 0$. For $a_i \in A$ we denote the image of $a_0 \otimes a_1 \otimes \ldots \otimes a_n$ in $\Omega^n\ A$ by

$$(a_0, \ldots, a_n).$$

Consider the vector space $\Omega\, A = \oplus_{n \in \mathbb{Z}}\, \Omega^n\, A$ and define a product on it by

$$(a_0, \ldots, a_n)(a_{n+1}, \ldots, a_m) = \sum_{i=0}^{n} (-1)^{n-i}(a_0, \ldots, a_{i-1}, a_i a_{i+1}, a_{i+2}, \ldots, a_m).$$

Further, define an operator d of degree one

$$\cdots \xrightarrow{\;d\;} \Omega^{n-1}\, A \xrightarrow{\;d\;} \Omega^n\, A \xrightarrow{\;d\;} \Omega^{n+1}\, A \xrightarrow{\;d\;} \cdots$$

by the rule

$$d(a_0, \ldots, a_n) = (1, a_0, \ldots, a_n)$$

THEOREM 7.7
These formulas define the unique `dgalg` *structure on* $\Omega\, A$ *such that*

$$a_0 da_1 \ldots da_n = (a_0, a_1, \ldots, a_n).$$

PROOF   In *any* $R = \oplus_i R_i \in$ `dgalg` containing A as an even degree subalgebra we have the following identities

$$d(a_0 da_1 \ldots da_n) = \qquad\qquad da_0 da_1 \ldots da_n$$
$$(a_0 da_1 \ldots da_n)(a_{n+1} da_{n+2} \ldots da_m) = \qquad (-1)^n a_0 a_1 da_2 \ldots da_m$$
$$+ \sum_{i=1}^{n}(-1)^{n-i} a_0 da_1 \ldots d(a_i a_{i+1}) \ldots da_m$$

which proves uniqueness.

To prove existence, we define d on $\Omega\, A$ as above making the $\mathbb{Z}$-graded $\mathbb{C}$-vector space $\Omega\, A$ into a complex as $d \circ d = 0$. Consider the *graded endomorphism ring* of the complex

$$\mathbf{End} = \oplus_{n \in \mathbb{Z}} \mathbf{End}_n = \oplus_{n \in \mathbb{Z}} \mathrm{Hom}_{complex}(\Omega^\bullet\, A, \Omega^{\bullet+n}\, A)$$

With the composition as multiplication, `End` is a $\mathbb{Z}$-graded $\mathbb{C}$-algebra and we make it into an object in `dgalg` by defining a differential

$$\cdots \xrightarrow{\;D\;} \mathbf{End}_{n-1} \xrightarrow{\;D\;} \mathbf{End}_n \xrightarrow{\;D\;} \mathbf{End}_{n+1} \xrightarrow{\;D\;} \cdots$$

by the formula on any homogeneous ϕ

$$D\phi = d \circ \phi - (-1)^{deg\ \phi}\phi \circ d$$

Now define the morphism $A \xrightarrow{\;l\;} \mathbf{End}_0$, which assigns to $a \in A$ the left multiplication operator

$$la(a_0, \ldots, a_n) = (aa_0, \ldots, a_n)$$

and extend it to a map

$$\Omega \; A \xrightarrow{\; l_* \;} \text{End} \qquad \text{by} \qquad l_*(a_0, \ldots, a_n) = l a_0 \circ D \; l a_1 \circ \ldots \circ D \; l a_n$$

Applying the general formulae given at the beginning of the proof to the subalgebra $l(A) \hookrightarrow \text{End}$ we see that the image of l_* is a differential graded subalgebra of End and is the differential graded subalgebra generated by $l(A)$.

Define an evaluation map $\text{End} \xrightarrow{\; ev \;} \Omega \; A$ by $ev(\phi) = \phi(1)$. Because

$$
\begin{aligned}
D \; l a_i (1, a_{i+1}, \ldots, a_n) &= d(a_i, a_{i-1}, \ldots, a_n) - l a_i d(1, a_{i+1}, \ldots, a_n) \\
&= \qquad\qquad (1, a_i, \ldots, a_n)
\end{aligned}
$$

we have that

$$ev(l a_0 \circ D \; l a_1 \circ \ldots \circ D \; l a_n) = (a_0, \ldots, a_n)$$

showing that ev is a left inverse for l_* whence l_* in injective.

Hence we can use the isomorphism $\Omega \; A \simeq Im(l_*)$ to transport the dgalg structure to $\Omega \; A$ finishing the proof. ∎

Example 7.10 Noncommutative differential forms of $\mathbb{C} \times \mathbb{C}$
Let $A = \mathbb{C} \times \mathbb{C}$ and e and f the idempotents corresponding to the two factors. The quotient space $\overline{A} = A/\mathbb{C}1$ can be identified with $\mathbb{C}\overline{e}$ and therefore

$$\Omega^n \; \mathbb{C} \times \mathbb{C} = (\mathbb{C} \times \mathbb{C}) \otimes \mathbb{C}\overline{e}^{\otimes n} = (\mathbb{C} \times \mathbb{C})de^n$$

The differential d is defined by the formula

$$d((\alpha e + \beta f)de^n) = (\alpha - \beta)de^{n+1}$$

and the product of $\Omega \; \mathbb{C} \times \mathbb{C}$ is defined by the rule

$$(\alpha e + \beta f)de^n (\gamma e + \delta f)de^m = \begin{cases} (\alpha\gamma e + \beta\delta f)de^{n+m} & \text{when } n \text{ is even} \\ (\alpha\delta e + \beta\gamma f)de^{n+m} & \text{when } n \text{ is odd.} \end{cases}$$

∎

We will relate the algebra structure of $\Omega \; A$ to that of A. The trick is to define *another* multiplication on $\Omega \; A$ making it only into a *filtered* algebra. We then prove that this filtered algebra is isomorphic to the I-adic filtration of an algebra constructed from A and we recover the dgalg multiplication on $\Omega \; A$ by taking the associated graded algebra.

We introduce the universal algebra $\mathbb{L}_A$ with respect to *based linear maps* from A to $\mathbb{C}$-algebras. A based linear map is a $\mathbb{C}$-linear map

$$A \xrightarrow{\; \rho \;} R$$

where R is a $\mathbb{C}$-algebra and $\rho(1) = 1$. The *curvature* of ρ is then defined to be the bilinear map $A \times A \xrightarrow{\omega} R$ defined by

$$\omega(a, a') = \rho(aa') - \rho(a)\rho(a')$$

that is, it is a measure for the failure of ρ to be an algebra map. Observe that ω vanishes if either a or a' is 1 so it can be viewed as a linear map

$$\overline{A} \otimes \overline{A} \xrightarrow{\omega} R$$

Let $T(A) = \oplus_{n \geq 0} A^{\otimes n}$ be the *tensor algebra* of the vector space A and define

$$\mathbb{L}_A = \frac{T(A)}{T(A)(1 - 1_A)T(A)}$$

where 1_A is the identity of A consider as a 1-tensor in $T(A)$, then we have a based linear map

$$A \xrightarrow{\rho^{un}} \mathbb{L}_A \qquad a \mapsto \overline{a}$$

where $\overline{a}$ is the image in $\mathbb{L}_A$ of the 1-tensor a in $T(A)$. The map ρ^{un} is universal for based linear maps $A \xrightarrow{\rho} R$, that is, there is a unique *algebra morphism* $\mathbb{L}_A \xrightarrow{\phi_\rho} R$ making the diagram commute

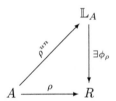

In particular, there is a canonical algebra map $\mathbb{L}_A \xrightarrow{\phi_{id}} A$ corresponding to the identity map on A. We define

$$\mathbb{I}_A = \mathrm{Ker}\ \phi_{id} \lhd \mathbb{L}_A$$

and equip $\mathbb{L}_A$ with the $\mathbb{I}_A$-adic filtration.

For an arbitrary $R \in \mathrm{dgalg}$ we define the *Fedosov product* on R to be the one induced by defining on homogeneous $r, s \in R$ the product

$$r.s = rs - (-1)^{deg\ r} drds$$

One easily checks that the Fedosov product is associative. Observe that if we decompose $R = R^{ev} \oplus R^{odd}$ into its homogeneous components of even (resp. odd) degree, then this new multiplication is compatible with this decomposition and makes R into a $\mathbb{Z}/2\mathbb{Z}$-graded algebra.

We will now investigate the Fedosov product on $\Omega\ A$. Let ω^{un} be the curvature of the universal based linear map $A \xrightarrow{\rho^{un}} \mathbb{L}_A$.

THEOREM 7.8

There is an isomorphism of algebras

$$\mathbb{L}_A \simeq (\Omega^{ev} A , .)$$

between $\mathbb{L}_A$ and the even forms $\Omega^{ev} A$ equipped with the Fedosov product given by

$$\rho^{un}(a_0)\omega^{un}(a_1, a_2)\ldots\omega^{un}(a_{2n-1}, a_{2n}) \longrightarrow a_0 da_1 \ldots da_{2n}$$

Under this isomorphism we have the correspondence

$$\mathbb{I}_A^n \simeq \oplus_{k \geq n}\Omega^{2k} A$$

The associated graded algebra gives an isomorphism

$$\mathbf{gr}_{\mathbb{I}_A} \mathbb{L}_A = \oplus \frac{\mathbb{I}_A^n}{\mathbb{I}_A^{n+1}} \simeq \Omega^{ev} A$$

with even forms equipped with the `dgalg` *structure.*

PROOF   Consider the based linear map $A \xrightarrow{\rho} \Omega^{ev} A$ given by inclusion, then its curvature is given by

$$\omega(a, a') = aa' - a.a' = dada'$$

By the universal property of $\mathbb{L}_A$ there is an algebra morphism

$$\mathbb{L}_A \xrightarrow{\phi} (\Omega^{ev} A , .)$$

such that $\phi(\rho^{un}(a)) = a$ and $\phi(\omega^{un}(a, a')) = dada'$. Observe that the Fedosov product coincides with the usual `dgalg` product when one of the terms is *closed* that is $dr = 0$. Therefore, we have

$$\phi(\rho^{un}(a_0)\omega^{un}(a_1, a_2)\ldots\omega^{un}(a_{2n-1}, a_{2n})) = a_0 da_1 \ldots da_{2n}$$

On the other hand, as $\Omega^{2n} A = A \otimes \overline{A}^{\otimes 2n}$ we have a well defined map $\Omega^{ev} A \xrightarrow{\psi} \mathbb{L}_A$ given by

$$\psi(a_0 da_1 \ldots da_{2n}) = \rho^{un}(a_0)\omega^{un}(a_1, a_2)\ldots\rho^{un}(a_{2n-1}, a_{2n})$$

and it remains to prove that this map is surjective. The image of ψ is closed under left multiplication as it is closed under left multiplication by elements $\rho^{un}(a)$ (and they generate $\mathbb{L}_A$) as

$$\rho^{un}(a).\rho^{un}(a_0)\omega^{un}(a_1, a_2)\ldots\omega^{un}(a_{2n-1}, a_{2n})$$
$$= \rho^{un}(aa_0)\omega^{un}(a_1, a_2)\ldots\omega^{un}(a_{2n-1}, a_{2n}) -$$
$$\omega^{un}(a, a_0)\omega^{un}(a_1, a_2)\ldots\omega^{un}(a_{2n-1}, a_{2n})$$

Because the image contains 1 this proves the claim and the isomorphism.

Identify via this isomorphism $\mathbb{L}_A$ with $\Omega^{ev}\,A$. Because $dada' \in \mathbb{I}_A$ we have $\Omega^{2k}\,A \hookrightarrow \mathbb{I}_A^n$ for all $k \geq n$. Thus, $F_n = \oplus_{k \geq n}\Omega^{2k}\,A \hookrightarrow \mathbb{I}_A$. Conversely, $\mathbb{I}_A = F_1$ and hence

$$\mathbb{I}_A^n = (F_1)^n \hookrightarrow F_n$$

by the definition of the Fedosov product. Therefore, $\mathbb{I}_A^n = F_n$ and the claim over the associated graded follows. $\square$

Example 7.11 Even differential forms of $\mathbb{C} \times \mathbb{C}$

As before, let e and f be the idempotents of $A = \mathbb{C} \times \mathbb{C}$ corresponding to the two components. By definition

$$\mathbb{L}_{\mathbb{C} \times \mathbb{C}} = \frac{T(\mathbb{C}e + \mathbb{C}f)}{(1 - e - f)} = \frac{\mathbb{C}\langle E, F\rangle}{(1 - E - F)} \simeq \mathbb{C}[E]$$

The universal based linear map is given by

$$\mathbb{C} \times \mathbb{C} \xrightarrow{\rho^{un}} \mathbb{C}[E] \qquad \begin{cases} e & \mapsto E \\ f & \mapsto 1 - E \end{cases}$$

and the curvature on $\overline{A} = \mathbb{C}\overline{e}$ is given by

$$w^{un}(e, e) = E - E^2$$

Therefore the isomorphism between $\Omega^{ev}\,A$ and $\mathbb{L}_A = \mathbb{C}[E]$ is given by

$$(\alpha e + \beta f)de^{2n} \xrightarrow{\psi} (\alpha E + \beta(1 - E))(E - E^2)^n$$

The Fedosov product on $\Omega^{ev}\,A$ is given by the formula (using the multiplication formulas we found above)

$$(\alpha e + \beta f)de^{2n}.(\gamma e + \delta f)de^{2m} = (\alpha\gamma e + \beta\delta f)de^{2n+2m} - (\alpha - \beta)(\gamma - \delta)de^{2n+2m+2}$$

In order to check that ψ is indeed an algebra morphism we need to verify that in $\mathbb{C}[E]$ we have the equality

$$(\alpha E + \beta(1 - E))(E - E^2)^n(\gamma E + \delta(1 - E))(E - E^2)^m$$
$$= (\alpha\gamma E + \beta\delta(1 - E))(E - E^2)^{n+m} - (\alpha - \beta)(\gamma - \delta)(E - E^2)^{n+m+1}$$

which is indeed the case.

Further, $\mathbb{I}_A = \mathbb{C}[E](E - E^2)$ and indeed $\frac{\mathbb{C}[E]}{(E-E^2)} \simeq \mathbb{C} \times \mathbb{C}$. Finally, under the identification ψ we obtain the usual multiplication of noncommutative differential forms from $\Omega^{2n}\,A \times \Omega^{2m}\,A =$

$$\frac{(E - E^2)^n}{(E - E^2)^{n+1}} \times \frac{(E - E^2)^m}{(E - E^2)^{m+1}} \longrightarrow \frac{(E - E^2)^{n+m}}{(E - E^2)^{n+m+1}} = \Omega^{2n+2m}\,A$$

We now turn to *all* noncommutative differential forms $\Omega\, A$. Observe that this algebra has an involution σ, which is the identity on even forms and is minus the identity on odd forms. σ is an algebra automorphism both for the usual `dgalg`-algebra structure as for the Fedosov product. Algebras with an involution are called *superalgebras* .

We want to construct an algebra universal for algebra morphisms from A to a superalgebra. Consider the *free product* $A * A$, which is defined as follows. Let $\mathcal{B}_1$ be a vector space basis for $A - \mathbb{C}.1$ and $\mathcal{B}_2$ a duplicate of it. As a $\mathbb{C}$-vector space $A * A$ has a basis consisting of words

$$w = a_1 b_1 a_2 b_2 \ldots a_k b_k \quad \text{or} \quad w = a_1 b_1 a_2 b_2 \ldots a_k$$

for some k where the a_i's all belong to $\mathcal{B}_1$ or all to $\mathcal{B}_2$ and the b_j's all belong to the other base set. On this vector space one defines a $\mathbb{C}$-algebra structure in the obvious way, that is by concatenating words and if necessary (if the end term of the first word lies in the same base-set as the beginning term of the second) use the multiplication table in A to reduce to a linear combination of allowed words.

The algebra $A * A$ is universal with respect to *pairs of algebra maps* $A \overset{f}{\underset{g}{\rightrightarrows}} R$ from A to R. That is, there is a unique algebra map γ

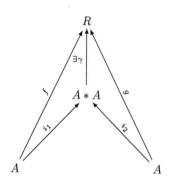

making the diagram commute. Here, i_1 is the inclusion of A in $A * A$ using only syllables in $\mathcal{B}_1$ and i_2 is defined similarly. The construction of γ clearly is induced by sending $a \in \mathcal{B}_1$ to $f(a)$ and $b \in \mathcal{B}_2$ to $g(b)$.

Interchanging the bases $\mathcal{B}_1 \overset{\tau}{\longrightarrow} \mathcal{B}_2$ equips $A * A$ with an involution, or if you prefer, makes $A * A$ a superalgebra. Now, let S be a superalgebra with involution σ_S and let $A \overset{f}{\longrightarrow} S$ be an algebra morphism, then there is a unique morphism of superalgebras ψ making the diagram commute

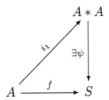

ψ is the universal map corresponding to the pair of algebra maps $A \underset{\sigma_S \circ f}{\overset{f}{\rightrightarrows}} S$.

For any $a \in A$ we define the elements in $A * A$

$$\begin{cases} p(a) &= \frac{1}{2}(i_1(a) + i_2(a)) \\ q(a) &= \frac{1}{2}(i_1(a) - i_2(a)) \end{cases}$$

and we define $\mathbb{Q}_A \lhd A * A$ to be the ideal of $A * A$ generated by the elements $q(a)$ for $a \in A$, then clearly

$$A \simeq \frac{A * A}{\mathbb{Q}_A}$$

We now have an analog of the previous theorem for all differential forms.

THEOREM 7.9

There is an isomorphism of superalgebras

$$A * A \simeq (\Omega \, A \, , \, . \,)$$

between $A * A$ and the noncommutative differential forms $\Omega \, A$ equipped with the Fedosov product given by

$$p(a_0)q(a_1) \ldots q(a_n) \longrightarrow a_0 da_1 \ldots da_n$$

Under this isomorphism we have the correspondence

$$\mathbb{Q}_A^n \simeq \oplus_{k \geq n} \Omega^n \, A$$

and the associated graded algebra is isomorphic to $\Omega \, A$ with the usual dgalg *structure.*

PROOF We have an algebra map $A \overset{u}{\longrightarrow} \Omega \, A$ equipped with the Fedosov product given by $a \mapsto a + da$ because

$$(a + da).(a' + da') = aa' - dada' + ada' + daa' + dada'$$
$$= aa' + d(aa')$$

By the universal property of $A * A$ there is a superalgebra morphism

$$A * A \overset{\psi}{\longrightarrow} \Omega \, A \qquad \psi(p(a)) = a \quad \text{and} \quad \psi(q(a)) = da$$

But then, because the Fedosov product coincides with the usual product when one of the forms is closed, we have

$$\psi(p(a_0)q(a_1)\ldots q(a_n)) = a_0 da_1 \ldots da_n$$

Conversely, we have a section to ψ defined by

$$\Omega\,A \xrightarrow{\ \phi\ } A * A \qquad a_0 da_1 \ldots da_n \mapsto p(a_0)q(a_1)\ldots q(a_n)$$

and we only have to prove that ϕ is surjective. The image $\text{Im}\,\phi$ is closed under left multiplication by $p(a)$ and $q(a)$ as $p(1) = 1$ and

$$\begin{cases} p(a)p(a_0)q(a_1)\ldots q(a_n) = & p(aa_0)q(a_1)\ldots q(a_n) - q(a)q(a_0)q(a_1)\ldots q(a_n) \\ q(a)p(a_0)q(a_1)\ldots q(a_n) = & q(aa_0)q(a_1)\ldots q(a_n) - p(a)q(a_0)q(a_1)\ldots q(a_n) \end{cases}$$

Because the elements $p(a)$ and $q(a)$ generate $A * A$, the image $\text{Im}\,\phi$ is a left ideal containing 1, whence ψ is surjective.

The claims about the ideals $\mathbb{Q}_A^n$ and about the associated graded algebra follow as in the proof for even forms. $\quad\Box$

Example 7.12 Noncommutative differential forms of $\mathbb{C}\langle x, y\rangle$
The noncommutative free algebra in two variables $\mathbb{C}\langle x, y\rangle$ is the path algebra of the quiver

Clearly we have $\mathbb{C}\langle x, y\rangle * \mathbb{C}\langle x, y\rangle = \mathbb{C}\langle x_1, y_1, x_2, y_2\rangle$ and the maps

$$\begin{cases} p(x) = \tfrac{1}{2}(x_1 + x_2) \qquad q(x) = \tfrac{1}{2}(x_1 - x_2) \\ p(y) = \tfrac{1}{2}(y_1 + y_2) \qquad q(y) = \tfrac{1}{2}(y_1 - y_2) \end{cases}$$

It is easy to compute the maps p and q on any monomial in x and y using the formulae holding in any $A * A$

$$\begin{cases} p(aa') = & p(a)p(a') + q(a)q(a') \\ q(aa') = & p(a)q(a') + q(a)p(a') \end{cases}$$

Further note that it follows from this that $\mathbb{Q}_{\mathbb{C}\langle x,y\rangle} = (x_1 - x_2, y_1 - y_2)$ and we have all the required tools to calculate (in principle) with $\Omega\,\mathbb{C}\langle x, y\rangle$. $\quad\Box$

Example 7.13 Noncommutative differential forms of $\mathbb{C} \times \mathbb{C}$
The *infinite dihedral group* D_∞ is the group with presentation

$$D_\infty = \langle a, b \mid a^2 = 1 = b^2\rangle$$

that is, an arbitrary element in D_∞ is a word of the form

$$a^i babab \ldots abab^j$$

where $i, j = 0$ or 1. Multiplication is given by concatenation of words, using the relations $a^2 = 1 = b^2$ when necessary.

The *group algebra* $\mathbb{C}[D_\infty]$ is the vector space with basis D_∞ and with multiplication induced by the group multiplication in D_∞. We now claim that

$$(\mathbb{C} \times \mathbb{C}) * (\mathbb{C} \times \mathbb{C}) \simeq \mathbb{C}[D_\infty]$$

Indeed, $\mathbb{C} \times \mathbb{C} \simeq \mathbb{C}[\mathbb{Z}_2]$ the group algebra of the cyclic group of order two, that is, $\mathbb{C}[\mathbb{Z}_2] = \mathbb{C}[x]/(x^2 - 1)$, the isomorphism being given by

$$e \longrightarrow \frac{1}{2}(1 + x) \qquad f \longrightarrow \frac{1}{2}(1 - x)$$

One also has the obvious notion of a free product in the category of groups and from the definition it is clear that

$$\mathbb{Z}_2 * \mathbb{Z}_2 \simeq D_\infty$$

and therefore also on the level of group algebras

$$\mathbb{C}[\mathbb{Z}_2] * \mathbb{C}[\mathbb{Z}_2] \simeq \mathbb{C}[D_\infty]$$

The relevant maps $\mathbb{C} \times \mathbb{C} \xrightarrow[q]{p} \mathbb{C}[D_\infty]$ are given by

$$\begin{cases} p(e) = \frac{1}{2} + \frac{1}{4}(a + b) & q(e) = \frac{1}{4}(a - b) \\ p(f) = \frac{1}{2} - \frac{1}{4}(a + b) & q(f) = -\frac{1}{4}(a - b) \end{cases}$$

and so $\mathbb{Q}_{\mathbb{C} \times \mathbb{C}} = (a - b) \triangleleft \mathbb{C}[D_\infty]$. Again, this information allows us to calculate with $\Omega \ \mathbb{C} \times \mathbb{C}$ by referring all computations to the more familiar group algebra $\mathbb{C}[D_\infty]$. $\Box$

The above definitions and results are valid for every $\mathbb{C}$-algebra A. We will indicate a few extra properties provided the algebra A is Quillen-smooth.

We have the universal lifting algebra $\mathbb{L}_A$ for based *linear* maps from A to $\mathbb{C}$-algebras and the ideal $\mathbb{I}_A$ such that

$$A \xleftarrow[\simeq]{\overline{\phi_{id}}} \frac{\mathbb{L}_A}{\mathbb{I}_A}$$

The $\mathbb{I}_A$-*adic completion* of $\mathbb{L}_A$ is by definition the inverse limit

$$\hat{\mathbb{L}}_A = \varprojlim_n \frac{\mathbb{L}_A}{\mathbb{I}_A}$$

Assume that A is Quillen-smooth, then for every k we have an algebra map lifting $\overline{\phi_{id}}^{-1}$

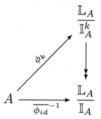

These compatible lifts define an algebra lift $A \xrightarrow{l^{un}} \hat{\mathbb{L}}_A$. This map can be used to construct algebra lifts modulo nilpotent ideals in a systematic way. Assume $I \lhd R$ is such that $I^k = 0$ and there is an algebra map $A \xrightarrow{\mu} \frac{R}{I}$. We can lift μ to R as a based linear map, say, ρ. Now we have the following situation

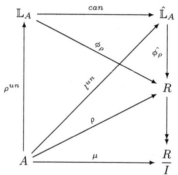

Here, ϕ_ρ is the algebra map coming from the universal lifting property of $\mathbb{L}_A$ and $\hat{\phi}_\rho$ is its extension to the completion. But then, $\tilde{\mu} = \hat{\phi}_\rho \circ l^{un}$ is an *algebra* lift of μ, as shown in the text that follows.

PROPOSITION 7.3

A is Quillen-smooth if and only if there is an algebra section $A \longrightarrow \hat{\mathbb{L}}_A$ to the projection $\hat{\mathbb{L}}_A \longrightarrow A$ defined by mapping out $\mathbb{I}_A$.

We will give an explicit construction of the embedding $A \xrightarrow{l^{un}} \hat{\mathbb{L}}_A$. By Quillen-smoothness we have an algebra lift

$$
\begin{array}{ccc}
 & A \oplus \Omega^2 A & = & \dfrac{\mathbb{L}_A}{\mathbb{I}_A^2} \\[4ex]
\overset{l_2}{\nearrow} & & & \downarrow \\[2ex]
A \xrightarrow{\quad id \quad} A & = & \dfrac{\mathbb{L}_A}{\mathbb{I}_A}
\end{array}
$$

which is of the form $l_2(a) = a - \phi(a)$ for a linear map $A \xrightarrow{\phi} \Omega^2 A$. As $\mathbb{L}_A$ is freely generated by the $a \in A - \mathbb{C}1$, we can define a *derivation* on $\mathbb{L}_A$ defined by

$$\mathbb{L}_A \xrightarrow{D} \mathbb{L}_A \qquad D(a) = \phi(a) \quad \forall a \in A$$

This derivation is called the *Yang-Mills derivation* of A.

Clearly $D(\mathbb{L}_A) \hookrightarrow \mathbb{I}_A$ and we have

$$
\begin{aligned}
D(dada') &= D(aa' - a.a') \\
&= D(aa') - D(a).a' - a.D(a') \\
&= \phi(aa') - \phi(a).a' - a.\phi(a') \\
&\equiv aa' - a.a' \bmod \mathbb{I}_A^2 \\
&\equiv dada' \bmod \mathbb{I}_A^2
\end{aligned}
$$

the next to last equality coming from the fact that l_2 is an algebra map. Hence, $D = id$ on $\frac{\mathbb{I}_A}{\mathbb{I}_A^2} = \Omega^2 A$.

Further, $D(\mathbb{I}_A^n) \hookrightarrow \mathbb{I}_A^n$ and so D induces a derivation on the associated graded $gr_{\mathbb{I}_A} \mathbb{L}_A$. As this derivation is zero on $A = \frac{\mathbb{L}_A}{\mathbb{I}_A}$ and one on $\frac{\mathbb{I}_A}{\mathbb{I}_A^2}$ it is n on $\frac{\mathbb{I}_A^n}{\mathbb{I}_A^{n+1}}$. But then we have by induction

$$(D - n)...(D - 1)D(\mathbb{L}_A) \hookrightarrow \mathbb{I}_A^{n+1}$$

Therefore, $\frac{\mathbb{L}_A}{\mathbb{I}_A^{n+1}}$ decomposes into eigenspaces of D corresponding to the eigenvalues $0, 1, \ldots, n$ and because D is a derivation this decomposition defines a grading compatible with the product.

Hence, we obtain an isomorphism of $\frac{\mathbb{L}_A}{\mathbb{I}_A^{n+1}}$ with its associated graded algebra by lifting $\frac{\mathbb{I}_A^k}{\mathbb{I}_A^{k+1}}$ to the eigenspace of D on $\frac{\mathbb{L}_A}{\mathbb{I}_A^{n+1}}$ corresponding to the eigenvalue k.

Taking the inverse limit as $n \longrightarrow \infty$ we obtain an algebra isomorphism of $\hat{\mathbb{L}}_A$ with the completion of its associated graded algebra, that is

$$\hat{\Omega^{ev}} A = \prod_n \Omega^{2n} A \simeq \hat{\mathbb{L}}_A$$

In particular, the kernel of D is a subalgebra of $\hat{\mathbb{L}}_A$ mapped isomorphically onto A by the canonical surjection $\hat{\mathbb{L}}_A \longrightarrow A$. Hence, this subalgebra gives the desired universal lift $A \xrightarrow{l^{un}} \hat{\mathbb{L}}_A$.

We can even give an explicit formula for l^{un}. Let L be the degree two operator on $\Omega^{ev} A$ defined by

$$L(a_0 da_1 \ldots da_{2n}) = \phi(a_0) da_1 \ldots da_{2n} + \sum_{j=1}^{2n} a_0 da_1 \ldots da_{j-1} d\phi(a_j) da_{j+1} \ldots da_{2n}$$

and let H denote the degree zero operator on even forms, which is multiplication by n on $\Omega^{2n} A$. Then, we have the relations

$$[H, L] = L \qquad \text{and} \qquad D = H + L$$

whence we have on $\hat{\Omega}^{ev} A$ that

$$e^{-L} H e^L = H + e^{-L}[H, e^L] = H + \int_0^1 e^{-tL}[H, L]e^{tL} dt = D$$

Therefore, the universal lift for all $a \in A$ is given by

$$l^{un}(a) = e^{-L} a = a - \phi(a) + \frac{1}{2} L\phi(a) - \cdots$$

Example 7.14 The universal lift for $\mathbb{C} \times \mathbb{C}$
Recall the correspondence between $\Omega^{ev} \mathbb{C} \times \mathbb{C}$ and $\mathbb{L}_{\mathbb{C} \times \mathbb{C}} = \mathbb{C}[E]$ given by

$$(\alpha e + \beta f)de^{2n} \longrightarrow (\alpha E + \beta(1 - E))(E - E^2)^n$$

Lifting e to $\frac{L}{l^2}$ we have to compute

$$(2 - E)^2 E^2 = E + (2E - 1)(E - E^2) + (E - E^2)^2$$

whence $\phi(e) = (1 - 2E)(E - E^2)$ and as $f = 1 - e$ we have $\phi(f) = (2E - 1)(E - E^2)$. The Yang-Mills derivation D on $\mathbb{C}[E]$ is hence the one determined by

$$\mathbb{C}[E] \xrightarrow{D} \mathbb{C}[E] \qquad D(E) = (1 - 2E)(E - E^2)$$

To determine the universal lift of e we have to compute

$$l^{un}(e) = e - Le + \frac{1}{2}L^2 e - \frac{1}{6}L^3 e + \cdots$$

and we have

$$\begin{aligned}
L(e) &= \phi(e) = (f - e)de^2 \\
L^2(e) &= L(f - e)de^2 = -6(f - e)de^4 \\
L^3(e) &= -6L(f - e)de^4 = 60(f - e)de^6 \\
L^4(e) &= \cdots
\end{aligned}$$

and therefore

$$l^{un}(e) = E + (2E-1)(E-E^2) + 3(2E-1)(E-E^2)^2 + 10(2E-1)(E-E^2)^3 + \cdots$$

$\square$

Another characteristic feature of Quillen-smooth algebras is the existence of *connections* on $\Omega^1 A$. If E is an A-bimodule, then a connection on E consists of two operators

- A *right connection* : $E \xrightarrow{\nabla_r} E \otimes_A \Omega^1 A$ satisfying

$$\nabla_r(aea') = a(\nabla_r e)a' + aeda'$$

- A *left connection* : $E \xrightarrow{\nabla_l} \Omega^1 A \otimes_A E$ satisfying

$$\nabla_l(aea') = a(\nabla_l e)a' + daea'$$

Given a right connection ∇_r there is a bimodule splitting s_r of the right multiplication map m_r

$$E \otimes_A A \underset{s_r}{\overset{m_r}{\rightleftarrows}} E$$

given by the formula

$$s_r(e) = e \otimes 1 - j(\nabla_r e) \quad \text{where} \quad j(e \otimes da) = ea \otimes 1 - e \otimes a$$

Similarly, a left connection gives a bimodule splitting s_l to the left multiplication map. Consequently, if a connection exists on E, then E must be a *projective* bimodule.

Consider the special bimodule of noncommutative 1-forms $\Omega^1 A$, then as $\Omega^1 A \otimes_A \Omega^1 A = \Omega^2 A$ a connection on $\Omega^1 A$ is the datum of three maps

$$\Omega^1 A \xrightarrow[\nabla_r]{\overset{\nabla_l}{\underset{d}{\longrightarrow}}} \Omega^2 A$$

satisfying the following properties

$$\begin{aligned}
\nabla_l(aea') &= a\nabla_l(e)a' &+(da)ea' \\
d(aea') &= a(de)a' &+(da)ea' -ae(da') \\
\nabla_r(aea') &= a\nabla_r(e)a' & +ae(da')
\end{aligned}$$

Hence, if ∇_r is a right connection then $d + \nabla_r$ is a left connection and if ∇_l is a left connection then $\nabla_l - d$ is a right connection. Therefore, one sided connections exist on $\Omega^1 A$ if and only if connections exist and hence if and only if $\Omega^1 A$ is a projective bimodule.

But then we have an A-bimodule splitting of the exact sequence

$$0 \longrightarrow \Omega^2 A \xrightarrow{j} \Omega^1 A \otimes A \xrightarrow{m} \Omega^1 A \longrightarrow 0$$

where $j(\omega da) = \omega a \otimes 1 - \omega \otimes a$ and $m(\omega \otimes a) = \omega a$.

PROPOSITION 7.4
A connection exists on $\Omega^1 A$ if and only if A is Quillen-smooth.

PROOF   A bimodule splitting of the above map is determined by a retraction bimodule map p for j. As $\Omega^1 \, A \otimes A \simeq A \otimes \overline{A} \otimes A$, a bimodule map p

$$\Omega^i \, A \otimes A \xrightarrow{\;p\;} \Omega^2 \, A$$

is equivalent to a map $\overline{A} \xrightarrow{\;\phi\;} \Omega^2 \, A$ via $p(a_0 da_1 \otimes a_2) = a_0 \phi(a_1 a_2)$. But then we have

$$\begin{aligned}
pj(da_1 da_2) &= p((da_1)a_2 \otimes 1 - da_1 \otimes da_2)\\
&= p(d(a_1 a_2) \otimes 1 - a_1(da_2) \otimes 1 - da_1 \otimes a_2)\\
&= \phi(a_1 a_2) - a_1\phi(a_2) - \phi(a_1)a_2
\end{aligned}$$

and splitting of the map means $pj = id$, that is, that ϕ satisfies

$$\phi(aa') = a\phi(a') + \phi(a)a' + dada'$$

which is equivalent to an algebra lift

$$A \xrightarrow{\;\phi^*\;} \frac{\mathbb{L}_A}{\mathbb{I}_A} = A \oplus \Omega^2 \, A$$

Now, assume we have an algebra morphism

$$A \xrightarrow{\;f\;} \frac{R}{I} \qquad \text{with} \qquad I^2 = 0$$

and lift f to a based linear map $A \xrightarrow{\;\rho\;} R$. By the universal property of $\mathbb{L}_A$ we have an algebra lift

$$\mathbb{L}_A \xrightarrow{\;\rho^*\;} R$$

living over f. Therefore $\rho^*(\mathbb{I}_A) \subset I$ and therefore ρ^* is zero on $\mathbb{I}_A^2$ giving an algebra morphism

$$\frac{\mathbb{L}_A}{\mathbb{I}_A^2} \xrightarrow{\;f^*\;} R$$

living over f. But then the existence of an algebra map ϕ^* as above gives a desired lifting $f^* \circ \phi^*$ of f, finishing the proof.   ☐

For a map $\overline{A} \xrightarrow{\;\phi\;} \Omega^2 \, A$ as above, a connection is given by the formulae

$$\nabla_r(ada') = a\phi(a') \quad \text{and} \quad \nabla_r(ada') = a\phi(a') + dada'$$

Example 7.15 Connection on $\mathbb{C}\langle x, y\rangle$
Clearly we have $\Omega^1 \, \mathbb{C}\langle x,y\rangle = \mathbb{C}\langle x,y\rangle \otimes \mathbb{C}x + \mathbb{C}y \otimes \mathbb{C}\langle x,y\rangle$, which is the free bimodule generated by dx and dy. There is a canonical connection with

$$\begin{cases} \phi(x) = 0 \quad \text{and} \quad \nabla_l(dx) = \nabla_r(dx) = 0\\ \phi(y) = 0 \quad \text{and} \quad \nabla_l(dy) = \nabla_r(dy) = 0 \end{cases}$$

The image of ϕ on any word $z_1 \ldots z_n$ with $z_i = x$ or y is given by the formula

$$\phi(z_1 \ldots z_n) = \nabla_r d(z_1 \ldots z_n)$$

$$= \nabla_r \left(\sum_{i=1}^{n} z_1 \ldots z_{i-1}(dz_i) z_{i+1} \ldots z_n \right)$$

$$= \sum_{i=1}^{n-1} z_1 \ldots z_{i-1}(dz_i) d(z_{i+1} \ldots z_n)$$

□

Example 7.16 Connection on $\mathbb{C} \times \mathbb{C}$

We have calculated above that the lifting map ϕ is determined by

$$\phi(e) = (1 - 2E)(E - E^2) = (f - e)de^2$$

Therefore the corresponding left and right connections are given by

$$\begin{cases} \nabla_r((\alpha e + \beta f)de) = & (\beta f - \alpha e)de^2 \\ \nabla_l((\alpha e + \beta f)de) = & (\alpha f - \beta e)de^2 \end{cases}$$

□

7.6 deRham cohomology

In this section we will compute various sorts of *noncommutative deRham cohomology* . We have for an arbitrary $\mathbb{C}$-algebra A the complex of *noncommutative* differential forms

$$A = \Omega^0 \, A \xrightarrow{d} \Omega^1 \, A \xrightarrow{d} \ldots \xrightarrow{d} \Omega^n \, A \xrightarrow{d} \Omega^{n+1} \, A \xrightarrow{d} \ldots$$

A first attempt to define noncommutative de Rham cohomology is to take the homology groups of this complex, we call these the *big* noncommutative de Rham cohomology

$$H_{big}^n \, A = \frac{Ker \, \Omega^n \, A \xrightarrow{d} \Omega^{n+1} \, A}{Im \, \Omega^{n-1} \, A \xrightarrow{d} \Omega^n \, A}$$

Example 7.17 Big de Rham cohomology of $\mathbb{C} \times \mathbb{C}$

We have seen before that $\Omega^n \, \mathbb{C} \times \mathbb{C} = (\mathbb{C} \times \mathbb{C})de^n$ and that the differential is given by

$$\begin{array}{ccc} \Omega^n \, \mathbb{C} \times \mathbb{C} & \xrightarrow{d} & \Omega^{n+1} \, \mathbb{C} \times \mathbb{C} \\ (\alpha e + \beta f)de^n & \mapsto & (\alpha - \beta)de^{n+1} \end{array}$$

From which it is immediately clear that

$$\begin{cases} H^0_{big} \; \mathbb{C} \times \mathbb{C} = & \mathbb{C} \\ H^n_{big} \; \mathbb{C} \times \mathbb{C} = & 0 \end{cases}$$

for all $n \geq 1$. This is not quite the answer $H^0 \, \mathbb{C} \times \mathbb{C} = \mathbb{C} \oplus \mathbb{C}$ we would expect from the commutative case.   ❏

For a general $\mathbb{C}$-algebra A it is usually very difficult to compute these co-homology groups. In case of free algebras we can use the graded structure of the complex together with the *Euler derivation* to compute them, a trick we will use later in greater generality.

Example 7.18 Big de Rham cohomology of $\mathbb{C}\langle x, y \rangle$
Define the *Euler derivation* E on $\mathbb{C}\langle x, y \rangle$ by

$$E(x) = x \qquad \text{and} \qquad E(y) = y$$

Observe that if w is a *word* in x and y of degree k, then we have the Eulerian property that

$$E(w) = kw$$

as one easily verifies.

We can define a degree preserving *derivation* L_E on the differentially graded algebra $\Omega \; \mathbb{C}\langle x, y \rangle$ by the rules

$$L_E(a) = E(a) \qquad \text{and} \qquad L_E(da) = dE(a) \qquad \forall a \in \mathbb{C}\langle x, y \rangle$$

Further we introduce the degree -1 *contraction operator* i_E which is the *super-derivation* on $\Omega \; \mathbb{C}\langle x, y \rangle$, that is

$$i_E(\omega \omega') = i_E(\omega)\omega' + (-1)^i \omega i_E(\omega') \qquad \text{for } \omega \in \Omega^i \; \mathbb{C}\langle x, y \rangle$$

defined by the rules

$$i_E(a) = 0 \qquad i_E(da) = E(a) \qquad \forall a \in \mathbb{C}\langle x, y \rangle$$

That is, we have the following situation

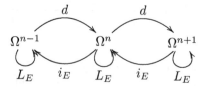

These operators satisfy the equation

$$L_E = i_E \circ d + d \circ i_E$$

as both sides are *derivations* on $\Omega\,\mathbb{C}\langle x,y\rangle$ and coincide on the generators a and da for $a \in \mathbb{C}\langle x,y\rangle$ of this differentially graded algebra.

We claim that L_E is a total degree preserving linear automorphism on

$$\Omega^n\,\mathbb{C}\langle x,y\rangle \qquad \text{for } n \geq 1.$$

For if w_i for $0 \leq i \leq n$ are words in x and y of degree k_i with $k_i \geq 1$ for $i \geq 1$, then we have

$$L_E(w_0 dw_1 \ldots dw_n) = (k_0 + \ldots + k_n)w_0 dw_1 \ldots dw_n$$

Using the words in x and y as a basis for $\overline{A}$ we see that the kernel and image of the differential d must be homogeneous. But then, if ω is a multihomogeneous element in $\Omega^n\,\mathbb{C}\langle x,y\rangle$ and in $Ker\ d$ we have for some integer $k \neq 0$ that

$$k\omega = L_E(\omega) = (i_E \circ d + d \circ i_E)\omega = d(i_E\ \omega)$$

and hence ω lies in $Im\ d$. Therefore, we have proved

$$\begin{cases} H^0_{big}\,\mathbb{C}\langle x,y\rangle = \mathbb{C} \\ H^n_{big}\,\mathbb{C}\langle x,y\rangle = 0 \end{cases}$$

for all $n \geq 1$. $\qquad\qquad\qquad\qquad\qquad\qquad\qquad\qquad\qquad\qquad\qquad$ □

The examples show that the differentially graded algebra $\Omega\,A$ is *formal* for $A = \mathbb{C} \times \mathbb{C}$ or $\mathbb{C}\langle x,y\rangle$. Recall that for an arbitrary A_∞-algebra Ω (in particular for $\Omega \in \mathtt{dgalg}$), the homology algebra $H^*\,\Omega$ has a *canonical A_∞-structure*. That is, we have $m_1 = 0$, m_2 is induced by the "multiplication" m_2 on Ω and there is a quasi-isomorphism of A_∞-algebras $H^*\,\Omega \longrightarrow \Omega$ lifting the identity of $H^*\,\Omega$.

The A_∞-algebra Ω is said to be *formal* if the canonical structure makes $H^*\,\Omega$ into an ordinary associative graded algebra (that is, such that all $m_n = 0$ for $n \geq 3$). In particular, if $\Omega = \Omega\,A$ and if the big deRham cohomology is concentrated in degree zero, then the degree properties of m_n imply that $m_n = 0$ for $n \geq 3$ and hence that $\Omega\,A$ is formal.

Let A be an arbitrary $\mathbb{C}$-algebra and $\theta \in Der_\mathbb{C}\,A$, the Lie algebra of $\mathbb{C}$-algebra *derivations* of A, then we define a degree preserving derivation L_θ and a degree -1 super-derivation i_θ on $\Omega\,A$

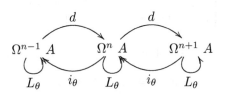

defined by the rules

$$\begin{cases} L_\theta(a) = \theta(a) & L_\theta(da) = d\,\theta(a) \\ i_\theta(a) = 0 & i_\theta(da) = \theta(a) \end{cases}$$

for all $a \in A$. In this generality we again have the fundamental identity

$$L_\theta = i_\theta \circ d + d \circ i_\theta$$

as both sides are degree preserving derivations on $\Omega\,A$ and they agree on all the generators a and da for $a \in A$.

LEMMA 7.2
Let $\theta, \gamma \in Der_{\mathbb{C}}\, A$, then we have on $\Omega\,A$ the following identities of operators

$$\begin{cases} L_\theta \circ i_\gamma - i_\gamma \circ L_\theta = [L_\theta, i_\gamma] & = i_{[\theta,\gamma]} = i_{\theta \circ \gamma - \gamma \circ \theta} \\ L_\theta \circ L_\gamma - L_\gamma \circ L_\theta = [L_\theta, L_\gamma] & = L_{[\theta,\gamma]} = L_{\theta \circ \gamma - \gamma \circ \theta} \end{cases}$$

PROOF   Consider the first identity. By definition both sides are degree -1 superderivations on $\Omega\,A$ so it suffices to check that they agree on generators. Clearly, both sides give 0 when evaluated on $a \in A$ and for da we have

$$(L_\theta \circ i_\gamma - i_\gamma \circ L_\theta)da = L_\theta\,\gamma(a) - i_\gamma\,d\,\theta(a) = \theta\,\gamma(a) - \gamma\,\theta(a) = i_{[\theta,\gamma]}(da)$$

A similar argument proves the second identity.   □

Let Q be a quiver on k vertices $\{v_1, \ldots, v_k\}$, then we can define an *Euler derivation* E on $\mathbb{C}Q$ by the rules that

$$E(v_i) = 0 \ \forall 1 \leq i \leq k \qquad \text{and} \qquad E(a) = a \ \forall a \in Q_a$$

By induction on the length $l(p)$ of an oriented path p in the quiver Q one easily verifies that $E(p) = l(p)p$. By the lemma above we have all the necessary ingredients to redo the argument in example 7.18.

THEOREM 7.10
For a quiver Q on k vertices, the noncommutative differential forms $\Omega\,\mathbb{C}Q$ is formal. In fact, we have for the big deRham cohomology

$$\begin{cases} H^0_{big}\,\mathbb{C}Q & \simeq \mathbb{C} \times \ldots \times \mathbb{C} \ (k \text{ factors}) \\ H^n_{big}\,\mathbb{C}Q & \simeq 0 \qquad \forall n \geq 1 \end{cases}$$

For $\omega \in \Omega^i\,A$ and $\omega' \in \Omega^j\,A$ we define the *supercommutator* to be

$$[\omega, \omega'] = \omega\omega' - (-1)^{ij}\omega'\omega$$

That is, it is the usual commutator unless both i and j are odd in which case it is the sum $\omega\omega' + \omega'\omega$.

As the differential d is a super-derivation on $\Omega\, A$ we have that

$$d([\omega, \omega']) = [d\omega, \omega'] + (-1)^i [\omega, d\omega']$$

and therefore the differential maps the *subspaces* of supercommutators to subspaces of supercommutators. Therefore, if we define

$$\text{DR}^n\, A = \frac{\Omega^n\, A}{\sum_{i=0}^{n}[\Omega^i\, A, \Omega^{n-i}\, A]}$$

Then the `dgalg`-structure on $\Omega\, A$ induces one on the complex

$$\text{DR}^0\, A \xrightarrow{\;d\;} \text{DR}^1\, A \xrightarrow{\;d\;} \text{DR}^2\, A \xrightarrow{\;d\;} \cdots$$

which is called the *Karoubi complex* of A.

We define the *noncommutative de Rham cohomology groups* of A to be the homology of the Karoubi complex, that is

$$H^n_{dR}\, A = \frac{Ker\ \text{DR}^n\, A \xrightarrow{\;d\;} \text{DR}^{n+1}\, A}{Im\ \text{DR}^{n-1}\, A \xrightarrow{\;d\;} \text{DR}^n\, A}$$

Example 7.19 Noncommutative de Rham cohomology of $\mathbb{C} \times \mathbb{C}$
Recall that the product on $\Omega\, \mathbb{C} \times \mathbb{C}$ is given by the formula

$$(\alpha e + \beta f)de^n(\gamma e + \delta f)de^m = \begin{cases} (\alpha\gamma e + \beta\delta f)de^{n+m} & \text{when } n \text{ is even} \\ (\alpha\delta e + \beta\gamma f)de^{n+m} & \text{when } n \text{ is odd} \end{cases}$$

If m is odd, then we deduce from this that the commutator

$$[\alpha e + \beta f, (\gamma e + \delta f)de^m] = (\alpha - \beta)(\gamma e - \delta f)de^m$$

and hence we can write any element of $\Omega^m\, \mathbb{C} \times \mathbb{C} = (\mathbb{C} \times \mathbb{C})de^m$ as a (super) commutator, whence

$$\text{DR}^m\, \mathbb{C} \times \mathbb{C} = 0 \qquad \text{when } m \text{ is odd.}$$

On the other hand, if m is even then any commutator with k even

$$[(\alpha e + \beta f)de^k, (\gamma e + \delta f)de^{m-k}] = 0$$

whereas if k is odd we have

$$[(\alpha e + \beta f)de^k, (\gamma e + \delta f)de^{m-k}] = (\alpha\delta + \beta\gamma)de^m$$

As a consequence the space of super-commutators in $\Omega^m \, \mathbb{C} \times \mathbb{C}$ is one dimensional and therefore

$$\mathrm{DR}^m \, \mathbb{C} \times \mathbb{C} = \mathbb{C} \qquad \text{when } m \text{ is even and} > 0$$

Thus, the Karoubi complex of $\mathbb{C} \times \mathbb{C}$ has the following form

$$\mathbb{C} \times \mathbb{C} \xrightarrow{d} 0 \xrightarrow{d} \mathbb{C} \xrightarrow{d} 0 \xrightarrow{d} \mathbb{C} \xrightarrow{d} 0 \xrightarrow{d} \dots$$

and therefore we have for the noncommutative de Rham cohomology groups

$$\mathrm{H}_{dR}^n \, \mathbb{C} \times \mathbb{C} = \begin{cases} \mathbb{C} \times \mathbb{C} & \text{when } n = 0 \\ 0 & \text{when } n \text{ is odd} \\ \mathbb{C} & \text{when } n \text{ is even and} > 0 \end{cases}$$

$\square$

Example 7.20 Noncommutative de Rham cohomology of $\mathbb{C}\langle x, y \rangle$

Consider again the Eulerian derivation E on $\mathbb{C}\langle x, y \rangle$ and the operators L_E and i_E on $\Omega \, \mathbb{C}\langle x, y \rangle$. Repeating the above argument that d is compatible with the subspaces of supercommutators for i_E and L_E we see that we have induced operations

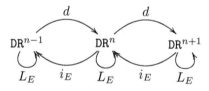

We have again that L_E is an isomorphism on $\mathrm{DR}^n \, \mathbb{C}\langle x, y \rangle$ whenever $n \geq 1$ and again we deduce from the equality $L_E = i_E \circ d + d \circ i_E$ that

$$\mathrm{H}_{dR}^n \, \mathbb{C}\langle x, y \rangle = \begin{cases} \mathbb{C} & \text{when } n = 0 \\ 0 & \text{when } n \geq 1 \end{cases}$$

$\square$

THEOREM 7.11

Let Q be a quiver on k vertices, then the Karoubi complex of $\mathbb{C}Q$ is acyclic. In particular

$$\begin{cases} \mathrm{H}_{dR}^0 \, \mathbb{C}Q & \simeq \mathbb{C} \times \dots \times \mathbb{C} \, (k \text{ factors}) \\ \mathrm{H}_{dR}^n \, \mathbb{C}Q & \simeq 0 \qquad \forall n \geq 1 \end{cases}$$

So far we have considered differential forms with respect to the basefield $\mathbb{C}$. Sometimes it is useful to consider only the *relative differential forms* on A with respect to a subalgebra B. These can be defined as follows.

Let $\overline{A}_B$ be the cokernel of the inclusion $B \hookrightarrow A$ in the category $B-\texttt{bimod}$ of *bimodules* over B. We define the space of relative differential forms of degree n with respect to B to be

$$\Omega_B^n \, A = A \otimes_B \underbrace{\overline{A}_B \otimes_B \ldots \otimes_B \overline{A}_B}_{n}$$

By definition $\Omega_B^n \, A$ is the quotient space of $\Omega^n \, A$ by the relations

$$(a_0, \ldots, a_{i-1}b, a_i, \ldots, a_n) = (a_0, \ldots, a_{i-1}, ba_i, \ldots, a_n)$$
$$(a_0, \ldots, a_{i-1}, s, a_{i+1}, \ldots, a_n) = 0$$

for all $b \in B$ and $1 \le i \le n$. One verifies that the multiplication and differential defined on $\Omega \, A$ are compatible with these relations, making $\Omega_B \, A$ an object in $\texttt{dgalg}$. Moreover, there is a canonical epimorphism

$$\Omega \, A \longrightarrow\!\!\!\rightarrow \Omega_B \, A \qquad \text{in } \texttt{dgalg}$$

We will now determine the kernel. First we give the universal property for $\Omega_B \, A$. Given $\Gamma = \oplus \Gamma^n$ in $\texttt{dgalg}$ and an algebra map $A \xrightarrow{f} \Gamma^0$ such that $d(f \, B) = 0$, then there is a unique morphism in $\texttt{dgalg}$ making the diagram commute

$$
\begin{array}{ccc}
\Omega_B \, A & \xrightarrow{\quad \exists f_* \quad} & \Gamma \\
\uparrow & & \uparrow \\
A & \xrightarrow{\quad f \quad} & \Gamma^0
\end{array}
$$

Indeed, by the universal property of $\Omega \, A$ there is a unique morphism $\Omega \, A \xrightarrow{f_*} \Gamma$ in $\texttt{dgalg}$ extending f given by

$$f_*(a_0 da_1 \ldots da_n) = f(a_0)d(f(a_1)) \ldots d(f(a_n)).$$

If $d(f \, B) = 0$ then one verifies that f_* is compatible with the relations defining $\Omega_B \, A$, proving the universal property.

PROPOSITION 7.5
For a subalgebra B of A we have an isomorphism in $\texttt{dgalg}$

$$\Omega_B \, A = \frac{\Omega \, A}{\Omega \, A \, d(B) \, \Omega \, A}$$

PROOF   The ideal generated by $d(B)$ is closed under d and therefore the quotient is an object in **dgalg** with the same universal property as $\Omega_B\ A$.   ▯

An important special case is when $B = \mathbb{C} \times \ldots \times \mathbb{C}$ is the subalgebra of $\mathbb{C}Q$ generated by the vertexidempotents. In this case we will denote

$$\Omega_{rel}\ \mathbb{C}Q = \Omega_B\ \mathbb{C}Q$$

and call it the relative differential forms on Q.

LEMMA 7.3
Let Q be a quiver on k vertices, then a basis for $\Omega^n_{rel}\ \mathbb{C}Q$ is given by the elements

$$p_0 dp_1 \ldots dp_n$$

where p_i is an oriented path in the quiver such that length $p_0 \geq 0$ and length $p_i \geq 1$ for $1 \leq i \leq n$ and such that the starting point of p_i is the endpoint of p_{i+1} for all $1 \leq i \leq n-1$.

PROOF   Clearly $l(p_i) \geq 1$ when $i \geq 1$ or p_i would be a vertexidempotent whence in B. Let v be the starting point of p_i and w the end point of p_{i+1} and assume that $v \neq w$, then

$$p_i \otimes_B p_{i+1} = p_i v \otimes_B w p_{i+1} = p_i vw \otimes_B p_{i+1} = 0$$

from which the assertion follows.   ▯

We define the *big relative de Rham* cohomology groups of A with respect to B to be the cohomology of the complex

$$\Omega^0_B\ A \xrightarrow{\ d\ } \Omega^1_B\ A \xrightarrow{\ d\ } \Omega^2_B\ A \xrightarrow{\ d\ } \ldots$$

that is,

$$H^n_B\ A = \frac{Ker\ \Omega^n\ A \xrightarrow{\ d\ } \Omega^{n+1}\ A}{Im\ \Omega^{n-1}\ A \xrightarrow{\ d\ } \Omega^n\ A}$$

In the case of path algebras of quivers, we can use the grading by length of paths and the Eulerian derivation to compute these relative de Rham groups.

Example 7.21 Big relative de Rham cohomology
Let $\mathbb{M}$ (resp. $\mathbb{C} \times \mathbb{C}$) be the path algebras of the quivers

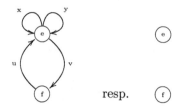

resp.

The Eulerian derivation E on $\mathbb{M}$ is defined by

$$E(e) = E(f) = 0 \quad E(x) = x \quad E(y) = y \quad E(u) = u \quad \text{and} \quad E(v) = v.$$

Observe that E respects all relations holding in $\mathbb{M}$ and so is indeed a $\mathbb{C} \times \mathbb{C}$-derivation of $\mathbb{M}$.

As before we define a degree preserving derivation L_E and a degree -1 superderivation i_E on $\Omega_{rel} \, \mathbb{M} = \Omega_{\mathbb{C} \times \mathbb{C}} \, \mathbb{M}$ by the rules

$$\begin{cases} L_E(a) = E(a) & L_E(da) = dE(a) \\ i_E(a) = 0 & i_E(da) = E(a) \end{cases}$$

for all $a \in \mathbb{M}$. We have the equality

$$L_E = i_E \circ d + d \circ i_E$$

and arguing as before we obtain that

$$H^n_{rel} \, \mathbb{M} = \begin{cases} \mathbb{C} \times \mathbb{C} & \text{when } n = 0, \\ 0 & \text{when } n \geq 1 \end{cases}$$

$\square$

THEOREM 7.12
Let Q be a quiver on k vertices, then the relative differential forms $\Omega_{rel} \, \mathbb{C}Q$ is a formal differentially graded algebra. In fact

$$\begin{cases} H^0_{rel} \, \mathbb{C}Q \simeq \mathbb{C} \times \ldots \times \mathbb{C} \ (k \ factors) \\ H^n_{rel} \, \mathbb{C}Q \simeq 0 \qquad \forall n \geq 1 \end{cases}$$

We can repeat the construction of the Karoubi complex verbatim for relative differential operators and define a *relative Karoubi complex*

$$\mathrm{DR}^0_B \, A \xrightarrow{d} \mathrm{DR}^1_B \, A \xrightarrow{d} \mathrm{DR}^2_B \, A \xrightarrow{d} \ldots$$

where

$$\mathrm{DR}^n_B \, A = \frac{\Omega^n_B \, A}{\sum_{i=0}^{n} [\, \Omega^i_B \, A, \Omega^{n-i}_B \, A \,]}$$

Clearly, we then define the *noncommutative relative de Rham cohomology groups* of A with respect to B to be the homology of this complex

$$H^n_{B,dR} \, A = \frac{Ker \ \mathrm{DR}^n_B \, A \xrightarrow{d} \mathrm{DR}^{n+1}_B \, A}{Im \ \mathrm{DR}^{n-1}_B \, A \xrightarrow{d} \mathrm{DR}^n_B \, A}$$

Let $\theta \in Der_B$ A, that is θ is a $\mathbb{C}$-derivation on A such that $\theta(b) = 0$ for every $b \in B$. Then, as

$$L_\theta(db) = d\,\theta(b) = 0 \qquad \text{and} \qquad i_\theta(db) = \theta(b) = 0$$

we see that the operators L_θ and i_θ can be defined on the relative forms

$$\Omega_B\ A = \frac{\Omega\ A}{\Omega\ A\ dB\ \Omega\ A}$$

and also on the relative Karoubi complex. Again, these induced operators satisfy the identities of lemma 7.2. In the special case of the Eulerian derivation E on the path algebra $\mathbb{C}Q$ we see that $E \in Der_B\ \mathbb{C}Q$ and hence we have the following result.

THEOREM 7.13

Let Q be a quiver on k vertices. Then, the relative Karoubi complex is acyclic. That is,

$$\begin{cases} H^0_{rel,dR}\ \mathbb{C}Q\ \simeq \mathbb{C} \times \ldots \times \mathbb{C}\ (k\ \text{factors}) \\ H^n_{rel,dR}\ \mathbb{C}Q\ \simeq 0 \qquad \forall n \geq 1 \end{cases}$$

7.7   Symplectic structure

Let Q be a quiver on k vertices $\{v_1, \ldots, v_k\}$. We will determine the first terms in the relative Karoubi complex. Define

$$\mathrm{dR}^n_{rel}\ \mathbb{C}Q = \frac{\Omega^n_{rel}\ \mathbb{C}Q}{\sum_{i=0}^n [\,\Omega^i_{rel}\ \mathbb{C}Q, \Omega^{n-i}\ \mathbb{C}Q\,]}$$

In the commutative case, dR^0 are the functions on the manifold and dR^1 the 1-forms. We will characterize the *noncommutative functions* and *noncommutative 1-forms* in the case of quivers.

Recall that a *necklace word* w in the quiver Q is an equivalence class of an oriented cycle $c = a_1 \ldots a_l$ of length $l \geq 0$ in Q, where $c \sim c'$ if c' is obtained from c by cyclically permuting the composing arrows a_i.

LEMMA 7.4

A $\mathbb{C}$-basis for the noncommutative functions

$$\mathrm{dR}^0_{rel}\ \mathbb{C}Q \simeq \frac{\mathbb{C}Q}{[\,\mathbb{C}Q, \mathbb{C}Q\,]}$$

are the necklace words in the quiver Q.

PROOF Let $\mathbb{W}$ be the $\mathbb{C}$-space spanned by all necklace words w in Q and define a linear map

$$\mathbb{C}Q \xrightarrow{\ n\ } \mathbb{W} \qquad \begin{cases} p \mapsto w_p & \text{if } p \text{ is a cycle} \\ p \mapsto 0 & \text{if } p \text{ is not} \end{cases}$$

for all oriented paths p in the quiver Q, where w_p is the necklace word in Q determined by the oriented cycle p. Because $w_{p_1 p_2} = w_{p_2 p_1}$ it follows that the commutator subspace $[\mathbb{C}Q, \mathbb{C}Q]$ belongs to the kernel of this map. Conversely, let

$$x = x_0 + x_1 + \ldots + x_m$$

be in the kernel where x_0 is a linear combination of noncyclic paths and x_i for $1 \leq i \leq m$ is a linear combination of cyclic paths mapping to the same necklace word w_i, then $n(x_i) = 0$ for all $i \geq 0$. Clearly, $x_0 \in [\mathbb{C}Q, \mathbb{C}Q]$ as we can write every noncyclic path $p = a.p' = a.p' - p'.a$ as a commutator. If $x_i = a_1 p_1 + a_2 p_2 + \ldots + a_l p_l$ with $n(p_i) = w_i$, then $p_1 = q.q'$ and $p_2 = q'.q$ for some paths q, q' whence $p_1 - p_2$ is a commutator. But then, $x_i = a_1(p_1 - p_2) + (a_2 - a_1)p_2 + \ldots + a_l p_l$ is a sum of a commutator and a linear combination of strictly fewer elements. By induction, this shows that $x_i \in [\mathbb{C}Q, \mathbb{C}Q]$. $\qquad \square$

If we fix a dimension vector α, then taking traces defines a map

$$dR^0 \, \mathbb{C}Q \xrightarrow{\ tr\ } \mathbb{C}[\mathbf{rep}_\alpha \, Q]$$

whence noncommutative functions determine $GL(\alpha)$-invariant commutative functions on the representation space $\mathbf{rep}_\alpha \, Q$ and hence commutative functions on the quotient varieties $\mathbf{iss}_\alpha \, Q$. In fact, we have seen that the image $tr(dR^0 \, \mathbb{C}Q)$ generates the ring of polynomial invariants $\mathbb{C}[\mathbf{rep}_\alpha \, Q]^{GL(\alpha)} = \mathbb{C}[\mathbf{iss}_\alpha \, Q]$.

LEMMA 7.5
$dR^1_{rel} \, \mathbb{C}Q$ *is isomorphic as $\mathbb{C}$-space to*

PROOF If $p.q$ is not a cycle, then $pdq = [p, dq]$ and so vanishes in $dR^1_{rel} \, \mathbb{C}Q$ so we only have to consider terms pdq with $p.q$ an oriented cycle in Q. For any three paths p, q and r in Q we have the equality

$$[p.qdr] = pqdr - qd(rp) + qrdp$$

whence in dR^1_{rel} $\mathbb{C}Q$ we have relations allowing reduction of the length of the differential part

$$qd(rp) = pqdr + qrdp$$

so dR^1_{rel} $\mathbb{C}Q$ is spanned by terms of the form pda with $a \in Q_a$ and $p.a$ an oriented cycle in Q. Therefore, we have a surjection

$$\Omega^1_{rel} \ \mathbb{C}Q \longrightarrow \bigoplus_{\substack{j \xleftarrow{\ a\ } i}} v_i.\mathbb{C}Q.v_j \ da$$

By construction, it is clear that $[\Omega^0_{rel} \ \mathbb{C}Q, \Omega^1_{rel} \ \mathbb{C}Q]$ lies in the kernel of this map and using an argument as in the lemma above one shows also the converse inclusion. $\qquad\qquad\qquad\qquad\qquad\qquad\qquad\qquad\qquad\qquad\qquad\quad$ ◻

Example 7.22 dR^i_{rel} $\mathbb{M}$

Take the path algebra $\mathbb{M}$ of the quiver of example 7.21. Noncommutative *functions* on $\mathbb{M}$ are the 0-forms, which is by definition the quotient space

$$dR^0_{rel} \ \mathbb{M} = \frac{\mathbb{M}}{[\ \mathbb{M}, \mathbb{M}\ ]}$$

If p is an oriented path of length ≥ 1 in the quiver with different begin- and endpoint, then we can write p as a concatenation $p = p_1 p_2$ with p_i an oriented path of length ≥ 0 such that $p_2 p_1 = 0$ in $\mathbb{M}$. As $[p_1, p_2] = p_1 p_2 - p_2 p_1 = 0$ in dR^0_{rel} $\mathbb{M}$ we deduce that the space of noncommutative functions on $\mathbb{M}$ has as $\mathbb{C}$-basis the *necklace words w*

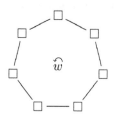

where each bead is one of the elements

$$\boxed{\bullet} = x \qquad \boxed{\circ} = y \quad \text{and} \quad \boxed{\blacktriangledown} = uv$$

together with the necklace words of length zero e and f. Each necklace word w corresponds to the equivalence class of the words in $\mathbb{M}$ obtained from multiplying the beads in the indicated orientation and and two words in $\{x, y, u, v\}$ in $\mathbb{M}$ are said to be equivalent if they are identical up to cyclic permutation of the terms.

Substituting each bead with the $n \times n$ matrices specified before and taking traces we get a map

$$dR^0_{rel} \; \mathbb{M} = \frac{\mathbb{M}}{[\,\mathbb{M},\mathbb{M}\,]} \xrightarrow{\;tr\;} \mathbb{C}[\mathbf{rep}_\alpha \; \mathbb{M}]$$

Hence, noncommutative functions on $\mathbb{M}$ induce ordinary functions on *all* the representation spaces $\mathbf{rep}_\alpha \; \mathbb{M}$ and these functions are $GL(\alpha)$-invariant. Moreover, the image of this map generates the ring of polynomial invariants as we mentioned before.

Next, we consider *noncommutative 1-forms* on $\mathbb{M}$ that are by definition elements of the space

$$dR^1_{rel} \; \mathbb{M} = \frac{\Omega^1_{rel} \; \mathbb{M}}{[\,\mathbb{M},\Omega^1_{rel} \; \mathbb{M}\,]}$$

Recall that $\Omega^1_{rel} \; \mathbb{M}$ is spanned by the expressions $p_0 dp_1$ with p_0 resp. p_1 oriented paths in the quiver of length ≥ 0 resp. ≥ 1 and such that the starting point of p_0 is the endpoint of p_1. To form $dR^1_{rel} \; \mathbb{M}$ we have to divide out expressions such as

$$[\,p, p_0 dp_1\,] = pp_0 dp_1 + p_0 p_1 dp - p_0 d(p_1 p)$$

That is, if we have connecting oriented paths p_2 and p_1 both of length ≥ 1 we have in $dR^1_{rel} \; \mathbb{M}$

$$p_0 d(p_1 p_2) = p_2 p_0 dp_1 + p_0 p_1 dp_2$$

and by iterating this procedure whenever the differential term is a path of length ≥ 2 we can represent each class in $dR^1_{rel} \; \mathbb{M}$ as a combination from

$$\mathbb{M} e \; dx + \mathbb{M} e \; dy + \mathbb{M} e \; du + \mathbb{M} f \; dv$$

Now, $\mathbb{M} e = e \mathbb{M} e + f \mathbb{M} e$ and let $p \in f \mathbb{M} e$. Then, we have in $dR^1_{rel} \; \mathbb{M}$

$$d(xp) = p \; dx + x \; dp$$

but by our description of $\Omega^1 \; \mathbb{M}$ the left hand term is zero as is the second term on the right, whence $p \; dx = 0$. A similar argument holds replacing x by y. As for u, let $p \in e \mathbb{M} e$, then we have in $dR^1_{rel} \; \mathbb{M}$

$$d(up) = p \; du + u \; dp$$

and again the left-hand and the second term on the right are zero whence $p \; du = 0$. An analogous result holds for v and $p \in f \mathbb{M} f$. Therefore, we have the description of noncommutative 1-forms on $\mathbb{M}$

$$dR^1_{rel} \; \mathbb{M} = e \mathbb{M} e \; dx + e \mathbb{M} e \; dy + f \mathbb{M} e \; du + e \mathbb{M} f \; dv$$

That is, in graphical terms

$$
dR^1_{rel}\ \mathbb{M}\qquad=\quad \text{(graph)}\ d\ \text{(graph)}\ +\ \text{(graph)}\ d\ \text{(graph)}\ +
$$

$$
\text{(graph)}\ d\ \text{(graph)}\ +\ \text{(graph)}\ d\ \text{(graph)}
$$

Using the above descriptions of $dR^i_{rel}\ \mathbb{C}Q$ for $i = 0, 1$ and the differential $dR^0_{rel}\ \mathbb{C}Q\ \xrightarrow{\ d\ }\ dR^1_{rel}\ \mathbb{C}Q$ we can define *partial differential operators* associated to any arrow $\text{(j)} \xleftarrow{\ a\ } \text{(i)}$ in Q.

$$
\frac{\partial}{\partial a}\ :\ dR^0_{rel}\ \mathbb{C}Q\ \longrightarrow\ v_i \mathbb{C}Q v_j\qquad \text{by}\qquad df = \sum_{a \in Q_a} \frac{\partial f}{\partial a} da
$$

To take the partial derivative of a necklace word w with respect to an arrow a, we run through w and each time we encounter a we open the necklace by removing that occurrence of a and then take the sum of all the paths obtained.

Example 7.23

For the path algebra $\mathbb{M}$ we have the partial differential operators

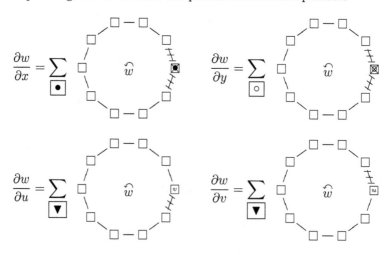

Recall that a *symplectic structure* on a (commutative) manifold M is given by a closed differential 2-form. The nondegenerate 2-form ω gives a canonical

isomorphism

$$T\,M \simeq T^*\,M$$

that is, between vector fields on M and differential 1-forms. Further, there is a unique $\mathbb{C}$-linear map from functions f on M to vector fields ξ_f by the requirement that $-df = i_{\xi_f}\omega$ where i_ξ is the contraction of n-forms to $n-1$-forms using the vector field ξ. We can make the functions on M into a *Poisson algebra* by defining

$$\{f,g\} = \omega(\xi_f,\xi_g)$$

and one verifies that this bracket satisfies the Jacobi and Leibnitz identities.

The *Lie derivative* L_ξ with respect to ξ is defined by the Cartan homotopy formula

$$L_\xi\,\varphi = i_\xi d\varphi + d i_\xi\varphi$$

for any differential form φ. A vector field ξ is said to be *symplectic* if it preserves the symplectic form, that is, $L_\xi\omega = 0$. In particular, for any function f on M we have that ξ_f is symplectic. Moreover the assignment

$$f \longrightarrow \xi_f$$

defines a Lie algebra morphism from the functions $\mathcal{O}(M)$ on M equipped with the Poisson bracket to the Lie algebra of symplectic vector fields, $Vect_\omega\,M$. Moreover, this map fits into the exact sequence

$$0 \longrightarrow \mathbb{C} \longrightarrow \mathcal{O}(M) \longrightarrow Vect_\omega\,M \longrightarrow H^1_{dR}\,M \longrightarrow 0$$

Recall the definition of the double quiver Q^d of a quiver Q given in section 5.5 by assigning to every arrow $a \in Q_a$ an arrow a^* in Q^d in the opposite direction.

DEFINITION 7.3 *The canonical* noncommutative symplectic structure *on the double quiver Q^d is given by the element*

$$\omega = \sum_{a \in Q_a} da\,da^* \in dR^2_{rel}\,\mathbb{C}Q^d$$

We will use ω to define a correspondence between the noncommutative 1-forms $dR^1_{rel}\,\mathbb{C}Q^d$ and the *noncommutative vector fields*, which are defined to be $B = \mathbb{C}^{Q_v}$-derivations of the path algebra $\mathbb{C}Q^d$. Recall that if $\theta \in Der_B\,\mathbb{C}Q^d$ we define operators L_θ and i_θ on $\Omega\,\mathbb{C}Q^d$ and on $dR\,\mathbb{C}Q^d$ by the rules

$$\begin{cases} L_\theta(a) = \theta(a) & L_\theta(da) = d\theta(a) \\ i_\theta(a) = 0 & i_\theta(da) = \theta(a) \end{cases}$$

and that the following identities are satisfied for all $\theta, \gamma \in Der_B\,\mathbb{C}Q^d$

$$[L_\theta, L_\gamma] = L_{[\theta,\gamma]} \qquad \text{and} \qquad [i_\theta, i_\gamma] = i_{[\theta,\gamma]}$$

These operators allow us to define a linear map

$$Der_B \ \mathbb{C}Q \xrightarrow{\ \tau\ } dR^1_{rel} \ \mathbb{C}Q \qquad \text{by} \qquad \tau(\theta) = i_\theta(\omega)$$

We claim that this is an isomorphism. Indeed, every B-derivation θ on $\mathbb{C}Q^d$ is fully determined by its image on the arrows in Q^d that satisfy if $a = \underset{j}{\bigcirc} \xleftarrow{\ a\ } \underset{i}{\bigcirc}$

$$\theta(a) = \theta(v_j a v_i) = v_j \theta(a) v_i \in v_j \mathbb{C}Q^d v_i$$

so determines an element $\theta(a)da^* \in dR^1_{rel} \ \mathbb{C}Q^d$. Further, we compute

$$
\begin{aligned}
i_\theta(\omega) &= \sum_{a \in Q_a} i_\theta(da)da^* - i_\theta(da^*)da \\
&= \sum_{a \in Q_a} \theta(a)da^* - \theta(a^*)da
\end{aligned}
$$

which lies in $dR^1_{rel} \ \mathbb{C}Q^d$. As both B-derivations and 1-forms are determined by their coefficients, τ is indeed bijective.

Example 7.24

For the path algebra of the double quiver $\mathbb{M}$, the analog of the classical isomorphism $T \ M \simeq T^* \ M$ is the isomorphism

$$Der_{\mathbb{C} \times \mathbb{C}} \ \mathbb{M} \xrightarrow{\ i.\omega\ } dR^1_{rel} \ \mathbb{M}$$

as for any $\mathbb{C} \times \mathbb{C}$-derivation θ we have

$$
\begin{aligned}
i_\theta \ \omega &= i_\theta(dx)dy - dx i_\theta(dy) + i_\theta(du)dv - du i_\theta(dv) \\
&= \theta(x)dy - dx\theta(y) + \theta(u)dv - du\theta(v) \\
&\equiv \theta(x)dy - \theta(y)dx + \theta(u)dv - \theta(v)du
\end{aligned}
$$

and using the relations in $\mathbb{M}$ we can easily prove that any $\mathbb{C} \times \mathbb{C}$ derivation on $\mathbb{M}$ must satisfy

$$\theta(x) \in e\mathbb{M}e \quad \theta(y) \in e\mathbb{M}e \quad \theta(u) \in e\mathbb{M}f \quad \theta(v) \in f\mathbb{M}e$$

so the above expression belongs to $dR^1_{rel} \ \mathbb{M}$. Conversely, any θ defined by its images on the generators x, y, u and v by

$$-\theta(y)dx + \theta(x)dy - \theta(v)du + \theta(u)dv \in dR^1_{rel} \ \mathbb{M}$$

induces a derivation on $\mathbb{M}$. ▯

In analogy with the commutative case we define a derivation $\theta \in Der_B \ \mathbb{C}Q^d$ to be *symplectic* if and only if $L_\theta \omega = 0 \in dR^2_{rel} \ \mathbb{C}Q^d$. We will denote the subspace of symplectic derivations by $Der_\omega\ \mathbb{C}Q$. It follows from the noncommutative analog of the Cartan homotopy equality

$$L_\theta = i_\theta \circ d + d \circ i_\theta$$

and the fact that ω is a closed form, that $\theta \in Der_\omega\ \mathbb{C}Q^d$ implies

$$L_\theta\omega = di_\theta\omega = \tau(\theta) = 0$$

That is, $\tau(\theta)$ is a closed form which by the acyclicity of the Karoubi complex shows that it must be an exact form. That is, we have an isomorphism of exact sequences of $\mathbb{C}$-vector spaces

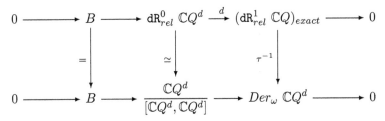

n the next section we will show that this is in fact an exact sequence of Lie algebras.

7.8 Necklace Lie algebras

Let Q be a quiver on k vertices, Q^d its double and $\omega = \sum_{a\in Q_a} dada^*$ the canonical symplectic form on $\mathbb{C}Q^d$. Recall from last section the definition of the partial differential operators $\frac{\partial}{\partial a}$ for an arrow a in Q^d.

DEFINITION 7.4   *The* Kontsevich bracket *on the necklace words in Q^d,* $dR^0_{rel}\ \mathbb{C}Q^d$ *is defined to be*

$$\{w_1, w_2\}_K = \sum_{a\in Q_a} \left(\frac{\partial w_1}{\partial a}\frac{\partial w_2}{\partial a^*} - \frac{\partial w_1}{\partial a^*}\frac{\partial w_2}{\partial a}\right) \bmod [\mathbb{C}Q^d, \mathbb{C}Q^d]$$

That is, to compute $\{w_1, w_2\}_K$ we consider for every arrow $a \in Q_a$ all occurrences of a in w_1 and a^* in w_2. We then open up the necklaces removing these factors and gluing the open ends together to form a new necklace word. We then replace the roles of a^* and a and redo this operation (with a minus sign), see figure 7.4. Finally, we add all the obtained necklace words.

Using this graphical description of the Kontsevich bracket, it is an enjoyable exercise to verify that the bracket turns $dR^0_{rel}\ \mathbb{C}Q^d$ into a Lie algebra. That

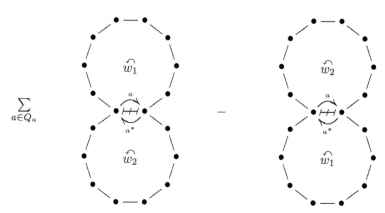

FIGURE 7.4: Kontsevich bracket $\{w_1, w_2\}_K$.

is, for all necklace words w_i, the bracket satisfies the *Jacobi identity*

$$\{\{w_1, w_2\}_K, w_3\}_K + \{\{w_2, w_3\}_K, w_1\}_K + \{\{w_3, w_1\}_K, w_2\}_K = 0$$

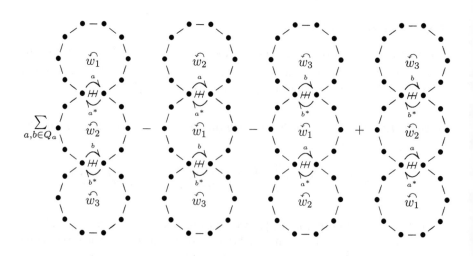

$$\{\{w_1, w_2\}_K, w_3\}_K$$

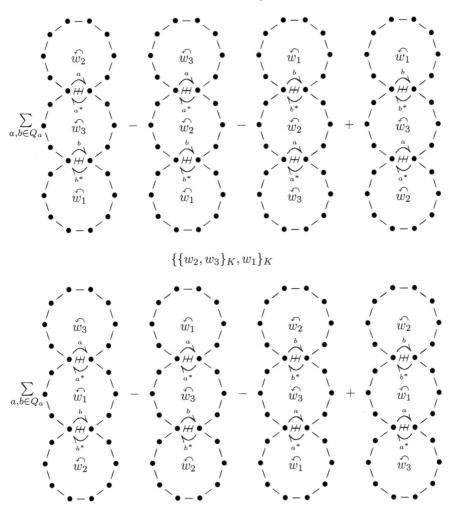

$$\{\{w_2, w_3\}_K, w_1\}_K$$

Term 1a vanishes against 2c, term 1b against 3d, 1c against 3a, 1d against 2b, 2a against 3c and 2d against 3b.

Recall the exact commutative diagram from last section

$$
\begin{array}{ccccccccc}
0 & \longrightarrow & B & \longrightarrow & \mathrm{dR}^0_{rel}\,\mathbb{C}Q^d & \xrightarrow{d} & (\mathrm{dR}^1_{rel}\,\mathbb{C}Q)_{exact} & \longrightarrow & 0 \\
& & \Big\| & & \Big\downarrow{\simeq} & & \Big\downarrow{\tau^{-1}} & & \\
0 & \longrightarrow & B & \longrightarrow & \dfrac{\mathbb{C}Q^d}{[\mathbb{C}Q^d, \mathbb{C}Q^d]} & \longrightarrow & Der_\omega\,\mathbb{C}Q^d & \longrightarrow & 0
\end{array}
$$

Clearly, the symplectic derivations $Der_\omega\,\mathbb{C}Q^d$ are equipped with a Lie algebra structure via $[\theta_1, \theta_2] = \theta_1 \circ \theta_2 - \theta_2 \circ \theta_1$.

For every necklace word w we have a derivation $\theta_w = \tau^{-1} dw$, which is defined by

$$\begin{cases} \theta_w(a) &= \frac{\partial w}{\partial a^*} \\ \theta_w(a^*) &= -\frac{\partial w}{\partial a} \end{cases}$$

With this notation we get the following interpretations of the Kontsevich bracket

$$\{w_1, w_2\}_K = i_{\theta_{w_1}}(i_{\theta_{w_2}}\omega) = L_{\theta_{w_1}}(w_2) = -L_{\theta_{w_2}}(w_1)$$

where the next to last equality follows because $i_{\theta_{w_2}}\omega = dw_2$ and the fact that $i_{\theta_{w_1}}(dw) = L_{\theta_{w_1}}(w)$ for any w. More generally, for any B-derivation θ and any necklace word w we have the equation

$$i_\theta(i_{\theta_w}\omega) = L_\theta(w)$$

By the commutation relations for the operators L_θ and i_θ we have for all B-derivations θ_i the equalities

$$\begin{aligned} L_{\theta_1} i_{\theta_2} i_{\theta_3} \omega - i_{\theta_2} i_{\theta_3} L_{\theta_1} \omega &= [L_{\theta_1}, i_{\theta_2}] i_{\theta_3} \omega + i_{\theta_2} L_{\theta_1} i_{\theta_3} \omega \\ &\quad - i_{\theta_2} L_{\theta_1} i_{\theta_3} \omega + i_{\theta_2}[L_{\theta_1}, i_{\theta_3}]\omega \\ &= i_{[\theta_1, \theta_2]} i_{\theta_3} \omega + i_{\theta_2} i_{[\theta_1, \theta_3]} \omega \end{aligned}$$

By the homotopy formula we have $L_{\theta_w}\omega = 0$ for every necklace word w, whence we get

$$L_{\theta_{w_1}} i_{\theta_2} i_{\theta_3} \omega = i_{[\theta_{w_1}, \theta_2]} i_{\theta_3} \omega + i_{\theta_2} i_{[\theta_{w_1}, \theta_3]} \omega$$

Take $\theta_2 = \theta_{w_2}$, then the left hand side is equal to

$$\begin{aligned} L_{\theta_{w_1}} i_{\theta_{w_2}} i_{\theta_3} \omega &= -L_{\theta_{w_1}} i_{\theta_3} i_{\theta_{w_2}} \omega \\ &= -L_{\theta_{w_1}} L_{\theta_3} w_2 \end{aligned}$$

whereas the last term on the right equals

$$\begin{aligned} i_{\theta_{w_2}} i_{[\theta_{w_1}, \theta_3]}\omega &= -i_{[\theta_{w_1}, \theta_3]} i_{\theta_{w_2}} \omega \\ &= -L_{[\theta_{w_1}, \theta_3]} w_2 = -L_{theta_{w_1}} L_{\theta_3} w_2 + L_{\theta_3} L_{\theta_{w_1}} w_2 \end{aligned}$$

and substituting this we obtain that

$$\begin{aligned} i_{[\theta_{w_1}, \theta_{w_2}]} i_{\theta_3} \omega &= -L_{\theta_{w_1}} L_{\theta_3} w_2 + L_{\theta_{w_1}} L_{\theta_3} w_2 - L_{\theta_3} L_{\theta_{w_1}} w_2 \\ &= -L_{\theta_3} L_{\theta_{w_1}} w_2 = -L_{\theta_3}\{w_1, w_2\}_K \\ &= -i_{\theta_3} i_{\theta_{\{w_1, w_2\}_K}} \omega = i_{\theta_{\{w_1, w_2\}_K}} i_{\theta_3} \omega \end{aligned}$$

Finally, if we take $\theta = [\theta_{w_1}, \theta_{w_2}] - \theta_{\{w_1, w_2\}_K}$ we have that $i_\theta \omega$ is a closed 1-form and that $i_\theta i_{\theta_3} \omega = -i_{\theta_3} i_\theta \omega = 0$ for all θ_3. But then by the homotopy formula $L_{\theta_3} i_\theta \omega = 0$ whence $i_\theta \omega = 0$, which finally implies that $\theta = 0$. The following concludes the proof.

THEOREM 7.14
With notations as before, the necklace words $\mathrm{dR}^0_{rel}\ \mathbb{C}Q^d$ *is a Lie algebra for the Kontsevich bracket, and the sequence*

$$0 \longrightarrow B \longrightarrow \mathrm{dR}^0_{rel}\ \mathbb{C}Q^d \xrightarrow{\ \tau^{-1}d\ } Der_\omega\ \mathbb{C}Q^d \longrightarrow 0$$

is an exact sequence (hence a central extension) of Lie algebras.

This result will be crucial in the study of coadjoint orbits in the final chapter.

References

The formal structure on smooth affine varieties is due to M. Kapranov [50] and its extension to module varieties is due to L. Le Bruyn [79]. The semi-invariants of quivers have been obtained independently by various people, among which A. Schofield, M. Van den Bergh [95], H. Derksen, J. Weyman [30] and M. Domokos, A. Zubkov [31]. We follow here the approach of [95]. The results of section 7.3 are due to A. Schofield [92] or based on discussions with him. The results of section 7.4 are due to L. Le Bruyn [73] and is inspired by prior work of B. Westbury [106]. The results of section 7.5 are due to J. Cuntz and D. Quillen [29]. The acyclicity results of section 7.6 for the free algebra is due to M. Kontsevich [58] and in the quiver case to R. Bockland, L. Le Bruyn [12] and V. Ginzburg [37] independently as is the description of the necklace Lie algebra.

Chapter 8

Moduli Spaces

So far, the more interesting applications of the theory developed in the previous chapter have not been to noncommutative manifolds but to families $(Y_n)_n$ of varieties in which the role of Quillen-smooth algebras is replaced by Cayley-smooth algebras and where the sum-maps are replaced by gluing into a larger space. In this chapter we give the details of Ginzburg's coadjoint-orbit result for Calogero-Moser phase space, which was the first instance of such a situation.

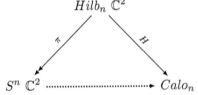

Here, $Hilb_n\ \mathbb{C}^2$ is the *Hilbert scheme* of n points in the complex plane $\mathbb{C}^2$, which is a desingularization of the symmetric power $S^n\ \mathbb{C}^2$. On the other hand, $S^n\ \mathbb{C}^2$ can be viewed as the special fiber of a family of which the general fiber is isomorphic to $Calo_n$, the phase space of Calogero-Moser particles. $Calo_n$ is a smooth affine variety and we will see that it is isomorphic to $\mathtt{triss}_n\ A_n$ for some Cayley-smooth order $A_n \in \mathtt{alg@n}$. Surprisingly, forgetting the complex structure, $Calo_n$ itself is diffeomorphic (as a C^∞-manifold) to $Hilb_n\ \mathbb{C}^2$ via rotations of hyper-Kähler structures.

George Wilson has shown that the varieties $Calo_n$ can be glued together to form an infinite dimensional manifold, the *adelic Grassmannian*

$$\bigsqcup_n Calo_n = Gr^{ad}$$

The adelic Grassmannian can be identified with the isomorphism classes of right ideals in the first Weyl algebra $A_1(\mathbb{C})$ and as the automorphism group of the Weyl algebra acts on this set with countably many orbits it was conjectured that every $Calo_n$ might be a coadjoint orbit. This fact was proved by Victor Ginzburg who showed that, indeed

$$Calo_n \hookrightarrow \mathfrak{g}^*$$

for some infinite dimensional Lie algebra $\mathfrak{g}$, which is nothing but the necklace Lie algebra of the path algebra of a double quiver naturally associated to the situation. After reading through this chapter, the reader will have no problem to prove for herself that every quiver variety, in the sense of Nakajima, is diffeomorphic to a coadjoint orbit of a necklace Lie algebra.

8.1 Moment maps

In section 2.8 we have studied in some detail the *real moment map* of m-tuples of $n \times n$ matrices. In this section we will first describe the obvious extension to representation spaces of quivers and then to prove the properties of the *real moment map* for moduli spaces of θ-semistable representations.

We fix a quiver Q on k vertices $\{v_1, \ldots, v_k\}$ and a dimension vector $\alpha = (a_1, \ldots, a_k) \in \mathbb{N}^k$. We take the standard Hermitian inproduct on each of the vertex spaces $\mathbb{C}^{\oplus a_i}$ and this induces the standard *operator inner product* on every arrow-component of $\mathbf{rep}_\alpha Q$. That is, for every arrow

$$\textcircled{j} \xleftarrow{\quad a \quad} \textcircled{i} \qquad \text{we define} \qquad (V_a, W_a) = tr(V_a W_a^*)$$

on the component $Hom_{\mathbb{C}}(\mathbb{C}^{\oplus a_i}, \mathbb{C}^{\oplus a_j})$ for all $V, W \in \mathbf{rep}_\alpha$ and where W_a^* is the *adjoint matrix* $(\overline{w_{ji}})_{i,j}$ of $W_a = (w_{ij})_{i,j}$. The *Hermitian inproduct* on $\mathbf{rep}_\alpha Q$ is defined to be

$$(V, W) = \sum_{a \in Q_a} tr(V_a W_a^*)$$

The *maximal compact subgroup* of the base change group $GL(\alpha) = \prod_{i=1}^k GL_{a_i}$ is the *multiple unitary group*

$$U(\alpha) = \prod_{i=1}^k U_{a_i}$$

which preserves the Hermitian inproduct under the base change action as subgroup of $GL(\alpha)$. The Lie algebra $Lie\ U(\alpha)$ is the algebra of multiple skew-Hermitian matrices

$$Lie\ U(\alpha) = \bigoplus_{j=1}^k iHerm_{a_j} = \{\ h = (h_1, \ldots, h_k)\ |\ h_j = -h_j^*\ \}$$

and the induced action of $Lie\ U(\alpha)$ on $\mathbf{rep}_\alpha Q$ is given by the rule

$$(h.V)_a = h_j V_a - V_a h_i \qquad \text{for} \qquad \textcircled{j} \xleftarrow{\quad a \quad} \textcircled{i}$$

for all $V \in \mathbf{rep}_\alpha Q$. This action allows us to define the *real moment map* μ for the action of $U(\alpha)$ on the representation space $\mathbf{rep}_\alpha Q$ by the assignment

$$\mathbf{rep}_\alpha Q \xrightarrow{\quad \mu \quad} (iLie\ U(\alpha))^* \qquad V \longrightarrow (h \mapsto i(h.V, V))$$

That is, the moment map is determined by

$$(h.V, V) = \sum_{\substack{\text{①}\xleftarrow{a}\text{①}}} tr(h_j V_a V_a^* - V_a h_i V_a^*)$$

$$= \sum_{v_i \in Q_v} tr(h_i(\sum_{\substack{\text{①}\xleftarrow{a}\text{①}}} V_a V_a^* - \sum_{\substack{\text{①}\xleftarrow{a}\text{①}}} V_a^* V_a))$$

Using the nondegeneracy of the Killing form on *Lie U(α)* we have the identification

$$\mu^{-1}(\underline{0}) = \{V \in \mathbf{rep}_\alpha Q \mid \sum_{\substack{\text{①}\xleftarrow{a}\text{①}}} V_a V_a^* = \sum_{\substack{\text{①}\xleftarrow{a}\text{①}}} V_a^* V_a \quad \forall v_i \in Q_v\}$$

The *real moment map* $\mu_{\mathbb{R}}$ is then defined to be

$$\mathbf{rep}_\alpha Q \xrightarrow{\mu_{\mathbb{R}}} Lie\ U(\alpha) \qquad V \mapsto i[V, V^*] = i(\sum_{\substack{\text{①}\xleftarrow{a}\text{①}}} V_a V_a^* - \sum_{\substack{\text{①}\xleftarrow{a}\text{①}}} V_a^* V_a)_j$$

Reasoning as in section 2.8 we can prove the following moment map description of the isomorphism classes of semisimple α-dimensional representations of Q.

THEOREM 8.1
There are natural one-to-one correspondences between

1. *points of* $\mathbf{iss}_\alpha Q$, *and*

2. $U(\alpha)$-*orbits in* $\mu_{\mathbb{R}}^{-1}(\underline{0})$.

Next, we will prove a similar result to describe the points of $M_\alpha^{ss}(Q, \theta)$, the moduli space of θ-semistable α-dimensional representations of Q, introduced and studied in section 4.8. Fix, an integral k-tuple $\theta = (t_1, \ldots, t_k) \in \mathbb{Z}^k$ with associated *character*

$$GL(\alpha) \xrightarrow{\chi_\theta} \mathbb{C}^* \qquad g = (g_1, \ldots, g_k) \mapsto \prod_{i=1}^{k} det(g_i)^{t_i}$$

We have seen in section 4.8 that in order to describe $M_\alpha^{ss}(Q, \theta)$ we consider the extended representation space $\mathbf{rep}_\alpha Q \oplus \mathbb{C}$. We introduce a function N on this extended space replacing the norm in the above discussion.

$$\mathbf{rep}_\alpha Q \oplus \mathbb{C} \xrightarrow{N} \mathbb{R}_+ \qquad (V, z) \mapsto |z| e^{\frac{1}{2}\|V\|^2}$$

where $\|V\|$ is the norm coming from the Hermitian inproduct on $\mathbf{rep}_\alpha Q$. Sometimes, the function N is called the *Kähler potential* for the inproduct on $\mathbf{rep}_\alpha Q$. We will investigate the properties of N.

LEMMA 8.1
Let X be a closed subvariety of $\mathbf{rep}_\alpha Q \oplus \mathbb{C}$ disjoint from $rep'_\alpha Q = \{(V,0) \mid V \in \mathbf{rep}_\alpha Q\} \hookrightarrow \mathbf{rep}_\alpha Q \oplus \mathbb{C}$. Then, the restriction of N to X is proper and therefore achieves its minimum.

PROOF   Because X and $rep'_\alpha Q$ are disjoint closed subvarieties of $\mathbf{rep}_\alpha Q \oplus \mathbb{C}$, there is a polynomial $f \in \mathbb{C}[\mathbf{rep}_\alpha Q \oplus \mathbb{C}] = \mathbb{C}[\mathbf{rep}_\alpha Q][z]$ such that $f \mid X = 1$ and $f \mid rep'_\alpha Q = 0$. That is, X is contained in the hypersurface

$$\mathbb{V}(f-1) = \mathbb{V}(zP_1(V) + \ldots + z^n P_n(V) - 1) \hookrightarrow \mathbf{rep}_\alpha Q \oplus \mathbb{C}$$

where the $P_i \in \mathbb{C}[\mathbf{rep}_\alpha Q]$.
Now, N is *proper* if the inverse images $N^{-1}([0,r])$ are compact for all $r \in \mathbb{R}_+$, that is, there exist constants r_1 and r_2 depending on X and r such that

$$N(z,V) \leq r \qquad \text{implies} \qquad |z| \leq r_1 \text{ and } \|V\| \leq r_2.$$

We can always take $r_1 = r$ so we only need to bound $\|V\|$. If $|z| \leq re^{-\frac{1}{2}\|V\|^2}$, then we have that

$$|zP_1(V) + \ldots + z^n P_n(V)| \leq r|P_1(V)|e^{-\frac{1}{2}\|V\|^2} + \ldots + r^n|P_n(V)|e^{-\frac{n}{2}\|V\|^2}$$

Choose r_2, depending on r and P_i such that the condition

$$\|V\| > r_2 \qquad \text{implies that} \qquad |P_i(V)| < \frac{1}{n}r^{-i}e^{\frac{i}{2}\|V\|^2} \qquad \forall 1 \leq i \leq n$$

But then if $\|V\| > r_2$, we have $|zP_1(V) + \ldots + z^n P_n(V)| < 1$ and so (V,z) does not belong to X. $\square$

Recall that $GL(\alpha)$ acts on the extended representation space $\mathbf{rep}_\alpha Q \oplus \mathbb{C}$ via

$$g.(V,z) = (g.V, \chi_\theta^{-1}(g)z)$$

LEMMA 8.2
Let $\mathcal{O}$ be a $GL(\alpha)$-orbit in the extended representation space $\mathbf{rep}_\alpha Q \oplus \mathbb{C}$ which is disjoint from $rep'_\alpha Q$. Then, if the restriction of N to $\mathcal{O}$ achieves its minimum, then $\mathcal{O}$ is a closed orbit.

PROOF   Assume that N achieves its minimum in the point $V_z = (V,z) \in \mathcal{O}$. If $\mathcal{O}$ is not a closed orbit we can by the Hilbert criterion find a one-parameter subgroup λ of $GL(\alpha)$ such that

$$\lim_{t \to 0} \lambda(t).V_z \notin \mathcal{O}$$

and the limit exists in $\mathbf{rep}_\alpha \, Q \oplus \mathbb{C}$. Decompose the representation $V = \sum_{n \in \mathbb{Z}} V_n$ into eigenspaces with respect to the one-parameter subgroup λ, that is

$$\lambda(t).V = \sum_{n \in \mathbb{Z}} t^n V_n$$

Because the limit exists, we have that $V_n = 0$ whenever $n < 0$ and $\theta(\lambda) \leq 0$. Because the limit is not contained in $\mathcal{O}$ we have that $V_n \neq 0$ for some $n > 0$. Further, by conjugating λ if necessary we may assume that the weight space decomposition $V = \sum_n V_n$ is orthogonal with respect to the inproduct in $\mathbf{rep}_\alpha \, Q$.

Using these properties we then have that

$$N(\lambda(t).(V,z)) = |z|e^{\frac{1}{2}|V_0|^2}|t|^{-\theta(\lambda)}e^{\frac{1}{2}\sum_{n>0}|t|^n \|V_n\|^2}$$

This expression will decrease when t approaches zero, contradicting the assumption that the minimum of $N \mid \mathcal{O}$ was achieved in (V,z). This contradiction implies that $\mathcal{O}$ must be a closed orbit. $\qquad \square$

Recall from section 4.8 that an orbit $\mathcal{O}(V,z)$ is closed and disjoint from $rep'_\alpha \, Q$ for some $z \in \mathbb{C}^*$ if and only if V is the direct sum of θ-stable representations of Q. Recall the real moment map

$$\mathbf{rep}_\alpha \, Q \xrightarrow{\mu} (iLie \, U(\alpha))^*$$

And consider the special real valued function $d\chi_\theta$ on $Lie \, U(\alpha)$, which is the restriction to $Lie \, U(\alpha)$ of the differential of $GL(\alpha) \xrightarrow{\chi_\theta} \mathbb{C}^*$ at the identity element (which takes real values). In fact, for any $m = (m_1, \ldots, m_k) \in Lie \, GL(\alpha) = M_\alpha(\mathbb{C})$ we have that

$$d\chi_\theta(m) = \sum_{v_j \in Q_v} t_j tr(m_j) = \sum_{v_j \in Q_v} tr(m_j t_j \mathbb{1}_{a_j})$$

With these notations we have the promised extension to moduli spaces of θ-semistable representations.

THEOREM 8.2

There are natural one-to-one correspondences between

1. *points of $M_\alpha^{ss}(Q, \theta)$, and*

2. *$U(\alpha)$-orbits in $\mu^{-1}(d\chi_\theta)$*

PROOF Let $V_z = (V, z) \in \mathbf{rep}_\alpha \, Q \oplus \mathbb{C}$ with $z \neq 0$. For any $h = (h_1, \ldots, h_k)$ in $iLie \, U(\alpha)$ we define the functions

$$m_V(h) = \frac{d}{dt}\Big|_{t=0} \, log \, N(e^{th}.V_z)$$
$$= (h.V, V) - d\chi_\theta(h)$$

$$m_V^{(2)}(h) = \frac{d}{dt}\,|_{t=0}\, log\, N(e^{th}.V_z)$$
$$= 2\|h.V\|^2$$

The function m_V is the zero map if and only if the restriction of N to the orbit $\mathcal{O}(V_z)$ has a critical point at V_z. As the base change action of $U(\alpha)$ on the extended representation space $\mathbf{rep}_\alpha\, Q \oplus \mathbb{C}$ preserves the Kähler potential N, N induces a function on the quotient $\mathcal{O}(V_z)/U(\alpha)$. The formula for $m_V^{(2)}$ shows that this function is strictly convex (except in directions along the fibers $\{(V, c) \mid c \in \mathbb{C}\}$ where it is linear). Hence, a critical point is a minimum and there can be at most one such critical point. From the lemmas above we have that N has a minimum on $\mathcal{O}(V_z)$ if and only if $\mathcal{O}(V_z)$ is a closed orbit, which in its turn is equivalent to V being the direct sum of θ-stable representations, whence determining a point of $M_\alpha^{ss}(Q, \theta)$.   □

Finally, for any $h \in iLie\, U(\alpha)$ we have the formulas

$$\mu(V)(h) = i \sum_{v_i \in Q_v} tr(h_i(\sum_{\text{ⓘ}\xleftarrow{a}\text{ⓙ}} V_a V_a^* - \sum_{\text{ⓙ}\xleftarrow{a}\text{ⓘ}} V_a^* V_a))$$

$$d\chi_\theta(h) = \sum_{v_i \in Q_v} tr(h_i t_i \mathbb{1}_{a_i})$$

whence by nondegeneracy of the Killing form, the equality $\mu(V) = d\chi_\theta$ is equivalent to the conditions

$$\sum_{\text{ⓙ}\xleftarrow{a}\text{ⓘ}} V_a V_a^* - \sum_{\text{ⓙ}\xleftarrow{a}\text{ⓘ}} V_a^* V_a = it_j \mathbb{1}_{a_j} \qquad \forall v_j \in Q_v$$

We can assign to $\theta = (t_1, \dots, t_k) \in \mathbb{Z}^k$ the element $i\theta \mathbb{1}_\alpha = (it_1 \mathbb{1}_{a_1}, \dots, it_k \mathbb{1}_{a_k}) \in Lie\, U(\alpha)$. We then can rephrase the results of this section as follows.

THEOREM 8.3
There are natural identifications between the spaces

$$\mathbf{iss}_\alpha\, Q \longleftrightarrow \mu_\mathbb{R}^{-1}(\underline{0})/U(\alpha) \quad and \quad M_\alpha^{ss}(Q, \theta) \longleftrightarrow \mu_\mathbb{R}^{-1}(i\theta \mathbb{1}_\alpha)/U(\alpha)$$

8.2   Dynamical systems

In this chapter we will illustrate what we have learned on the simplest *wild* quiver Q, which is neither Dynkin nor extended Dynkin

In this section we will show that the representation theory of this quiver is of importance in system theory.

A *linear time invariant dynamical system* Σ is determined by the following system of differential equations

$$\begin{cases} \dfrac{dx}{dt} & = Bx + Au \\ y & = Cx \end{cases} \qquad (8.1)$$

Here, $u(t) \in \mathbb{C}^m$ is the *input* or *control* of the system at tome t, $x(t) \in \mathbb{C}^n$ the *state* of the system and $y(t) \in \mathbb{C}^p$ the *output* of the system Σ. *Time invariance* of Σ means that the matrices $A \in M_{n \times m}(\mathbb{C})$, $B \in M_n(\mathbb{C})$ and $C \in M_{p \times n}(\mathbb{C})$ are constant, that is, $\Sigma = (A, B, C)$ is a representation of the quiver $\overset{\circ}{Q}$

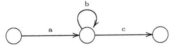

of dimension vector $\alpha = (m, n, p)$. The system Σ can be represented as a *black box*

which is in a certain state $x(t)$ that we can try to change by using the input controls $u(t)$. By reading the output signals $y(t)$ we can try to determine the state of the system.

Recall that the *matrix exponential* e^B of any $n \times n$ matrix B is defined by the infinite series

$$e^B = \mathbb{1}_n + B + \frac{B^2}{2!} + \ldots + \frac{B^m}{m!} + \cdots$$

The importance of this construction is clear from the fact that e^{Bt} is the *fundamental matrix* for the homogeneous differential equation $\frac{dx}{dt} = Bx$. That is, the columns of e^{Bt} are a basis for the n-dimensional space of solutions of the equation $\frac{dx}{dt} = Bx$.

Motivated by this, we look for a solution to equation (8.1) as the form $x(t) = e^{Bt}g(t)$ for some function $g(t)$. Substitution gives the condition

$$\frac{dg}{dt} = e^{-Bt}Au \quad \text{whence} \quad g(\tau) = g(\tau_0) + \int_{\tau_0}^{\tau} e^{-Bt}Au(t)dt$$

Observe that $x(\tau_0) = e^{B\tau_0}g(\tau_0)$ and we obtain the solution of the linear dynamical system $\Sigma = (A, B, C)$:

$$\begin{cases} x(\tau) & = e^{(\tau-\tau_0)B}x(\tau_0) + \int_{\tau_0}^{\tau} e^{(\tau-t)B}Au(t)dt \\ y(\tau) & = Ce^{B(\tau-\tau_0)}x(\tau_0) + \int_{\tau_0}^{\tau} Ce^{(\tau-t)B}Au(t)dt \end{cases}$$

Differentiating we see that this is indeed a solution and it is the unique one having a prescribed starting state $x(\tau_0)$. Indeed, given another solution $x_1(\tau)$ we have that $x_1(\tau) - x(\tau)$ is a solution to the homogeneous system $\frac{dx}{dt} = Bt$, but then

$$x_1(\tau) = x(\tau) + e^{\tau B} e^{-\tau_0 B}(x_1(\tau_0) - x(\tau_0))$$

We call the system Σ *completely controllable* if we can steer any starting state $x(\tau_0)$ to the zero state by some control function $u(t)$ in a finite time span $[\tau_0, \tau]$. That is, the equation

$$0 = x(\tau_0) + \int_{\tau_0}^{\tau} e^{(\tau_0 - t)B} Au(t)dt$$

has a solution in τ and $u(t)$. As the system is time-invariant we may always assume that $\tau_0 = 0$ and have to satisfy the equation

$$0 = x_0 + \int_0^{\tau} e^{tB} Au(t)dt \quad \text{for every} \quad x_0 \in \mathbb{C}^n \tag{8.2}$$

Consider the *control matrix* $c(\Sigma)$, which is the $n \times mn$ matrix

$$c(\Sigma) = \boxed{\;A\;}\;\boxed{\;BA\;}\;\boxed{\;B^2A\;}\quad \cdots \quad \boxed{\;B^{n-1}A\;}$$

Assume that $rk\ c(\Sigma) < n$ then there is a nonzero state $s \in \mathbb{C}^n$ such that $s^{tr}c(\Sigma) = 0$, where s^{tr} denotes the transpose (row column) of s. Because B satisfies the characteristic polynomial $\chi_B(t)$, B^n and all higher powers B^m are linear combinations of $\{\mathbb{1}_n, B, B^2, \ldots, B^{n-1}\}$. Hence, $s^{tr}B^m A = 0$ for all m. Writing out the power series expansion of e^{tB} in equation (8.2) this leads to the contradiction that $0 = s^{tr}x_0$ for all $x_0 \in \mathbb{C}^n$. Hence, if $rk\ c(\Sigma) < n$, then Σ is not completely controllable.

Conversely, let $rk\ c(\Sigma) = n$ and assume that Σ is not completely controllable. That is, the space of all states

$$s(\tau, u) = \int_0^{\tau} e^{-tB} Au(t)dt \qquad {}_!$$

is a proper subspace of $\mathbb{C}^n$. But then, there is a nonzero state $s \in \mathbb{C}^n$ such that $s^{tr}s(\tau, u) = 0$ for all τ and all functions $u(t)$. Differentiating this with respect to τ we obtain

$$s^{tr}e^{-\tau B} Au(\tau) = 0 \quad \text{whence} \quad s^{tr}e^{-\tau B}A = 0 \tag{8.3}$$

for any τ as $u(\tau)$ can take on any vector. For $\tau = 0$ this gives $s^{tr}A = 0$. If we differentiate (8.3) with respect to τ we get $s^{tr}Be^{-\tau B}A = 0$ for all τ and for $\tau = 0$ this gives $s^{tr}BA = 0$. Iterating this process we show that $s^{tr}B^m A = 0$ for any m, whence

$$s^{tr}\begin{bmatrix} A\ BA\ B^2A \ldots B^{n-1}A \end{bmatrix} = 0$$

contradicting the assumption that $rk\ c(\Sigma) = n$. The proof follows.

PROPOSITION 8.1
A linear time-invariant dynamical system Σ determined by the matrices (A, B, C) is completely controllable if and only if $rk\ c(\Sigma)$ is maximal.

We say that a state $x(\tau)$ at time τ is *unobservable* if $Ce^{(\tau-t)B}x(\tau) = 0$ for all t. Intuitively this means that the state $x(\tau)$ cannot be detected uniquely from the output of the system Σ. Again, if we differentiate this condition a number of times and evaluate at $t = \tau$ we obtain the conditions

$$Cx(\tau) = CBx(\tau) = \ldots = CB^{n-1}x(\tau) = 0.$$

We say that Σ is *completely observable* if the zero state is the only unobservable state at any time τ. Consider the *observation matrix* $o(\Sigma)$ of the system Σ that is the $pn \times n$ matrix

$$o(\Sigma) = \left[C^{tr}\ (CB)^{tr} \cdots (CB^{n-1})^{tr} \right]^{tr}$$

An analogous argument as in the proof of proposition 8.1 gives us that a linear time-invariant dynamical system Σ determined by the matrices (A, B, C) is completely observable if and only if $rk\ o(\Sigma)$ is maximal.

Assume we have two systems Σ and Σ', determined by matrix triples from $\text{rep}_\alpha\ Q = M_{n\times m}(\mathbb{C}) \times M_n(\mathbb{C}) \times M_{p\times n}(\mathbb{C})$ producing the same output $y(t)$ when given the same input $u(t)$, for all possible input functions $u(t)$. We recall that the output function y for a system $\Sigma = (A, B, C)$ is determined by

$$y(\tau) = Ce^{B(\tau-\tau_0)}x(\tau_0) + \int_{\tau_0}^{\tau} Ce^{(\tau-t)B}Au(t)dt$$

Differentiating this a number of times and evaluating at $\tau = \tau_0$ as in the proof of proposition 8.1 equality of input/output for Σ and Σ' gives the conditions

$$CB^i A = C'B'^i A' \quad \text{for all} \quad i$$

But then, we have for any $v \in \mathbb{C}^{mn}$ that $c(\Sigma)(v) = 0 \Leftrightarrow c(\Sigma')(v) = 0$ and we can decompose $\mathbb{C}^{pn} = V \oplus W$ such that the restriction of $c(\Sigma)$ and $c(\Sigma')$ to V are the zero map and the restrictions to W give isomorphisms with $\mathbb{C}^n$. Hence, there is an invertible matrix $g \in GL_n$ such that $c(\Sigma') = gc(\Sigma)$ and from the commutative diagram

$$
\begin{array}{ccccc}
\mathbb{C}^{mn} & \xrightarrow{c(\Sigma)} & \mathbb{C}^n & \xleftarrow{o(\Sigma)} & \mathbb{C}^{pn} \\
\| & & \downarrow{g} & & \| \\
\mathbb{C}^{mn} & \xrightarrow{c(\Sigma')} & \mathbb{C}^n & \xleftarrow{o(\Sigma')} & \mathbb{C}^{pn}
\end{array}
$$

we obtain that also $o(\Sigma') = o(\Sigma)g^{-1}$.

Consider the system $\Sigma_1 = (A_1, B_1, C_1)$ *equivalent* with Σ under the base-change matrix g. That is, $\Sigma_1 = g.\Sigma = (gA, gBg^{-1}, Cg^{-1})$. Then

$$\left[A_1, B_1 A_1, \ldots, B_1^{n-1} A_1\right] = gc(\Sigma) = c(\Sigma') = \left[A', B'A', \ldots, B'^{n-1}A'\right]$$

and so $A_1 = A'$. Further, as $B_1^{i+1} A_1 = B'^{i+1} A'$ we have by induction on i that the restriction of B_1 on the subspace of $B'^i Im(A')$ is equal to the restriction of B' on this space. Moreover, as $\sum_{i=0}^{n-1} B'^i Im(A') = \mathbb{C}^n$ it follows that $B_1 = B'$. Because $o(\Sigma') = o(\Sigma)g^{-1}$ we also have $C_1 = C'$. The following completes the proof.

PROPOSITION 8.2
Let $\Sigma = (A, B, C)$ and $\Sigma' = (A', B', C')$ be two completely controllable and completely observable dynamical systems. The following are equivalent

1. *The input/output behavior of Σ and Σ' are equal*

2. *The systems Σ and Σ' are* equivalent, *that is, there exists an invertible matrix $g \in GL_n$ such that*

$$A' = gA, \quad B' = gBg^{-1} \quad \text{and} \quad C' = Cg^{-1}$$

Hence, in *system identification* it is important to classify completely controllable and observable systems $\Sigma \in \text{rep}_\alpha \tilde{Q}$ under this restricted base change action. We will concentrate on the input part and consider *completely controllable minisystems*, that is, representations $\Sigma = (A, B) \in \text{rep}_\alpha Q$ where $\alpha = (m, n)$ such that $c(\Sigma)$ is of maximal rank. First, we relate the system theoretic notion to that of θ-semistability for $\theta = (-n, m)$ (observe that $\theta(\alpha) = 0$).

LEMMA 8.3
If $\Sigma = (A, B) \in \text{rep}_\alpha Q$ is θ-semistable, then Σ is completely controllable and $m \leq n$.

PROOF   If $m > n$ then $(Ker\ A, 0)$ is a proper subrepresentation of Σ of dimension vector $\beta = (dim\ Im\ A - m, 0)$ with $\theta(\beta) < 0$ so Σ cannot be θ-semistable. If Σ is not completely controllable then the subspace W of $\mathbb{C}^{\oplus n}$ spanned by the images of $A, BA, \ldots, B^{n-1}A$ has dimension $k < n$. But then, Σ has a proper subrepresentation of dimension vector $\beta = (m, k)$ with $\theta(\beta) < 0$, contradicting the θ-semistability assumption.   ∎

We introduce a combinatorial gadget: the *Kalman code* . It is an array consisting of $(n + 1) \times m$ boxes each having a position label (i, j) where

$0 \le i \le n$ and $1 \le j \le m$. These boxes are ordered *lexicographically*, that is, $(i', j') < (i, j)$ if and only if either $i' < i$ or $i' = i$ and $j' < j$. Exactly n of these boxes are painted black subject to the rule that if box (i, j) is black, then so is box (i', j) for all $i' < i$. That is, a Kalman code looks like

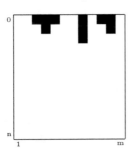

We assign to a completely controllable couple $\Sigma = (A, B)$ its Kalman code $K(\Sigma)$ as follows: let $A = \begin{bmatrix} A_1 \ A_2 \ \dots \ A_m \end{bmatrix}$, that is A_i is the i-th column of A. Paint the box (i, j) black if and only if the column vector $B^i A_j$ is linearly independent of the column vectors $B^k A_l$ for all $(k, l) < (i, j)$.

The painted array $K(\Sigma)$ is indeed a Kalman code. Assume that box (i, j) is black but box (i', j) white for $i' < i$, then

$$B^{i'} A_j = \sum_{(k,l)<(i',j)} \alpha_{kl} B^k A_l \quad \text{but then,} \quad B^i A_j = \sum_{(k,l)<(i',j)} \alpha_{kl} B^{k+i-i'} A_l$$

and all $(k + i - i', l) < (i, l)$, a contradiction. Moreover, $K(\Sigma)$ has exactly n black boxes as there are n linearly independent columns of the control matrix $c(\Sigma)$ when Σ is completely controllable.

The Kalman code is a discrete invariant of the orbit $\mathcal{O}(\Sigma)$ under the restricted base change action by GL_n. This follows from the fact that $B^i A_j$ is linearly independent of the $B^k A_l$ for all $(k, l) < (i, j)$ if and only if $gB^i A_j$ is linearly independent of the $gB^k A_l$ for any $g \in GL_n$ and the observation that $gB^k A_l = (gBg^{-1})^k (gA)_l$.

With $\mathbf{rep}_\alpha^c \ Q$ we will denote the open subset of $\mathbf{rep}_\alpha \ Q$ of all completely controllable couples (A, B). We consider the map

$$\mathbf{rep}_\alpha \ Q \xrightarrow{\quad \psi \quad} M_{n \times (n+1)m}(\mathbb{C})$$

$$(A, B) \quad \mapsto \quad \begin{bmatrix} A \ BA \ B^2 A \ \dots \ B^{n-1} A \ B^n A \end{bmatrix}$$

The matrix $\psi(A, B)$ determines a linear map $\psi_{(A,B)} : \mathbb{C}^{(n+1)m} \longrightarrow \mathbb{C}^n$ and (A, B) is a completely controllable couple if and only if the corresponding linear map $\psi_{(A,B)}$ is surjective. Moreover, there is a linear action of GL_n on $M_{n \times (n+1)m}(\mathbb{C})$ by left multiplication and the map ψ is GL_n-equivariant.

The Kalman code induces a *bar code* on $\psi(A, B)$, that is, the $n \times n$ minor of $\psi(A, B)$ determined by the columns corresponding to black boxes in the

$$\psi(A, B)$$

FIGURE 8.1: Kalman code and bar code.

Kalman code, see figure 8.1 By construction this minor is an invertible matrix $g^{-1} \in GL_n$. We can choose a canonical point in the orbit $\mathcal{O}(\Sigma)$: $g.(A, B)$. It does have the characteristic property that the $n \times n$ minor of its image under ψ, determined by the Kalman code is the identity matrix $\mathbb{1}_n$. The matrix $\psi(g.(A, B))$ will be denoted by $b(A, B)$ and is called bar code of the completely controllable pair $\Sigma = (A, B)$. We claim that the bar code determines the orbit uniquely.

The map ψ is injective on the open set $\mathbf{rep}^c_\alpha Q$. Indeed, if

$$\begin{bmatrix} A & BA & \dots & B^n A \end{bmatrix} = \begin{bmatrix} A' & B'A' & \dots & B'^n A' \end{bmatrix}$$

then $A = A'$, $B \mid Im(A) = B' \mid Im(A)$ and hence by induction also

$$B \mid B^i Im(A) = B' \mid B'^i Im(A') \quad \text{for all } i \leq n - 1$$

But then, $B = B'$ as both couples (A, B) and (A', B') are completely controllable. Hence, the bar code $b(A, B)$ determines the orbit $\mathcal{O}(\Sigma)$ and is a point in the *Grassmannian* $Grass_n(m(n + 1))$. We have

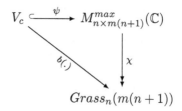

where ψ is a GL_n-equivariant embedding and χ the orbit map. Observe that the bar code matrix $b(A, B)$ shows that the stabilizer of (A, B) is trivial. Indeed, the minor of $g.b(A, B)$ determined by the Kalman code is equal to g. Moreover, continuity of b implies that the orbit $\mathcal{O}(\Sigma)$ is closed in $\mathbf{rep}^c_\alpha Q$.

Consider the differential of ψ. For all $(A, B) \in \mathbf{rep}_\alpha Q$ and $(X, Y) \in T_{(A,B)} \mathbf{rep}_\alpha Q \simeq \mathbf{rep}_\alpha Q$ we have

$$(B + \epsilon Y)^j (A + \epsilon X) = B^n A + \epsilon \left(B^n X + \sum_{i=0}^{j-1} B^i Y B^{n-1-i} A \right)$$

Therefore the differential of ψ in $(A, B) \in \mathbf{rep}_\alpha Q$, $d\psi_{(A,B)}(X, Y)$ is equal to

$$\begin{bmatrix} X & BX + YA & B^2 X + BYA + YBA & \dots & B^n X + \sum_{i=0}^{n-1} B^i Y B^{n-1-i} A \end{bmatrix}$$

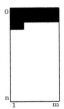

FIGURE 8.2: Generic Kalman code.

Assume $d\psi_{(A,B)}(X,Y)$ is the zero matrix, then $X = 0$ and substituting in the next term also $YA = 0$. Substituting in the third gives $YBA = 0$, then in the fourth $YB^2A = 0$ and so on until $YB^{n-1}A = 0$. But then

$$Y\left[A\ BA\ B^2A \ldots B^{n-1}A\right] = 0$$

If (A, B) is a completely controllable pair, this implies that $Y = 0$ and hence shows that $d\psi_{(A,B)}$ is injective for all $(A, B) \in \mathbf{rep}^c_\alpha\ Q$. By the implicit function theorem, ψ induces a GL_n-equivariant diffeomorphism between $\mathbf{rep}^c_\alpha\ Q$ and a locally closed submanifold of $M^{max}_{n\times(n+1)m}(\mathbb{C})$. The image of this submanifold under the orbit map χ is again a manifold as all fibers are equal to GL_n. This concludes the difficult part of the *Kalman theorem*.

THEOREM 8.4
The orbit space $O_c = \mathbf{rep}^c_\alpha\ Q/GL_n$ of equivalence classes of completely controllable couples is a locally closed submanifold of dimension $m.n$ of the Grassmannian $Grass_n(m(n+1))$. In fact $\mathbf{rep}^c_\alpha\ Q \xrightarrow{\ b\ } O_c$ is a principal GL_n-bundle.

To prove the dimension statement, consider $\mathbf{rep}^c_\alpha(K)$ the set of completely controllable pairs (A, B) having Kalman code K and let $O_c(K)$ be the image under the orbit map. After identifying $\mathbf{rep}^c_\alpha(K)$ with its image under ψ, the bar code matrix $b(A, B)$ gives a section $O_c(K) \xrightarrow{\ s\ } \mathbf{rep}^c_\alpha(K)$. In fact,

$$GL_n \times O_c(K) \longrightarrow V_c(K) \qquad (g, x) \mapsto g.s(x)$$

is a GL_n-equivariant diffeomorphism because the $n \times n$ minor determined by K of $g.b(A, B)$ is g. Consider the *generic* Kalman code K^g of figure 8.2 obtained by painting the top boxes black from left to right until one has n black boxes. Clearly $\mathbf{rep}^c_\alpha(K^g)$ is open in $\mathbf{rep}^c_\alpha$ and one deduces

$$dim\ O_c = dim\ O_c(K^g) = dim\ V_c(K^g) - dim\ GL_n = mn + n^2 - n^2 = mn$$

The Kalman orbit space also naturally defines an order over the moduli space $M^{ss}_\alpha(Q, \theta)$. First, observe that whenever $m \leq n$ we have θ-stable representations of dimension vector $\alpha = (m, n)$ for $\theta = (-n, m)$. Then, $dim\ M^{ss}_\alpha(Q, \theta) =$

$$dim\ \mathbf{rep}_\alpha\ Q - dim\ GL(\alpha) + 1 = n^2 + mn - n^2 - m^2 + 1 = m(n - m) + 1$$

By the lemma we have that $\mathbf{rep}_\alpha^{ss} Q$ is an open subset of $\mathbf{rep}_\alpha^c Q$ and let O_{ss} be the open subset of O_c it determines. Then, the natural quotient map

$$O_{ss} \longrightarrow\!\!\!\!\!\rightarrow M_\alpha^{ss}(Q, \theta)$$

is generically a principal PGL_m-fibration, so determines a central simple algebra over the function field of $M_\alpha^{ss}(Q, \theta)$.

In particular, if $m = 1$ then $O_{ss} \simeq M_\alpha^{ss}(Q, \theta)$ and both are isomorphic to $\mathbb{A}^n$ and the orbits are parameterized by an old acquaintance, the *companion matrix* and its canonical *cyclic vector*

$$A = \begin{bmatrix} 1 \\ 0 \\ \dots \\ 0 \\ 0 \end{bmatrix} \qquad B = \begin{bmatrix} 0 & & & & x_n \\ -1 & 0 & & & x_{n-1} \\ & \ddots & \ddots & & \vdots \\ & & -1 & 0 & x_2 \\ & & & -1 & x_1 \end{bmatrix}$$

Trivial as this case seems, we will see that it soon gets interesting if we consider its extension to the double quiver Q^d and to deformed preprojective algebras.

8.3   Deformed preprojective algebras

Recall the construction of *deformed preprojective algebras* given in section 5.5. Let Q be a quiver on k vertices and Q^d its *double quiver*, that is to each arrow $a \in Q_a$ we add an arrow a^* with the reverse orientation in Q_a^d and define the *commutator element* $c = \sum_{a \in Q_a} [a, a^*]$ in the path algebra $\mathbb{C}Q^d$. For a weight $\lambda = (\lambda_1, \dots, \lambda_k) \in \mathbb{C}^k$ we define the *deformed preprojective algebra*

$$\Pi_\lambda = \frac{\mathbb{C}Q^d}{c - \lambda}$$

In this section we will give an outline of the determination of the dimension vectors of simple Π_λ-representations due to W. Crawley-Boevey [26].

We know already that a dimension vector $\alpha = (a_1, \dots, a_k)$ can be the dimension vector of a Π_λ-representation only if $\lambda.\alpha = 0$, so we will denote this subset of $\mathbb{N}^k$ by $\mathbb{N}_\lambda^k$. With Δ_λ^+ we will denote the subset of *positive roots* α of Q lying in $\mathbb{N}_\lambda^k$ and with $\mathbb{N}\Delta_\lambda^+$ the additive semigroup they generate.

If v_i is a loop-free vertex of Q we have defined the *reflexion* $\mathbb{Z}^k \xrightarrow{r_i} \mathbb{Z}^k$ by

$$r_i(\alpha) = \alpha - T_Q(\alpha, \epsilon_i)$$

and we define its *dual reflexion* $\mathbb{C}^k \xrightarrow{s_i} \mathbb{C}^k$ by the formula

$$s_i(\lambda)_j = \lambda_j - T_Q(\epsilon_i, \epsilon_j)\lambda_i$$

Clearly, we have $s_i(\lambda).\alpha = \lambda.r_i(\alpha)$. We say that a loop-free vertex v_i in Q is *admissible* for (λ, α) (or for λ) if $\lambda_i \neq 0$. We define an equivalence relation $\sim$ on pairs $(\lambda, \alpha) \in \mathbb{C}^k \times \mathbb{Z}^k$ induced by $(\lambda, \alpha) \sim (s_i(\lambda), r_i(\alpha))$ whenever v_i is an admissible vertex for (λ, α). We want to relate the representation theory of Π_λ to that of $\Pi_{s_i(\lambda)}$.

THEOREM 8.5

If v_i is an admissible vertex for λ, then there is an equivalence of categories

$$\Pi_\lambda - \texttt{rep} \xrightarrow{\ E_i\ } \Pi_{s_i(\lambda)} - \texttt{rep}$$

that acts as the reflection r_i on the dimension vectors.

PROOF Because the definition of Π_λ does not depend on the orientation of the quiver Q, we may assume that there are no arrows in Q having starting vertex v_i. Let $V \in \texttt{rep}_\alpha \Pi_\lambda$ and consider V as a representation of the double quiver Q^d. Consider the vectorspace

$$V_\oplus = \bigoplus_{\overset{a}{\textcircled{i} \leftarrow \textcircled{j}}} V_j$$

where the sum is taken over all arrows $a \in Q_a$ terminating in v_i. Let μ_a and π_a be the inclusion and projection between V_j and $V_\oplus$ and define maps $V_i \xrightarrow{\ \mu\ } V_\oplus$ and $V_\oplus \xrightarrow{\ \pi\ } V_i$ by the formulas

$$\pi = \frac{1}{\lambda_i} \sum_{\overset{a}{\textcircled{i} \leftarrow \textcircled{}}} V_a \circ \pi_a \qquad \text{and} \qquad \mu = \sum_{\overset{a}{\textcircled{i} \leftarrow \textcircled{}}} \mu_a \circ V_a$$

then $\pi \circ \mu = \mathbb{1}_{V_i}$ whence $\mu \circ \pi$ is an idempotent endomorphism on $V_\oplus$.

We define the representation V' of Q^d by the following data : $V'_j = V_j$ for $j \neq i$, $V'_a = V_a$ and $V'_{a^*} = V_{a^*}$ whenever the terminating vertex of a is not v_i. Further

$$V'_i = Im\ \mathbb{1} - \mu \circ \pi = Ker\ \pi$$

and for an arrow $\textcircled{i} \xleftarrow{\quad a \quad} \textcircled{j}$ in Q we define

$$\begin{cases} V'_a &= -\lambda_i(\mathbb{1} - \mu \circ \pi) \circ \mu_a \ : \ V'_j \longrightarrow V'_i \\ V'_{a^*} &= \pi_a \mid V'_i \ : \ V'_i \longrightarrow V'_j \end{cases}$$

We claim that V' is a representation of $\Pi_{s_i(\lambda)}$. Indeed, for a vertex v_i we have

$$\sum_{\overset{a}{\textcircled{i} \leftarrow \textcircled{}}} V'_a V'_{a^*} = \sum_{\overset{a}{\textcircled{i} \leftarrow \textcircled{}}} -\lambda_i(\mathbb{1} - \mu \circ \pi) \circ \mu_a \circ \pi_a \mid V'_i = -\lambda_i(\mathbb{1} - \mu \circ \pi) \mid V'_i = -\lambda_i \mathbb{1}_{V'_i}$$

and $(s_i\lambda)_i = -\lambda_i$. Further, for an arrow $\underset{\textstyle i}{\bigcirc} \xleftarrow{\quad a \quad} \underset{\textstyle j}{\bigcirc}$ in Q then

$$V'_{a^*} V'_a = \pi_a \circ (-\lambda_i(\mathbb{1} - \mu\circ\pi)) \circ \mu_a = -\lambda_i \pi_a \circ \mu_a + \lambda_i \pi_a \circ \mu \circ \pi \circ \mu_a = -\lambda_i \mathbb{1}_{V_j} + V_{a^*} V_a$$

but then, whenever $j \neq i$ we have the equality

$$\sum_{\underset{\textstyle a}{\underset{\textstyle i \leftarrow \bigcirc}{}}} V'_a V'_{a^*} - \sum_{\underset{\textstyle a}{\underset{\textstyle \bigcirc \to i}{}}} V'_{a^*} V'_a = \sum_{\underset{\textstyle a}{\underset{\textstyle i \leftarrow \bigcirc}{}}} V_a V_{a^*} - \sum_{\underset{\textstyle a}{\underset{\textstyle \bigcirc \to i}{}}} V_{a^*} V_a - T_Q(\epsilon_j, \epsilon_i)\lambda_i \mathbb{1}_{V_j}$$

because there are $-T_Q(\epsilon_j, \epsilon_i)$ arrows from v_j to v_i. Then, this reduces to

$$\lambda_j \mathbb{1}_{V_j} - T_Q(\epsilon_j, \epsilon_i)\lambda_i \mathbb{1}_{V_i} = (s_i\lambda)_j \mathbb{1}_{V_j}$$

The assignment $V \mapsto V'$ extends to a functor E_i and the exact sequence

$$0 \longrightarrow V'_i \longrightarrow V_\oplus \xrightarrow{\;\pi\;} V_i \longrightarrow 0$$

shows that it acts as r_i on the dimension vectors. Finally, the reflection also defines a functor $E'_i : \Pi_{s_i(\lambda)} - \mathbf{rep} \longrightarrow \Pi_\lambda - \mathbf{rep}$ and one shows that there is a natural equivalence $V \longrightarrow E'_i(E_i(V))$ finishing the proof. $\quad\Box$

Recall from section 5.5 that for a fixed dimension vector α we have the complex moment map

$$\mathbf{rep}_\alpha\, Q^d \xrightarrow{\;\mu_\alpha\;} M_\alpha \qquad \mu_\alpha(V)_i = \sum_{\underset{\textstyle a}{\underset{\textstyle \bigcirc \to i}{}}} V_a V_{a^*} - \sum_{\underset{\textstyle a}{\underset{\textstyle i \leftarrow \bigcirc}{}}} V_{a^*} V_a$$

and that we have the identification $\underline{rep}_\alpha\, \Pi_\lambda = \mu_\alpha^{-1}(\lambda)$. A geometric interpretation of the proof of the foregoing theorem tells us that the schemes $\mu_\alpha^{-1}(\lambda)$ and $\mu_{r_i(\alpha)}^{-1}(s_i(\lambda))$ have the same number of irreducible components and that

$$dim\ \mu_\alpha^{-1}(\lambda) - \alpha.\alpha = dim\ \mu_{r_i(\alpha)}^{-1}(s_i(\lambda)) - r_i(\alpha).r_i(\alpha)$$

see [26, lemma 1.2] for full details. The set of λ-Schur roots S_λ was defined to be the set of $\alpha \in \mathbb{N}^k$ such that

$$p_Q(\alpha) \geq p_Q(\beta_1) + \ldots + p_Q(\beta_r)$$

for all decompositions $\alpha = \beta_1 + \ldots + \beta_r$ with the $\beta_i \in \Delta_\lambda^+$. If we demand a proper inequality $>$ for all decompositions we get a subset Σ_λ and call it the *set of λ-simple roots* . Recall that S_λ and hence Σ_λ consists of Schur roots of Q.

As in the case of Kac's theorem where one obtains the set of all roots from the subsets $\Pi = \{\epsilon_i \mid v_i$ has no looops$\}$ and the *fundamental set of roots* $F_Q = \{\alpha \in \mathbb{N}^k - \underline{0} \mid T_Q(\alpha, \epsilon_i) \leq 0$ and $supp(\alpha)$ is connected $\}$, we can use the above reflection functors E_i to reduce pairs (λ, α) under the equivalence relation $\sim$ to a particularly nice form, see [26, Thm. 4.8].

THEOREM 8.6

If $\alpha \in \Sigma_\lambda$, then $(\lambda, \alpha) \sim (\lambda', \alpha')$ with

$$\begin{cases} \alpha' \in \Pi & \text{if } \alpha \text{ is a real root,} \\ \alpha' \in F_Q & \text{if } \alpha \text{ is an imaginary root} \end{cases}$$

PROPOSITION 8.3

If (λ, α) is such that $\alpha \in \Sigma_\lambda$, then $\underline{rep}_\alpha \, \Pi_\lambda = \mu_\alpha^{-1}(\lambda)$ is irreducible and

$$dim \, \mu_\alpha^{-1}(\lambda) = \alpha.\alpha - 1 + 2p_Q(\alpha)$$

In particular, $\mu_\alpha^{-1}(\lambda)$ is a complete intersection.

PROOF   If $\alpha \in \Sigma_\lambda$, then we know by theorem 5.18 that

$$dim \, \mu_\alpha^{-1}(\lambda) = \alpha.\alpha - \chi_Q(\alpha, \alpha) + p_Q(\alpha) = \alpha.\alpha - 1 + 2p_Q(\alpha)$$

as $p_Q(\alpha) = 1 - \chi_Q(\alpha, \alpha)$. Moreover, this number is also the relative dimension of the complex moment map μ_α. Therefore, $\mu_\alpha^{-1}(\lambda)$ is equidimensional and we only have to prove that it is irreducible.

By theorem 8.6 and the geometric interpretation of the reflexion functor equivalence we may reduce to the case where α is either a coordinate vector or lies in the fundamental region. The former case being trivial, we assume $\alpha \in F_Q$. Consider the projection map

$$\mu_\alpha^{-1}(\lambda) \xrightarrow{\ \pi\ } \mathbf{rep}_\alpha \, Q$$

then the image of π is described in theorem 5.17 and any nonempty fiber $\pi^{-1}(V) \simeq (Ext^1_{\mathbb{C}Q}(V, V))^*$ is irreducible. As in the proof of theorem 5.18 we can decompose $\mathbf{rep}_\alpha \, Q$ according to representation types in $\mathbf{rep}_\alpha(\tau)$. Because $\alpha \in \Sigma_\lambda$ we have that $dim \, \pi^{-1}(\mathbf{rep}_\alpha(\tau)) < d = \alpha.\alpha - 1 + 2p_Q(\alpha)$. for all $\tau \neq (1, \alpha)$.

Because α is a Schur root, $\mathbf{rep}_\alpha(1, \alpha)$ is an open set and $\pi^{-1}(\mathbf{rep}_\alpha \, Q - \mathbf{rep}_\alpha(1, \alpha))$ has a dimension less than d, whence it is sufficient to prove that $\pi^{-1}(\mathbf{rep}_\alpha(1, \alpha))$ is irreducible. Because it is an open subset of $\mu_\alpha^{-1}(\lambda)$ it is equidimensional of dimension d and every fiber is irreducible. But, if $X \longrightarrow Y$ is a dominant map with Y irreducible and all fibers irreducible of the same dimension, then X is irreducible, finishing the proof.   ▯

The term λ-simple roots for Σ_λ is justified by the following result.

THEOREM 8.7

Let (λ, α) be such that $\alpha \in \Sigma_\lambda$. Then, $\underline{rep}_\alpha \, \Pi_\lambda = \mu_\alpha^{-1}(\lambda)$ is a reduced and irreducible complete intersection of dimension $d = \alpha.\alpha - 1 + 2p_Q(\alpha)$ and the general element of $\mu_\alpha^{-1}(\lambda)$ is a simple representation of Π_λ.

In particular, $\mathrm{iss}_\alpha \, \Pi_\lambda$ is an irreducible variety of dimension $2p_Q(\alpha)$.

PROOF   We know that $\mu_\alpha^{-1}(\lambda)$ is irreducible of dimension d. By the type stratification, it is enough to prove the existence of one simple representation of dimension vector α. The reflection functors being equivalences of categories, we may assume that α is either in Π or in F_Q. Clearly, for α a dimension vector, there is a simple representation, whence assume $\alpha \in F_Q$.

Assume there is no simple α-dimensional representation of Π_λ. Because $\underline{rep}_\alpha \, \Pi_\lambda$ is irreducible, there is a dimension vector $\beta < \alpha$ and an open subset of representations containing a subrepresentation of dimension vector β. As the latter condition is closed, every α-dimensional representation of Π_λ contains a β-dimensional subrepresentation.

Because α is a Schur root for Q, the general α-dimensional representation of Q extends to Π_λ and hence contains a subrepresentation of dimension vector β, that is $\beta \xhookrightarrow{Q} \alpha$. Applying the same argument to the quiver Q^o we also have $\beta \xhookrightarrow{Q^o} \alpha$.

If we now consider duals, this implies that the general α-dimensional representation of Q has a subrepresentation of dimension vector $\alpha - \beta$. But then, by the results of section 4.7 we have $ext(\beta, \alpha - \beta) = 0 = ext(\alpha - \beta, \beta)$ whence a general α-dimensional representation of Q decomposes as a direct sum of representations of dimension β and $\alpha - \beta$, contradicting the fact that α is a Schur root. Hence, there are α-dimensional simple representations of Π_λ.

Let V be a simple representation in $\mu_\alpha^{-1}(\lambda)$, then computing differentials it follows that μ_α is smooth at V, whence $\mu_\alpha^{-1}(\lambda)$ is generically reduced. But then, being a complete intersection, it is Cohen-Macaulay and therefore reduced.   ☐

This finishes the proof of the *easy* part of the characterization of simple roots for Π_λ due to W. Crawley-Boevey [26].

THEOREM 8.8

The following are equivalent

1. *Π_λ has α-dimensional simple representations*

2. *$\alpha \in \Sigma_\lambda$*

The proof of [26] involves a lengthy case-by-case study and awaits a more transparent argument, perhaps along the lines of hyper-Kähler reduction as in section 8.5.

If $\alpha \in \Sigma_\lambda$, then $\Pi_\lambda(\alpha)$ is an order in a central simple algebra over the functionfield of $\mathtt{iss}_\alpha \, \Pi_\lambda$.

8.4 Hilbert schemes

In this section we will illustrate some of the foregoing results in the special case of the quiver Q coming from the study of linear dynamical systems, and its double quiver Q^d

In order to avoid heavy use of stars, we denote as in the previous chapters, $a = u$, $a^* = v$, $b = x$ and $b^* = y$, so the path algebra of the double Q^d

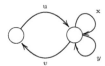

is the algebra $\mathbb{M}$ considered before. We fix the dimension vector $\alpha = (1, n)$ and the character $\theta = (-n, 1)$ and recall from section 8.2 that the moduli space $M_\alpha^{ss}(Q, \theta) \simeq \mathbb{C}^n$.

We say that u is a *cyclic vector* for the matrix-couple $(X, Y) \in M_n(\mathbb{C}) \oplus M_n(\mathbb{C})$ if there is no proper subspace of $\mathbb{C}^n$ containing u, which is stable under left multiplication by X and Y.

LEMMA 8.4
A representation $V = (X, Y, u, v) \in \mathtt{rep}_\alpha \, \mathbb{M}$ is θ-semistable if and only if u is a cyclic vector for (X, Y). Moreover, in this case V is even θ-stable.

PROOF   If there is a proper subspace of $\mathbb{C}^n$ of dimension k containing u and stable under the multiplication with X and Y then V contains a subrepresentation of dimension $\beta = (1, k)$ and $\theta(\beta) < 0$. If u is cyclic for (X, Y) then the only proper subrepresentations of V are of dimension $(0, k)$ for some k, but for those $\theta(\beta) > 0$ whence V is θ-stable.   ▯

The complex moment map $\mu = \mu_\alpha$ for this situation is

$$\mathbf{rep}_\alpha\ Q^d = M_n(\mathbb{C}) \oplus M_n(\mathbb{C}) \oplus \mathbb{C}^n \oplus \mathbb{C}^{n*} \xrightarrow{\quad \mu \quad} \mathbb{C} \oplus M_n(\mathbb{C})$$

$$(X, Y, u, v) \qquad\qquad \mapsto \qquad (-v.u, [Y, X] + u.v)$$

Observe that the image is contained in $M_\alpha^0(\mathbb{C}) = \{(c, M) \mid c + tr(M) = 0\}$. The differential $d\mu$ in the point (X, Y, u, v) is equal to

$$d\mu_{(X,Y,u,v)}\ (A, B, c, d) = (-v.c - d.u, [B, X] + [Y, A] + u.d + c.v)$$

LEMMA 8.5
The second component of the differential $d\mu$ is surjective in (X, Y, u, v) if u is a cyclic vector for (X, Y).

PROOF   Consider the *nondegenerate* symmetric bilinear form $tr(MN)$ on $M_n(\mathbb{C})$ With respect to this inproduct on $M_n(\mathbb{C})$ the space orthogonal to the image of (the second component of) $d\mu_{(X,Y,u,v)}$ is equal to

$$\{M \in M_n(\mathbb{C}) \mid tr([B, X]M + [Y, A]M + u.dM + c.vM) = 0, \forall(A, B, c, d)\}$$

Because the trace does not change under cyclic permutations and is nondegenerate we see that this space is equal to

$$\{M \in M_n(\mathbb{C}) \mid [M, X] = 0 \quad [Y, M] = 0 \quad Mu = 0 \quad \text{and} \quad vM = 0\}$$

But then, the kernel *ker* M is a subspace of $\mathbb{C}^n$ containing u and stable under left multiplication by X and Y. By the cyclicity assumption this implies that *ker* $M = \mathbb{C}^n$ or equivalently that $M = 0$. As $d\mu_{(X,Y,u,v)}^\perp = 0$ and *tr* is nondegenerate, this implies that the differential is surjective.   □

Let $\mathbf{rep}_\alpha^{ss}\ Q^d = \mathbf{rep}_\alpha^s\ Q^d = \mathbf{rep}_\alpha^s\ \mathbb{M}$ be the open variety of θ-(semi)stable representations.

PROPOSITION 8.4
For every matrix $(c, M) \in M_\alpha^0(\mathbb{C})$ in the image of the map

$$\mathbf{rep}_\alpha^s\ \mathbb{M} \xrightarrow{\quad \mu \quad} M_\alpha^0(\mathbb{C})$$

the inverse image $\mu^{-1}(M)$ is a submanifold of $\mathbf{rep}_\alpha\ \mathbb{M}$ of dimension $n^2 + 2n$.

This is a special case of theorem 5.19. Observe that for the quiver Q we have $p_Q(m, n) = mn + 1 - m^2$. As any decomposition of $\alpha = (1, n)$ is of the form

$$(1, n) = (1, a_1) + (0, a_2) + \ldots + (0, a_k) \qquad \text{with} \qquad \sum_i a_i = n$$

we have that $p_Q(\alpha) = n \geq \sum_i p_Q(\beta_i) = a_1 + 1 + \ldots + 1$ and equality only occurs for $(1,1) + (0,1) + \ldots + (0,1)$. Therefore $\alpha \in S_{\underline{0}}$.

We now turn to the description of the moduli space $M_\alpha^{ss}(Q^d, \theta)$. In this particular case we clearly have.

LEMMA 8.6
For $\alpha = (1, n)$ and $\theta = (-n, 1)$ there is a natural one-to-one correspondence between

1. $GL(\alpha)$-orbits in $\mathbf{rep}_\alpha^s \, \mathbb{M}$, and

2. GL_n-orbits in $\mathbf{rep}_\alpha^s \, \mathbb{M}$ under the induced action

For the investigation of the $GL_n(\mathbb{C})$-orbits on $\mathbf{rep}_\alpha^s \, \mathbb{M}$ we introduce a combinatorial gadget : the *Hilbert n-stair*. This is the lower triangular part of a square $n \times n$ array of boxes

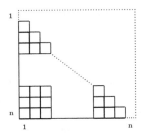

filled with *go-stones* according to the following two rules :

- each row contains exactly one stone, and

- each column contains at most one stone of each color

For example, the set of all possible Hilbert 3-stairs is given below

To every Hilbert stair σ we will associate a sequence of monomials $W(\sigma)$ in the free noncommutative algebra $\mathbb{C}\langle x, y \rangle$, that is, $W(\sigma)$ is a sequence of words in x and y.

At the top of the stairs we place the identity element 1. Then, we descend the stairs according to the following rule.

- Every go-stone has a *top word* T, which we may assume we have constructed before and a *side word* S and they are related as indicated

below

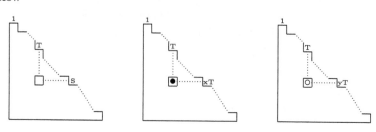

For example, for the Hilbert 3-stairs we have the following sequences of noncommutative words

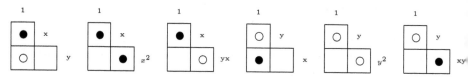

We will evaluate a Hilbert n-stair σ with associated sequence of noncommutative words $W(\sigma) = \{1, w_2(x, y), \ldots, w_n(x, y)\}$ on

$$\mathbf{rep}_\alpha \ \mathbb{M} = M_n(\mathbb{C}) \oplus M_n(\mathbb{C}) \oplus \mathbb{C}^n \oplus \mathbb{C}^{n*}$$

For a quadruple (X, Y, u, v) we replace every occurrence of x in the word $w_i(x, y)$ by X and every occurrence of y by Y to obtain an $n \times n$ matrix $w_i = w_i(X, Y) \in M_n(\mathbb{C})$ and by left multiplication on u a column vector $w_i.v$. The *evaluation of σ on* (X, Y, u, v) is the determinant of the $n \times n$ matrix

$$\sigma(X, Y, u, v) = \det \begin{array}{|c|c|c|c|c|} \hline u & w_2.u & w_3.u & \cdots & w_n.u \\ \hline \end{array}$$

For a fixed Hilbert n-stair σ we denote with $rep(\sigma)$ the subset of quadruples (X, Y, u, v) in $\mathbf{rep}_\alpha \ \mathbb{M}$ such that the evaluation $\sigma(v, X, Y) \neq 0$.

THEOREM 8.9
For every Hilbert n-stair, rep $(\sigma) \neq \emptyset$

PROOF   Let u be the basic column vector

$$e_1 = \begin{bmatrix} 1 \\ 0 \\ \vdots \\ 0 \end{bmatrix}$$

Let every black stone in the Hilbert stair σ fix a column of X by the rule

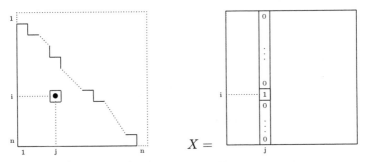

$$X =$$

That is, one replaces every black stone in σ by 1 at the same spot in X and fills the remaining spots in the same column by zeroes. The same rule applies to Y for white stones. We say that such a quadruple (X, Y, u, v) is in σ-*standard form*.

With these conventions one easily verifies by induction that

$$w_i(X, Y)e_1 = e_i \quad \text{for all } 2 \leq i \leq n$$

Hence, filling up the remaining spots in X and Y arbitrarily one has that $\sigma(X, Y, u, v) \neq 0$ proving the claim. $\square$

Hence, $rep\,(\sigma)$ is an open subset of $\mathbf{rep}_\alpha\,\mathbb{M}$ (and even of $\mathbf{rep}_\alpha^s\,\mathbb{M}$) for every Hilbert n-stair σ. Further, for every word (monomial) $w(x, y)$ and every $g \in GL_n(\mathbb{C})$ we have that

$$w(gXg^{-1}, gYg^{-1})gv = gw(X, Y)v$$

and therefore the open sets $rep\,(\sigma)$ are stable under the $GL_n(\mathbb{C})$-action on $\mathbf{rep}_\alpha\,\mathbb{M}$. We will give representatives of the orbits in $rep\,(\sigma)$.

Let $W_n = \{1, x, \dots, x^n, xy, \dots, y^n\}$ be the set of all words in the non-commuting variables x and y of length $\leq n$, ordered lexicographically.

For every quadruple $(X, Y, u, v) \in \mathbf{rep}_\alpha\,\mathbb{M}$ consider the $n \times m$ matrix

$$\psi(X, Y, u, v) = [u\ Xu\ X^2u \ \dots\ Y^n u]$$

where $m = 2^{n+1} - 1$ and the j-th column is the column vector $w(X, Y)v$ with $w(x, y)$ the j-th word in W_n.

Hence, $(X, Y, u, v) \in rep\,(\sigma)$ if and only if the $n \times n$ minor of $\psi(X, Y, u, v)$ determined by the word-sequence $\{1, w_2, \dots, w_n\}$ of σ is invertible. Moreover, as

$$\psi(gXg^{-1}, gYg^{-1}, gu, vg^{-1}) = g\psi(v, X, Y)$$

we deduce that the $GL_n(\mathbb{C})$-orbit of $(X, Y, u, v) \in \mathbf{rep}_\alpha\,\mathbb{M}$ contains a *unique* quadruple (X_1, Y_1, u_1, v_1) such that the corresponding minor of $\psi(X_1, Y_1, u_1, v_1) = \mathbb{1}_n$.

Hence, each $GL_n(\mathbb{C})$-orbit in *rep* (σ) contains a unique representant in σ-standard form. See discussion below.

PROPOSITION 8.5

The action of $GL_n(\mathbb{C})$ on rep (σ) is free and the orbit space

$$rep\ (\sigma)/GL_n(\mathbb{C})$$

is an affine space of dimension $n^2 + 2n$.

PROOF   The dimension is equal to the number of nonforced entries in X, Y and v. As we fixed $n-1$ columns in X or Y this dimension is equal to

$$k = 2n^2 - (n-1)n + n = n^2 + 2n$$

The argument above shows that every $GL_n(\mathbb{C})$-orbit contains a unique quadruple in σ-standard form so the orbit space is an affine space.   ☐

THEOREM 8.10

For $\alpha = (1, n)$ and $\theta = (-n, 1)$, the moduli space

$$M_\alpha^{ss}(Q^d, \theta) = M_\alpha^{ss}(\mathbb{M}, \theta)$$

is a complex manifold of dimension $n^2 + 2n$ and is covered by the affine spaces rep (σ).

PROOF   Recall that $\mathbf{rep}_\alpha^s\ \mathbb{M}$ is the open submanifold consisting of quadruples (x, Y, u, v) such that u is a cyclic vector of (X, Y) or equivalently such that

$$\mathbb{C}\langle X, Y \rangle u = \mathbb{C}^n$$

where $\mathbb{C}\langle X, Y \rangle$ is the not necessarily commutative subalgebra of $M_n(\mathbb{C})$ generated by the matrices X and Y.

Hence, clearly *rep* $(\sigma) \subset \mathbf{rep}_n\ \mathbb{M}$ for any Hilbert n-stair σ. Conversely, we claim that a quadruple $(X, Y, u, v) \in \mathbf{rep}_\alpha^s\ \mathbb{M}$ belongs to at least one of the open subsets *rep* (σ).

Indeed, either $Xu \notin \mathbb{C}u$ or $Yu \notin \mathbb{C}u$ as otherwise the subspace $W = \mathbb{C}u$ would contradict the cyclicity assumption. Fill the top box of the stairs with the corresponding stone and define the 2-dimensional subspace $V_2 = \mathbb{C}u_1 + \mathbb{C}u_2$ where $u_1 = u$ and $u_2 = w_2(X, Y)u$ with w_2 the corresponding word (either x or y).

Assume by induction we have been able to fill the first i rows of the stairs with stones leading to the sequence of words $\{1, w_2(x, y), \ldots, w_i(x, y)\}$ such that the subspace $V_i = \mathbb{C}u_1 + \ldots + \mathbb{C}u_i$ with $u_i = w_i(X, Y)v$ has dimension i.

Then, either $Xu_j \notin V_i$ for some j or $Yu_j \notin V_i$ (if not, V_i would contradict cyclicity). Then, fill the j-th box in the $i+1$-th row of the stairs with the corresponding stone. Then, the top $i+1$ rows of the stairs form a Hilbert $i+1$-stair as there can be no stone of the same color lying in the same column. Define $w_{i+1}(x,y) = xw_i(x,y)$ (or $yw_i(x,y)$) and $u_{i+1} = w_{i+1}(X,Y)u$. Then, $V_{i+1} = \mathbb{C}u_1 + \ldots + \mathbb{C}u_{i+1}$ has dimension $i+1$.

Continuing we end up with a Hilbert n-stair σ such that $(X,Y,u,v) \in$ $rep\,(\sigma)$. This concludes the proof. $\qquad\qquad\qquad\qquad\qquad\qquad\qquad$ 🗌

Example 8.1 The moduli space $M_\alpha^{ss}(Q^d,\theta)$ when $n=3$

Representatives for the $GL_3(\mathbb{C})$-orbits in $rep\,(\sigma)$ are given by the following quadruples for σ a Hilbert 3-stair

$$X\quad \begin{bmatrix} 0 & a & b \\ 1 & c & d \\ 0 & e & f \end{bmatrix} \begin{bmatrix} 0 & 0 & a \\ 1 & 0 & b \\ 0 & 1 & c \end{bmatrix} \begin{bmatrix} 0 & a & b \\ 1 & c & d \\ 0 & e & f \end{bmatrix} \begin{bmatrix} 0 & a & b \\ 0 & c & d \\ 1 & e & f \end{bmatrix} \begin{bmatrix} a & b & c \\ d & e & f \\ g & h & i \end{bmatrix} \begin{bmatrix} a & 0 & b \\ c & 0 & d \\ e & 1 & f \end{bmatrix}$$

$$Y\quad \begin{bmatrix} 0 & g & h \\ 0 & i & j \\ 1 & k & l \end{bmatrix} \begin{bmatrix} d & e & f \\ g & h & i \\ j & k & l \end{bmatrix} \begin{bmatrix} g & 0 & h \\ i & 0 & j \\ k & 1 & l \end{bmatrix} \begin{bmatrix} 0 & g & h \\ 1 & i & j \\ 0 & k & l \end{bmatrix} \begin{bmatrix} 0 & 0 & j \\ 1 & 0 & k \\ 0 & 1 & l \end{bmatrix} \begin{bmatrix} 0 & g & h \\ 1 & i & j \\ 0 & k & l \end{bmatrix}$$

$$u\quad \begin{bmatrix} 1 \\ 0 \\ 0 \end{bmatrix} \begin{bmatrix} 1 \\ 0 \\ 0 \end{bmatrix} \begin{bmatrix} 1 \\ 0 \\ 0 \end{bmatrix} \begin{bmatrix} 1 \\ 0 \\ 0 \end{bmatrix} \begin{bmatrix} 1 \\ 0 \\ 0 \end{bmatrix} \begin{bmatrix} 1 \\ 0 \\ 0 \end{bmatrix}$$

$$v\quad [m\ n\ o]\ [m\ n\ o]\ [m\ n\ o]\ [m\ n\ o]\ [m\ n\ o]\ [m\ n\ o]$$

$\qquad\qquad\qquad\qquad\qquad\qquad\qquad\qquad\qquad\qquad\qquad\qquad\qquad$ 🗌

We now turn to the deformed preprojective algebras. Let $\lambda = (-n\lambda, \lambda \mathbb{1}_n) \in$ $M_\alpha^0(\mathbb{C})$ for $\lambda \in \mathbb{C}$. Then

$$\Pi_\lambda = \frac{\mathbb{M}}{(v.u + \lambda v_1, [Y,X] + u.v - \lambda v_2)}$$

then if we denote by $M_\alpha^{ss}(\Pi_\lambda, \theta)$ the moduli space of θ-semistable representations of Π_λ, then we have the following situation

$$
\begin{array}{ccc}
\mu^{-1}(\lambda) \cap \mathbf{rep}_\alpha^s \; \mathbb{M} & \lhook\joinrel\longrightarrow & \mathbf{rep}_\alpha^s \; \mathbb{M} \\
\downarrow & & \downarrow \\
M_\alpha^{ss}(\Pi_\lambda, \theta) & \lhook\joinrel\longleftarrow & M_\alpha^{ss}(\mathbb{M}, \theta)
\end{array}
$$

and from the theorem above we obtain the following theorem.

THEOREM 8.11
For a $\lambda \in M_\alpha^0(\mathbb{C})$, the orbit space of θ-semistable representations of the deformed preprojective algebra

$$M_\alpha^{ss}(\Pi_\lambda, \theta)$$

is a submanifold of $M_\alpha^{ss}(\mathbb{M}, \theta)$ of dimension $2n$.

We will identify the special case of the preprojective algebra (that is, $\lambda = \underline{0}$) with the *Hilbert scheme of n points in the plane* .

Consider a codimension n ideal $i \lhd \mathbb{C}[x, y]$ and fix a basis $\{v_1, \ldots, v_n\}$ of the quotient space

$$V_i = \frac{\mathbb{C}[x, y]}{i} = \mathbb{C}v_1 + \ldots + \mathbb{C}v_n$$

Multiplication by x on $\mathbb{C}[x, y]$ induces a linear operator on the quotient V_i and hence determines a matrix $X_i \in M_n(\mathbb{C})$ with respect to the chosen basis $\{v_1, \ldots, v_n\}$. Similarly, multiplication by y determines a matrix $Y_i \in M_n(\mathbb{C})$.

Moreover, the image of the unit element $1 \in \mathbb{C}[x, y]$ in V_i determines with respect to the basis $\{v_1, \ldots, v_n\}$ a column vector $u \in \mathbb{C}^n = V_i$. Clearly, this vector and matrices satisfy

$$[X_i, Y_i] = 0 \quad \text{and} \quad \mathbb{C}[X_i, Y_i]u = \mathbb{C}^n$$

Here, $\mathbb{C}[X_i, Y_i]$ is the n-dimensional subalgebra of $M_n(\mathbb{C})$ generated by the two matrices X_i and Y_i. In particular, u is a cyclic vector for the matrix-couple (X, Y).

Conversely, if $(X, Y, u) \in M_n(\mathbb{C}) \oplus M_n(\mathbb{C}) \oplus \mathbb{C}^n$ is a cyclic triple such that $[X, Y] = 0$, then $\mathbb{C}\langle X, Y \rangle = \mathbb{C}[X, Y]$ is an n-dimensional commutative subalgebra of $M_n(\mathbb{C})$, then the kernel of the natural epimorphism

$$\mathbb{C}[x, y] \longrightarrow\!\!\!\!\!\rightarrow \mathbb{C}[X, Y] \qquad x \mapsto X \quad y \mapsto Y$$

is a codimension n ideal i of $\mathbb{C}[x, y]$.

However, there is some redundancy in the assignment $i \longrightarrow (X_i, Y_i, u_i)$ as it depends on the choice of basis of V_i. If we choose a different basis $\{v'_1, \ldots, v'_n\}$ with base change matrix $g \in GL_n(\mathbb{C})$, then the corresponding triple is

$$(X'_i, Y'_i, u'_i) = (g.X_i.g^{-1}, g.Y_i.g^{-1}, gu_i)$$

The above discussion shows that there is a one-to-one correspondence between

- codimension n ideals i of $\mathbb{C}[x,y]$, and

- $GL_n(\mathbb{C})$-orbits of cyclic triples (X,Y,u) in $M_n(\mathbb{C}) \oplus M_n(\mathbb{C}) \oplus \mathbb{C}^n$ such that $[X,Y] = 0$

Example 8.2 The Hilbert scheme $Hilb_2$
Consider a triple $(X,Y,u) \in M_2(\mathbb{C}) \oplus M_2(\mathbb{C}) \oplus \mathbb{C}^2$ and assume that either X or Y has distinct eigenvalues (type a). As

$$[\begin{bmatrix} \nu_1 & 0 \\ 0 & \nu_2 \end{bmatrix}, \begin{bmatrix} a & b \\ c & d \end{bmatrix}] = \begin{bmatrix} 0 & (\nu_1 - \nu_2)b \\ (\nu_2 - \nu_1)c & 0 \end{bmatrix}$$

we have a representant in the orbit of the form

$$(\begin{bmatrix} \lambda_1 & 0 \\ 0 & \lambda_2 \end{bmatrix}, \begin{bmatrix} \mu_1 & 0 \\ 0 & \mu_2 \end{bmatrix}, \begin{bmatrix} u_1 \\ u_2 \end{bmatrix})$$

where cyclicity of the column vector implies that $u_1 u_2 \neq 0$.

The stabilizer subgroup of the matrix-pair is the group of diagonal matrices $\mathbb{C}^* \times \mathbb{C}^* \hookrightarrow GL_2(\mathbb{C})$, hence the orbit has a unique representant of the form

$$(\begin{bmatrix} \lambda_1 & 0 \\ 0 & \lambda_2 \end{bmatrix}, \begin{bmatrix} \mu_1 & 0 \\ 0 & \mu_2 \end{bmatrix}, \begin{bmatrix} 1 \\ 1 \end{bmatrix})$$

The corresponding ideal i $\lhd \mathbb{C}[x,y]$ is then

$$i = \{ f(x,y) \in \mathbb{C}[x,y] \mid f(\lambda_1, \mu_1) = 0 = f(\lambda_2, \mu_2) \}$$

hence these orbits correspond to sets of two *distinct* points in $\mathbb{C}^2$.

The situation is slightly more complicated when X and Y have only one eigenvalue (type b). If (X,Y,u) is a cyclic commuting triple, then either X or Y is not diagonalizable. But then, as

$$[\begin{bmatrix} \nu & 1 \\ 0 & \nu \end{bmatrix}, \begin{bmatrix} a & b \\ c & d \end{bmatrix}] = \begin{bmatrix} c & d - a \\ 0 & c \end{bmatrix}$$

we have a representant in the orbit of the form

$$(\begin{bmatrix} \lambda & \alpha \\ 0 & \lambda \end{bmatrix}, \begin{bmatrix} \mu & \beta \\ 0 & \mu \end{bmatrix}, \begin{bmatrix} u_1 \\ u_2 \end{bmatrix})$$

with $[\alpha : \beta] \in \mathbb{P}^1$ and $u_2 \neq 0$. The stabilizer of the matrixpair is the subgroup

$$\{ \begin{bmatrix} c & d \\ 0 & c \end{bmatrix} \mid c \neq 0 \} \hookrightarrow GL_2(\mathbb{C})$$

and hence we have a unique representant of the form

$$\left(\begin{bmatrix} \lambda & \alpha \\ 0 & \lambda \end{bmatrix}, \begin{bmatrix} \mu & \beta \\ 0 & \mu \end{bmatrix}, \begin{bmatrix} 0 \\ 1 \end{bmatrix}\right)$$

The corresponding ideal $i \lhd \mathbb{C}[x, y]$ is

$$i = \{f(x, y) \in \mathbb{C}[x, y] \mid f(\lambda, \mu) = 0 \ \text{ and } \ \alpha\frac{\partial f}{\partial x}(\lambda, \mu) + \beta\frac{\partial f}{\partial y}(\lambda, \mu) = 0\}$$

as one proves by verification on monomials because

$$\begin{bmatrix} \lambda & \alpha \\ 0 & \lambda \end{bmatrix}^k \begin{bmatrix} \mu & \beta \\ 0 & \mu \end{bmatrix}^l \begin{bmatrix} 0 \\ 1 \end{bmatrix} = \begin{bmatrix} k\alpha\lambda^{k-1}\mu^l + l\beta\lambda^k\mu^{l-1} \\ \lambda^k\mu^l \end{bmatrix}$$

Therefore, i corresponds to the set of two points at $(\lambda, \mu) \in \mathbb{C}^2$ infinitesimally attached to each other in the direction $\alpha\frac{\partial}{\partial x} + \beta\frac{\partial}{\partial y}$. For each point in $\mathbb{C}^2$ there is a $\mathbb{P}^1$ family of such *fat points*.

Thus, points of $Hilb_2$ correspond to either of the following two situations

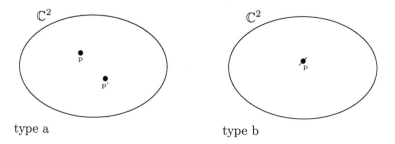

type a   type b

The Hilbert-Chow map $Hilb_2 \xrightarrow{\ \pi\ } S^2\,\mathbb{C}^2$ (where $S^2\,\mathbb{C}^2$ is the symmetric power of $\mathbb{C}^2$, that is $S_2 = \mathbb{Z}/2\mathbb{Z}$ orbits of couples of points from $\mathbb{C}^2$) sends a point of type a to the formal sum $[p] + [p']$ and a point of type b to $2[p]$. Over the complement of (the image of) the diagonal, this map is a one-to-one correspondence.

However, over points on the diagonal the fibers are $\mathbb{P}^1$ corresponding to the directions in which two points can approach each other in $\mathbb{C}^2$. As a matter of fact, the symmetric power $S^2\,\mathbb{C}^2$ has singularities and the Hilbert-Chow map $Hilb_2 \xrightarrow{\ \pi\ } S^2\,\mathbb{C}^2$ is a *resolution of singularities*. ∎

THEOREM 8.12

Let $\mathbf{rep}_\alpha\,\mathbb{M} \xrightarrow{\ \mu\ } M_\alpha^0(\mathbb{C})$ be the complex moment map, then

$$Hilb_n \simeq M_\alpha^{ss}(\Pi_0, \theta)$$

and is therefore a complex manifold of dimension $2n$.

PROOF   We identify the triples $(X, Y, u) \in M_n(\mathbb{C}) \oplus M_n(\mathbb{C}) \oplus \mathbb{C}^n$ such that u is a cyclic vector of (X, Y) and $[X, Y] = 0$ with the subspace

$$\{(X, Y, u, \underline{0}) \mid [X, Y] = 0 \text{ and } u \text{ is cyclic }\} \hookrightarrow \mathbf{rep}^s_\alpha \, \mathbb{M}$$

which is clearly contained in $\mu^{-1}(0)$. To prove the converse inclusion assume that (X, Y, u, v) is a cyclic quadruple such that

$$[X, Y] + uv = 0$$

Let $m(x, y)$ be any word in the noncommuting variables x and y. We claim that

$$v.m(X, Y).u = 0$$

We will prove this by induction on the length $l(m)$ of the word $m(x, y)$. When $l(m) = 0$ then $l(x, y) = 1$ and we have

$$v.l(X, Y).u = v.u = tr(u.v) = tr([X, Y]) = 0$$

Assume we proved the claim for all words of length $< l$ and take a word of the form $m(x, y) = m_1(x, y)yxm_2(x, y)$ with $l(m_1) + l(m_2) + 2 = l$. Then, we have

$$
\begin{aligned}
wm(X, Y) = & \quad wm_1(X, Y)YXm_2(X, Y) \\
= & \quad wm_1(X, Y)([Y, X] + XY)m_2(X, Y) \\
= & \, (wm_1(X, Y)v).wm_2(X, Y) + wm_1(X, Y)XYm_2(X, Y) \\
= & \quad wm_1(X, Y)XYm_2(X, Y)
\end{aligned}
$$

where we used the induction hypotheses in the last equality (the bracketed term vanishes).

Hence we can reorder the terms in $m(x, y)$ if necessary and have that $wm(X, Y) = wX^{l_1}Y^{l_2}$ with $l_1 + l_2 = l$ and l_1 the number of occurrences of x in $m(x, y)$. Hence, we have to prove the claim for $X^{l_1}Y^{l_2}$

$$
\begin{aligned}
wX^{l_1}Y^{l_2}v = & \quad tr(X^{l_1}Y^{l_2}vw) \\
= & \quad -tr(X^{l_1}Y^{l_2}[X, Y]) \\
= & \quad -tr([X^{l_1}Y^{l_2}, X]Y) \\
= & \quad -tr(X^{l_1}[Y^{l_2}, X]Y) \\
= & \, -\sum_{i=0}^{l_2-1} tr(X^{l_1}Y^i[Y, X]Y^{l_2-i}) \\
= & \, -\sum_{i=0}^{l_2-1} tr(Y^{l_2-i}X^{l_1}Y^i[Y, X] \\
= & \, -\sum_{i=0}^{l_2-1} tr(Y^{l_2-i}X^{l_1}Y^iv.w) \\
= & \, -\sum_{i=0}^{l_2-1} wY^{m_2-i}X^{l_1}Y^iv
\end{aligned}
$$

But we have seen that $wY^{l_2-i}X^{l_1}Y^i = wX^{l_1}Y^{l_2}$ hence the above implies that $wX^{l_1}Y^{l_2}v = -l_2 wX^{l_1}Y^{l_2}v$. But then $wX^{l_1}Y^{l_2}v = 0$, proving the claim.

Consequently, $w.\mathbb{C}\langle X, Y \rangle.v = 0$ and by the cyclicity condition we have $w.\mathbb{C}^n = 0$ hence $w = 0$. Finally, as $v.w + [X, Y] = 0$ this implies that $[X, Y] = 0$ and we can identify the fiber $\mu^{-1}(0)$ with the indicated subspace. From this the result follows. $\square$

We can use the affine covering of $M_\alpha^{ss}(\mathbb{M}, \theta)$ by Hilbert stairs, to cover the Hilbert scheme $Hilb_n$ by the intersections $Hilb(\sigma) = rep(\sigma) \cap Hilb_n$.

Example 8.3 The Hilbert scheme $Hilb_2$
Consider $Hilb_2$ ($\boxed{\bullet}$). Because

$$[\begin{bmatrix} 0 & a \\ 1 & b \end{bmatrix}, \begin{bmatrix} c & d \\ e & f \end{bmatrix}] = \begin{bmatrix} ae - d & af - ac - bd \\ c + be - f & d - ae \end{bmatrix}$$

this subset can be identified with $\mathbb{C}^4$ using the equalities

$$d = ar \quad \text{and} \quad f = c + be$$

Similarly, $Hilb_2$ ($\boxed{\circ}$) $\simeq \mathbb{C}^4$. $\square$

THEOREM 8.13
The Hilbert scheme $Hilb_n$ of n points in $\mathbb{C}^2$ is a complex connected manifold of dimension $2n$.

PROOF   The symmetric power $S^n \mathbb{C}^1$ parameterizes *sets* of n-points on the line $\mathbb{C}^1$ and can be identified with $\mathbb{C}^n$. Consider the map

$$Hilb_n \xrightarrow{\pi} S^n \mathbb{C}^1$$

defined by mapping a cyclic triple (X, Y, u) with $[X, Y] = 0$ in the orbit corresponding to the point of $Hilb_n$ to the set $\{\lambda_1, \ldots, \lambda_n\}$ of eigenvalues of X. Observe that this map does not depend on the point chosen in the orbit.

Let Δ be the *big diagonal* in $S^n \mathbb{C}^1$, that is, $S^n \mathbb{C}^1 - \Delta$ is the space of all sets of n distinct points from $\mathbb{C}^1$. Clearly, $S^n \mathbb{C}^1 - \Delta$ is a connected n-dimensional manifold. We claim that

$$\pi^{-1}(S^n \mathbb{C}^1 - \Delta) \simeq (S^n \mathbb{C}^1 - \Delta) \times \mathbb{C}^n$$

and hence is connected.

Indeed, take a matrix X with n distinct eigenvalues $\{\lambda_1, \ldots, \lambda_n\}$. We may diagonalize X. But then, as

$$[\begin{bmatrix} \lambda_1 & & \\ & \ddots & \\ & & \lambda_n \end{bmatrix}, \begin{bmatrix} y_{11} & \cdots & y_{1n} \\ \vdots & & \vdots \\ y_{n1} & \cdots & y_{nn} \end{bmatrix}] = \begin{bmatrix} (\lambda_1 - \lambda_1)y_{11} & \cdots & (\lambda_1 - \lambda_n)y_{1n} \\ \vdots & & \vdots \\ (\lambda_n - \lambda_1)y_{n1} & \cdots & (\lambda_n - \lambda_n)y_{nn} \end{bmatrix}$$

we see that also Y must be a diagonal matrix with entries $(\mu_1, \ldots, \mu_n) \in \mathbb{C}^n$ where $\mu_i = y_{ii}$. But then the cyclicity condition implies that all coordinates of v must be nonzero.

Now, the stabilizer subgroup of the commuting (diagonal) matrix-pair (X, Y) is the *maximal torus* $T_n = \mathbb{C}^* \times \ldots \times \mathbb{C}^*$ of diagonal invertible $n \times n$ matrices. Using its action we may assume that all coordinates of v are equal to 1. That is, the points in $\pi^{-1}(\{\lambda_1, \ldots, \lambda_n\})$ with $\lambda_i \neq \lambda_j$ have unique (up to permutation as before) representatives of the form

$$\left(\begin{bmatrix} \lambda_1 & & & \\ & \lambda_2 & & \\ & & \ddots & \\ & & & \lambda_n \end{bmatrix}, \begin{bmatrix} \mu_1 & & & \\ & \mu_2 & & \\ & & \ddots & \\ & & & \mu_n \end{bmatrix}, \begin{bmatrix} 1 \\ 1 \\ \vdots \\ 1 \end{bmatrix} \right)$$

that is, $\pi^{-1}(\{\lambda_1, \ldots, \lambda_n\}$ can be identified with $\mathbb{C}^n$, proving the claim.

Next, we claim that *all* the fibers of π have dimension at most n. Let $\{\lambda_1, \ldots, \lambda_n\} \in S^n \, \mathbb{C}^1$, then there are only finitely many X in Jordan normal-form with eigenvalues $\{\lambda_1, \ldots, \lambda_n\}$. Fix such an X, then the subset $T(X)$ of cyclic triples (X, Y, u) with $[X, Y] = 0$ has dimension at most $n + dim \, C(X)$ where $C(X)$ is the centralizer of X in $M_n(\mathbb{C})$, that is

$$C(X) = \{Y \in M_n(\mathbb{C}) \mid XY = YX\}$$

The stabilizer subgroup $Stab(X) = \{g \in GL_n(\mathbb{C}) \mid gXg^{-1} = X\}$ is an open subset of the vectorspace $C(X)$ and acts freely on the subset $T(X)$ because the action of $GL_n(\mathbb{C})$ on $\mu^{-1}(0) \cap \mathbf{rep}_\alpha^s \, \mathbb{M}$ has trivial stabilizers.

But then, the orbit space for the $Stab(X)$-action on $T(X)$ has a dimension at most

$$n + dim \, C(X) - dim \, Stab(X) = n.$$

As we only have to consider finitely many X this proves the claim. The diagonal Δ has dimension $n - 1$ in $S^n \, \mathbb{C}^1$ and hence by the foregoing we know that the dimension of $\pi^{-1}(\Delta)$ is at most $2n - 1$. Let H be the connected component of $Hilb_n$ containing the connected subset $\pi^{-1}(S^n \, \mathbb{C}^1 - \Delta)$. If $\pi^{-1}(\Delta)$ were not entirely contained in H, then $Hilb_n$ would have a component of dimension less than $2n$, which we proved not to be the case. This finishes the proof. □

We can give a representation theoretic interpretation of the *resolution of singularities Hilbert-Chow morphism*

$$Hilb_n \xrightarrow{\pi} S^n \, \mathbb{C}^2$$

$\Sigma_0 = \{(1, 0), (0, 1)\}$, that is, the only simple Π_0-representations are one-dimensional. Any semisimple representation of Π_0 of dimension vector $\alpha = (1, n)$ therefore decomposes as $T_0 \oplus S_1^{\oplus e_1} \oplus \ldots \oplus S_r^{\oplus e_r}$ with T_0 the unique simple

$(1, 0)$-dimensional representation and the S_i in the two-dimensional family of $(0, 1)$-simple representations of Π_0 (corresponding to couples $(\lambda_i, \mu_i) \in \mathbb{C}^2$). Therefore we have the projective bundle morphism

$$Hilb_n = M_\alpha^{ss}(\Pi_0, \theta) \xrightarrow{\pi'} \text{iss}_\alpha \, \Pi_0 = S^n \, \mathbb{C}^2$$

where the mapping sends a point of $Hilb_n$ determined by a cyclic triple (X, Y, u) to the n-tuple of eigenvalues (λ_i, μ_i) of X and Y.

8.5 Hyper Kähler structure

Again, Q is a quiver on k vertices and Q^d its double. We fix a dimension vector $\alpha = (a_1, \dots, a_k) \in \mathbb{N}^k$ and a character $\theta = (t_1, \dots, t_k) \in \mathbb{Z}^k$ and a weight $\lambda = (\lambda_1, \dots, \lambda_k) \in \mathbb{C}^k$ such that the numerical conditions

$$\theta(\alpha) = \sum_{i=1}^{k} t_i a_i = 0 \qquad \text{and} \qquad \lambda(\alpha) = \sum_{i=1}^{n} \lambda_i a_i = 0$$

are satisfied. The first is required to have θ-semistable representations, the second for λ to lie in the image of the complex moment map

$$\text{rep}_\alpha \, Q^d \xrightarrow{\mu_{\mathbb{C}}} M_\alpha^0(\mathbb{C}) \qquad V \mapsto \sum_{\substack{a \in Q_a}} [V_a, V_{a^*}]$$

where a^* is the arrow in Q_a^d corresponding to $a \in Q_a$ (that is, with the opposite direction).

Recall that the *quaternion algebra* $\mathbb{H}$ is the 4-dimensional division algebra over $\mathbb{R}$ defined by

$$\mathbb{H} = \mathbb{R}.1 \oplus \mathbb{R}.i \oplus \mathbb{R}.j \oplus \mathbb{R}.k \qquad i^2 = j^2 = k^2 = -1 \quad k = ij = -ji$$

DEFINITION 8.1   *A C^∞ (real) manifold M is said to be a hyper-Kähler manifold if $\mathbb{H}$ acts on H by diffeomorphisms.*

LEMMA 8.7

For any quiver Q, the representation space $\text{rep}_\alpha \, Q^d$ is a hyper-Kähler manifold.

PROOF   We have to specify the actions. They are defined as follows, for $V \in \mathtt{rep}_\alpha Q^d$ for all arrows $b \in Q_a^d$ and all arrows $a \in Q_a$

$$(i.V)_b = iV_b$$

$$(j.V)_a = -V_{a*}^\dagger \quad (j.V)_{a*} = V_a^\dagger$$

$$(k.V)_a = -iV_{a*}^\dagger \quad (k.V)_{a*} = iV_a^\dagger$$

where this time we denote the *Hermitian adjoint* of a matrix M by $M^\dagger$ to distinguish it from the star-operation on the arrows of Q^d. A calculation shows that these operations satisfy the required relations. $\square$

In section 8.1 we introduced the *real moment map* for quiver representations. If we apply this to the double quiver Q^d we can take

$$\mathtt{rep}_\alpha Q^d \xrightarrow{\mu_\mathbb{R}} Lie\ U(\alpha) \qquad V \mapsto \sum_{b \in Q_a^d} \frac{i}{2}[V_b, V_b^\dagger]$$

We will use the action by nonzero elements of $\mathbb{H}$ to obtain C^∞-diffeomorphisms between certain subsets of $\mathtt{rep}_\alpha Q^d$. Let $h = \frac{i-k}{\sqrt{2}}$ then we have

$$\mu_\mathbb{C}(h.V) = \frac{1}{2} \sum_{a \in Q_a} [iV_a + iV_{a*}^\dagger, iV_{a*} - iV_a^\dagger]$$

$$= \frac{1}{2} \sum_{a \in Q_a} (-[V_a, V_{a*}] + [V_a, V_a^\dagger] - [V_{a*}^\dagger, V_{a*}] + [V_{a*}^\dagger, V_a^\dagger])$$

$$= \frac{1}{2} \sum_{a \in Q_a} ([V_a, V_{a*}]^\dagger - [V_a, V_{a*}]) + \frac{1}{2} \sum_{a \in Q_a} ([V_a, V_a^\dagger] + [V_{a*}, V_{a*}^\dagger])$$

$$= \frac{1}{2}(\mu_\mathbb{C}(V)^\dagger - \mu_\mathbb{C}(V)) - i\mu_\mathbb{R}(V)$$

and

$$\mu_\mathbb{R}(h.V) = \frac{i}{4}(\sum_{a \in Q_a} [iV_a + iV_{a*}^\dagger, -iV_a^\dagger - iV_{a*}] + \sum_{a \in Q_a} [iV_{a*} - iV_a^\dagger, -iV_{a*}^\dagger + iV_a])$$

$$= \frac{i}{4} \sum_{a \in Q_a} ([V_a, V_a^\dagger] + [V_a, V_{a*}] + [V_{a*}^\dagger, V_a^\dagger] + [V_{a*}^\dagger, V_{a*}]$$

$$+ [V_{a*}, V_{a*}^\dagger] - [V_{a*}, V_a] - [V_a^\dagger, V_{a*}^\dagger] + [V_a^\dagger, V_a])$$

$$= \frac{i}{4}(2\mu_\mathbb{C}(V) + 2\mu_\mathbb{C}(V)^\dagger)$$

In particular we have the following.

PROPOSITION 8.6

If $\lambda \in \mathbb{R}^k$, then we have a homeomorphism between the real varieties

$$\mu_{\mathbb{C}}^{-1}(\lambda \mathbb{1}_\alpha) \cap \mu_{\mathbb{R}}^{-1}(\underline{0}) \overset{h.}{\longrightarrow} \mu_{\mathbb{C}}^{-1}(\underline{0}) \cap \mu_{\mathbb{R}}^{-1}(i\lambda \mathbb{1}_\alpha)$$

Moreover, the hyper-Kähler structure commutes with the base-change action of $U(\alpha)$, whence we have a natural one-to-one correspondence between the quotient spaces

$$(\mu_{\mathbb{C}}^{-1}(\lambda \mathbb{1}_\alpha) \cap \mu_{\mathbb{R}}^{-1}(\underline{0}))/U(\alpha) \overset{h.}{\longrightarrow} (\mu_{\mathbb{C}}^{-1}(\underline{0}) \cap \mu_{\mathbb{R}}^{-1}(i\lambda \mathbb{1}_\alpha))/U(\alpha)$$

By the results of section 8.1 we can identify both sides. To begin, by definition of the complex moment map $\mu_{\mathbb{C}}$ we have that

$$\mu_{\mathbb{C}}^{-1}(\underline{0}) = \underline{rep}_\alpha \Pi_0 \quad \text{and} \quad \mu_{\mathbb{C}}^{-1}(\lambda \mathbb{1}_\alpha) = \underline{rep}_\alpha \Pi_\lambda$$

Moreover, applying theorem 8.3 to the double quiver Q^d we have

$$\mathtt{iss}_\alpha \, Q^d \simeq \mu_{\mathbb{R}}^{-1}(\underline{0})/U(\alpha) \quad \text{and} \quad M_\alpha^{ss}(Q^d, \lambda) \simeq \mu_{\mathbb{R}}^{-1}(i\lambda \mathbb{1}_\alpha)/U(\alpha)$$

when $\lambda \in \mathbb{Z}^k$, concluding with the following proof.

THEOREM 8.14

For a character $\theta = (t_1, \ldots, t_k) \in \mathbb{Z}^k$ such that $\theta(\alpha) = 0$, there is a natural one-to-one correspondence between

$$\mathtt{iss}_\alpha \, \Pi_\theta \overset{h.}{\longrightarrow} M_\alpha^{ss}(\Pi_0, \theta)$$

which is an homeomorphism in the (induced) real topology.

Note however that this bijection does *not* respect the complex structures of these varieties. This is already clear from the fact that $\mathtt{iss}_\alpha \, \Pi_\theta$ is an affine complex variety and $M_\alpha^{ss}(\Pi_0, \theta)$ is a projective bundle over $\mathtt{iss}_\alpha \, \Pi_0$.

If $V \in \mathtt{rep}_\alpha \, Q^d$ belongs to $\mu_{\mathbb{R}}^{-1}(\underline{0})$ we know that V is a semisimple representation, that is

$$V = S_1^{\oplus e_1} \oplus \ldots \oplus S_r^{\oplus e_r}$$

with the S_i simple representations of dimension vector β_i. Further, if $W \in \mathtt{rep}_\alpha^{-1}(i\theta \mathbb{1}_\alpha)$, then W is a direct sum of θ-stable representations, that is

$$W = T_1^{\oplus f_1} \oplus \ldots \oplus T_s^{\oplus f_s}$$

with the T_i θ-stable representations of dimension vector γ_i. By the explicit form of the map, we have that if $W = h.V$ that $r = s$, $e_i = f_i$ and $\beta_i = \gamma_i$. See the discussion below.

PROPOSITION 8.7

Let θ be a character such that $\theta(\alpha) = 0$, then the deformed preprojective algebra Π_θ has semisimple representations of dimension vector α of representation type $\tau = (e_1, \beta_1; \ldots; e_r, \beta_r)$ if and only if the preprojective algebra Π_0 has θ-stable representations of dimension vectors β_i for all $1 \leq i \leq r$.

In particular, Π_θ has a simple representation of dimension vector α if and only if Π_0 has a θ-stable representation of dimension vector α.

The variety $M_\alpha^{ss}(\Pi_0, \theta)$ is locally controlled by noncommutative algebras. Indeed, as in the case of moduli spaces of θ-semistable quiver-representations, it is locally isomorphic to $\text{iss}_\alpha (\Pi_0)_\Sigma$ for some universal localization of Π_0. We can determine the α-smooth locus of the corresponding sheaf of Cayley-Hamilton algebras.

PROPOSITION 8.8

Let $\alpha \in \Sigma_\theta$, then the α-smooth locus of $M_\alpha^{ss}(\Pi_0, \theta)$ is the open subvariety $M_\alpha^s(\Pi_0, \theta)$ of θ-stable representations of Π_0.

In particular, if the sheaf of Cayley-Hamilton algebras over $M_\alpha^{ss}(\Pi_0, \theta)$ is a sheaf of α-smooth algebras if and only if α is a minimal dimension vector in Σ_θ.

PROOF As $\alpha \in \Sigma_\theta$ we know that $\text{iss}_\alpha \Pi_\theta$ has dimension $2p_Q(\alpha) = 2 - T_Q(\alpha, \alpha)$. By the hyper-Kähler correspondence so is the dimension of $M_\alpha^{ss}(\Pi_0, \theta)$, whence the open subset of $\mu_{\mathbb{C}}^{-1}(\underline{0})$ consisting of θ-semistable representations has dimension

$$\alpha.\alpha - 1 + 2p_Q(\alpha)$$

as there are θ-stable representations in it. Take a $GL(\alpha)$-closed orbit $\mathcal{O}(V)$ in this open set. That is, V is the direct sum of θ-stable subrepresentations

$$V = S_1^{\oplus e_1} \oplus \ldots \oplus S_r^{\oplus e_r}$$

with S_i a θ-stable representation of Π_0 of dimension vector β_i occurring in V with multiplicity e_i whence $\alpha = \sum_i e_i \beta_i$.

As all S_i are Π_0-representations we can determine the local quiver Q_V by the knowledge of all $Ext_{\Pi_0}^1(S_i, S_j)$ from proposition 5.12

$$Ext_{\Pi_0}^1(S_i, S_j) = 2\delta_{ij} - T_Q(\beta_i, \beta_j)$$

But then the dimension of the normal space to the orbit is

$$dim\ Ext_{\Pi_0}^1(V, V) = 2\sum_{i=1}^{r} e_i - T_Q(\alpha, \alpha)$$

whence the étale local structure in an n-smooth point is of the form

$$GL(\alpha) \times^{GL(\tau)} Ext^1(V, V)$$

where $\tau = (e_1, \ldots, e_r)$ and is therefore of dimension

$$\alpha.\alpha + \sum_{i=1}^{2} e_i^2 - T_Q(\alpha, \alpha)$$

This number is equal to the dimension of the subvariety of θ-semistable representations of Π_0, which has dimension $\alpha.\alpha - 1 + 2 - T_Q(\alpha, \alpha)$ if and only if $r = 1$ and $e_1 = 1$, that is, if V is θ-stable. $\qquad\square$

Even in points of $M_\alpha^{ss}(\Pi_0, \theta)$, which are not in the α-smooth locus, we can use the local quiver to deduce combinatorial properties of the set of dimension vectors Σ_θ of simple representations of Π_θ.

PROPOSITION 8.9
Let $\alpha, \beta \in \Sigma_\theta$, then

1. *If $T(\alpha, \beta) \leq -2$ then $\alpha + \beta \in \Sigma_\theta$*

2. *If $T(\alpha, \beta) \geq -1$ then $\alpha + \beta \notin \Sigma_\theta$*

PROOF   The property that α and β are Schur roots of Q such that $T_Q(\alpha, \beta) \leq -2$ ensures that $\gamma = \alpha + \beta$ is a Schur root of Q and hence that $\mu_{\mathbb{C}}^{-1}(\theta 1\!\!1_\gamma$ has dimension $\gamma.\gamma - 1 + 2p_Q(\gamma)$, whence so is the subvariety of θ-semistable γ-dimensional representations of Π_0. We have to prove that Π_0 has a θ-stable γ-dimensional representation.

Let $V = S \oplus T$ with S resp. T a θ-stable representation of Π_0 of dimension vector α resp. β (they exist by the hyper-Kähler correspondence). But then the local quiver Q_V has the following form

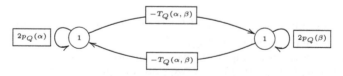

and by a calculation similar to the one in the foregoing proof we see that the image of the slice morphism in the space $GL(\gamma) \times^{\mathbb{C}^* \times \mathbb{C}^*} \mathbf{rep}_{(1,1)} Q_V$ has codimension 1. However, as $T_Q(\alpha, \beta) \leq -2$ there are at least 3 algebraically independent new invariants coming from the nonloop cycles in Q_V, so they cannot all vanish on the image. This means that $(1, \alpha; 1, \beta)$ cannot be the generic type for θ-semistables of dimension γ, so by the stratification result, there must exist θ-stables of dimension γ.

For the second assertion, assume that $\gamma = \alpha + \beta$ is the dimension vector of a simple representation of Π_θ, then $\text{iss}_\gamma \Pi_\theta$ has dimension $2p_Q(\gamma) = 2 - T_Q(\alpha, \beta, \alpha + \beta) = 2p_Q(\alpha) + 2p_Q(\beta)$ whence so is the dimension of $M_\gamma^{ss}(\Pi_0, \theta)$. By assumption $(1, \alpha; 1, \beta)$ cannot be the generic type for θ-semistable representations, but the stratum consisting of direct sums $S \oplus T$ with $S \in M_\alpha^s(\Pi_0, \theta)$ and $T \in M_\beta^s(Q, \theta)$ has the same dimension as the total space, a contradiction. $\square$

The first part of the foregoing proof can also be used to show that usually the moduli spaces $M_\alpha^{ss}(\Pi), \theta)$ and the quotient varieties $\text{iss}_\alpha \Pi_\theta$ have lots of singularities.

PROPOSITION 8.10
Let $\alpha \in |sigma_\theta$ such that $\alpha = \beta + \gamma$ with $\beta, \gamma \in \Sigma_\theta$. Then

$$M_\alpha^{ss}(\Pi_0, \theta) \qquad \text{and} \qquad \text{iss}_\alpha \Pi_\theta$$

is singular along the stratum of points of type $(1, \beta; 1, \gamma)$.

PROOF The quotient space of the local quiver situation (as in the foregoing proof) contains singularities at the trivial representations that remain singularities in any codimension one subvariety. $\square$

Still, if α is a minimal dimension vector in Σ_θ, the varieties $M_\alpha^{ss}(\Pi_0, \theta)$ and $\text{iss}_\alpha \Pi_\theta$ are smooth. In fact, we will show in section 8.7 that the affine smooth variety $\text{iss}_\alpha \Pi_\theta$ is in fact a coadjoint orbit.

8.6 Calogero particles

The *Calogero system* is a classical particle system of n particles on the real line with inverse square potential

That is, if the i-th particle has position x_i and velocity (momentum) y_i, then the Hamiltonian is equal to

$$H = \frac{1}{2} \sum_{i=1}^{n} y_i^2 + \sum_{i<j} \frac{1}{(x_i - x_j)^2}$$

The Hamiltonian *equations of motions* is the system of $2n$ differential equations

$$\begin{cases} \dfrac{dx_i}{dt} = \dfrac{\partial H}{\partial y_i} \\[2ex] \dfrac{dy_i}{dt} = -\dfrac{\partial H}{\partial x_i} \end{cases}$$

This defines a dynamical system that is *integrable* .

A convenient way to study this system is as follows. Assign to a position defined by the $2n$ vector $(x_1, y_1; \ldots, x_n, y_n)$ the couple of *Hermitian* $n \times n$ matrices

$$X = \begin{bmatrix} x_1 & & \\ & \ddots & \\ & & x_n \end{bmatrix} \quad \text{and} \quad Y = \begin{bmatrix} y_1 & \frac{i}{x_1-x_2} & \cdots & \cdots & \frac{i}{x_1-x_n} \\ \frac{i}{x_2-x_1} & y_2 & \ddots & & \vdots \\ \vdots & & \ddots & \ddots & \vdots \\ \vdots & & \ddots & \ddots & \frac{i}{x_{n-1}-x_n} \\ \frac{i}{x_n-x_1} & \cdots & \cdots & \frac{i}{x_n-x_{n-1}} & y_n \end{bmatrix}$$

Physical quantities are given by invariant polynomial functions under the action of the unitary group $U_n(\mathbb{C})$ under simultaneous conjugation. In particular one considers the functions

$$F_j = tr \, \frac{Y^j}{j}$$

For example

$$\begin{cases} tr(Y) = \sum y_i & \text{the total momentum} \\ \frac{1}{2} tr(Y^2) = \frac{1}{2} \sum y_i^2 - \sum_{i<j} \frac{1}{(x_i-x_j)^2} & \text{the Hamiltonian} \end{cases}$$

We can now consider the $U_n(\mathbb{C})$-translates of these matrix couples. This is shown to be a manifold with a free action of $U_n(\mathbb{C})$ such that the orbits are in one-to-one correspondence with points $(x_1, y_1; \ldots; x_n, y_n)$ in the phase space (that is, we agree that two such $2n$ tuples are determined only up to permuting the couples (x_i, y_i). The n-functions F_j give a completely integrable system on the phase space via *Liouville's theorem*, see, for example [1].

In the classical case, all points are assumed to lie on the real axis and the potential is repulsive so that collisions do not appear. G. Wilson [108] considered an alternative where the points are assumed to lie in the complex numbers and such that the potential is attractive (to allow for collisions), that is, the Hamiltonian is of the form

$$H = \frac{1}{2} \sum_i y_i^2 - \sum_{i<j} \frac{1}{(x_i - x_j)^2}$$

again giving rise to a dynamical system via the equations of motion. One recovers the classical situation back if the particles are assumed only to move on the imaginary axis

In general, we want to extend the phase space of n distinct points analytically to allow for collisions.

When all the points are distinct, that is, if all eigenvalues of X are distinct we will see in a moment that there is a unique $GL_n(\mathbb{C})$-orbit of couples of $n \times n$ matrices (up to permuting the n couples (x_i, y_i)).

$$X = \begin{bmatrix} x_1 & & \\ & \ddots & \\ & & x_n \end{bmatrix} \quad \text{and} \quad Y = \begin{bmatrix} y_1 & \frac{1}{x_1-x_2} & \cdots & & \cdots & \frac{1}{x_1-x_n} \\ \frac{1}{x_2-x_1} & y_2 & \ddots & & & \vdots \\ \vdots & & \ddots & \ddots & \ddots & \vdots \\ \vdots & & & \ddots & \ddots & \frac{1}{x_{n-1}-x_n} \\ \frac{1}{x_n-x_1} & \cdots & & \cdots & \frac{1}{x_n-x_{n-1}} & y_n \end{bmatrix}$$

For matrix couples in this standard form one verifies that

$$[Y, X] + \begin{bmatrix} 1 & \cdots & 1 \\ \vdots & \ddots & \vdots \\ 1 & \cdots & 1 \end{bmatrix} = \mathbb{1}_n$$

This equality suggests an approach to extend the phase space of n distinct complex Calogero particles to allow for collisions.

Assign the representation $(X, Y, u, v) \in \mathbf{rep}_\alpha \, \mathbb{M}$ where $\alpha = (1, n)$ and $\mathbb{M}$ is the path algebra of the quiver Q^d is

where X and Y are the matrices above and where

$$u = \begin{bmatrix} 1 \\ 1 \\ \vdots \\ 1 \end{bmatrix} \qquad v = \begin{bmatrix} 1 & 1 & \cdots & 1 \end{bmatrix}$$

Recall that the complex moment map for this quiver-setting is defined to be

$$\mathbf{rep}_\alpha\ Q^d = M_n(\mathbb{C}) \oplus M_n(\mathbb{C}) \oplus \mathbb{C}^n \oplus \mathbb{C}^{n*} \xrightarrow{\quad\mu\quad} \mathbb{C} \oplus M_n(\mathbb{C})$$

$$(X, Y, u, v) \qquad\qquad\qquad \mapsto \qquad (-v.u, [Y, X] + u.v)$$

Therefore, the above equation entails that $(X, Y, u, v) \in \mu_{\mathbb{C}}^{-1}(\theta 1_\alpha)$ where $\theta = (-n, 1)$, that is $(X, Y, u, v) \in \mathbf{rep}_\alpha\ \Pi_\theta$. Observe that $\alpha = (1, n) \in \Sigma_\theta$ (in fact, α is a minimal element in Σ_θ), whence theorem 8.7, $\mathbf{rep}_\alpha\ \Pi_\theta$ is an irreducible complete intersection of dimension $d = n^2 + 2n$ and there are α-dimensional simple representations of Π_θ. In particular, $\mathbf{iss}_\alpha\ \Pi_\theta$ is an irreducible variety of dimension $2n$.

We can define the *phase space for Calogero collisions* of n particles to be the quotient space

$$Calo_n = \mathbf{iss}_\alpha\ \Pi_\theta$$

In the following we will show that this is actually an orbit-space.

THEOREM 8.15

The phase space $Calo_n$ of Calogero collisions of n-particles is a connected complex manifold of dimension $2n$.

THEOREM 8.16

Let $\mathbf{rep}_\alpha\ \mathbb{M} \xrightarrow{\mu_{\mathbb{C}}} M_\alpha^0(\mathbb{C})$ be the complex moment map, then any $V = (X, Y, u, v) \in \mathbf{rep}_\alpha\ \Pi_\theta$ is a θ-stable representation. Therefore $Calo_n =$

$$\mu_{\mathbb{C}}^{-1}(\theta 1_\alpha)/GL(\alpha) = \mathbf{iss}_\alpha\ \Pi_\theta \simeq (\mu_{\mathbb{C}}^{-1}(\theta 1_\alpha) \cap \mathbf{rep}_\alpha^s\ \mathbb{M})/GL(\alpha) = M_\alpha^s(\Pi_\theta, \theta)$$

and is therefore a complex manifold of dimension $2n$, which is connected by theorem 8.7.

PROOF   The result will follow if we can prove that any Calogero quadruple (X, Y, u, v) has the property that u is a cyclic vector, that is, lies in $\mathbf{rep}_\alpha^s\ \mathbb{M}$.

Assume that U is a subspace of $\mathbb{C}^n$ stable under X and Y and containing u. U is then also stable under left multiplication with the matrix

$$A = [X, Y] + 1_n$$

and we have that $tr(A \mid U) = tr(1_n \mid U) = dim\ U$. On the other hand, $A = u.v$ and therefore

$$A.\begin{bmatrix} c_1 \\ \vdots \\ c_n \end{bmatrix} = \begin{bmatrix} u_1 \\ \vdots \\ u_n \end{bmatrix} \cdot [v_1\ \dots\ v_n] \cdot \begin{bmatrix} c_1 \\ \vdots \\ c_n \end{bmatrix} = (\sum_{i=1}^{n} v_i c_i) \begin{bmatrix} u_1 \\ \vdots \\ u_n \end{bmatrix}$$

Hence, if we take a basis for U containing u, then we have that

$$tr(A \mid U) = a$$

where $A.u = au$, that is $a = \sum u_i v_i$.

But then, $tr(A \mid U) = \dim U$ is independent of the choice of U. Now, $\mathbb{C}^n$ is clearly a subspace stable under X and Y and containing u, so we must have that $a = n$ and so the only subspace U possible is $\mathbb{C}^n$ proving cyclicity of u with respect to the matrix-couple (X, Y). □

Again, it follows that we can cover the phase space $Calo_n$ by open subsets

$$Calo_n \ (\sigma) = \{(X, Y, u, v) \quad \text{in } \sigma\text{-standard form such that } [Y, X] + u.v = \mathbb{1}_n \}$$

where σ runs over the Hilbert n-stairs.

Example 8.4 The phase space $Calo_2$
Consider $Calo_2$ ($\boxed{\circ}$). Because

$$\left[\begin{bmatrix} 0 & a \\ 1 & b \end{bmatrix}, \begin{bmatrix} c & d \\ e & f \end{bmatrix}\right] + \begin{bmatrix} 1 \\ 0 \end{bmatrix} \cdot [g \ h] - \mathbb{1}_2 = \begin{bmatrix} g - d + ae - 1 & h + af - ac - bd \\ c - f + be & d - ae - 1 \end{bmatrix}$$

We obtain after taking Groebner bases that the defining equations are

$$\begin{cases} g & = 2 \\ h & = b \\ f & = c + eh \\ d & = 1 + ae \end{cases}$$

In particular we find

$$Calo_2 \ (\boxed{\circ}) = \{(\begin{bmatrix} 0 & a \\ 1 & b \end{bmatrix}, \begin{bmatrix} c & 1+ae \\ e & c+be \end{bmatrix}, \begin{bmatrix} 1 \\ 0 \end{bmatrix}, [2 \ b]) \mid a, b, c, e \in \mathbb{C}\} \simeq \mathbb{C}^4$$

and a similar description holds for $Calo_2$ ($\boxed{\bullet}$). □

Example 8.5 The phase space $Calo_3$
We claim that

$$Calo_3 \ (\begin{smallmatrix} \boxed{\circ} \\ \boxed{\bullet}\boxed{}\end{smallmatrix}) \simeq \mathbb{C}^6$$

For, if we compute the 3×3 matrix

$$[\begin{bmatrix} 0 & a & b \\ 1 & c & d \\ 0 & e & f \end{bmatrix}, \begin{bmatrix} 0 & g & h \\ 0 & i & j \\ 1 & k & l \end{bmatrix}] + \begin{bmatrix} 1 \\ 0 \\ 0 \end{bmatrix} \cdot [m \ n \ o] - \mathbb{1}_3$$

then the Groebner basis for its entries gives the following defining equations

$$
\begin{cases}
m &= 3 \\
n &= c + k \\
o &= i + l \\
f &= k \\
d &= o - l \\
g &= 2 + b \\
l &= g - ej - kl + ko \\
h &= 2jk + 2l^2 - jn - 3lo + o^2 \\
a &= 2k^2 - 2el - kn + eo
\end{cases}
$$

In a similar manner one can show that

$$
Calo_3 \; (\; \boxed{} \;) \simeq \mathbb{C}^6 \qquad \text{but} \qquad Calo_3 \; (\; \boxed{} \;)
$$

is again more difficult to describe.   ⬚

We can identify the classical Calogero situations as an open subset of $Calo_n$.

PROPOSITION 8.11
Let $(X, Y, u, v) \in \mathbf{rep}_\alpha \; \Pi_\theta$ and suppose that X is diagonalizable. Then

1. all eigenvalues of X are distinct, and

2. the $GL(\alpha)$-orbit contains a representative such that

$$
X = \begin{bmatrix} \lambda_1 & & \\ & \ddots & \\ & & \lambda_n \end{bmatrix}
\qquad
Y = \begin{bmatrix}
\alpha_1 & \frac{1}{\lambda_1 - \lambda_2} & \cdots & \cdots & \frac{1}{\lambda_1 - \lambda_n} \\
\frac{1}{\lambda_2 - \lambda_1} & \alpha_2 & \ddots & & \vdots \\
\vdots & \ddots & \ddots & \ddots & \vdots \\
\vdots & & \ddots & \ddots & \frac{1}{\lambda_{n-1} - \lambda_n} \\
\frac{1}{\lambda_n - \lambda_1} & \cdots & \cdots & \frac{1}{\lambda_n - \lambda_{n-1}} & \alpha_n
\end{bmatrix}
$$

$$
u = \begin{bmatrix} 1 \\ 1 \\ \vdots \\ 1 \end{bmatrix}
\qquad
v = \begin{bmatrix} 1 & 1 & \ldots & 1 \end{bmatrix}
$$

and this representative is unique up to permutation of the n couples (λ_i, α_i).

PROOF   Choose a representative with X a diagonal matrix as indicated. Equating the diagonal entries in $[Y, X] + u.v = 1_n$ we obtain that for all $1 \leq i \leq n$ we have $u_i v_i = 1$. Hence, none of the entries of

$$[X, Y] + 1_n = u.v$$

is zero. Consequently, by equating the (i, j)-entry it follows that $\lambda_i \neq \lambda_j$ for $i \neq j$.

The representative with X a diagonal matrix is therefore unique up to the action of a diagonal matrix D and of a permutation. The freedom in D allows us to normalize u and v as indicated, the effect of the permutation is described in the last sentence.

Finally, the precise form of Y can be calculated from the normalized forms of X, u and v and the equation $[Y, X] + u.v = 1_n$.   ⬜

Invoking the hyper-Kähler structure on $\mathtt{rep}_\alpha$ M we have by theorem 8.14 a homeomorphism, in fact in this case a C^∞-diffeomorphism between the Calogero phase-space and the Hilbert scheme

$$Calo_n = \mathtt{iss}_\alpha \ \Pi_\theta \xrightarrow{\ h. \ } M_\alpha^{ss}(\Pi_0, \theta) = Hilb_n$$

8.7   Coadjoint orbits

In this section we will give an important application of noncommutative geometry@n developed in the foregoing chapter. If α is a minimal dimension vector in Σ_θ we will prove that the quotient variety $\mathtt{iss}_\alpha \ \Pi_\theta$ is smooth and a coadjoint orbit for the dual of the necklace algebra. In particular, the phase space of Calogero particles is a coadjoint orbit.

We fix a quiver Q on k vertices, a dimension vector $\alpha \in \mathbb{N}^k$ and a character $\theta \in \mathbb{Z}^k$ such that $\theta(\alpha) = 0$ with corresponding weight $\theta 1_\alpha$. Recall that Σ_θ is the subset of dimension vectors α such that

$$p_Q(\alpha) > p_Q(\beta_1) + \ldots + p_Q(\beta_r)$$

for all decompositions $\alpha = \beta_1 + \ldots + \beta_r$ with the $\beta_i \in \Delta^+\theta$, that is, β_i is a positive root for the quiver Q and $\theta(\beta_i) = 0$. With Σ_θ^{min} we will denote the subset of *minimal dimension vectors* in Σ_θ, that is, such that for all $\beta < \alpha$ we have $\beta \notin \Sigma_\theta$.

PROPOSITION 8.12
If $\alpha \in \Sigma_\theta^{min}$, then the deformed preprojective algebra Π_θ is α-smooth, that is, $\mathtt{rep}_\alpha \ \Pi_\theta$ is a smooth $GL(\alpha)$-variety of dimension $d = \alpha.\alpha - 1 + 2p_Q(\alpha)$.

Moreover, the quotient variety $\mathrm{iss}_\alpha\ \Pi_\theta$ *is a smooth variety of dimension* $2p_Q(\alpha)$, *and the quotient map*

$$\mathrm{rep}_\alpha\ \Pi_\theta \longrightarrow\!\!\!\!\rightarrow \mathrm{iss}_\alpha\ \Pi_\theta$$

is a principal $PGL(\alpha)$*-fibration, so determines a central simple algebra.*

PROOF   Let $V \in \mathrm{rep}_\alpha\ \Pi_\theta$ and let V^{ss} be its semisimplification. As Σ_θ is the set of simple dimension vectors of Π_θ by theorem 8.8 and α is a minimal dimension vector in this set, V^{ss} must be simple. As V^{ss} is the direct sum of the Jordan-Hölder components of V, it follows that $V \simeq V^{ss}$ is simple and hence its orbit $\mathcal{O}(V)$ is closed. As the stabilizer subgroup of V is $\mathbb{C}^*\mathbb{1}_\alpha$ computing the differential of the complex moment map shows that V is a smooth point of $\mu_\mathbb{C}^{-1}(\theta\mathbb{1}_\alpha = \mathrm{rep}_\alpha\ \Pi_\theta$.

Therefore, $\mathrm{rep}_\alpha\ \Pi_\theta$ is a smooth $GL(\alpha)$-variety whence Π_θ is α-smooth. Because each α-dimensional representation is simple, the quotient map

$$\mathrm{rep}_\alpha\ \Pi_\theta \xrightarrow{\ \pi\ }\!\!\!\!\rightarrow \mathrm{iss}_\alpha\ \Pi_\theta$$

is a principal $PGL(\alpha)$-fibration in the étale topology. The total space being smooth, so is the base space $\mathrm{iss}_\alpha\ \Pi_\theta$.   ▯

The trace pairing identifies $\mathrm{rep}_\alpha\ Q^d$ with the *cotangent bundle* $T^*\ \mathrm{rep}_\alpha\ Q$ and as such it comes equipped with a *canonical symplectic structure* . More explicit, for every arrow $\mathbb{j} \xleftarrow{\quad a \quad} \mathbb{i}$ in Q we have an $a_j \times a_i$ matrix of coordinate functions A_{uv} with $1 \le u \le a_j$ and $1 \le v \le a_i$ and for the adjoined arrow $\mathbb{j} \xrightarrow{\quad\ \ } \mathbb{i}$ in Q^d an $a_i \times a_j$ matrix of coordinate functions A_{vu}^*. The canonical symplectic structure on $\mathrm{rep}_\alpha\ Q^d$ is then induced by the closed 2-form

$$\omega = \sum_{\mathbb{j} \xleftarrow{a} \mathbb{i}}^{\substack{1\le v\le a_i \\ 1\le u\le a_j}} dA_{uv} \wedge dA_{vu}^*$$

This symplectic structure induces a *Poisson bracket* on the coordinate ring $\mathbb{C}[\mathrm{rep}_\alpha\ Q^d]$ by the formula

$$\{f,g\} = \sum_{\mathbb{j} \xleftarrow{a} \mathbb{i}}^{\substack{1\le v\le a_i \\ 1\le u\le a_j}} \left(\frac{\partial f}{\partial A_{uv}} \frac{\partial g}{\partial A_{vu}^*} - \frac{\partial f}{\partial A_{vu}^*} \frac{\partial g}{\partial A_{uv}} \right)$$

The base change action of $GL(\alpha)$ on the representation space $\mathrm{rep}_\alpha\ Q^d$ is *symplectic*, which means that for all *tangent vectors* $t, t' \in T\ \mathrm{rep}_\alpha\ Q^d$ we have for the induced $GL(\alpha)$ action that

$$\omega(t, t') = \omega(g.t, g.t')$$

for all $g \in GL(\alpha)$.

The infinitesimal $GL(\alpha)$ action gives a Lie algebra homomorphism

$$Lie\ PGL(\alpha) \longrightarrow Vect_\omega\ \mathbf{rep}_\alpha\ Q^d$$

which factorizes through a Lie algebra morphism H to the coordinate ring making the diagram below commute

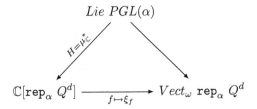

where $\mu_{\mathbb{C}}$ is the complex moment map introduced before. We say that the action of $GL(\alpha)$ on $\mathbf{rep}_\alpha\ \mathbb{M}$ is *Hamiltonian* .

This makes the ring of polynomial invariants $\mathbb{C}[\mathbf{rep}_\alpha\ Q^d]^{GL(\alpha)}$ into a *Poisson algebra* and we will write

$$\mathfrak{lie} = (\mathbb{C}[\mathbf{rep}_\alpha\ Q^d]^{GL(\alpha)}, \{-,-\})$$

for the corresponding abstract *infinite dimensional Lie algebra.*

The dual space of this Lie algebra $\mathfrak{lie}^*$ is then a *Poisson manifold* equipped with the *Kirillov-Kostant bracket* .

Evaluation at a point in the quotient variety $\mathbf{iss}_\alpha\ Q^d$ defines a linear function on $\mathfrak{lie}$ and therefore evaluation gives an embedding

$$\mathbf{iss}_\alpha\ Q^d \hookrightarrow \mathfrak{lie}^*$$

as Poisson varieties. That is, the induced map on the polynomial functions is a morphism of Poisson algebras.

Let us return to the setting of deformed preprojective algebras. So let θ be a character with $\theta(\alpha) = 0$ and corresponding weight $\theta \mathbb{1}_\alpha \in Lie\ PGL(\alpha)$.

THEOREM 8.17

Let $\alpha \in \Sigma_\theta^{min}$, then $\mathbf{iss}_\alpha\ \Pi_\theta$ is an affine symplectic manifold and the Poisson embeddings

$$\mathbf{iss}_\alpha\ \Pi_\theta \hookrightarrow \mathbf{iss}_\alpha\ Q^d \hookrightarrow \mathfrak{lie}^*$$

make $\mathbf{iss}_\alpha\ \Pi_\theta$ into a closed coadjoint orbit of the infinite dimensional Lie algebra $\mathfrak{lie}^*$.

PROOF We know from proposition 8.12 that $\mathbf{iss}_\alpha\ \Pi_\theta$ is a smooth affine variety and that $PGL(\alpha)$ acts freely on $\mu_{\mathbb{C}}^{-1}(\theta \mathbb{1}_\alpha) = \mathbf{rep}_\alpha\ \Pi_\theta$. Moreover, the infinitesimal coadjoint action of $\mathfrak{lie}$ on $\mathfrak{lie}^*$ preserves $\mathtt{iss}_\alpha \ \Pi_\theta$ and therefore $\mathbb{C}[\mathtt{iss}_\alpha \ \Pi_\theta]$ is a quotient Lie $\overline{\mathfrak{lie}}$ algebra (for the induced bracket) of $\mathfrak{lie}$.

In general, if X is a smooth affine variety, then the differentials of polynomial functions on X span the tangent spaces at all points x of X. Therefore, if X is in addition symplectic, the infinitesimal Hamiltonian action of the Lie algebra $\mathbb{C}[X]$ (with the natural Poisson bracket) on X is infinitesimally transitive. But then, the evaluation map makes X a coadjoint orbit of the dual Lie algebra $\mathbb{C}[X]^*$.

Hence, the quotient variety $iss_\alpha \ \Pi_\theta$ is a coadjoint orbit in $\overrightarrow{\mathfrak{lie}}^*$. Therefore, the infinite dimensional group Ham generated by all *Hamiltonian flows* on $\mathtt{iss}_\alpha \ \Pi_\theta$ acts with *open* orbits.

By proposition 8.12 $\mathtt{iss}_\alpha \ \Pi_\theta$ is an irreducible variety, whence is a single Ham-orbit, finishing the proof. $\qquad\qquad\qquad\qquad\qquad\qquad\qquad\qquad\qquad\quad\square$

The Lie algebra $\mathfrak{lie}$ depends on the dimension vector α. By the general principle of $\mathtt{noncommutative\ geometry@n}$ we would like to construct a noncommutative variety from a family of coadjoint quotient spaces of deformed preprojective algebras. For this reason we need a larger Lie algebra, the *necklace Lie algebra*.

Recall that the necklace Lie algebra introduced in section 7.8

$$\mathtt{necl} = \mathrm{dR}^0_{rel} \ \mathbb{C}Q^d = \frac{\mathbb{C}Q^d}{[\mathbb{C}Q^d, \mathbb{C}Q^d]}$$

is the vector space with basis all the necklace words w in the quiver Q^d, that is, all equivalence classes of oriented cycles in the quiver Q^d, equipped with the Kontsevich bracket

$$\{w_1, w_2\}_K = \sum_{a \in Q_a} \left(\frac{\partial w_1}{\partial a} \frac{\partial w_2}{\partial a^*} - \frac{\partial w_1}{\partial a^*} \frac{\partial w_2}{\partial a} \right) \quad \mathrm{mod} \ [\mathbb{C}Q^d, \mathbb{C}Q^d]$$

We recall that the algebra of polynomial quiver invariants $\mathbb{C}[\mathtt{iss}_\alpha \ Q^d] = \mathbb{C}[\mathtt{rep}_\alpha \ Q^d]^{GL(\alpha)}$ is generated by traces of necklace words. That is, we have a map

$$\mathtt{necl} = \frac{\mathbb{C}Q^d}{[\mathbb{C}Q^d, \mathbb{C}Q^d]} \xrightarrow{\ tr\ } \mathfrak{lie} = \mathbb{C}[\mathtt{iss}_\alpha \ Q^d]$$

Recalling the definition of the Lie bracket on $\mathfrak{lie}$ we see that this map is actually a Lie algebra map, that is, for all necklace words w_1 and w_2 in Q^d we have the identity

$$tr \ \{w_1, w_2\}_K = \{tr(w_1), tr(w_2)\}$$

Now, the image of tr contains a set of *algebra* generators of $\mathbb{C}[\mathtt{iss}_\alpha \ Q^d]$, so the elements $tr \ \mathtt{necl}$ are enough to separate points in $\mathtt{iss}_\alpha \ Q^d$ and in the closed subvariety $\mathtt{iss}_\alpha \ \Pi_\theta$. That is, the composition

$$\mathtt{iss}_\alpha \ \Pi_\theta \hookrightarrow \mathtt{iss}_\alpha \ Q^d \xrightarrow{\ tr^*\ } \mathtt{necl}^*$$

is injective. Again, the differentials of functions on $\text{iss}_\alpha \ \Pi_\theta$ obtained by restricting traces of necklace words span the tangent spaces at all points if the affine variety $\text{iss}_\alpha \ \Pi_\theta$ is smooth. A summary follows.

THEOREM 8.18
Let $\alpha \in \Sigma_\theta^{min}$. Then, the quotient variety of the preprojective algebra $\text{iss}_\alpha \ \Pi_\theta$ is an affine smooth manifold and the embeddings

$$\text{iss}_\alpha \ \Pi_\theta \hookrightarrow \text{iss}_\alpha \ Q^d \hookrightarrow \mathfrak{lie}^* \hookrightarrow \mathfrak{neck}^*$$

make $\text{iss}_\alpha \ \Pi_\theta$ into a closed coadjoint orbit in the dual of the necklace Lie algebra $\mathfrak{neck}$.

We have proved in section 7.8 that there is an exact sequence of Lie algebras

$$0 \longrightarrow \underbrace{\mathbb{C} \oplus \ldots \oplus C}_{k} \longrightarrow \mathfrak{neck} \longrightarrow Der_\omega \ \mathbb{C}Q^d \longrightarrow 0$$

That is, the necklace Lie algebra $\mathfrak{neck}$ is a central extension of the Lie algebra of symplectic derivations of $\mathbb{C}Q^d$. This Lie algebra corresponds to the automorphism group of all $B = \mathbb{C} \times \ldots \times \mathbb{C}$-automorphisms of the path algebra $\mathbb{C}Q^d$ preserving the *moment map element*, the commutator

$$c = \sum_{a \in Q_a} [a, a^*]$$

That is, we expect a transitive action of an extension of this automorphism group on the quotient varieties of deformed preprojective algebras $\text{iss}_\alpha \ \Pi_\theta$ when $\alpha \in \Sigma_\theta^{min}$. Further, it should be observed that these coadjoint cases are precisely the situations were the preprojective algebra Π_θ is α-smooth. That is, whereas the Lie algebra of vector fields of the smooth noncommutative variety corresponding to $\mathbb{C}Q^d$ has rather unpredictable behavior on the *singular* noncommutative closed subvariety corresponding to the quotient algebra Π_θ, it behaves as expected on those α-dimensional components where Π_θ is α-smooth.

8.8 Adelic Grassmannian

At the moment of this writing it is unclear which coadjoint orbits $\text{iss}_\alpha \ \Pi_\theta$ should be taken together to form an object in noncommutative geometry@n, for a general quiver Q. In this section we will briefly recall how the phase spaces $Calo_n$ of Calogero particles can be assembled together to form an infinite dimensional cellular complex, the *adelic Grassmannian* Gr^{ad}.

Let $\lambda \in \mathbb{C}$, a subset $V \subset \mathbb{C}[x]$ is said to be λ-*primary* if there is some power $r \in \mathbb{N}_+$ such that

$$(x - \lambda)^r \mathbb{C}[x] \subset V \subset \mathbb{C}[x]$$

A subset $V \subset \mathbb{C}[x]$ is said to be *primary decomposable* if it is the finite intersection

$$V = V_{\lambda_1} \cap \ldots \cap V_{\lambda_r}$$

with $\lambda_i \neq \lambda_j$ if $i \neq j$ and V_{λ_i} is a λ_i-primary subset. Let k_{λ_i} be the codimension of V_{λ_i} in $\mathbb{C}[x]$ and consider the polynomial

$$p_V(x) = \prod_{i=1}^{r} (x - \lambda_i)^{k_{\lambda_i}}$$

Finally, take $W = p_V(x)^{-1} V$, then W is a vectorsubspace of the rational functionfield $\mathbb{C}(x)$ in one variable.

DEFINITION 8.2   *The adelic Grassmannian Gr^{ad} is the se of subspaces $W \subset \mathbb{C}(x)$ that arise in this way.*

We can decompose Gr^{ad} in affine cells as follows. For a fixed $\lambda \in \mathbb{C}$ we define

$$Gr_\lambda = \{W \in Gr^{ad} \mid \exists k, l \in \mathbb{N} \; : \; (x - \lambda)^k \mathbb{C}[x] \subset W \subset (x - \lambda)^{-l} \mathbb{C}[x]\}$$

Then, we can write every element $w \in W$ as a *Laurent series*

$$w = \alpha_s (x - \lambda)^s + \text{ higher terms}$$

Consider the increasing set of integers $S = \{s_0 < s_1 < \ldots\}$ consisting of all *degrees s* of elements $w \in W$. Now, define natural numbers

$$v_i = i - s_i \qquad \text{then} \qquad v_0 \geq v_1 \geq \ldots \geq v_z = 0 = v_{z+1} = \ldots$$

That is, to $W \in Gr_\lambda$ we can associate a *partition*

$$p(W) = (v_0, v_1, \ldots, v_{z-1})$$

Conversely, if p is a partition of some n, then the set of all $W \in Gr_\lambda$ with associated partition $p_W = p$ form an affine space $\mathbb{A}^n$ of dimension n. Hence, Gr_λ has a cellular structure indexed by the set of all partitions.

As $Gr^{ad} = \prod'_{\lambda \in \mathbb{C}} Gr_\lambda$ because for any $W \in Gr^{ad}$ there are uniquely determined $W(\lambda_i) \in Gr_{\lambda_i}$ such that $W = W(\lambda_1) \cap \ldots \cap W(\lambda_r)$, there is a natural number n associated to W where $n = |p_i|$ where $p_i = p(W(\lambda_i))$ is the partition determined by $W(\lambda_i)$. Again, all $W \in Gr^{ad}$ with corresponding $(\lambda_1, p_1; \ldots; \lambda_r, p_r)$ for an affine cell $\mathbb{A}^n$ of dimension n. In his way, the adelic Grassmannian Gr^{ad} becomes an infinite cellular space with the cells indexed by r-tuples of complex numbers and partitions for all $r \geq 0$. The adelic Grassmannian is an important object in the theory of dynamical systems as it parameterizes rational solutions of the *KP hierarchy*. A surprising connection between Gr^{ad} and the Calogero system was discovered by G. Wilson in [108].

THEOREM 8.19
Let $Gr^{ad}(n)$ be the collection of all cells of dimension n in gr^{ad}, then there is a set-theoretic bijection
$$Gr^{ad}(n) \longleftrightarrow Calo_n$$
between $Gr^{ad}(n)$ and the phase space of n Calogero particles.

The adelic Grassmannian also appears in the study of right ideals of the first *Weyl algebra*
$$A_1(\mathbb{C}) = \frac{\mathbb{C}\langle x, y \rangle}{(xy - yx - 1)}$$
which is an infinite dimensional simple $\mathbb{C}$-algebra, having no finite dimensional representations. Consider right ideals of $A_1(\mathbb{C})$ under isomorphism, that is

$$\mathfrak{p} \simeq \mathfrak{p}' \qquad \text{iff} \qquad f.\mathfrak{p} = g.\mathfrak{p}' \quad \text{for some } f, g \in A_1(\mathbb{C}).$$

If we denote with $D_1(\mathbb{C})$ the *Weyl skew field*, that is, the field of fractions of $A_1(\mathbb{C})$, then the foregoing can also be expressed as

$$\mathfrak{p} \simeq \mathfrak{p}' \qquad \text{iff} \qquad \mathfrak{p} = h.\mathfrak{p}' \quad \text{for some } h \in D_1(\mathbb{C})$$

The set of isomorphism classes will be denoted by $Weyl$.

The connection between right ideals of $A_1(\mathbb{C})$ and gr^{ad} is contained (in disguise) in the paper of R. Cannings and M. Holland [19]. $A_1(\mathbb{C})$ acts as differential operators on $\mathbb{C}[x]$ and for every right ideal I of $A_1(\mathbb{C})$ they show that $I.\mathbb{C}[x]$ is primary decomposable. Conversely, if $V \subset \mathbb{C}[x]$ is primary decomposable, they associate the right ideal

$$I_V = \{\theta \in A_1(\mathbb{C}) \mid \theta.\mathbb{C}[x] \subset V\}$$

of $A_1(\mathbb{C})$ to it. Moreover, isomorphism classes of right ideals correspond to studying primary decomposable subspaces under multiplication with polynomials. Hence
$$Gr^{ad} \simeq Weyl$$

The group $Aut\, A_1(\mathbb{C})$ of $\mathbb{C}$-algebra automorphisms of $A_1(\mathbb{C})$ acts on the set of right ideals of $A_1(\mathbb{C})$ and respects the notion of isomorphism whence acts on $Weyl$. The group $Aut\, A_1(\mathbb{C})$ is generated by automorphisms σ_i^f defined by

$$\begin{cases} \sigma_1^f(x) &= x + f(y) \\ \sigma_1^f(y) &= y \end{cases} \quad \text{with } f \in \mathbb{C}[y], \qquad \begin{cases} \sigma_2^f(x) &= x \\ \sigma_2^f(y) &= y + f(x) \end{cases} \quad \text{with } f \in \mathbb{C}[x]$$

We claim that for any polynomial in one variable $f(z) \in \mathbb{C}[z]$ we have that

$$f(xy).x^n = x^n.f(xy - n) \qquad \text{and} \qquad f(xy).y^n = y^n.f(xy + n)$$

Indeed, we have $(xy).x = x.(yx) = x.(xy - 1)$ and therefore

$$f(xy).x = x.f(xy - 1)$$

from which the claim follows by recursion. In particular, as $x^n.y^n = x^{n-1}(xy)y^{n-1} = x^{n-1}y^{n-1}(xy + n - 1)$ we get by recurrence that

$$x^n y^n = xy(xy + 1)(xy + 2) \ldots (xy + n - 1)$$

In calculations with the Weyl algebra it is often useful to decompose $A_1(\mathbb{C})$ in weight spaces. For $t \in \mathbb{Z}$ let us define

$$A_1(\mathbb{C})(t) = \{ \, f \in A_1(\mathbb{C}) \mid [xy, f] = tf \, \}$$

then the foregoing asserts that $A_1(\mathbb{C}) = \oplus_{t \in \mathbb{Z}} A_1(\mathbb{C})(t)$ where $A_1(\mathbb{C})(t)$ is equal to

$$\begin{cases} y^t \mathbb{C}[xy] = \mathbb{C}[xy]y^t & \text{for } t \geq 0 \\ x^{-t} \mathbb{C}[xy] = \mathbb{C}[xy]x^{-t} & \text{for } t < 0 \end{cases}$$

For a natural number $n \geq 1$ we define the *n-th canonical right ideal* of $A_1(\mathbb{C})$ to be

$$\mathfrak{p}_n = x^{n+1} A_1(\mathbb{C}) + (xy + n) A_1(\mathbb{C})$$

LEMMA 8.8
The weight space decomposition of $\mathfrak{p}_n$ is given for $t \in \mathbb{Z}$

$$\mathfrak{p}_n(t) = x^{n+1} A_1(\mathbb{C})(t + n + 1) + (xy + n) A_1(\mathbb{C})(t)$$

which is equal to

$$\begin{cases} (xy + n)\mathbb{C}[xy]y^t & \text{for } t \geq 0, \\ (xy + n)\mathbb{C}[xy]x^{-t} & \text{for } -n \leq t < 0, \\ \mathbb{C}[xy]x^{-t} & \text{for } t < -n \end{cases}$$

PROOF   Let $t = -1$, then $\mathfrak{p}_n(-1)$ is equal to

$$x^{n+1}\mathbb{C}[xy]y^n + (xy + n)\mathbb{C}[xy]x$$

Using $x^{n+1}y^{n+1} = xy(xy + 1) \ldots (xy + n)$ this is equal to

$$xy(xy + 1) \ldots (xy + n)\mathbb{C}[xy]y^{-1} + (xy + n)\mathbb{C}[xy]x$$

The first factor is $(xy + 1) \ldots (xy + n)\mathbb{C}[xy]x$ from which the claim follows. For all other t the calculations are similar.   ▯

One can show that $\mathfrak{p}_n \not\simeq \mathfrak{p}_m$ whenever $n \neq m$ so the isomorphism classes $[\mathfrak{p}_n]$ are distinct points in $Weyl$ for all n. We define

$$Weyl_n = Aut\, A_1(\mathbb{C}).[\mathfrak{p}_n] = \{\, [\sigma(\mathfrak{p}_n)]\ \ \forall \sigma \in Aut\, A_1(\mathbb{C})\}$$

the orbit in $Weyl$ of the point $[\mathfrak{p}_n]$ under the action of the automorphism group.

Example 8.6 The Weyl right ideals $Weyl_1$
For a point $(a,b) \in \mathbb{C}^2$ we define a right ideal of $A_1(\mathbb{C})$ by

$$\mathfrak{p}_{a,b} = (x+a)^2 A_1(\mathbb{C}) + ((x+a)(y+b)+1)A_1(\mathbb{C})$$

Observe that $\mathfrak{p}_1 = \mathfrak{p}_{0,0}$. Consider the action of the automorphism σ_2^f on these right ideals. As $f \in \mathbb{C}[x]$ we can write

$$f = f(-a) + (x+a)f_1 \qquad \text{with } f_1 \in \mathbb{C}[x]$$

Then, recalling the definition of σ_2^f we have

$$\begin{aligned}
\sigma_2^f(\mathfrak{p}_{a,b}) &= (x+a)^2 A_1(\mathbb{C}) + ((x+a)(y+b+f(-a)+(x+a)f_1)+1)A_1(\mathbb{C}) \\
&= (x+a)^2 A_1(\mathbb{C}) + ((x+a)(y+b+f(-a))+1)A_1(\mathbb{C}) = \mathfrak{p}_{a,b+f(-a)}
\end{aligned}$$

Now, consider the action of an automorphism σ_1^f. We claim that

$$\mathfrak{p}_{a,b} = A_1(\mathbb{C}) \cap (y+b)^{-1}(x+a)A_1(\mathbb{C})$$

This is easily verified on the special case $\mathfrak{p}_1$ using the above lemma, the arbitrary case follows by changing variables. We have

$$\begin{aligned}
\mathfrak{p}_{a,b} &= A_1(\mathbb{C}) \cap (y+b)^{-1}(x+a)A_1(\mathbb{C}) \\
&\simeq (x+a)^{-1}(y+b)A_1(\mathbb{C}) \cap A_1(\mathbb{C}) \quad \text{(multiply with } h = (x+a)^{-1}(y+b)) \\
&= (y+b)^2 A_1(\mathbb{C}) + ((y+b)(x+a)-1)A_1(\mathbb{C}) \overset{def}{=} \mathfrak{q}_{b,a}
\end{aligned}$$

Writing $f = f(-b) + (y+b)f_1$ with $f_1 \in \mathbb{C}[y]$ we then obtain by mimicking the foregoing

$$\begin{aligned}
\sigma_1^f(\mathfrak{p}_{a,b}) &\simeq \sigma_1^f(\mathfrak{q}_{b,a}) \\
&= \mathfrak{q}_{b,a+f(-b)} \\
&\simeq \mathfrak{p}_{a+f(-b),b}
\end{aligned}$$

and therefore there is an $h \in D_1(\mathbb{C})$ such that $\sigma_1^f(\mathfrak{p}_{a,b}) = h\mathfrak{p}_{a+f(-b),b}$.

As the group $Aut\, A_1(\mathbb{C})$ is generated by the automorphisms σ_1^f and σ_2^f we see that

$$Weyl_1 = Aut\, A_1(\mathbb{C}).[\mathfrak{p}_1] \longrightarrow \{\, [\mathfrak{p}_{a,b}\ \mid\ a,b \in \mathbb{C}\,\}$$

Moreover, this inclusion is clearly surjective by the above arguments. Finally, we claim that $Weyl_1 \simeq \mathbb{C}^2$. That is, we have to prove that if

$$\mathfrak{p}_{a,b} = h.\mathfrak{p}_{a',b'} \quad \Rightarrow \quad (a,b) = (a',b').$$

Observe that $A_1(\mathbb{C}) \hookrightarrow \mathbb{C}(x)[y,\delta]$ where this algebra is the differential polynomial algebra over the field $\mathbb{C}(x)$ and is hence a right principal ideal domain. That is, we may assume that the element $h \in D_1(\mathbb{C})$ actually lies in $\mathbb{C}(x)[y,\delta]$. Now, induce the filtration by y-degree on $\mathbb{C}(x)[y,\delta]$ to the subalgebra $A_1(\mathbb{C})$. This is usually called the *Bernstein filtration* . Because $A_1(\mathbb{C})$ and $\mathbb{C}(x)[y,\delta]$ are domains we have for all $f \in A_1(\mathbb{C})$ that

$$deg(h.f) = deg(h) + deg(f)$$

Now, as both $\mathfrak{p}_{a,b}$ and $\mathfrak{p}_{a'.b'}$ contain elements of degree zero x^2+a resp. x^2+a' we must have that $h \in \mathbb{C}(x)$.

View y as the differential operator $-\frac{\partial}{\partial x}$ on $\mathbb{C}[x]$ and define for every right ideal $\mathfrak{p}$ of $A_1(\mathbb{C})$ its *evaluation* to be the *subspace* of polynomials

$$ev(\mathfrak{p}) = \{ \ D.f \ \mid \ D \in \mathfrak{p} \ , \ f \in \mathbb{C}[x] \ \}$$

where $D.f$ is the evaluation of the differential operator on f. One calculates that

$$ev(\mathfrak{p}_{a,b}) = \mathbb{C}(1 + b(x + a)) + (x + a)^2 \mathbb{C}[x]$$

and as from $\mathfrak{p}_{a,b} = h.\mathfrak{p}_{a',b'}$ and $h \in \mathbb{C}(x)$ follows that

$$ev(\mathfrak{p}_{a,b}) = hev(\mathfrak{p}_{a',b'})$$

we deduce that $h \in \mathbb{C}^*$ and hence that $\mathfrak{p}_{a,b} = \mathfrak{p}_{a',b'}$ and $(a,b) = (a',b')$. $\Box$

Yu. Berest and G. Wilson proved in [7] that the Cannings-Holland correspondence respects the automorphism orbit decomposition.

THEOREM 8.20
We have $Weyl = \bigsqcup_n Weyl_n$ and there are set-theoretic bijections

$$Weyl_n \longleftrightarrow Gr^{ad}(n)$$

whence also with $Calo_n$.

Example 8.7
Consider the special case $n = 1$. As $\square$ is the only partition of 1, for every $\lambda \in \mathbb{C}$, gr_λ is a one-dimensional cell $\mathbb{A}^1$, whence $Gr^{ad}(1) \simeq \mathbb{A}^2$. In fact we

have

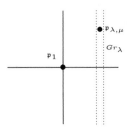

where the origin corresponds to the canonical right ideal $\mathfrak{p}_1$ and the right ideal corresponding to (λ, μ) is $\mathfrak{p}_{\lambda,\mu} = (x - \lambda)^2 A_1(\mathbb{C}) + ((x - \lambda)(y - \mu) + 1)A_1(\mathbb{C})$.

Finally, let us verify that $\mathfrak{p}_n$ should correspond to a point in $Gr^{ad}(n)$. As $\mathfrak{p}_n = x^{n+1}A_1(\mathbb{C}) + (xy + n)A_1(\mathbb{C})$ we have that

$$\mathfrak{p}_n.\mathbb{C}[x] = \mathbb{C} + \mathbb{C}x + \ldots + \mathbb{C}x^{n-1} + (x^{n+1})\mathbb{C}[x]$$

whence $(x^{n+1})\mathbb{C}[x] \subset \mathfrak{p}_n.\mathbb{C}[x] \subset \mathbb{C}[x]$ and converting this to Gr^{ad} the corresponding subspace is

$$(x^n)\mathbb{C}[x] \subset x^{-1}\mathfrak{p}_n.\mathbb{C}[x] \subset x^{-1}\mathbb{C}[x]$$

The associated sequence of degrees is $(-1, 0, 1, \ldots, n - 2, n, \ldots)$ giving rise to the partition $p = \underbrace{(1, 1, \ldots, 1)}_{n}$ proving the claim. □

If we trace the action of $Aut\, A_1(\mathbb{C})$ on $Weyl_n$ through all the identifications, we get a transitive action of $Aut\, A_1(\mathbb{C})$ on $Calo_n$. However, this action is non-differentiable, hence highly nonalgebraic. Berest and Wilson asked whether it is possible to identify $Calo_n$ with a coadjoint orbit in some infinite dimensional Lie algebra. We have seen before that this is indeed the case if we consider the necklace Lie algebra.

References

The results of section 8.1 are due to A. King [54]. Section 8.2 is based on the lecture notes of H.P. Kraft [62] and those of A. Tannenbaum [100]. Theorem 8.5 is due to W. Crawley-Boevey and M. Holland [28], the other results of section 8.3 are due to W. Crawley-Boevey [26]. The description of Hilbert schemes via Hilbert stairs is my interpretation of a result of M. Van den Bergh [101]. Example 8.2 is taken from the lecture notes of H. Nakajima [83]. The hyper-Kähler correspondence is classical, see, for example, the notes of Nakajima [83], here we follow W. Crawley-Boevey [25]. The applications to the hyper-Kähler correspondence are due to R. Bockland and L. Le Bruyn

[12]. The treatment of Calogero particles is taken from G. Wilson's paper [108]. The fact that the Calogero phase space is a coadjoint orbit is due to V. Ginzburg [36]. The extension to deformed preprojective algebras is independently due to V. Ginzburg [37] and R. Bockland and L. Le Bruyn [12]. The results on adelic Grassmannians are due to G. Wilson [108] and Y. Berest and G. Wilson [7]. More details on the connection with right ideals of the Weyl algebra can be found in the papers of R. C. Cannings and M. Holland [19] and L. Le Bruyn [69].

References

[1] V. I. Arnol'd, *Mathematical methods of classical mechanics*, Graduate Texts in Mathematics, vol. 60, Springer-Verlag, New York, 1989, Translated from the 1974 Russian original by K. Vogtmann and A. Weinstein, Corrected reprint of the second (1989) edition. MR MR1345386 (96c:70001)

[2] M. Artin, *On Azumaya algebras and finite dimensional representations of rings.*, J. Algebra **11** (1969), 532–563. MR MR0242890 (39 #4217)

[3] ———, *Brauer-Severi varieties*, Brauer groups in ring theory and algebraic geometry (Wilrijk, 1981), Lecture Notes in Math., vol. 917, Springer, Berlin, 1982, pp. 194–210. MR MR657430 (83j:14015)

[4] ———, *Local structure of maximal orders on surfaces*, Brauer groups in ring theory and algebraic geometry (Wilrijk, 1981), Lecture Notes in Math., vol. 917, Springer, Berlin, 1982, pp. 146–181. MR MR657428 (83k:14003)

[5] ———, *Maximal orders of global dimension and Krull dimension two*, Invent. Math. **84** (1986), no. 1, 195–222. MR MR830045 (88e:16008a)

[6] M. Artin and D. Mumford, *Some elementary examples of unirational varieties which are not rational*, Proc. London Math. Soc. (3) **25** (1972), 75–95. MR MR0321934 (48 #299)

[7] Yuri Berest and George Wilson, *Automorphisms and ideals of the Weyl algebra*, Math. Ann. **318** (2000), no. 1, 127–147. MR MR1785579 (2001m:16036)

[8] George M. Bergman and Lance W. Small, *P.I. degrees and prime ideals*, J. Algebra **33** (1975), 435–462. MR MR0360682 (50 #13129)

[9] David Birkes, *Orbits of linear algebraic groups*, Ann. of Math. (2) **93** (1971), 459–475. MR MR0296077 (45 #5138)

[10] Raf Bocklandt, *Smooth quiver representation spaces*, J. Algebra **253** (2002), no. 2, 296–313. MR MR1929191 (2003i:16020)

[11] ———, *Symmetric quiver settings with a regular ring of invariants*, Linear Algebra Appl. **365** (2003), 25–43, Special issue on linear algebra methods in representation theory. MR MR1987326 (2004k:16031)

[12] Raf Bocklandt and Lieven Le Bruyn, *Necklace Lie algebras and noncommutative symplectic geometry*, Math. Z. **240** (2002), no. 1, 141–167. MR MR1906711 (2003g:16013)

[13] Raf Bocklandt, Lieven Le Bruyn, and Stijn Symens, *Isolated singularities, smooth orders, and Auslander regularity*, Comm. Algebra **31** (2003), no. 12, 6019–6036. MR MR2014913 (2004j:16019)

[14] Raf Bocklandt, Lieven Le Bruyn, and Geert Van de Weyer, *Smooth order singularities*, J. Algebra Appl. **2** (2003), no. 4, 365–395. MR MR2020934 (2005e:16029)

[15] Nicolas Bourbaki, *Éléments de mathématique: groupes et algèbres de Lie*, Masson, Paris, 1982, Chapitre 9. Groupes de Lie réels compacts. [Chapter 9. Compact real Lie groups]. MR MR682756 (84i:22001)

[16] Kenneth A. Brown and Iain Gordon, *The ramification of centres: Lie algebras in positive characteristic and quantised enveloping algebras*, Math. Z. **238** (2001), no. 4, 733–779. MR MR1872572 (2002i:17027)

[17] ———, *The ramifications of the centres: quantised function algebras at roots of unity*, Proc. London Math. Soc. (3) **84** (2002), no. 1, 147–178. MR MR1863398 (2002i:20072)

[18] ———, *Poisson orders, symplectic reflection algebras and representation theory*, J. Reine Angew. Math. **559** (2003), 193–216. MR MR1989650 (2004i:16025)

[19] R. C. Cannings and M. P. Holland, *Right ideals of rings of differential operators*, J. Algebra **167** (1994), no. 1, 116–141. MR MR1282820 (95f:16028)

[20] Neil Chriss and Victor Ginzburg, *Representation theory and complex geometry*, Birkhäuser Boston Inc., Boston, MA, 1997. MR MR1433132 (98i:22021)

[21] Paul Moritz Cohn, *Skew field constructions*, Cambridge University Press, Cambridge, 1977, London Mathematical Society Lecture Note Series, No. 27. MR MR0463237 (57 #3190)

[22] Jean-Louis Colliot-Thélène, Raymond T. Hoobler, and Bruno Kahn, *The Bloch-Ogus-Gabber theorem*, Algebraic K-theory (Toronto, 1996), Fields Inst. Commun., vol. 16, Amer. Math. Soc., Providence, RI, 1997, pp. 31–94. MR MR1466971 (98j:14021)

[23] Alastair Craw, *The mckay correspondence and representations of the mckay quiver*, Warwick Ph. D. thesis, 2001.

[24] William Crawley-Boevey, *Geometry of representations of algebras*, Lecture Notes, 1993.

[25] William Crawley-Boevey, *On the exceptional fibres of Kleinian singularities*, Amer. J. Math. **122** (2000), no. 5, 1027–1037. MR MR1781930 (2001f:14009)

[26] ———, *Geometry of the moment map for representations of quivers*, Compositio Math. **126** (2001), no. 3, 257–293. MR MR1834739 (2002g:16021)

[27] ———, *Decomposition of Marsden-Weinstein reductions for representations of quivers*, Compositio Math. **130** (2002), no. 2, 225–239. MR MR1883820 (2003j:16019)

[28] William Crawley-Boevey and Martin P. Holland, *Noncommutative deformations of Kleinian singularities*, Duke Math. J. **92** (1998), no. 3, 605–635. MR MR1620538 (99f:14003)

[29] Joachim Cuntz and Daniel Quillen, *Algebra extensions and nonsingularity*, J. Amer. Math. Soc. **8** (1995), no. 2, 251–289. MR MR1303029 (96c:19002)

[30] Harm Derksen and Jerzy Weyman, *On the canonical decomposition of quiver representations*, Compositio Math. **133** (2002), no. 3, 245–265. MR MR1930979 (2003h:16017)

[31] M. Domokos and A. N. Zubkov, *Semi-invariants of quivers as determinants*, Transform. Groups **6** (2001), no. 1, 9–24. MR MR1825166 (2002d:16015)

[32] Edward Formanek, *The polynomial identities and invariants of $n \times n$ matrices*, CBMS Regional Conference Series in Mathematics, vol. 78, Published for the Conference Board of the Mathematical Sciences, Washington, DC, 1991. MR MR1088481 (92d:16031)

[33] Edward Formanek, Patrick Halpin, and Wen Ch'ing Winnie Li, *The Poincaré series of the ring of 2×2 generic matrices*, J. Algebra **69** (1981), no. 1, 105–112. MR MR613860 (82i:16020)

[34] Ofer Gabber, *Some theorems on Azumaya algebras*, The Brauer group (Sem., Les Plans-sur-Bex, 1980), Lecture Notes in Math., vol. 844, Springer, Berlin, 1981, pp. 129–209. MR MR611868 (83d:13004)

[35] P. Gabriel and A. V. Roiter, *Representations of finite-dimensional algebras*, Springer-Verlag, Berlin, 1997, Translated from the Russian, with a chapter by B. Keller, Reprint of the 1992 English translation. MR MR1475926 (98e:16014)

[36] Victor Ginzburg, *Non-commutative symplectic geometry and Calogero-Moser space*, preprint Chicago, preliminary version, 1999.

[37] Victor Ginzburg, *Non-commutative symplectic geometry, quiver varieties, and operads*, Math. Res. Lett. **8** (2001), no. 3, 377–400. MR MR1839485 (2002m:17020)

[38] Jonathan S. Golan, *Structure sheaves over a noncommutative ring*, Lecture Notes in Pure and Applied Mathematics, vol. 56, Marcel Dekker Inc., New York, 1980. MR MR582091 (81k:16002)

[39] K. R. Goodearl and R. B. Warfield, Jr., *An introduction to noncommutative Noetherian rings*, second ed., London Mathematical Society Student Texts, vol. 61, Cambridge University Press, Cambridge, 2004. MR MR2080008 (2005b:16001)

[40] Alexander Grothendieck, *Le groupe de Brauer. II. Théorie cohomologique [MR0244270 (39 #5586b)]*, Séminaire Bourbaki, Vol. 9, Soc. Math. France, Paris, 1995, pp. Exp. No. 297, 287–307. MR MR1608805

[41] G. B. Gurevich, *Foundations of the theory of algebraic invariants*, Translated by J. R. M. Radok and A. J. M. Spencer, P. Noordhoff Ltd., Groningen, 1964. MR MR0183733 (32 #1211)

[42] Dieter Happel and Claus Michael Ringel, *Tilted algebras*, Trans. Amer. Math. Soc. **274** (1982), no. 2, 399–443. MR MR675063 (84d:16027)

[43] Robin Hartshorne, *Algebraic geometry*, Graduate Texts in Mathematics, vol. 52, Springer-Verlag, Berlin, Heidelberg, New York, 1977.

[44] Michiel Hazewinkel, *On representations of the symmetric groups, nilpotent matrices, systems, vectorbundles and Schubert cells*, Representations of algebras (Puebla, 1980), Lecture Notes in Math., vol. 903, Springer, Berlin, 1981, pp. 145–167. MR MR654708 (83g:14028)

[45] Wim H. Hesselink, *Desingularizations of varieties of nullforms*, Invent. Math. **55** (1979), no. 2, 141–163. MR MR553706 (81b:14025)

[46] Klaus Hulek, *On the classification of stable rank-r vector bundles over the projective plane*, Vector bundles and differential equations (Proc. Conf., Nice, 1979), Progr. Math., vol. 7, Birkhäuser Boston, Mass., 1980, pp. 113–144. MR MR589223 (82b:14012)

[47] A. V. Jategaonkar, *Localization in Noetherian rings*, London Mathematical Society Lecture Note Series, vol. 98, Cambridge University Press, Cambridge, 1986. MR MR839644 (88c:16005)

[48] V. G. Kac, *Infinite root systems, representations of graphs and invariant theory*, Invent. Math. **56** (1980), no. 1, 57–92. MR MR557581 (82j:16050)

[49] _____, *Infinite root systems, representations of graphs and invariant theory. II*, J. Algebra **78** (1982), no. 1, 141–162. MR MR677715 (85b:17003)

[50] M. Kapranov, *Noncommutative geometry based on commutator expansions*, J. Reine Angew. Math. **505** (1998), 73–118. MR MR1662244 (2000b:14003)

References

509

[51] Bernhard Keller, *Introduction to A-infinity algebras and modules*, Homology Homotopy Appl. **3** (2001), no. 1, 1–35 (electronic). MR MR1854636 (2004a:18008a)

[52] G. Kempf, Finn Faye Knudsen, D. Mumford, and B. Saint-Donat, *Toroidal embeddings. I*, Springer-Verlag, Berlin, 1973, Lecture Notes in Mathematics, Vol. 339. MR MR0335518 (49 #299)

[53] George Kempf and Linda Ness, *The length of vectors in representation spaces*, Algebraic geometry (Proc. Summer Meeting, Univ. Copenhagen, Copenhagen, 1978), Lecture Notes in Math., vol. 732, Springer, Berlin, 1979, pp. 233–243. MR MR555701 (81i:14032)

[54] A. D. King, *Moduli of representations of finite-dimensional algebras*, Quart. J. Math. Oxford Ser. (2) **45** (1994), no. 180, 515–530. MR MR1315461 (96a:16009)

[55] Frances Clare Kirwan, *Cohomology of quotients in symplectic and algebraic geometry*, Mathematical Notes, vol. 31, Princeton University Press, Princeton, 1984. MR MR766741 (86i:58050)

[56] Friedrich Knop, *Der Scheibensatz für algebraische Transformationsgruppen*, Algebraische Transformationsgruppen und Invariantentheorie, DMV Sem., vol. 13, Birkhäuser, Basel, 1989, pp. 100–113. MR MR1044587

[57] Max-Albert Knus and Manuel Ojanguren, *Théorie de la descente et algèbres d'Azumaya*, Springer-Verlag, Berlin, 1974, Lecture Notes in Mathematics, Vol. 389. MR MR0417149 (54 #5209)

[58] Maxim Kontsevich, *Formal (non)commutative symplectic geometry*, The Gel'fand Mathematical Seminars, 1990–1992, Birkhäuser Boston, Boston, MA, 1993, pp. 173–187. MR MR1247289 (94i:58212)

[59] Maxim Kontsevich, *Triangulated categories and geometry*, Course at the Ecole Normale Sup., Paris, 1998.

[60] Maxim Kontsevich, *Operads and motives in deformation quantization*, Lett. Math. Phys. **48** (1999), no. 1, 35–72, Moshé Flato (1937–1998). MR MR1718044 (2000j:53119)

[61] H. Kraft and Ch. Riedtmann, *Geometry of representations of quivers*, Representations of algebras (Durham, 1985), London Math. Soc. Lecture Note Ser., vol. 116, Cambridge Univ. Press, Cambridge, 1986, pp. 109–145. MR MR897322 (88k:16028)

[62] Hanspeter Kraft, *Geometric methods in representation theory*, Representations of algebras (Puebla, 1980), Lecture Notes in Math., vol. 944, Springer, Berlin, 1982, pp. 180–258. MR MR672117 (84c:14007)

[63] _____, *Geometrische Methoden in der Invariantentheorie*, Aspects of Mathematics, D1, Friedr. Vieweg & Sohn, Braunschweig, 1984. MR MR768181 (86j:14006)

[64] _____, *Klassische Invariantentheorie. Eine Einführung*, Algebraische Transformationsgruppen und Invariantentheorie, DMV Sem., vol. 13, Birkhäuser, Basel, 1989, pp. 41–62. MR MR1044584

[65] Lieven Le Bruyn, *The Artin-Schofield theorem and some applications*, Comm. Algebra **14** (1986), no. 8, 1439–1455. MR MR859443 (87k:16020)

[66] _____, *A cohomological interpretation of the reflexive Brauer group*, J. Algebra **105** (1987), no. 1, 250–254. MR MR871757 (88j:16009)

[67] _____, *Trace rings of generic 2 by 2 matrices*, Mem. Amer. Math. Soc. **66** (1987), no. 363, vi+100. MR MR878906 (88b:16032)

[68] _____, *Central singularities of quantum spaces*, J. Algebra **177** (1995), no. 1, 142–153. MR MR1356364 (96k:16051)

[69] _____, *Moduli spaces for right ideals of the Weyl algebra*, J. Algebra **172** (1995), no. 1, 32–48. MR MR1320617 (96b:16025)

[70] _____, *Etale cohomology in non-commutative geometry*, TMR-Lecture Notes, 1997.

[71] _____, *Nilpotent representations*, J. Algebra **197** (1997), no. 1, 153–177. MR MR1480781 (98k:14070)

[72] _____, *Orbits of matrix tuples*, Algèbre non commutative, groupes quantiques et invariants (Reims, 1995), Sémin. Congr., vol. 2, Soc. Math. France, Paris, 1997, pp. 245–261. MR MR1601151 (99a:16019)

[73] _____, *Noncommutative compact manifolds constructed from quivers*, AMA Algebra Montp. Announc. (1999), Paper 1, 5 pp. (electronic). MR MR1734896 (2001m:14004)

[74] _____, *Local structure of Schelter-Procesi smooth orders*, Trans. Amer. Math. Soc. **352** (2000), no. 10, 4815–4841. MR MR1695028 (2001b:16018)

[75] _____, *Optimal filtrations on representations of finite-dimensional algebras*, Trans. Amer. Math. Soc. **353** (2001), no. 1, 411–426. MR MR1707199 (2001c:16025)

[76] Lieven Le Bruyn and Claudio Procesi, *Semisimple representations of quivers*, Trans. Amer. Math. Soc. **317** (1990), no. 2, 585–598. MR MR958897 (90e:16048)

[77] Lieven Le Bruyn and Aidan Schofield, *Rational invariants of quivers and the ring of matrixinvariants*, Perspectives in ring theory (Antwerp,

1987), NATO Adv. Sci. Inst. Ser. C Math. Phys. Sci., vol. 233, Kluwer Acad. Publ., Dordrecht, 1988, pp. 21–29. MR MR1048393 (91f:14047)

[78] Lieven Le Bruyn and Michel Van den Bergh, *Regularity of trace rings of generic matrices*, J. Algebra **117** (1988), no. 1, 19–29. MR MR955588 (90b:16024)

[79] Lieven LeBruyn, *Noncommutative geometry @n*, math.AG/9904171, 1999.

[80] Thierry Levasseur, *Some properties of noncommutative regular graded rings*, Glasgow Math. J. **34** (1992), no. 3, 277–300. MR MR1181768 (93k:16045)

[81] Domingo Luna, *Slices étales*, Sur les groupes algébriques, Soc. Math. France, Paris, 1973, pp. 81–105. Bull. Soc. Math. France, Paris, Mémoire 33. MR MR0342523 (49 #7269)

[82] James S. Milne, *Étale cohomology*, Princeton Mathematical Series, vol. 33, Princeton University Press, 1980. MR MR559531 (81j:14002)

[83] Hiraku Nakajima, *Lectures on Hilbert schemes of points on surfaces*, University Lecture Series, vol. 18, American Mathematical Society, Providence, RI, 1999. MR MR1711344 (2001b:14007)

[84] Richard S. Pierce, *Associative algebras*, Graduate Texts in Mathematics, vol. 88, Springer-Verlag, New York, 1982, Studies in the History of Modern Science, 9. MR MR674652 (84c:16001)

[85] C. Procesi, *The invariant theory of $n \times n$ matrices*, Advances in Math. **19** (1976), no. 3, 306–381. MR MR0419491 (54 #7512)

[86] Claudio Procesi, *Rings with polynomial identities*, Marcel Dekker Inc., New York, 1973, Pure and Applied Mathematics, 17. MR MR0366968 (51 #3214)

[87] ———, *A formal inverse to the Cayley-Hamilton theorem*, J. Algebra **107** (1987), no. 1, 63–74. MR MR883869 (88b:16033)

[88] Y. P. Razmyslov, *Trace identities of full matrix algebras over a field of characteristic zero*, Math. USSR-Izv. **8** (1974), 727–760.

[89] Miles Reid, *La correspondance de McKay*, Astérisque (2002), no. 276, 53–72, Séminaire Bourbaki, Vol. 1999/2000. MR MR1886756 (2003h:14026)

[90] I. Reiner, *Maximal orders*, London Mathematical Society Monographs. New Series, vol. 28, The Clarendon Press Oxford University Press, 2003, Corrected reprint of the 1975 original, With a foreword by M. J. Taylor. MR MR1972204 (2004c:16026)

[91] William F. Schelter, *Smooth algebras*, J. Algebra **103** (1986), no. 2, 677–685. MR MR864437 (88a:16034)

[92] A. H. Schofield, *Representation of rings over skew fields*, London Mathematical Society Lecture Note Series, vol. 92, Cambridge University Press, 1985. MR MR800853 (87c:16001)

[93] Aidan Schofield, *General representations of quivers*, Proc. London Math. Soc. (3) **65** (1992), no. 1, 46–64. MR MR1162487 (93d:16014)

[94] ———, *Birational classification of moduli spaces of representations of quivers*, Indag. Math. (N.S.) **12** (2001), no. 3, 407–432. MR MR1914089 (2003k:16028)

[95] Aidan Schofield and Michel van den Bergh, *Semi-invariants of quivers for arbitrary dimension vectors*, Indag. Math. (N.S.) **12** (2001), no. 1, 125–138. MR MR1908144 (2003e:16016)

[96] Gerald W. Schwarz, *Lifting smooth homotopies of orbit spaces*, Inst. Hautes Études Sci. Publ. Math. (1980), no. 51, 37–135. MR MR573821 (81h:57024)

[97] Stephen S. Shatz, *Profinite groups, arithmetic, and geometry*, Princeton University Press, 1972, Annals of Mathematics Studies, No. 67. MR MR0347778 (50 #279)

[98] K. S. Siberskii, *Algebraic invariants for a set of matrices*, Siberian Math. J. **9** (1968), 115–124.

[99] Peter Slodowy, *Der Scheibensatz für algebraische Transformationsgruppen*, Algebraische Transformationsgruppen und Invariantentheorie, DMV Sem., vol. 13, Birkhäuser, Basel, 1989, pp. 89–113. MR MR1044587

[100] Allen Tannenbaum, *Invariance and system theory: algebraic and geometric aspects*, Lecture Notes in Mathematics, vol. 845, Springer-Verlag, Berlin, 1981. MR MR611155 (83g:93009)

[101] Michel Van den Bergh, *The Brauer-Severi scheme of the trace ring of generic matrices*, Perspectives in ring theory (Antwerp, 1987), NATO Adv. Sci. Inst. Ser. C Math. Phys. Sci., vol. 233, Kluwer Acad. Publ., Dordrecht, 1988, pp. 333–338. MR MR1048420 (91g:13001)

[102] Michel van den Bergh, *Non-commutative crepant resolutions*, The legacy of Niels Henrik Abel, Springer, Berlin, 2004, pp. 749–770. MR MR2077594 (2005e:14002)

[103] F. van Oystaeyen, *Prime spectra in non-commutative algebra*, Springer-Verlag, Berlin, 1975, Lecture Notes in Mathematics, Vol. 444. MR MR0419497 (54 #7518)

[104] Freddy Van Oystaeyen, *Algebraic geometry for associative algebras*, Monographs and Textbooks in Pure and Applied Mathematics, vol. 232, Marcel Dekker Inc., New York, 2000. MR MR1807463 (2002i:14002)

[105] Freddy M. J. Van Oystaeyen and Alain H. M. J. Verschoren, *Noncommutative algebraic geometry*, Lecture Notes in Mathematics, vol. 887, Springer-Verlag, Berlin, 1981, An introduction. MR MR639153 (85i:16006)

[106] Bruce W. Westbury, *On the character varieties of the modular group*, preprint, Nottingham, 1995.

[107] Hermann Weyl, *The Classical Groups. Their Invariants and Representations*, Princeton University Press, 1939. MR MR0000255 (1,42c)

[108] George Wilson, *Collisions of Calogero-Moser particles and an adelic Grassmannian*, Invent. Math. **133** (1998), no. 1, 1–41, With an appendix by I. G. Macdonald. MR MR1626461 (99f:58107)

Index

Milton Keynes UK
Ingram Content Group UK Ltd.
UKHW021930071024
449327UK00022B/1749